Lecture Notes in Mathematics

Edited by A. Dold and B. Eckmann

1153

Probability in Banach Spaces V

Proceedings of the International Conference held in Medford, USA, July 16–27, 1984

Edited by A. Beck, R. Dudley, M. Hahn, J. Kuelbs and M. Marcus

Springer-Verlag
Berlin Heidelberg New York Tokyo

Editors

Anatole Beck
Department of Mathematics, University of Wisconsin
Madison, WI 53706, USA

Richard Dudley
Department of Mathematics, Massachusetts Institute of Technology
Cambridge, MA 02139, USA

Marjorie Hahn
Department of Mathematics, Tufts University
Medford, MA 02155, USA

James Kuelbs
Department of Mathematics, University of Wisconsin
Madison, WI 53706, USA

Michael Marcus
Department of Mathematics, Texas A&M University
College Station, TX 77843, USA

Mathematics Subject Classification (1980): Primary: 60Bxx, 60Gxx, 60Fxx
Secondary: 43A46, 46Bxx, 46Cxx, 60D05, 60Exx, 60J30

ISBN 3-540-15704-2 Springer-Verlag Berlin Heidelberg New York Tokyo
ISBN 0-387-15704-2 Springer-Verlag New York Heidelberg Berlin Tokyo

Library of Congress Cataloging-in-Publication Data. Main entry under title: Probability in Banach spaces V. (Lecture notes in mathematics; 1153) Bibliography: p. 1. Probabilities–Congresses. 2. Banach spaces–Congresses. I. Beck, Anatole, 1930–. II. Series: Lecture notes in mathematics (Springer-Verlag); 1153.
QA3.L28 no. 1153 [QA273.43] 510 s [519.2] 85-22080
ISBN 0-387-15704-2 (U.S.)

Printing and binding: Beltz Offsetdruck, Hemsbach/Bergstr.
2146/3140-543210

Introduction

The first international conference on Probability in Banach Spaces was held at Oberwolfach, West Germany in 1975. It brought together European researchers who, under the inspiration of the Schwartz seminar in Paris, were using probabilistic methods in the study of the geometry of Banach spaces, a rather small number of probabilists who were already studying classical limit laws on Banach spaces and a larger number of probabilists, specialists in various aspects of the study of Gaussian processes, whose results and techniques were of interest to the members of the first two groups. This first conference was very fruitful. It fostered a continuing relationship amongst fifty to seventy-five probabilists and analysts working on probability on infinite-dimensional spaces, the geometry of Banach spaces and the use of random methods in harmonic analysis. Three more international conferences were held, two at Oberwolfach in 1978 and 1982 and one at Tufts University in 1980. This volume contains a selection of papers by the participants of the Fifth International Conference held at Medford, Massachusetts, July 16-27, 1984. This exciting and provocative conference was attended by eighty mathematicians, thirty of whom were from outside the United States. These papers demonstrate the range of interests of the conference participants. The distinction between probabilists and analysts has diminished during the ten years since the first conference. During the same time a number of workers have been considering problems in statistics. It is this interaction between probabilists, statisticians, functional analysts and harmonic analysts which enriches and propels this field. Arrangements have already been made to hold the sixth and seventh international conferences in 1986 and 1988. We look forward with great enthusiasm to further developments in this field.

The Editors.

Table of Contents

ON LARGE DEVIATIONS OF SUMS OF INDEPENDENT RANDOM VECTORS

A. de Acosta
Case Western Reserve University

1. Introduction. The object of this paper is to give essentially self-contained proofs of some basic results on large deviations of sums of independent, identically distributed random vectors taking values in a separable Banach space. Although the main theorem that we prove is known, our approach differs in several aspects from the methods of proof in the literature. Also, several new results are presented.

In Section 2 (Theorem 2.1) we prove Donsker and Varadhan's [4] basic result on upper bounds for closed sets and lower bounds for open sets. Azencott's notes [2] give a fine exposition of this result, incorporating as well the contributions of Bahadur and Zabell [3]. In the present paper, the proof of the upper bound is taken from [1] and the lower bound is obtained by combining an adaptation of Cramér's classical one-dimensional argument with a density result, Proposition 2.3, which appears to be of independent interest.

As a simple consequence of Theorem 2.1, we give in Section 3 geometric conditions under which the upper and lower bounds coalesce. A result of Bahadur and Zabell [3] is obtained as a corollary.

2. Upper and lower bounds for large deviations. Let B be a separable Banach space, and let X, $\{X_j, j \in \mathbb{N}\}$ be independent B-valued random vectors with common distribution μ. We will assume that the following integrability condition is satisfied:

$$\text{(2.1)} \qquad \text{for every } t > 0, \quad \int e^{t\|x\|}\mu(dx) < \infty .$$

The Laplace transform of μ will be denoted

$$\hat{\mu}(\xi) = \int e^{\xi(x)}\mu(dx), \quad \text{for} \quad \xi \in B' \quad ,$$

where B' is the dual space of B. The Cramér transform of μ, denoted λ, is the convex conjugate of $\log \hat{\mu}$; that is,

$$\lambda(x) = \sup_{\xi \in B'} [\xi(x) - \log\hat{\mu}(\xi)] , \quad \text{for} \quad x \in B.$$

Obviously $\lambda \geq 0$ and λ is lower semicontinuous and convex. We define $\mathrm{Dom}\,\lambda = \{x \in B : \lambda(x) < \infty\}$; the set functional Λ is defined by

$$\Lambda(A) = \inf_{x \in A} \lambda(x) , \quad \text{for} \quad A \subset B.$$

In this section we prove

Theorem 2.1 (Donsker-Varadhan [4]).

(1) For every closed set $F \subset B$,

$$\overline{\lim}_n \, n^{-1} \log P\{S_n/n \in F\} \leq - \Lambda(F) .$$

(2) For every open set $G \subset B$,

$$\underline{\lim}_n \, n^{-1} \log P\{S_n/n \in G\} \geq - \Lambda(G) .$$

(3) For every $a \geq 0$, $\{x : \lambda(x) \leq a\}$ is compact.

Remark. Statement (1) for compact sets holds true under a weaker condition, involving only the finiteness of $\hat{\mu}$; this is clear from the proof. Statement (2) may also be proved under the weaker condition (see [2], [3]). However, the full strength of (1) as well as (3) require condition (2.1) for their proof. For this reason we have adopted (2.1) as a blanket assumption.

The proof of statement (1) is carried out in two steps: first it is proved for compact F and then for closed F. The passage to closed sets is based on an integrability result - Lemma 2.2 below - which is taken from [1]. Statement (3) also follows from Lemma 2.2.

We recall that the Minkowski functional of a convex, symmetric set $A \subset B$ is defined $q_A(x) = \inf\{\alpha > 0 : x \in \alpha A\}$, $x \in B$.

Lemma 2.2. Assume that the probability measure μ satisfies (2.1). Then there exists a compact convex symmetric set K such that

$$\int \exp(q_K)\, d\mu < \infty \quad . \tag{2.2}$$

Proof. Let $\tau(t) = \mu(\{x:\|x\| > t\})$. We claim

$$\lim_{t\to\infty} (\tau(t))^{1/t} = 0 \quad . \tag{2.3}$$

In fact, given $\varepsilon > 0$, choose $a > 0$ so that $e^{-a} < \varepsilon$. Then for all $t > 0$,

$$\tau(t) \le e^{-at} \int \exp(a\|x\|)\, \mu(dx) \quad ,$$

which implies $\overline{\lim_{t\to\infty}} (\tau(t))^{1/t} \le e^{-a} < \varepsilon$, proving (2.3).

We prove now the following statement, which clearly implies the conclusion of the lemma: for every $\beta \in (0,1)$, there exist $c > 0$ and a compact, convex, symmetric set K such that

$$\mu(\{x:q_K(x) > t\}) \le c\, \beta^t \qquad \text{for all} \quad t \ge 1 \quad . \tag{2.4}$$

Choose $\beta \in (0,1)$ and set $t_m = \inf\{t > 0:\tau(t) < \beta^m\}$, $m \in \mathbb{N}$; it follows that $\tau(t_m) \le \beta^m$. Let $B_m = \{x:\|x\| \le t_m\}$.

Let K_m be a compact set such that $\mu(K_m^c) < \beta^m$; we may assume that K_m is convex and symmetric (see e.g. [6], p. 50). We may also assume $K_m \subset K_{m+1}$ for all m. Now let

$$A_m = m^{-1}(K_m \cap B_m),$$

$$K = \text{closed convex symmetric hull of } \{\cup_m A_m\}.$$

We claim that K is compact. To prove this, we first observe that

$$m^{-1}\, t_m \to 0 \qquad \text{as} \quad m \to \infty \; . \tag{2.5}$$

For, given $\varepsilon > 0$, choose m_o so that $m_o^{-1/2} < \varepsilon/2$ and for all $t > m_o^{1/2}$,

$$\log \beta/(t^{-1} \log \tau(t)) < \varepsilon/2 \quad ;$$

this is possible by (2.3). Let $m > m_o$. If $t_m/2 > m^{1/2}$, we have

$$\tau(t_m/2) \geq \beta^m \quad ,$$

$$\frac{t_m}{2m} \leq \frac{\log \beta}{(\frac{2}{t_m}) \log \tau(t_m/2)} < \varepsilon/2 \quad ,$$

$$t_m/m < \varepsilon \ .$$

On the other hand, if $t_m/2 \leq m^{1/2}$, then

$$t_m/m \leq \frac{2}{m^{1/2}} < \frac{2}{m_o^{1/2}} < \varepsilon \ .$$

Thus for $m > m_o$, $m^{-1} t_m < \varepsilon$, proving (2.5). It follows that $\cup_m A_m$ is totally bounded; for, let $\varepsilon > 0$ and choose m_o so that $m^{-1} t_m < \varepsilon$ for $m \geq m_o$. Then

$$\cup_m A_m = (\bigcup_{m \leq m_o} A_m) \cup (\bigcup_{m > m_o} A_m)$$

$$\subset K_{m_o} \cup \{x : \| x \| \leq \varepsilon\} \quad ,$$

proving the total boundedness of $(\cup_m A_m)$. It follows that K is compact (see e.g. [6], p. 50), as stated.

To conclude the proof we verify (2.4). Given $t \geq 1$, let $m = [t]$. Then

$$\begin{aligned} \mu(\{x : q_K(x) > t\}) &= \mu((tK)^c) \\ &\leq \mu((mK^c) \\ &\leq \mu((K_m \cap B_m)^c) \\ &\leq \beta^m + \beta^m = 2\beta^m \ , \end{aligned}$$

and since $\beta^{m+1} < \beta^t$ we have

$$\mu(\{x:q_K(x) > t\}) \leq (2\beta^{-1})\beta^t \quad . \qquad \square$$

Proof of (1).

Assume first that F is compact. The result is trivially true if $\Lambda(F) = 0$. Suppose that $0 < \Lambda(F) < \infty$; the case $\Lambda(F) = \infty$ is handled similarly. For $\varepsilon > 0$, $\xi \in B'$, let $H(\xi) = \{x:\xi(x) - \log\hat{\mu}(\xi) > \Lambda(F) - \varepsilon\}$. Then

$$F \subset \{x:\lambda(x) > \Lambda(F) - \varepsilon\} = \bigcup_{\xi} H(\xi) ,$$

and hence $F \subset \bigcup_{i=1}^{k} H(\xi_i)$ for certain $\xi_1,\ldots,\xi_k$ in B'. Now if $b = \Lambda(F) - \varepsilon$,

$$\begin{aligned} P\{S_n/n \in F\} &\leq \sum_i P\{S_n/n \in H(\xi_i)\} \\ &= \sum_i P\{\xi_i(S_n) > n(\log\hat{\mu}(\xi_i) + b)\} \\ &\leq \sum_i e^{-n(\log\hat{\mu}(\xi_i) + b)} \left\{E\, e^{\xi_i(X)}\right\}^n \\ &= k\, e^{-nb} , \end{aligned}$$

and therefore $\overline{\lim}\, n^{-1} \log P\{S_n/n \in F\} \leq - b = -\Lambda(F) + \varepsilon$. Since ε is arbitrary, this proves (1) when F is compact.

Next, we show that (2.2) implies the following fact: for every $a > 0$, there exists a compact set K_a such that

$$P\{S_n/n \in K_a^c\} \leq e^{-na} \qquad \text{for } n \geq n_o(a) \quad . \tag{2.6}$$

In fact, if K is as in Lemma 2.2, $\beta = \int \exp(q_K)d\mu$ and $\delta > 0$,

$$\begin{aligned} P\{S_n/n \in (\delta K)^c\} &= P\{q_K(S_n) > n\delta\} \\ &\leq e^{-n\delta}\, E \exp(q_K(S_n)) \\ &\leq e^{-n\delta}\, \beta^n \\ &= e^{-n(\delta - \log\beta)} \quad . \end{aligned}$$

Take now $\delta \geq a + \log \beta$, $K_a = \delta K$.

Let F be a closed set and let K_a be as in (2.6). Then

$$P\{S_n/n \in F\} = P\{S_n/n \in K_a \cap F\} + P\{S_n/n \in K_a^c \cap F\}$$

$$\leq 2 \max(P\{S_n/n \in K_a \cap F\}, \ P\{S_n/n \in K_a^c\})$$

and since $K_a \cap F$ is compact,

$$\overline{\lim}\, n^{-1} \log P\{S_n/n \in F\} \leq \max(-\Lambda(K_a \cap F), -a)$$

$$\leq \max(-\Lambda(F), -a).$$

Since a is arbitrary, this finishes the proof of (1). □

Proof of (3).

Let $K^o = \{\xi \in B' : \xi(x) \leq 1 \text{ for all } x \in K\}$, the polar set of K. If $\xi \in K^o$, then

$$\hat{\mu}(\xi) = \int e^{\xi(x)} \mu(dx) \leq \int \exp(q_K) d\mu = \beta .$$

If $x \in L_a = \{x : \lambda(x) \leq a\}$, then for all $\xi \in B'$

$$\xi(x) \leq \log \hat{\mu}(\xi) + a \quad ,$$

and if $\xi \in K^o$, then $\xi(x) \leq c$, where $c = \log \beta + a$. Therefore

$$c^{-1} x \in K^{oo}.$$

Since $K^{oo} = K$ by the bipolar theorem (see, e.g. [6], p. 126), it follows that $L_a \subset c\, K$, which proves (3). □

The proof of statement (2) of Theorem 2.1 is based on Proposition 2.3. In order to make clear the structure of the proof, we shall state Proposition 2.3 and proceed to prove (2). The rest of the section will be devoted to the proof of Proposition 2.3.

In view of (2.1), it is easily seen that the Fréchet derivative $\psi(\xi) = (\log\hat{\mu})'(\xi)$ exists for every $\xi \in B'$, and

$$\psi(\xi) = \frac{\int x\, e^{\xi(x)}\mu(dx)}{\hat{\mu}(\xi)} \in B .$$

Moreover, since $\log \hat{\mu}$ is a convex function (by Hölder's inequality), it follows that if $x = \psi(\xi)$, then for all $\eta \in B'$,

$$\log \hat{\mu}(\eta) \geq \log \hat{\mu}(\xi) + (\eta-\xi)(x),$$

(2.7)

$$\xi(x) - \log \hat{\mu}(\xi) \geq \eta(x) - \log \hat{\mu}(\eta) .$$

Therefore $\lambda(x) = \xi(x) - \log \hat{\mu}(\xi)$ and $x \in \mathrm{Dom}\, \lambda$.

<u>Proposition 2.3</u>. For every $x \in \mathrm{Dom}\,\lambda$, $\varepsilon > 0$, there exists $y \in \psi(B')$ such that

$$\|x-y\| < \varepsilon, \quad |\lambda(x) - \lambda(y)| < \varepsilon .$$

In particular,

$$\psi(B') \subset \mathrm{Dom}\,\lambda \subset \overline{\psi(B')} .$$

Let us point out that since λ is not continuous in general, but merely lower semicontinuous, the closeness of $\lambda(x)$ and $\lambda(y)$ does not automatically follow from the closeness of x,y.

<u>Proof of (2)</u>.

Let G be an open subset of B. It is enough to prove: for every $x \in \mathrm{Dom}\,\lambda \cap G$,

$$\text{(2.8)} \qquad \underline{\lim}\, n^{-1} \log P\{S_n/n \in G\} \geq -\lambda(x) .$$

By Proposition 2.3, given $\varepsilon > 0$ there exists $\xi \in B'$ such that if $y = \psi(\xi)$, then $y \in G$ and $|\lambda(x) - \lambda(y)| < \varepsilon$. For $\delta > 0$, let $V = \{z \in B: |\xi(z) - \xi(y)| < \delta\}$, and let $\nu_n = \mathcal{L}(S_n/n)$. Then

$$\nu_n(G) \geq \nu_n(G \cap V)$$

$$\geq e^{-n\xi(y)}[\hat{\mu}(\xi)]^n e^{-n\delta} a_n \quad ,$$

where $a_n = \int_{G\cap V} [\hat{\mu}(\xi)]^{-n} e^{n\xi(z)} \nu_n(dz)$, and therefore

$$n^{-1} \log \nu_n(G) \geq [\log\hat{\mu}(\xi) - \xi(y)] - \delta + n^{-1} \log a_n$$

$$\geq - \lambda(y) - \delta + n^{-1} \log a_n$$

$$\geq - \lambda(x) - \varepsilon - \delta + n^{-1} \log a_n \quad .$$

It follows that in order to prove (2.8) it is enough to prove: for every $\delta > 0$,

(2.9) $$\lim a_n = 1 \ .$$

Let $\{Y_j, j \in \mathbb{N}\}$ be independent random vectors with common law $([\hat{\mu}(\xi)]^{-1} e^{\xi} d\mu)$, $T_n = \sum_{j=1}^n Y_j$. It is easily checked that

$$E\, Y_1 = y \qquad \mathcal{L}(T_n/n) = [\hat{\mu}(\xi)]^{-n} e^{n\xi}\, d\nu_n \quad .$$

Since $G \cap V$ is an open set containing y, (2.9) follows from the weak law of large numbers. □

The proof of Proposition 2.3 is based on several lemmas. We shall need a basic fact about duality of convex functions. Let $\phi : B' \to \mathbb{R}$ be a convex function, and assume that there exists $A \subset B$ and $f : A \to \mathbb{R}$ such that for every $\xi \in B'$,

(2.10) $$\phi(\xi) = \sup_{x\varepsilon A}[\xi(x) - f(x)] \quad .$$

Let $\phi^*(x) = \sup_{\xi\varepsilon B'} [\xi(x) - \phi(\xi)]$, $\phi^{**}(\xi) = \sup_{x\varepsilon B}[\xi(x) - \phi^*(x)]$. Then

(2.11) $$\phi^{**} = \phi \quad .$$

For, obviously $\phi^{**} \leq \phi$. On the other hand, it follows from (2.10) that $\phi^*(x) \leq f(x)$ for $x \in A$, and therefore $\phi \leq \phi^{**}$. It is an elementary fact that (2.10) is always satisfied if B is finite-dimensional (see [5], pp. 102-104). If $\phi = \log \hat{\mu}$, then

(2.7) implies that (2.10) is satisfied, and therefore (2.11) holds; that is,

$$(\log\hat{\mu})^* = \lambda \ , \quad (\log\hat{\mu})^{**} = \lambda^* = \log\hat{\mu} \ . \tag{2.12}$$

We recall that the relative interior of a convex set A in a finite-dimensional space, $ri(A)$, is the interior of A relative to its affine hull (see [5], p. 44). The convex hull of a set E is denoted $co(E)$.

<u>Lemma 2.4</u>. Let S be the topological support of μ. Then

(a) $\overline{\mathrm{Dom}\,\lambda} = \overline{co(S)}$.

(b) If B is finite-dimensional, then $ri(\mathrm{Dom}\,\lambda) = ri(co(S))$.

<u>Proof</u>. (a) Suppose $x \notin \overline{co(S)}$. Then there exist $\xi \in B'$, $\alpha,\beta \in \mathbb{R}$ such that $\xi(y) \leq \alpha < \beta \leq \xi(x)$ for all $y \in \overline{co(S)}$. Therefore for all $t \in \mathbb{R}^+$,

$$\hat{\mu}(t\xi) = \int_S e^{t\xi(y)}\mu(dy) \leq e^{t\alpha} \ ,$$

$$t\xi(x) - \log\hat{\mu}(t\xi) \geq t(\beta-\alpha) \ ,$$

implying $\lambda(x) = \infty$. Thus $\mathrm{Dom}\,\lambda \subset \overline{co(S)}$, and hence $\overline{\mathrm{Dom}\,\lambda} \subset \overline{co(S)}$.

Suppose $x \in S$, $x \notin \overline{\mathrm{Dom}\,\lambda}$. Then there exist $\xi \in B'$, $\alpha,\beta \in \mathbb{R}$ such that $\xi(x) > \beta > \alpha \geq \xi(y)$ for all $y \in \mathrm{Dom}\,\lambda$. By (2.12) and since $\lambda \geq 0$, we have for every $t \in \mathbb{R}^+$,

$$\log\hat{\mu}(t\xi) = \sup_{y \in \mathrm{Dom}\,\lambda} [t\xi(y) - \lambda(y)] \leq t\alpha \ . \tag{2.13}$$

On the other hand,

$$\hat{\mu}(t\xi) \geq \delta\, e^{t\beta} \qquad \text{with } \delta = \mu(\{z:\xi(z) > \beta\}) > 0 \tag{2.14}$$

because $x \in S \cap \{z:\xi(z) > \beta\}$. Therefore, from (2.13) and (2.14), for all $t \in \mathbb{R}^+$

$$t(\beta-\alpha) \leq -\log\delta \ , \quad \text{a contradiction} \ .$$

Thus $S \subset \overline{\mathrm{Dom}\,\lambda}$, and since $\mathrm{Dom}\,\lambda$ is convex, $\overline{co(S)} \subset \overline{\mathrm{Dom}\,\lambda}$.

(b) follows by an elementary argument (see [5], p. 46). □

Lemma 2.5. Let B be finite-dimensional. Then

(a) If $x \in \mathrm{ri}(\mathrm{Dom}\ \lambda)$, then there exists $\xi \in B'$ such that $\psi(\xi) = x$.

(b) If $x \in \mathrm{Dom}\ \lambda$, then for every $\varepsilon > 0$ there exists $y \in \psi(B')$ such that

$$\|x-y\| < \varepsilon , \quad |\lambda(x) - \lambda(y)| < \varepsilon .$$

Proof. Let $m = \int x\mu(dx)$, and define $\tilde{\mu} = \delta_{-m} * \mu$ and $\tilde{\lambda}, \tilde{\psi}, \tilde{S}$ in the obvious way. Then it is easily verified that $\int x\tilde{\mu}(dx) = 0$ and

$$\tilde{\lambda} = \lambda(\cdot + m), \quad \mathrm{Dom}\ \tilde{\lambda} = \mathrm{Dom}\ \lambda - m, \ \tilde{\psi} = \psi - m, \quad \tilde{S} = S - m .$$

It is clear from these relations that in order to prove (a) and (b) for μ it is enough to prove them for $\tilde{\mu}$; therefore we may assume that $m = 0$.

This implies that the common affine hull of $\mathrm{Dom}\ \lambda$ and S (Lemma 2.4) is a linear subspace, and therefore we may reduce the general case to the case span $S = B$, and hence by Lemma 2.4 $\mathrm{ri}(\mathrm{Dom}\ \lambda) = \mathrm{int}(\mathrm{Dom}\ \lambda)$ (here "int" is the usual interior).

Let $x \in \mathrm{int}(\mathrm{Dom}\ \lambda)$. Let $U = \{x : \|x\| \leq 1\}$ and let $r > 0$ be such that $x + r\,U \subset \mathrm{int}(\mathrm{Dom}\ \lambda)$. By an elementary property of convex functions, λ is continuous or $\mathrm{int}(\mathrm{Dom}\ \lambda)$, and therefore

$$b = \sup_{y \in x + r\,U} \lambda(y) < \infty .$$

Now for all $\xi \in B'$, $y \in x + r\,U$, say $y = x + rv$ with $v \in U$,

$$\xi(y) \leq \log\hat{\mu}(\xi) + \lambda(y),$$

$$r\xi(v) \leq \log\hat{\mu}(\xi) + b - \xi(x) ,$$

which implies

$$r\|\xi\| \leq \log\hat{\mu}(\xi) + b - \xi(x),$$

$$\xi(x) - \log\hat{\mu}(\xi) \leq b - r\|\xi\| ,$$

$$\lim_{\|\xi\|\to\infty} [\xi(x) - \log\hat{\mu}(\xi)] = -\infty .$$

Since B is locally compact, this implies that there exists $\xi \in B'$ such that $\lambda(x) = \xi(x) - \log\hat{\mu}(\xi)$. If $g(\eta) = \eta(x) - \log\hat{\mu}(\eta)$ for $\eta \in B'$, it follows that $g'(\xi) = 0$, and hence $x = \psi(\xi)$. This proves (a).

Let $y \in \text{Dom}\,\lambda$, $x \in \text{int}(\text{Dom}\,\lambda)$. Then for every $\alpha \in (0,1]$, $z_\alpha = \alpha x + (1-\alpha)y \in \text{int}(\text{Dom}\,\lambda)$. Therefore, by the convexity of λ,

$$\lambda(z_\alpha) \leq \alpha\lambda(x) + (1-\alpha)\lambda(y) \quad ,$$

$$\overline{\lim_{\alpha\to 0}}\, \lambda(z_\alpha) \leq \lambda(y) \quad .$$

On the other hand, the lower semicontinuity of λ implies

$$\underline{\lim}_{\alpha\to 0}\, \lambda(z_\alpha) \geq \lambda(y) \quad .$$

Hence $\lambda(y) = \lim_{\alpha\to 0} \lambda(z_\alpha)$. By (a), for each $\alpha > 0$ there exists $\xi_\alpha \in B'$ such that $\psi(\xi_\alpha) = z_\alpha$; (b) follows. □

<u>Lemma 2.6</u>. For $a \geq 0$, let $L_a = \{x:\lambda(x) \leq a\}$. Then for every $0 \leq a < b$,

$$L_a \subset \overline{\psi(B') \cap L_b} \quad .$$

<u>Proof</u>. Let $Q = \overline{\psi(B') \cap L_b}$. By statement (3) of Theorem 2.1, Q is compact. Let $x \in L_a$ and suppose $x \notin Q$. By the Hahn-Banach theorem, there exist $d \in \mathbb{N}$ and a continuous linear map $T:B \to \mathbb{R}^d$ such that $T(x) \notin T(Q)$.

Let $\tilde{\mu} = T(\mu)$, $u = T(x)$. Let $\tilde{\lambda}$ be the Cramér transform of $\tilde{\mu}$. Since $\hat{\tilde{\mu}} = \hat{\mu} \circ T'$, where $T':(\mathbb{R}^d)' \to B'$ is the transpose of T, it is easily checked that $\tilde{\lambda} \circ T \leq \lambda$, and therefore $\tilde{\lambda}(u) \leq \lambda(x) \leq a$.

Let $\rho = \inf\{\|u-z\|:z \in T(Q)\}$, where $\|\cdot\|$ is any norm in $\mathbb{R}^d$. By Lemma 2.5, there exists $v \in (\mathbb{R}^d)'$ such that

$$\|u - \tilde{\psi}(v)\| < \rho/2 \quad , \tag{2.14}$$

$$|\tilde{\lambda}(u) - \tilde{\lambda}(\tilde{\psi}(v))| < b - a \quad , \tag{2.15}$$

where $\widetilde{\psi}$ is defined for $\widetilde{\mu}$ in the obvious way. It is easily checked that $\widetilde{\psi} = T \circ \psi \circ T'$ and hence (2.14) may be rewritten

$$\| T(x) - T(\psi(T'(v))) \| < \rho/2 \quad . \tag{2.16}$$

We claim now that

$$\psi(T'(v)) \in L_b \quad . \tag{2.17}$$

It will then follow that $\|T(x) - T(\psi(T'(v)))\| \geq \rho$, contradicting (2.16) and thus proving that $x \in Q$.

To prove (2.17): by (2.7), setting $\xi = T'(v)$,

$$\begin{aligned}\lambda(\psi(\xi)) &= \xi(\psi(\xi)) - \log\hat{\mu}(\xi) \\ &= v(\widetilde{\psi}(v)) - \log\hat{\widetilde{\mu}}(v) \\ &= \widetilde{\lambda}(\widetilde{\psi}(v)) < a + (b-a) = b. \quad \square\end{aligned}$$

Proof of Proposition 2.3.

By (2.7), $\psi(B') \subset \mathrm{Dom}\,\lambda$. Now let $x \in \mathrm{Dom}\,\lambda$, $\varepsilon > 0$. By Lemma 2.6, there exists $x_\varepsilon \in \psi(B') \cap L_{a+\varepsilon}$ such that $\|x_\varepsilon - x\| < \varepsilon$, where $a = \lambda(x)$. Then $\lim_{\varepsilon\to 0} x_\varepsilon = x$, $\overline{\lim}_{\varepsilon\to 0} \lambda(x_\varepsilon) \leq a$. On the other hand, by the lower semicontinuity of λ,

$$a \leq \underline{\lim}_{\varepsilon\to 0} \lambda(x_\varepsilon) \quad .$$

Hence $\lim_{\varepsilon\to 0} \lambda(x_\varepsilon) = \lambda(x)$. $\square$

3. Existence of limits for certain sets.

The following result is an immediate consequence of Theorem 2.1.

Lemma 3.1. Let C be a Borel set in B such that $\Lambda(\mathrm{int}(C)) = \Lambda(\overline{C})$. Then

$$\lim_n n^{-1} \log P\{S_n/n \in C\} = -\Lambda(C) \quad . \tag{3.1}$$

Let $[x,y) = \{(1-\alpha)x + \alpha y : \alpha \in [0,1)\}$, for $x, y \in B$.

Theorem 3.2. Let C be a Borel set in B such that for every $y \in \partial C \cap \text{Dom}\,\lambda$, there exists $x \in C \cap \text{Dom}\,\lambda$ such that $[x,y) \subset \text{int}(C)$. Then (3.1) holds.

Proof. It is enough to show that $\Lambda(\text{int}(C)) \leq \Lambda(\overline{C})$. Let y, x be as above, and let $z_\alpha = (1-\alpha)x + \alpha y$. Then

$$\lambda(z_\alpha) \leq (1-\alpha)\,\lambda(x) + \alpha\,\lambda(y) \quad ,$$

and since $z_\alpha \in \text{int}(C)$ for $\alpha \in [0,1)$ and $\lambda(x) < \infty$,

$$\Lambda(\text{int}(C)) \leq \underline{\lim}_{\alpha \uparrow 1}\, \lambda(z_\alpha) \leq \lambda(y) \quad .$$

Since y is an arbitrary point of $\partial C \cap \text{Dom}\,\lambda$, this implies $\Lambda(\text{int}(C)) \leq \Lambda(\overline{C})$. □

Theorem 3.3. Let C be a convex Borel set in B such that $\text{int}(C) \cap \text{Dom}\,\lambda \neq \emptyset$. Then (3.1) holds.

Proof. Let $x \in \text{int}(C) \cap \text{Dom}\,\lambda$ and let $y \in \partial C$. Then by an elementary property of convex sets (see [6], p. 38), we have $[x,y) \subset \text{int}(C)$. The result follows now from Theorem 3.2. □

Corollary 3.4 (Bahadur and Zabell [3]). Let C be an open convex set. Then (3.1) holds.

Proof. If $C \cap \text{Dom}\,\lambda \neq \emptyset$, we apply Theorem 3.3. If $C \cap \text{Dom}\,\lambda = \emptyset$, then it follows from a standard separation theorem ([6], p. 64) that $C \cap \overline{\text{Dom}\,\lambda} = \emptyset$. Since $\overline{\text{Dom}\,\lambda} = \overline{\text{co}(S)}$ by Lemma 2.4 and obviously $P\{S_n/n \in \overline{\text{co}(S)}\} = 1$ for all n, we have $P\{S_n/n \in C\} = 0$ and hence (3.1) is satisfied with $\Lambda(C) = \infty$. □

Remarks (a) It is easily verified that finite unions of sets satisfying (3.1) also satisfy (3.1) ([2] or [3]).

(b) In [2] and [3] Corollary 3.4 is proved under a weaker integrability assumption.

Acknowledgement. I thank J. Kuelbs for some useful comments.

References

[1] de Acosta, A. (1984). Upper bounds for large deviations of dependent random vectors. To appear in Z. Wahrscheinlichkeitstheorie.

[2] Azencott, R. (1980). Grandes deviations et applications. Lecture Notes in Mathematics 774. Springer-Verlag, Berlin and New York.

[3] Bahadur, R. R. and Zabell, S. (1979). Large deviations of the sample mean in general vector spaces. Ann. Probability 7, 587-621.

[4] Donsker, M. D. and Varadhan, S. R. S. (1976). Asymptotic evaluation of certain Markov process expectations for large time III. Comm. Pure Appl. Math. 29, 389-461.

[5] Rockafellar, R. T. (1970). Convex Analysis. Princeton University Press.

[6] Schaefer, H. H. (1966). Topological vector spaces. MacMillan, New York.

The Non-Existence of a Universal Multiplier Moment for the Central Limit Theorem

*Kenneth S. Alexander**
Department of Mathematics
University of Washington
Seattle, Washington 98195

The following question was raised by J. Hoffman-Jørgensen. Does there exist a non-negative function ψ on $[0,\infty) \times [0,\infty)$, non-decreasing in each variable, with the following property: If B is a Banach space, X a B-valued random element satisfying the central limit theorem (CLT), and ξ a real random variable for which $E(\xi X) = 0$ and $E\psi(|\xi|, \|X\|) < \infty$, then ξX satisfies the CLT in B. We call such a function ψ a ***universal multiplier moment.*** In this note we show no universal multiplier moment exists.

Let $\mathbb{Z}_+$ denote the positive integers, let C be the class of all subsets of $\mathbb{Z}_+$, and let $\ell^\infty(C)$ be the Banach space of all bounded functions on C, endowed with the sup norm $\|\cdot\|_C$. Let ψ be a non-negative function on $[0,\infty)\times[0,\infty)$, non-decreasing in each variable. Let $\psi_1(x) := \psi(x,1)$. We may assume $\psi_1(O) = 0$, and by using a larger function if necessary, we may assume ψ_1 is continuous and strictly increasing, so ψ_1^{-1} is well-defined. Choose a sequence (b_m) satisfying

$$(1) \qquad b_m \searrow 0, \quad \Sigma\, b_m < \infty, \quad \text{and} \quad \Sigma\psi_1^{-1}(m)b_m = \infty.$$

Then $mb_m \to 0$, so $m \le b_m^{-1}$ eventually, and therefore by (1),

$$(2) \qquad \Sigma\, \psi_1^{-1}(b_m^{-1})b_m = \infty.$$

We may assume $\Sigma\, b_m^2 = 1$ and $b_m^2 \le 1/2$. Let P be the law on $\mathbb{Z}_+$ given by $P(m) = b_m^2$, let Y be an r.v. with law P, let ε be a Rademacher r.v. (i.e. $Pr[\varepsilon = 1] = Pr[\varepsilon = -1] = 1/2$) independent of Y, and let $X := \varepsilon\delta_Y$. Then X is a random element of $\ell^\infty(C)$. By (1) we have $\Sigma\, P(m)^{1/2} < \infty$, so a theorem of Durst and Dudley (1981) tells us that $\delta_Y - P$ satisfies the CLT in $\ell^\infty(C)$. In other words, the empirical process on C for the law P converges weakly to a sample-continuous Gaussian limit. By the arguments in Section 2 of Giné and Zinn (1984), especially Theorem 2.14 and its proof and Lemma 2.7, X therefore also satisfies the CLT in $\ell^\infty(C)$.

Now define φ on $\mathbb{Z}_+$ by

$$\varphi(m) := \psi_1^{-1}(b_m^{-1}) \wedge b_m^{-1},$$

and define the r.v.

$$\xi := \varphi(Y).$$

Then $E(\xi X) = 0$ and

* Research supported by an NSF Postdoctoral Fellowship, Grant No. MCS-83-11686.

$$E\psi(|\xi|, \|X\|) \le E\psi_1(|\xi|) = E\psi_1(\varphi(Y))$$
$$= \Sigma\psi_1(\varphi(m))P(m) = \Sigma(b_m^{-1} \wedge \psi_1(b_m^{-1}))b_m^2$$
$$\le \Sigma\, b_m < \infty,$$

but we will show that ξX is not pregaussian, hence *a fortiori* does not satisfy the CLT.

Suppose W is a mean-0 Gaussian r.v. in $\ell^\infty(\mathcal{C})$ with the same covariance as ξX. Then $E\, W(m)W(n) = 0$ for all $m \ne n$, so W is finitely additive a.s. Therefore using (2),

$$E\, \|W\|_{\mathcal{C}} \ge \frac{1}{2}\Sigma\, E\,|W(m)| = \frac{1}{(2\pi)^{1/2}}\Sigma(E\, W^2(m))^{1/2}$$
$$= \frac{1}{(2\pi)^{1/2}}\Sigma\,\{E(\xi X)^2(m)\}^{1/2}$$
$$\ge \frac{1}{(2\pi)^{1/2}}\Sigma\{P(m)(1-P(m))^2\varphi(m)^2\}^{1/2}$$
$$\ge \frac{1}{8}\Sigma\,\varphi(m)b_m$$
$$= \frac{1}{8}\Sigma(\psi_1^{-1}(b_m^{-1})b_m) \wedge 1 = \infty.$$

But we may assume the law of W has separable support (Marczewski and Sikorski, 1948), and a Gaussian r.v. in a separable Banach space always has a finite first moment (see e.g. Theorem 6.5 of Araujo and Giné, 1980) so this is a contradiction. Thus ξX does not satisfy the CLT.

The CLT for ξX is equivalent, again by the arguments in Section 2 of Giné and Zinn (1984), to the CLT for the empirical process indexed by $\{\varphi 1_C : C \in \mathcal{C}\}$. Thus our example shows that no moment condition on φ is sufficient to ensure that when the empirical process indexed by $\mathcal{C}$ satisfies the CLT, so does the empirical process on $\{\varphi 1_C : C \in \mathcal{C}\}$.

When ξ is independent of X, Giné and Zinn (personal communication) have observed that if $\int_0^\infty P[|\xi| > u]^{1/2} du < \infty$ and X satisfies the CLT, then ξX satisfies the CLT. This is slightly stronger than the condition $E\xi^2 < \infty$. The proof is similar to that of Lemma 2.9 of Giné and Zinn (1984). LeDoux and Talagrand (1984) have shown that this is best possible.

References

Araujo, A. and Giné, E. (1980). *The Central Limit Theorem for Real and Banach Valued Random Variables.* Wiley, New York.

Durst, M. and Dudley, R.M. (1981). Empirical processes, Vapnik-Červonenkis classes and Poisson processes. *Prob. Math. Statist.* (Wroclaw) 1 no. 2, 109-115.

Giné, E. and Zinn, J. (1984). Some limit theorems for empirical processes. *Ann. Probability* **12**, 929-989.

LeDoux, M. and Talagrand, M. (1984). Conditions d'intégrabilité pour les multiplicateurs dans les TLC Banachique. Preprint.

Marczewski, E. and Sikorski, P. (1948). Measures in nonseparable metric spaces. *Colloq. Math.* **1**, 133-139.

THE FATOU INEQUALITY REVISITED. - VARIATIONS ON A THEME BY A. DVORETZKY.

D. Austin, A. Bellow[1)] and N. Bouzar[2)]

0. Introduction

The purpose of this paper is to elaborate on the basic theme of Dvoretzky's paper [5]. In the process, we establish a natural connection with the ideas contained in [3].

1. Preliminaries

Throughout the paper $(\Omega, \mathcal{F}, P)$ is a *fixed probability* space, $(\mathcal{F}_n, n \in \mathbb{N})$ is a *fixed, increasing sequence of sub-σ-fields of* $\mathcal{F}$. We let $\mathcal{A} = \bigcup_{n\in\mathbb{N}} \mathcal{F}_n$. Clearly $\mathcal{A}$ is a *field*. We shall assume throughout the paper that $\mathcal{F} = \sigma(\mathcal{A})$, that is, $\mathcal{F}$ *is the σ-field generated by* $\mathcal{A}$. A *stopping time relative to* $(\mathcal{F}_n, n \in \mathbb{N})$ is a mapping $\tau: \Omega \to \mathbb{N}$ such that $[\tau = n] \in \mathcal{F}_n$ for every $n \in \mathbb{N}$. Let T be the set of bounded stopping times relative to $(\mathcal{F}_n, n \in \mathbb{N})$. For each $\tau \in T$ we define

$$\mathcal{F}_\tau = \{A \in \mathcal{F}: A \cap [\tau = n] \in \mathcal{F}_n \text{ for every } n \in \mathbb{N}\}.$$

We recall that $\mathcal{F}_\tau$ is a σ-field, τ is $\mathcal{F}_\tau$-measurable, and $\sigma \le \tau$ implies $\mathcal{F}_\sigma \subset \mathcal{F}_\tau$. For each $\tau \in T$ we also define

$$T(\tau) = \{\sigma \in T \mid \sigma \ge \tau\}.$$

A sequence $(X_n, n \in \mathbb{N})$ of (real valued) random variables is called *adapted* relative to $(\mathcal{F}_n, n \in \mathbb{N})$ if for each $n \in \mathbb{N}$ X_n is $\mathcal{F}_n$-measurable. If $(X_n, n \in \mathbb{N})$ is a sequence of random variables, then X_τ denotes the random variable defined by

$$X_\tau(\omega) = X_{\tau(\omega)}(\omega), \quad \omega \in \Omega.$$

Note that if $(X_n, n \in \mathbb{N})$ is adapted then X_τ is $\mathcal{F}_\tau$-measurable.

We denote by $L^1(\Omega, \mathcal{F}, P)$ the Banach space of integrable random variables with its usual norm: $\|X\|_1 = E|X|$.

If $(a_\tau; \tau \in T)$ is a net of real numbers we define

$$\underline{\lim}_{\tau\in T} a_\tau = \liminf_{\tau \in T} a_\tau = \sup_{\tau\in T}(\inf_{\sigma\ge\tau} a_\sigma).$$

Note that $\underline{\lim}_{\tau\in T} a_\tau = \sup_{n\in\mathbb{N}}(\inf_{\sigma\ge n} a_\sigma)$.

<u>Definition 1.1</u>: Let $(X_n, n \in \mathbb{N})$ be a sequence of random variables and $Y: \Omega \to \mathbb{R}$ be a mapping. Then Y is said to be a *measurable cluster point*, or simply a cluster point of the sequence $(X_n, n \in \mathbb{N})$ if Y is $\mathcal{F}$-measurable and for almost every $\omega \in \Omega$, $Y(\omega)$ is a cluster point of the sequence $(X_n(\omega), n \in \mathbb{N})$.

Equivalently, Y is a cluster point of $(X_n, n \in \mathbb{N})$ if and only if Y is measurable and $\underline{\lim}_n |X_n - Y| = 0$ a.s.

We now recall two results that will be used later.

<u>Lemma 1.1</u> (Fatou's Lemma). Let $(Y_n, n \in \mathbb{N})$ be a sequence of positive random variables. Then

$$E(\underline{\lim}_{n\in\mathbb{N}} Y_n) \le \underline{\lim}_{\tau\in T} EY_\tau.$$

Proof: We remark that $\underline{\lim}_n Y_n = \underline{\lim}_\tau Y_\tau$. The Lemma follows by applying the monotone convergence theorem to the sequence $(\inf_{\sigma \geq n} Y_\sigma,\ n \in \mathbb{N})$.

Proposition 1.1. Let $Y \in L^1(\Omega, \mathcal{F}, P)$. Then, for every increasing sequence $(\tau_n,\ n \in \mathbb{N})$ in T such that $\tau_n \to +\infty$, the sequence $(E(Y|\mathcal{F}_{\tau_n}),\ n \in \mathbb{N})$ converges to Y both a.s. and in $L^1(\Omega, \mathcal{F}, P)$. Moreover the net $(E(Y|\mathcal{F}_\tau),\ \tau \in T)$ converges to Y in the L^1-norm.

Proof: Note that by assumption $\mathcal{F} = \sigma(\bigcup_{n \in \mathbb{N}} \mathcal{F}_n) = \sigma(\bigcup_{\tau \in T} \mathcal{F}_\tau)$. The conclusion follows by applying Prop. II-2-11 and Prop. V-1-2 in [7].

Let $\mu: \mathcal{A} \to \mathbb{R}$ be a finitely additive measure. We define the measure $|\mu|: \mathcal{A} \to \mathbb{R}_+$, called the *variation* of μ, by

$$|\mu|(A) = \sup\{\sum_{B \in \pi} |\mu(B)| : \pi \text{ finite partition of } A \text{ in } \mathcal{A}\}.$$

μ is said to be of *finite variation* if $|\mu|(\Omega) < \infty$. The measure μ is said to be *absolutely continuous with respect to* P if $\mu(A) \to 0$ when $P(A) \to 0$. Note that μ is absolutely continuous with respect to P if and only if $|\mu|$ is absolutely continuous with respect to P. The measure μ is said to be *singular with respect to* P if the measure $|\mu| \wedge P$ defined by

$$|\mu| \wedge P(A) = \inf\{|\mu|(B) + P(A \setminus B) : B \in \mathcal{A},\ B \subset A\}$$

for $A \in \mathcal{A}$ is identically 0. Or equivalently, for every $\varepsilon > 0$, there exists $A \in \mathcal{A}$ such that $P(A) > 1 - \varepsilon$ and $|\mu|(A) < \varepsilon$.

2. The measures $(\Lambda_Y,\ Y \in L^1(\Omega, \mathcal{F}, P))$

Definition 2.1: Let $(X_n,\ n \in \mathbb{N})$ be an adapted sequence of integrable random variables and let $Y \in L^1(\Omega, \mathcal{F}, P)$. We define a set function $\Lambda_Y: \mathcal{A} \to [0, +\infty]$ by

$$\Lambda_Y(A) = \underline{\lim}_{\tau \in T} E|X_\tau - Y|\, I_A, \qquad A \in \mathcal{A}.$$

When it is necessary to exhibit the dependence of Λ_Y on the sequence $(X_n,\ n \in \mathbb{N})$ we will write $\Lambda_Y(\cdot, (X_n,\ n \in \mathbb{N}))$. Similarly for the set function ϕ_Y which will be introduced later.

Let $(X_n,\ n \in \mathbb{N})$ be an adapted sequence of integrable random variables and let $Y \in L^1(\Omega, \mathcal{F}, P)$. We define a sequence of set functions $\mu_n: \mathcal{F}_n \to \mathbb{R}_+$ as follows:

$$\mu_n(A) = \inf\{\int_A |(X_\tau - E(Y|\mathcal{F}_\tau)|dP : \tau \in T(n)\}, \qquad A \in \mathcal{F}_n.$$

For $\sigma \in T$, we define

$$\mu_\sigma(A) = \sum_{i=1}^{+\infty} \mu_i(A \cap [\sigma = i]), \qquad A \in \mathcal{F}_\sigma.$$

Note that the sum above is really finite.

Lemma 2.1. Let $(X_n,\ n \in \mathbb{N})$ be an adapted sequence of integrable random variables and let $Y \in L^1(\Omega, \mathcal{F}, P)$. Then

i) For each $n \in \mathbb{N}$, μ_n is finitely additive and absolutely continuous with respect to P.

ii) For each $n \in \mathbb{N}$ and $A \in \mathcal{F}_n$, $\mu_n(A) \le \mu_{n+1}(A)$.

iii) For each $\sigma \in T$ and each $A \in \mathcal{F}_\sigma$,

$$\mu_\sigma(A) = \inf\{\int_A |X_\tau - E(Y|\mathcal{F}_\tau)|dP: \tau \in T(\sigma)\}.$$

<u>Proof</u>: ii) is straightforward from the definition of $(\mu_n, n \in \mathbb{N})$. i) and iii) have already been proved in [3]. For the sake of completeness we give a proof here:

i) Let $A_1, A_2 \in \mathcal{F}_n$, $A_1 \cap A_2 = \emptyset$. By definition

$$\mu_n(A_1 \cup A_2) \ge \mu_n(A_1) + \mu_n(A_2).$$

For $\varepsilon > 0$, $\exists\ \tau_i \in T(n)$ such that

$$\mu_n(A_i) + \frac{\varepsilon}{2} \ge \int_{A_i} |X_{\tau_i} - E(Y|\mathcal{F}_{\tau_i})|dP, \qquad i = 1,2$$

Define $\tau \in T(n)$ by $\tau = \tau_1 I_{A_1} + \tau_2 I_{A_2} + n I_{(A_1 \cup A_2)^c}$.

$$\int_{A_1 \cup A_2} |X_\tau - E(Y|\mathcal{F}_\tau)|dP = \int_{A_1} |X_{\tau_1} - E(Y|\mathcal{F}_{\tau_1})|dP + \int_{A_2} |X_{\tau_2} - E(Y|\mathcal{F}_{\tau_2})|dP$$

$$\le \mu_n(A_1) + \mu_n(A_2) + \varepsilon.$$

Therefore $\mu_n(A_1 \cup A_2) \le \mu_n(A_1) + \mu_n(A_2)$.

The absolute continuity of μ_n with respect to P follows from the inequality

$$\mu_n(A) \le \int_A |X_n - E(Y|\mathcal{F}_n)|dP.$$

iii) Let $m = \max\{\sigma(\omega): \omega \in \Omega\}$, $\tau \in T(\sigma)$ and $A \in \mathcal{F}_\sigma$. Then $\int_A |X_\tau - E(Y|\mathcal{F}_\tau)|dP = \sum_{i=1}^m \int_{A\cap[\sigma=i]} |X_\tau - E(Y|\mathcal{F}_\tau)|dP$. For each $i \in \{1,\dots,m\}$ define $\tau_i \in T(i)$ by

$$\tau_i = \tau I_{[\sigma=i]} + i I_{[\sigma=i]^c}.$$

Therefore

$$\int_A |X_\tau - E(Y|\mathcal{F}_\tau)|dP = \sum_{i=1}^m \int_{A\cap[\sigma=i]} |X_{\tau_i} - E(Y|\mathcal{F}_{\tau_i})|dP$$

$$\ge \sum_{i=1}^m \mu_i(A\cap[\sigma=i]) = \mu_\sigma(A).$$

Now for each $\varepsilon > 0$ and $i \in \{1,\dots,m\}$ there exists $\tau_i \in T(i)$ such that

$$\mu_i(A\cap[\sigma=i]) + \frac{\varepsilon}{m} \ge \int_{A\cap[\sigma=i]} |X_{\tau_i} - E(Y|\mathcal{F}_{\tau_i})|dP.$$

Define $\tau \in T(\sigma)$ by $\tau = \sum_{i=1}^m \tau_i I_{[\sigma=i]}$. Therefore

$$\int_A |X_\tau - E(Y|\mathcal{F}_\tau)|dP = \sum_{i=1}^m \int_{A\cap[\sigma=i]} |X_{\tau_i} - E(Y|\mathcal{F}_{\tau_i})|dP$$

$$\leq \sum_{i=1}^{m} \mu_i(A \cap [\sigma = i]) + \varepsilon = \mu_\sigma(A) + \varepsilon.$$

Q.E.D.

Proposition 2.1. Let $(X_n, n \in \mathbb{N})$ be an adapted sequence of integrable random variables and let $Y \in L^1(\Omega, \mathcal{F}, P)$. Then

i) For every $A \in \mathcal{A}$,

$$\Lambda_Y(A) = \underline{\lim}_{\tau \in T} E|X_\tau - E(Y|\mathcal{F}_\tau)| I_A = \lim_{n \in \mathbb{N}} \mu_n(A).$$

ii) Λ_Y is finitely additive.

iii) If $(\sigma_n, n \in \mathbb{N})$ is an increasing sequence in T such that $\sigma_n \in T(n)$ for each $n \in \mathbb{N}$, then for every $A \in \mathcal{A}$

$$\Lambda_Y(A) = \lim_{n \in \mathbb{N}} (\mu_{\sigma_n}(A).$$

Proof: i) First note that $\lim_n \mu_n(A)$ always exists by Lemma 2.1(ii). The second equality is straightforward from the definitions. The first equality is a direct consequence of Prop. 1.1 and the definition of Λ_Y.

ii) By Lemma 2.1(i), μ_n is finitely additive. Therefore $\Lambda_Y = \lim_n \mu_n$ is finitely additive.

iii) Clearly $\mathcal{A} = \bigcup_{n \in \mathbb{N}} \mathcal{F}_{\sigma_n}$. Let $N_n = \max \sigma_n$. By Lemma 2.1(iii), $\mu_{\sigma_n}(A) \leq \mu_{\sigma_{n+1}}(A)$ for every $A \in \mathcal{F}_{\sigma_n}$. Fix $n \in \mathbb{N}$, $A \in \mathcal{F}_n$ and $k \geq n$. Since $k \leq \sigma_k \leq N_k$

$$\mu_k(A) \leq \mu_{\sigma_k}(A) \leq \mu_{N_k}(A).$$

Therefore

$$\Lambda_Y(A) = \lim_{\substack{k \\ k \geq n}} \mu_k(A) \leq \lim_k \mu_{\sigma_k}(A) \leq \lim_k \mu_{N_k}(A) = \Lambda_Y(A).$$

Q.E.D.

Remarks. 2.1. Λ_Y need not be countably additive on $\mathcal{A}$ as the following example shows. Let $\Omega = (0,1]$, $\mathcal{F}$ = Borel sets of $(0,1]$, P = Lebesgue measure on $(0,1]$. Let $Y \equiv 0$ and for each $n \in \mathbb{N}$ let $X_n = 2^n I_{(0,2^{-n})}$, $\mathcal{F}_n = \sigma(X_1, \ldots, X_n)$, $A_n = (2^{-n}, 2^{-n+1}]$. Then $\Lambda_Y(A_n) = 0$ for all $n \in \mathbb{N}$ but $\Lambda_Y(\Omega) = \Lambda_Y(\bigcup_n A_n) = 1$.

2.2. Λ_Y need not be σ-finite. In fact Λ_Y could be infinite on $\mathcal{A} \setminus \{\phi\}$ as the following example shows: Let $\Omega_0 = \{0,1\}$, $\mathcal{F}_0 = \mathcal{P}(\Omega)$, $P_0(\{0\}) = P_0(\{1\}) = 1/2$. Let $(\Omega, \mathcal{F}, P)$ be the probability space that is the infinite product of copies of $(\Omega_0, \mathcal{F}_0, P_0)$. Let $(pr_n, n \in \mathbb{N})$ be the sequence of projections from Ω to Ω_0. For every $n \in \mathbb{N}$ we set

$$A_n = pr_n^{-1}\{0\}, \qquad B_n = pr_n^{-1}\{1\}.$$

We also define $X_n = 2^{2n} I_{B_n}$ and $\mathcal{F}_n = \sigma(X_1, \ldots, X_n)$ for each $n \in \mathbb{N}$. Note that $(X_n, n \in \mathbb{N})$ is a sequence of integrable and independent random variables. We take $Y \equiv 0$. Let now $A \in \mathcal{A}$, then $A \in \mathcal{F}_{N_A}$ and A is a union of atoms,

$$A = \bigcup_i C_i, \qquad \text{each } C_i \in \mathcal{F}_{N_A}.$$

Take $k \in \mathbb{N}$ and let $\tau \in T$, $\tau \geq k + N_A$. Since τ is bounded, there is $m \geq k$ such that

$k + N_A \le \tau \le m + N_A$. Fix now $C = C_i$ one of the atoms of A. We know that

$$C \cap \Big(\bigcap_{j=k}^{m} B_{N_a + j}\Big) \neq \emptyset$$

and that the sets $C \cap [\tau = N_A + j]$, $(k \le j \le m)$, form a partition of C. Hence for some $j^* = j_i^*$, $k \le j^* \le m$,

$$C \cap [\tau = N_A + j^*] \cap \Big(\bigcap_{j=k}^{m} B_{N_A + j}\Big) \neq \emptyset$$

which implies that

$$C \cap [\tau = N_A + j^*] \cap B_{N_A + j^*} \neq \emptyset$$

and since this is in $\mathcal{F}_{N_A + j^*}$,

$$P(C \cap [\tau = N_A + j^*] \cap B_{N_A + j^*}) \ge \frac{1}{2^{N_A + j^*}} = P(C) \cdot \frac{1}{2^{j^*}}.$$

We deduce that

$$\int_A X_\tau dP = \sum_i \int_{C_i} X_\tau dP \ge \sum_i \int_{C_i \cap [\tau = N_A + j_i^*] \cap B_{N_A + j_i^*}} X_{N_A + j_i^*} dP$$

$$\ge \sum_i 2^{2(N_A + j_i^*)} P(C_i) \frac{1}{2^{j_i^*}} \ge 2^k P(A).$$

Hence $\Lambda_Y(A) = \varliminf_{\tau \in T} \int_A X_\tau \, dP = \infty$. Note also that by the Borel-Cantelli Lemma, $\varliminf_{n \in \mathbb{N}} X_n = 0$ a.s.

To avoid the situation of Remark 2.2 we introduce the following condition: an adapted sequence of integrable random variables $(X_n, n \in \mathbb{N})$ is said to satisfy *condition* (I) if

$$\varliminf_{\tau \in T} E|X_\tau| < \infty. \tag{I}$$

Clearly, if $(X_n, n \in \mathbb{N})$ satisfies condition (I), then for every $Y \in L^1(\Omega, \mathcal{F}, P)$, $\Lambda_Y(\Omega) < \infty$.

We now show that in the presence of condition (I) the measure Λ_Y is completely described by a martingale. First we need some more notation: if $\mu: \mathcal{Q} \to \mathbb{R}$ is a set function, then for every $n \in \mathbb{N}$ we let $R_n\mu$ denote the restriction of μ to $\mathcal{F}_n$.

Proposition 2.2. Let $(X_n, n \in \mathbb{N})$ be an adapted sequence of integrable random variables satisfying condition (I) and let $Y \in L^1(\Omega, \mathcal{F}, P)$. Then there exists a positive martingale $(M_n, n \in \mathbb{N})$ relative to $(\mathcal{F}_n, n \in \mathbb{N})$, unique up to indistinguishability, such that

$$\Lambda_Y(A) = \int_A M_n \, dP, \qquad A \in \mathcal{F}_n.$$

Proof: We prove the uniqueness assertion first. If $(M_n', n \in \mathbb{N})$ is another martingale such that $\Lambda_Y(A) = \int_A M'_n \, dP$, $A \in \mathcal{F}_n$, then $\int_A M_n \, dP = \int_A M'_n \, dP$, $A \in \mathcal{F}_n$, which implies $M_n = M'_n$ a.s. We now prove the existence. Fix $n \in \mathbb{N}$. We will show that $R_n \Lambda_Y$ is absolutely continuous with respect to P. For every $A \in \mathcal{F}_n$ we have $R_n \Lambda_Y(A) = \lim_{k \geq n} R_n \mu_k(A)$ by Proposition 2.1(i). Moreover by Lemma 2.1(i), $(R_n \mu_k), k \geq n)$ is a sequence of measures that are absolutely continuous with respect to P. Therefore by the Vitali-Hahn-Saks Theorem (Thm III-7-2, p. 158, [4]), $R_n \Lambda_Y$ is a countably additive measure that is absolutely continuous with respect to P. Since $R_n \Lambda_Y(\Omega) < \infty$, by the Radon-Nikodym Theorem (Thm 10-2, p. 176, [4]) there exists an integrable random variable M_n that is $\mathcal{F}_n$-measurable such that $R_n \Lambda_Y(A) = \int_A M_n \, dP$ for every $A \in \mathcal{F}_n$. Since $R_n \Lambda_Y \geq 0$, $M_n \geq 0$ a.s. That $(M_n, n \in \mathbb{N})$ is a martingale follows trivially from the defining relation of $(M_n, n \in \mathbb{N})$. Q.E.D.

Definition 2.2: Let $(X_n, n \in \mathbb{N})$ be an adapted sequence of integrable random variables satisfying condition (I) and let $Y \in L^1(\Omega, \mathcal{F}, P)$. Then the martingale $(M_n, n \in \mathbb{N})$ from Proposition 2.2 will be called the *martingale associated with* $(X_n, n \in \mathbb{N})$ *and* Y.

Remark 2.3. The existence of $(M_n, n \in \mathbb{N})$ in Proposition 2.2 can be established differently. We first prove the existence of a positive, L^1-bounded submartingale $(S_n, n \in \mathbb{N})$ such that $\Lambda_Y(A) = \lim_n \int_A S_n \, dP$ for $A \in \mathcal{A}$ (as was done in [3]). The sequence $(M_n, n \in \mathbb{N})$ is the martingale part of the Riesz decomposition of $(S_n, n \in \mathbb{N})$ (see for example [6], p. 31).

We need the following Lemma that is of interest in itself.

Lemma 2.2. Let $(M_n, n \in \mathbb{N})$ be a positive martingale. Then the set function $Q: \mathcal{A} \to \mathbb{R}_+$ defined by

$$Q(A) = \int_A M_n \, dP, \qquad A \in \mathcal{F}_n$$

is a well defined finitely additive measure. Moreover, Q is singular with respect to P if and only if $\lim_n M_n = 0$ a.s.

Proof: The first part of the Lemma is straightforward from the assumptions. Suppose now that $\lim_{n \in \mathbb{N}} M_n = 0$ a.s. Then, for $\varepsilon > 0$, $\exists\, n_0$ large enough such that $P(M_{n_0} > \varepsilon) < \varepsilon$. Let $A = [M_{n_0} \leq \varepsilon]$, $A \in \mathcal{F}_{n_0}$. Therefore $Q(A) = \int_A M_{n_0} \, dP < \varepsilon$. Hence Q is singular with respect to P. Conversely, suppose that Q is singular with respect to P. Then for $\varepsilon > 0$, $\delta > 0$, $\exists\, A \in \mathcal{A}$, $A \in \mathcal{F}_N$ say, such that $P(A) > 1 - \delta$ and $Q(A) < \varepsilon\delta$. Therefore,

$$\begin{aligned} P(\sup_{n \geq N} M_n > \varepsilon) &\leq P([\sup_{n \geq N} M_n > \varepsilon] \cap A) + P(A^c) \\ &\leq \frac{1}{\varepsilon} \sup_{n \geq N} E(M_n I_A) + \delta \\ &= \frac{1}{\varepsilon} E(M_N I_A) + \delta < \delta + \delta = 2\delta. \end{aligned}$$

The second inequality follows from the maximal inequality applied to the martingale

$(M_n I_A, n \geq N)$. Hence,

$$\lim_{N\to+\infty} P(\sup_{n\geq N} M_n > \varepsilon) = 0$$

which implies that $\lim_n M_n = 0$ a.s. Q.E.D.

We may now give some characterizations of measurable cluster points.

Proposition 2.3. Let $(X_n, n \in \mathbb{N})$ be an adapted sequence of integrable random variables satisfying condition (I). Let $Y \in L^1(\Omega,\mathcal{F},P)$. Let $(M_n, n \in \mathbb{N})$ be the martingale associated with $(X_n, n \in \mathbb{N})$ and Y. Then the following assertions are equivalent:

i) Y is a cluster point of $(X_n, n \in \mathbb{N})$.

ii) $\lim_n M_n = 0$ a.s.

iii) Λ_Y is singular with respect to P.

iv) There exists an increasing sequence $(\sigma_n, n \in \mathbb{N})$ in T such that $\sigma_n \in T(n)$ and such that X_{σ_n} converges to Y in probability.

Proof: The equivalence (i) <=> (iv) is well known and was proved in [1] (Lemma 1). The equivalence (ii) <=> (iii) follows from Proposition 2.2 and Lemma 2.2. Therefore we only have to show that (iii) <=> (iv).

(iii) => (iv). Let $\varepsilon_n \downarrow 0$. For each $n \in \mathbb{N}$ there exists $A_n \in \mathcal{A}$, $P(A_n) > 1 - \varepsilon_n$ and $\Lambda_Y(A_n) < \varepsilon_n$. By definition of Λ_Y, it implies that there exists a sequence $(\sigma_n, n \in \mathbb{N})$ in T such that for every $n \in \mathbb{N}$, $\sigma_{n+1} \geq \sigma_n$, $\sigma_n \in T(n)$ and

$$\int_{A_n} |X_{\sigma_n} - Y|dP < \varepsilon_n.$$

Now for $\varepsilon > 0$,

$$P(|X_{\sigma_n} - Y| > \varepsilon) \leq P(A_n^c) + \frac{1}{\varepsilon}\int_{A_n} |X_{\sigma_n} - Y|dP < \varepsilon_n + \frac{\varepsilon_n}{\varepsilon}.$$

Hence, $\lim_{n\in\mathbb{N}} P(|X_{\sigma_n} - Y| > \varepsilon) = 0$.

iv) => iii). By Proposition 1.1, $X_{\sigma_n} - E(Y|\mathcal{F}_{\sigma_n})$ converges to 0 in probability. Moreover by Proposition 2.1(iii) $\lim_n \mu_{\sigma_n}(\Omega) = \Lambda_Y(\Omega)$. Therefore for $\varepsilon > 0$, there exists n large enough such that $P(|X_{\sigma_n} - E(Y|\mathcal{F}_{\sigma_n})| < \frac{\varepsilon}{2}) > 1 - \frac{\varepsilon}{2}$ and $\mu_{\sigma_n}(\Omega) + \frac{\varepsilon}{2} > \Lambda_Y(\Omega)$. The latter inequality implies that for every $B \in \mathcal{F}_{\sigma_n}$, $\mu_{\sigma_n}(B) + \frac{\varepsilon}{2} > \Lambda_Y(B)$. For if not, there would exist $B \in \mathcal{F}_{\sigma_n}$ such that $\mu_{\sigma_n}(B) + \frac{\varepsilon}{2} \leq \Lambda_Y(B)$ and $\mu_{\sigma_n}(B^c) \leq \Lambda_Y(B^c)$ which implies $\Lambda_Y(\Omega) \geq \mu_{\sigma_n}(\Omega) + \frac{\varepsilon}{2}$, and hence a contradiction. Let $A = [|X_{\sigma_n} - E(Y|\mathcal{F}_{\sigma_n})| < \frac{\varepsilon}{2}]$. Then $A \in \mathcal{F}_{\sigma_n}$ and $P(A) > 1 - \frac{\varepsilon}{2}$. Moreover, by Lemma 2.1(iii)

$$\Lambda_Y(A) < \mu_{\sigma_n}(A) + \frac{\varepsilon}{2} \leq \int_A |X_{\sigma_n} - E(Y|\mathcal{F}_{\sigma_n})|dP + \frac{\varepsilon}{2} < \frac{\varepsilon}{2} + \frac{\varepsilon}{2} = \varepsilon.$$

Hence, Λ_Y is singular with respect to P. Q.E.D.

Remark 2.4: A similar equivalence to (ii) <=> (iv) of Proposition 2.3 was first proved in [3] (theorem 1) by a different method (see Remark 2.3 for the connection between the two results).

3. The Fatou discrepancy

Definition 3.1: Let $(X_n, n \in \mathbb{N})$ be an adapted sequence of integrable random variables satisfying condition (I). Let $Y \in L^1(\Omega,\mathcal{F},P)$. We define a set function $\Phi_Y: \mathcal{A} \to \mathbb{R}$ by

$$\Phi_Y(A) = \Lambda_Y(A) - E(\underline{\lim}_{n\in\mathbb{N}} |X_n - Y| I_A), \quad A \in \mathcal{A}.$$

When $Y = 0$ we write Φ instead of Φ_0, and Φ is called the *Fatou discrepancy* associated with $(X_n, n \in \mathbb{N})$.

We will show that $\Phi_Y = \Phi$ for any $Y \in L^1(\Omega,\mathcal{F},P)$. First we need some preliminary results.

Lemma 3.1. Let $(X_n, n \in \mathbb{N})$ be an adapted sequence of integrable random variables satisfying condition (I). Let $Y \in L^1(\Omega,\mathcal{F},P)$. Then,

i) $0 \leq \Phi_Y < \infty$.

ii) Φ_Y is finitely additive.

iii) If Y is a cluster point of $(X_n, n \in \mathbb{N})$ then $\Phi_Y = \Lambda_Y$.

Proof: (i) follows from Fatou's lemma and condition (I), (ii) follows from Proposition 2.1(ii), and (iii) is straightforward.

Lemma 3.2. Let $(X_n, n \in \mathbb{N})$ be an adapted sequence of integrable random variables satisfying condition (I). Let Y_1, Y_2 be in $L^1(\Omega,\mathcal{F},P)$. Then for every $A \in \mathcal{A}$,

$$|\Lambda_{Y_1}(A) - \Lambda_{Y_2}(A)| \leq E|Y_1 - Y_2| I_A.$$

Therefore the total variation $|\Lambda_{Y_1} - \Lambda_{Y_2}|$ of $\Lambda_{Y_1} - \Lambda_{Y_2}$ is absolutely continuous with respect to P.

Proof: Recall that $\Lambda_{Y_i}(A) = \underline{\lim}_{\tau\in T} E|X_\tau - E(Y_i|\mathcal{F}_\tau)| I_A$ for $A \in \mathcal{A}$ and $i \in \{1,2\}$. Now for $A \in \mathcal{A}$,

$$|X_\tau - E(Y_1|\mathcal{F}_\tau)| I_A \leq |X_\tau - E(Y_2|\mathcal{F}_\tau)| I_A + E(|Y_2 - Y_1||\mathcal{F}_\tau) I_A$$

which implies that $\Lambda_{Y_1}(A) - \Lambda_{Y_2}(A) \leq E|Y_2 - Y_1| I_A$. Likewise we show that $\Lambda_{Y_2}(A) - \Lambda_{Y_1}(A) \leq E|Y_2 - Y_1| I_A$ and hence $|\Lambda_{Y_2} - \Lambda_{Y_1}|(A) \leq E|Y_2 - Y_1| I_A$. Q.E.D.

Lemma 3.3. Let $(X_n, n \in \mathbb{N})$ be an adapted sequence of integrable random variables satisfying condition (I). Let $Y \in L^1(\Omega,\mathcal{F},P)$. Then Φ_Y is singular with respect to P.

Proof: For every $n \in \mathbb{N}$, let $\bar{X}_n = X_n - E(Y|\mathcal{F}_n)$. Note that $\bar{X}_\tau = X_\tau - E(Y|\mathcal{F}_\tau)$ for every $\tau \in T$, and that $\underline{\lim}_{n\in\mathbb{N}} |\bar{X}_n| = \underline{\lim}_{n\in\mathbb{N}} |X_n - Y|$ (by Proposition 1.1). Therefore

$$\phi_Y(A) = \lim_{\tau\in T} E|\bar{X}_\tau| I_A - E(\lim_{n\in \mathbb{N}} |\bar{X}_n| I_A), \quad A \in \mathcal{A}.$$

Hence,

$$\phi_Y(A) \le \lim_{\tau\in T} E| |\bar{X}_\tau| - \lim |\bar{X}_n| | I_A, \quad A \in \mathcal{A}.$$

By Proposition 2.3 applied to the sequence $(|\bar{X}_n|, n \in \mathbb{N})$ and the integrable cluster point $Z = \lim_{n\in\mathbb{N}} |\bar{X}_n|$, we deduce that $\Lambda_Z(\cdot, (|\bar{X}_n|, n \in \mathbb{N}))$ is singular with respect to P and therefore so is $\phi_Y(\cdot, (X_n, n \in \mathbb{N}))$. Q.E.D.

Proposition 3.1. Let $(X_n, n \in \mathbb{N})$ be an adapted sequence of integrable random variables satisfying condition (I). Then for any $Y_1, Y_2 \in L^1(\Omega,\mathcal{F},P)$, $\phi_{Y_1} = \phi_{Y_2}$.

Proof: Let $Z_i = \lim_{n\in\mathbb{N}} |X_n - Y_i|$ for $i = 1,2$. The total variation $|\phi_{Y_1} - \phi_{Y_2}|$ of $\phi_{Y_1} - \phi_{Y_2}$ satisfies

$$|\phi_{Y_1} - \phi_{Y_2}|(A) \le |\Lambda_{Y_1} - \Lambda_{Y_2}|(A) + E(Z_1 + Z_2)I_A$$

for every $A \in \mathcal{A}$. Therefore by Lemma 3.2 $|\phi_{Y_1} - \phi_{Y_2}|$ is absolutely continuous with respect to P. Since $|\phi_{Y_1} - \phi_{Y_2}| \le \phi_{Y_1} + \phi_{Y_2}$, we deduce, by Lemma 3.3, that $|\phi_{Y_1} - \phi_{Y_2}|$ is singular with respect to P. Therefore $|\phi_{Y_1} - \phi_{Y_2}| = 0$. Q.E.D.

Corollary 3.1. Let $(X_n, n \in \mathbb{N})$ be an adapted sequence of integrable random variables satisfying condition (I). Let ϕ be the Fatou discrepancy associated with $(X_n, n \in \mathbb{N})$. Then, for every $Y \in L^1(\Omega,\mathcal{F},P)$

i) $\phi_Y = \phi$.

ii) Y is a cluster point of $(X_n, n \in \mathbb{N})$ if and only if $\Lambda_Y = \phi$.

Proof: i) straightforward from Proposition 3.1.

ii) If Y is a cluster point then $\Lambda_Y = \phi_Y = \phi$ by (i). Conversely, if $\Lambda_Y = \phi$ then $\Lambda_Y = \phi_Y$ which implies that $E(\lim_{n\in\mathbb{N}} |X_n - Y|) = 0$ which in turn implies $\lim_{n\in\mathbb{N}} |X_n - Y| = 0$ a.s. Q.E.D.

Corollary 3.2 (Lebesgue decomposition). Let $(X_n, n \in \mathbb{N})$ be an adapted sequence of integrable random variables and let ϕ be the Fatou discrepancy associated with $(X_n, n \in \mathbb{N})$. Then, for any $Y \in L^1(\Omega,\mathcal{F},P)$ the Lebesgue decomposition of the measure Λ_Y is

$$\Lambda_Y(A) = \phi(A) + E(\lim_{n\in\mathbb{N}} |X_n - Y| I_A), \quad A \in \mathcal{A}.$$

Proof: The decomposition follows from Corollary 3.1(i). The fact that it is the Lebesgue decomposition of Λ_Y follows from the fact that ϕ is singular with respect to P (Lemma 3.3) and the fact that the measure $A \to E(\lim_{n\in\mathbb{N}} |X_n - Y| I_A)$ defined on $\mathcal{A}$ is absolutely continuous with respect to P. Q.E.D.

Proposition 3.2. Let $(X_n, n \in \mathbb{N})$ be an adapted sequence of integrable random variables satisfying condition (I). Let $Y \in L^1(\Omega,\mathcal{F},P)$ and let $(M_n, n \in \mathbb{N})$ be the martingale associated with $(X_n, n \in \mathbb{N})$ and Y. Let $M_\infty = \lim_{n\in\mathbb{N}} M_n$. Let ϕ be the Fatou discrepancy associated with $(X_n, n \in \mathbb{N})$. Then,

i) $M_\infty = \overline{\lim_{n\in\mathbb{N}}} |X_n - Y|$ a.s.

ii) For every $n \in \mathbb{N}$, $M_n \geq E(M_\infty|\mathcal{F}_n)$ a.s.

iii) For every $n \in \mathbb{N}$ and $A \in \mathcal{F}_n$,

$$\phi(A) = \int_A (M_n - E(M_\infty|\mathcal{F}_n))dP.$$

Proof: Set $Z = \overline{\lim_{n\in\mathbb{N}}} |X_n - Y|$. For $n \in \mathbb{N}$, we note that by Proposition 2.2, $\Lambda_Y(A) = \int_A M_n dP$ for each $A \in \mathcal{F}_n$. Therefore,

$$\phi_Y(A) = \Lambda_Y(A) - \int_A Z\, dP = \int_A (M_n - E(Z|\mathcal{F}_n))dP$$

for $n \in \mathbb{N}$ and $A \in \mathcal{F}_n$. By Fatou's Lemma, $\phi_Y \geq 0$. Hence $M_n - E(Z|\mathcal{F}_n) \geq 0$ a.s. for every $n \in \mathbb{N}$. Clearly $(M_n - E(Z|\mathcal{F}_n), n \in \mathbb{N})$ is a martingale. Moreover, ϕ is singular with respect to P. Therefore, by Lemma 2.2, $\lim_{n\to+\infty} [M_n - E(Z|\mathcal{F}_n)] = 0$ a.s. We deduce by Proposition 1.1 that $Z = \lim_{n\in\mathbb{N}} M_n$. Q.E.D.

Remarks. 3.1. Under the assumptions of Corollary 3.1, the inequality $0 \leq \phi \leq \Lambda_Y$ implies $\Lambda_Y = \phi$ if and only if $\Lambda_Y(\Omega) = \phi(\Omega)$. Therefore Corollary 3.1(ii) can be restated as: Y is a cluster point if and only if $\Lambda_Y(\Omega) = \phi(\Omega)$.

3.2. Note that Proposition 3.2(iii) is the martingale representation of ϕ. Therefore by uniqueness, $(M_n - E(M_\infty|\mathcal{F}_n), n \in \mathbb{N})$ has a version independent of Y.

3.3. The Fatou discrepancy need not be countably additive. For an example see Remark 2.1.

3.4. Using the notation of Proposition 3.2 we have: Φ is countably additive if and only if for any $Y \in L^1(\Omega,\mathcal{F},P)$, Λ_Y is countably additive if and only if for any $Y \in L^1(\Omega,\mathcal{F},P)$ and any finite stopping time σ (not necessarily bounded) $\Lambda_Y(\Omega) = \int_\Omega M_\sigma dP$. The second equivalence follows from Proposition III 1-1 p. 35 in [7].

3.5. Φ is absolutely continuous with respect to P if and only if $\Phi = 0$.

4. The cluster set

Definition 4.1: Let $(X_n, n \in \mathbb{N})$ be a sequence of random variables. The cluster set $\bar{C}$ $(= \bar{C}[X_n, n \in \mathbb{N}])$ of $(X_n, n \in \mathbb{N})$ is the set of all measurable cluster points of $(X_n, n \in \mathbb{N})$. The subset of $\bar{C}$ consisting of all integrable cluster points of $(X_n, n \in \mathbb{N})$ will be denoted by $C = (C[X_n, n \in \mathbb{N}])$. That is, $C = \{Y \in L^1(\Omega,\mathcal{F},P): \underline{\lim_{n\in\mathbb{N}}} |X_n - Y| = 0 \text{ a.s.}\}$.

We recall that if $Y \in L^1(\Omega,\mathcal{F},P)$ and if C is not empty, then $\rho(Y,C) = \inf\{\|X - Y\|_1: X \in C\}$ is the distance from Y to C. We also note that C is closed in $L^1(\Omega,\mathcal{F},P)$. Hence, $Y \in C$ if and only if $\rho(Y,C) = 0$. If $(X_n, n \in \mathbb{N})$ is an adapted sequence of integrable random variables satisfying condition (I) then, by Corollary 3.1(ii), for any $Y \in C$

$$\Lambda_Y(\Omega) = \phi(\Omega) + \rho(Y,C).$$

We will show that the above identity holds for any $Y \in L^1(\Omega,\mathcal{F},P)$. We need the following lemmas.

Lemma 4.1. Let $(X_n, n \in \mathbb{N})$ be an adapted sequence of random variables and let $Y \in \bar{C}[|X_n|, n \in \mathbb{N}]$, $Y \geq 0$. Then there exists $Y' \in \bar{C}[X_n, n \in \mathbb{N}]$ such that $|Y'| = Y$.

Proof: By Lemma 1 in [1], there exists an increasing sequence $(\tau_n, n \in \mathbb{N})$ in T, $\tau_n \in T(n)$ for each $n \in \mathbb{N}$, such that $Y = \lim_{n\in\mathbb{N}} |X_{\tau_n}|$. Define $Y' = Y$ on the set $[\limsup_{n\in\mathbb{N}} X_{\tau_n} \geq 0]$ and $Y' = -Y$ on the set $[\limsup_{n\in\mathbb{N}} X_{\tau_n} < 0]$. We easily check that Y' is a cluster point of $(X_n, n \in \mathbb{N})$ and that $|Y'| = Y$. Q.E.D.

Note that, in particular, Lemma 4.1 implies that $\bar{C}[X_n, n \in \mathbb{N}]$ is nonempty if and only if $P(\underline{\lim}_{n\in\mathbb{N}} |X_n| < \infty) = 1$ and that $C[X_n, n \in \mathbb{N}]$ is nonempty if and only if $\underline{\lim}_{n\in\mathbb{N}} |X_n| \in L^1(\Omega,\mathcal{F},P)$.

Lemma 4.2. Let $(X_n, n \in \mathbb{N})$ be an adapted sequence of random variables such that C is nonempty. Then, for every $Y \in L^1(\Omega,\mathcal{F},P)$ there exists $Z \in C$ such that

$$\rho(Y,C) = E|Y - Z| = E(\underline{\lim}_{n\in\mathbb{N}} |X_n - Y|).$$

Proof: First we note that by Proposition 1.1,

$$\underline{\lim}_{n\in\mathbb{N}} |X_n - Y| = \underline{\lim}_{n\in\mathbb{N}} |X_n - E(Y|\mathcal{F}_n)| \quad \text{a.s.}$$

By Lemma 4.1, applied to the sequence $(X_n - E(Y|\mathcal{F}_n), n \in \mathbb{N})$, there exists a cluster point Z' of $(X_n - E(Y|\mathcal{F}_n), n \in \mathbb{N})$ such that $|Z'| = \underline{\lim}_{n\in\mathbb{N}} |X_n - E(Y|\mathcal{F}_n)|$. Therefore $Z = Z' + Y$ is a cluster point of $(X_n, n \in \mathbb{N})$. Since C is nonempty, $\underline{\lim}_{n\in\mathbb{N}} |X_n| \in L^1(\Omega,\mathcal{F},P)$ which implies that $Z \in L^1(\Omega,\mathcal{F},P)$. Hence $Z \in C$ and

$$E(\underline{\lim}_{n\in\mathbb{N}} |X_n - E(Y|\mathcal{F}_n)|) = E|Z'| = E|Z - Y| \geq \rho(Y,C).$$

To prove the reverse inequality, let $X \in C$, then

$$|X_n - E(Y|\mathcal{F}_n)| \leq |X_n - X| + |X - E(Y|\mathcal{F}_n)|.$$

Now $\underline{\lim}_{n\in\mathbb{N}} |X_n - X| = 0$ and $\lim_{n\in\mathbb{N}} |X - E(Y|\mathcal{F}_n)| = |X - Y|$. Hence,

$$E(\underline{\lim}_{n\in\mathbb{N}} |X_n - E(Y|\mathcal{F}_n)|) \leq E|X - Y|.$$

Q.E.D.

Proposition 4.1. Let $(X_n, n \in \mathbb{N})$ be an adapted sequence of integrable random variables satisfying condition (I). Let ϕ be the Fatou discrepancy associated with $(X_n, n \in \mathbb{N})$. Then for every $Y \in L^1(\Omega,\mathcal{F},P)$,

$$\Lambda_Y(\Omega) = \phi(\Omega) + \rho(Y,C).$$

Proof: First we note that condition (I) implies that C is nonempty. Now by Corollary 3.1(i),

$$\Lambda_Y(\Omega) = \phi(\Omega) + E(\varliminf_{n\in\mathbb{N}} |X_n - Y|).$$

Therefore, by Lemma 4.2, $\Lambda_Y(\Omega) = \phi(\Omega) + \rho(Y,C)$. Q.E.D.

Let $(X_n, n \in \mathbb{N})$ be a sequence of random variables and let $Z \in L^1(\Omega,\mathcal{F},P)$. We define C_Z $(= C_Z[X_n, n \in \mathbb{N}])$ as the set

$$C_Z = \{U \in L^1(\Omega,\mathcal{F},P) : U + Z \in C\}.$$

Corollary 4.1. Let $(X_n, n \in \mathbb{N})$ be an adapted sequence of integrable random variables satisfying condition (I). Let ϕ be the Fatou discrepancy associated with $(X_n, n \in \mathbb{N})$. Then for any $Y, Z \in L^1(\Omega,\mathcal{F},P)$,

$$\Lambda_{Y+Z}(\Omega) = \phi(\Omega) + \rho(Y,C_Z) = \phi(\Omega) + \rho(Z,C_Y).$$

Proof: We prove only the first equation. The second one can be proved in the same way. For any $U \in L^1(\Omega,\mathcal{F},P)$

$$\varliminf_{n\in\mathbb{N}} |X_n - (U+Z)| = \varliminf_{n\in\mathbb{N}} |(X_n - E(Z|\mathcal{F}_n)) - U|.$$

Therefore $C_Z[X_n, n \in \mathbb{N}] = C[X_n - E(Z|\mathcal{F}_n), n \in \mathbb{N}]$. Also, $\Lambda_{Y+Z}(\Omega,(X_n, n \in \mathbb{N})) = \Lambda_Y(\Omega,(X_n - E(Z|\mathcal{F}_n), n \in \mathbb{N}))$. Moreover the Fatou discrepancy of $(X_n - E(Z|\mathcal{F}_n), n \in \mathbb{N})$ is $\phi_Z(X_n, n \in \mathbb{N})$ and therefore, by Corollary 3.1(i), it is equal to ϕ. Hence by Proposition 4.1 applied to the sequence $(X_n - E(Z|\mathcal{F}_n), n \in \mathbb{N})$,

$$\Lambda_{Y+Z}(\Omega,(X_n, n \in \mathbb{N})) = \phi(\Omega) + \rho(Y, C_Z(X_n, n \in \mathbb{N})).$$ Q.E.D.

Remark 4.1. We cannot hope for uniqueness of Z in Lemma 4.2 as the following trivial example shows: let $X_n = (-1)^n$. Then $Z_1 = 1$ and $Z_2 = -1$ are cluster points and if $Y = 0$,

$$E(\varliminf_{n\in\mathbb{N}} |X_n - Y|) = E|Z_1 - Y| = E|Z_2 - Y| = 1.$$

5. Simultaneous approximations

Let $(X_n, n \in \mathbb{N})$ be an adapted sequence of integrable random variables satisfying condition (I), and let $Y, Z \in C$. If $\phi(\Omega) = 0$, there exist sequences $(s_n, n \in \mathbb{N})$ and $(t_n, n \in \mathbb{N})$ in T, $s_n, t_n \in T(n)$ for each $n \in \mathbb{N}$, such that $X_{s_n} \to Y$ and $X_{t_n} \to Z$ in the L^1-norm. This, by Corollary 3.1(i). Hence $X_{s_i} - X_{t_j} \to Y - Z$ in the L^1-norm as $i, j \to +\infty$ independently of one another. The situation is quite different when $\phi(\Omega) > 0$. In the example given in Remark 2.1, $C = \{0\}$, and for any $s, t \in T$ with $\min t > \max s$ we have $E|X_t - X_s| > 1$. Thus there cannot exist sequences $(s_n, n \in \mathbb{N})$, $(t_n, n \in \mathbb{N})$ such that $E|X_{s_i} - X_{t_j}| \to 0$ as $i, j \to \infty$ independently of one another.

We do however have the following result:

Proposition 5.1. Let $(X_n, n \in \mathbb{N})$ be an adapted sequence of integrable random variables satisfying condition (I). Then with every $U \in C$ we can associate an increasing sequence $(t_n(U), n \in \mathbb{N})$ in T in such a way that for every $n \in \mathbb{N}$,

$t_n(U) \in T(n)$, $X_{t_n(U)} \to U$ in probability and that whenever $Y, Z \in C$ we have

$$\lim_{n\in\mathbb{N}} E|(X_{t_n(Y)} - X_{t_n(Z)}) - (Y - Z)| = 0.$$

Proof: By Lemma 3.3 the Fatou discrepancy Φ is singular with respect to P. Let $(\varepsilon_n, n \in \mathbb{N})$ be a sequence of positive numbers decreasing to 0. There exist then a sequence $(A_n, n \in \mathbb{N})$ in $\mathcal{A}$, an increasing sequence $(k_n, n \in \mathbb{N})$ in $\mathbb{N}$ with $k_n \geq n$, such that $A_n \in \mathcal{F}_{k_n}$, $\Phi(A_n) < \varepsilon_n$ and $P(A_n) > 1 - \varepsilon_n$ for each $n \in \mathbb{N}$.

Let now $U \in C$. By Corollary 3.1(ii), $\Lambda_U = \Phi$. Hence there exists an increasing sequence $(\tau_n(U), n \in \mathbb{N})$ in T such that for every $n \in \mathbb{N}$, $\tau_n(U) > k_n$, $E|X_{\tau_n(U)} - U|I_{A_n} < \varepsilon_n$. Define now

$$t_n(U) = \tau_n(U)I_{A_n} + k_n I_{A_n^c}, \quad \text{for every } n \in \mathbb{N}.$$

Then $(t_n(U), n \in \mathbb{N})$ is the desired sequence. In fact, given $\varepsilon > 0$, we have for every $n \in \mathbb{N}$

$$P(|X_{t_n(U)} - U| > \varepsilon) = P([|X_{t_n(U)} - U| > \varepsilon] \cap A_n) + P([|X_{t_n(U)} - U| > \varepsilon] \cap A_n^c)$$

$$\leq \frac{1}{\varepsilon} E|X_{\tau_n(U)} - U|I_{A_n} + P(A_n^c) < \frac{\varepsilon_n}{\varepsilon} + \varepsilon_n.$$

Hence, $\lim_{n\in\mathbb{N}} P(|X_{t_n(U)} - U| > \varepsilon) = 0$.

Let now $Y, Z \in C$. Since $t_n(Y) = \tau_n(Y)I_{A_n} + k_n I_{A_n^c}$, $t_n(Z) = \tau_n(Z|I_{A_n} + k_n I_{A_n^c}$, we have for every $n \in \mathbb{N}$,

$$E|(X_{t_n(Y)} - X_{t_n(Z)}) - (Y - Z)| \leq E|X_{\tau_n(Y)} - Y|I_{A_n} + E|X_{\tau_n(Z)} - Z|I_{A_n} + E|Y - Z|I_{A_n^c}$$

$$\leq 2\varepsilon_n + E|Y - Z|I_{A_n^c}.$$

Since $Y - Z \in L^1(\Omega, \mathcal{F}, P)$ and $P(A_n^c) \to 0$ as $n \to \infty$, we deduce

$$\lim_{n\in\mathbb{N}} E|(X_{t_n(Y)} - X_{t_n(Z)}) - (Y - Z)| = 0. \qquad \text{Q.E.D.}$$

Remarks 5.1. The sequence $(t_n(U), n \in \mathbb{N})$ associated with $U \in C$ depends only on U and the singularity of the Fatou discrepancy Φ.

5.2. J. R. Baxter [2] (see also A. Bellow [3]) proved that if $\underline{\lim}_{n\in\mathbb{N}} E|X_n| < \infty$ and $Y, Z \in C$ there exist sequences $(s_n, n \in \mathbb{N})$, $(t_n, n \in \mathbb{N})$ in T, increasing to ∞ such that $X_{s_n} - X_{t_n} \to Y - Z$ in the L^1-norm. These sequences were however dependent on the pair (Y,Z) making Proposition 5.1 a slightly stronger result.

Corollary 5.1. Let $(X_n, n \in \mathbb{N})$ be an adapted sequence of integrable random variables satisfying condition (I). Then to every $Y, Z \in L^1(\Omega, \mathcal{F}, P)$ there exist two increasing sequences $(t_n(Y), n \in \mathbb{N})$, $(t_n(Z), n \in \mathbb{N})$ such that $t_n(Y), t_n(Z) \in T(n)$ for every $n \in \mathbb{N}$ and such that

$$\limsup_{n\in\mathbb{N}} E|(X_{t_n(Y)} - X_{t_n(Z)}) - (Y - Z)| \leq \rho(Y, C) + \rho(Z, C).$$

Proof: By Lemma 4.2, there exists $Y', Z' \in C$ such that $\rho(Y,C) = E|Y - Y'|$ and $\rho(Z,C) = E|Z - Z'|$. By Proposition 5.1, there exist $(t_n(Y),\ n \in \mathbb{N})$ and $(t_n(Z),\ n \in \mathbb{N})$ in T such that $t_n(Y),\ t_n(Z) \in T(n)$ for each $n \in \mathbb{N}$ and

$$\lim_{n\in\mathbb{N}} E|(X_{t_n(Y)} - X_{t_n(Z)}) - (Y' - Z')| = 0.$$

Hence,

$$\limsup_{n\in\mathbb{N}} E|(X_{t_n(Y)} - X_{t_n(Z)}) - (Y - Z)| \le E|Y - Y'| + E|Z - Z'|$$
$$= \rho(Y,C) + \rho(Z,C).$$

Q.E.D.

6. The signed set functions $(\tilde{\Lambda}_Y,\ Y \in L^1(\Omega,\mathcal{F},P))$

Throughout this section $(X_n,\ n \in \mathbb{N})$ will be an adapted sequence of integrable random variables satisfying condition (I), and ϕ and C will be respectively the Fatou discrepancy associated with $(X_n,\ n \in \mathbb{N})$ and the set of integrable cluster points of $(X_n,\ n \in \mathbb{N})$. We recall from section 2 that for every $A \in \mathcal{Q}$ and $Y \in L^1(\Omega, \mathcal{F}, P)$

$$\Lambda_Y(A) = \varliminf_{t\in T} E|X_t - Y|I_A.$$

We start out with a lemma.

Lemma 6.1. Let $Y \in L^1(\Omega,\mathcal{F},P)$ and let $(t_n,\ n \in \mathbb{N})$ be an increasing sequence in T, $t_n \in T(n)$ for each $n \in \mathbb{N}$, such that $\Lambda_Y(\Omega) = \lim_{n\in\mathbb{N}} E|X_{t_n} - Y|$. Then,

$$\lim_{n\in\mathbb{N}}\Big(\sup_{A\in\mathcal{F}_{t_n}} |E|X_{t_n} - Y|I_A - \Lambda_Y(A)|\Big) = 0.$$

In particular, for any $A \in \mathcal{Q}$,

$$\Lambda_Y(A) = \lim_{n\in\mathbb{N}} E|X_{t_n} - Y|I_A.$$

Proof: For $n \in \mathbb{N}$ and $A \in \mathcal{F}_{t_n}$, we have by Proposition 2.1,

$$\Lambda_Y(A) = \lim_{k\in\mathbb{N}} \mu_{t_k}(A)$$

where

$$\mu_{t_k}(A) = \inf\{\textstyle\int_A |X_\tau - E(Y|\mathcal{F}_\tau)|dP,\quad \tau \in T,\quad \tau \ge t_k\}.$$

Note also that by assumption and by Proposition 1.1

$$\Lambda_Y(\Omega) = \lim_{n\in\mathbb{N}} E|X_{t_n} - E(Y|\mathcal{F}_{t_n})|.$$

Therefore,

$$\sup_{A\in\mathcal{F}_{t_n}} |E|X_{t_n} - Y|I_A - \Lambda_Y(A)| \le (\Lambda_Y(\Omega) - \mu_{t_n}(\Omega))$$
$$+ (E|X_{t_n} - E(Y|\mathcal{F}_{t_n})| - \mu_{t_n}(\Omega)) + E|E(Y|\mathcal{F}_{t_n}) - Y|.$$

The first two summands clearly tend to 0; the third summand tends to 0 by Proposition 1.1.

Q.E.D.

Definition 6.1: For $Y \in L^1(\Omega,\mathcal{F},P)$ and $A \in \mathcal{A}$ we define $\tilde{\Lambda}_Y^+(A) = \sup\{\lim_{n\in\mathbb{N}} E(X_{t_n} - Y)^+ I_A;\ \lim_{n\in\mathbb{N}} E|X_{t_n} - Y|I_A = \Lambda_Y(A),\ t_{n+1} \geq t_n,\ t_n \in T(n)\}$ (the supremum being taken over all sequences $(t_n, n \in \mathbb{N}) \subset T$ for which both limits exist).

We let $\tilde{\Lambda}_Y^- = \Lambda_Y - \tilde{\Lambda}_Y^+$ and $\tilde{\Lambda}_Y = \tilde{\Lambda}_Y^+ - \tilde{\Lambda}_Y^-$. Clearly $0 \leq \tilde{\Lambda}_Y^+ \leq \Lambda_Y$.

Lemma 6.2. Let $Y \in L^1(\Omega,\mathcal{F},P)$ and $A \in \mathcal{A}$. Then, there exists an increasing sequence $(t_n, n \in \mathbb{N})$ in T, $t_n \in T(n)$ such that

$$\tilde{\Lambda}_Y^+(A) = \lim_{n\in\mathbb{N}} E(X_{t_n} - Y)^+ I_A \quad\text{and}\quad \Lambda_Y(A) = \lim_{n\in\mathbb{N}} E|X_{t_n} - Y|I_A.$$

Proof: Let $(\varepsilon_n, n \in \mathbb{N})$ be a decreasing sequence in $\mathbb{R}_+$ such that $\varepsilon_n \to 0$. Then for every $n \in \mathbb{N}$, there exists an increasing sequence $(s_k^n, k \in \mathbb{N})$, $s_k^n \in T(k)$, such that

$$\lim_{k\in\mathbb{N}} E(X_{s_k^n} - Y)^+ I_A + \varepsilon_n > \tilde{\Lambda}_Y^+(A) \quad\text{and}\quad \lim_{k\in\mathbb{N}} E|X_{s_k^n} - Y|I_A = \Lambda_Y(A).$$

Therefore we can construct an increasing sequence $(\tau_k, k \in \mathbb{N})$ in T, $\tau_k \in T(k)$, such that

$$E(X_{\tau_k} - Y)^+ I_A + \varepsilon_k > \tilde{\Lambda}_Y^+(A)$$

and

$$|E|X_{\tau_k} - Y|I_A - \Lambda_Y(A)| < \varepsilon_k.$$

Hence, there exists a subsequence $(\tau_{k_n}, n \in \mathbb{N})$ of $(\tau_k, k \in \mathbb{N})$ such that $\tau_{k_n} \in T(n)$ and such that $\lim_{n\in\mathbb{N}} E(X_{\tau_{k_n}} - Y)^+ I_A \geq \tilde{\Lambda}_Y^+(A)$ and $\lim_{n\in\mathbb{N}} E|X_{\tau_{k_n}} - Y|I_A = \Lambda_Y(A)$. Let now $\tau_{k_n} = t_n$ for each $n \in \mathbb{N}$. Clearly by definition of $\tilde{\Lambda}_Y^+$, the sequence $(t_n, n \in \mathbb{N})$ satisfies the Lemma. Q.E.D.

Lemma 6.3. For every $Y \in L^1(\Omega,\mathcal{F},P)$, the set functions $\tilde{\Lambda}_Y^+$, $\tilde{\Lambda}_Y^-$ and Λ_Y are finitely additive on $\mathcal{A}$.

Proof: It is enough to show that $\tilde{\Lambda}_Y^+$ is finitely additive. Let $A_1, A_2 \in \mathcal{F}_k$ for some $k \in \mathbb{N}$ such that $A_1 \cap A_2 = \phi$. By Lemma 6.2 there exist increasing sequences $(t_n^i, n \in \mathbb{N})$ in T, $t_n^i \geq k$, $i = 1,2$, such that $\lim_{n\in\mathbb{N}} E(X_{t_n^i} - Y)^+ I_{A_i} = \tilde{\Lambda}_Y^+(A_i)$ and $\lim_{n\in\mathbb{N}} E|X_{t_n^i} - Y|I_{A_i} = \Lambda_Y(A_i)$ for $i = 1,2$. We define $\tau_n \in T$ by $\tau_n = t_n^1 I_{A_1} + t_n^2 I_{A_1^c}$ for each $n \in \mathbb{N}$. Then,

$$\lim_{n\in\mathbb{N}} E|X_{\tau_n} - Y|I_{A_1\cup A_2} = \lim_{n\in\mathbb{N}} (E|X_{t_n^1} - Y|I_{A_1} + E|X_{t_n^2} - Y|I_{A_2})$$
$$= \Lambda_Y(A_1) + \Lambda_Y(A_2) = \Lambda_Y(A_1 \cup A_2).$$

Likewise, $\lim E(X_{\tau_n} - Y)^+ I_{A_1\cup A_2} = \tilde{\Lambda}_Y^+(A_1) + \tilde{\Lambda}_Y^+(A_2)$. Therefore,

$$\tilde{\Lambda}_Y^+(A_1 \cup A_2) \geq \tilde{\Lambda}_Y^+(A_1) + \tilde{\Lambda}_Y^+(A_2).$$

To prove the reverse inequality, let $(\tau_n, n \in \mathbb{N})$ be an increasing sequence in T, $\tau_n \in \tau(n)$, such that

$$\lim_{n\in\mathbb{N}} E(X_{\tau_n} - Y)^+ I_{A_1 \cup A_2} = \tilde{\Lambda}_Y^+(A_1 \cup A_2) \quad \text{and} \quad \lim_{n\in\mathbb{N}} E|X_{\tau_n} - Y| I_{A_1 \cup A_2} = \Lambda_Y(A_1 \cup A_2).$$

By Lemma 6.1, $\lim_{n\in\mathbb{N}} E|X_{\tau_n} - Y| I_{A_i} = \Lambda_Y(A_i)$, $i = 1,2$. Moreover,

$$\tilde{\Lambda}_Y^+(A_1 \cup A_2) \le \overline{\lim_{n\in\mathbb{N}}}\, E(X_{\tau_n} - Y)^+ I_{A_1} + \overline{\lim_{n\in\mathbb{N}}}\, E(X_{\tau_n} - Y)^+ I_{A_2}.$$

Reverting to subsequences of $(\tau_n, n \in \mathbb{N})$ if necessary, we conclude

$$\tilde{\Lambda}_Y^+(A_1 \cup A_2) \le \tilde{\Lambda}_Y^+(A_1) + \tilde{\Lambda}_Y^+(A_2).$$

Q.E.D.

Lemma 6.4. Let $Y \in L^1(\Omega, \mathcal{F}, P)$ and let $(t_n, n \in \mathbb{N})$ be an increasing sequence in T such that $t_n \in T(n)$ and such that $\lim_{n\in\mathbb{N}} E(X_{t_n} - Y)^+ = \tilde{\Lambda}_Y^+(\Omega)$ and $\lim_{n\in\mathbb{N}} E|X_{t_n} - Y| = \Lambda_Y(\Omega)$. Then,

$$\lim_{n\in\mathbb{N}} \Big(\sup_{A\in\mathcal{F}_{t_n}} |E(X_{t_n} - Y)^+ I_A - \tilde{\Lambda}_Y^+(A)|\Big) = 0$$

$$\lim_{n\in\mathbb{N}} \Big(\sup_{A\in\mathcal{F}_{t_n}} |E(X_{t_n} - Y) I_A - \tilde{\Lambda}_Y(A)|\Big) = 0.$$

In particular, for any $A \in \mathcal{A}$,

$$\tilde{\Lambda}_Y^+(A) = \lim_{n\in\mathbb{N}} E(X_{t_n} - Y)^+ I_A \quad \text{and} \quad \tilde{\Lambda}_Y(A) = \lim_{n\in\mathbb{N}} E(X_{t_n} - Y) I_A.$$

Proof: In view of Lemma 6.1 and the definition of $\tilde{\Lambda}_Y$, it is enough to prove that

$$\lim_{n\in\mathbb{N}} \Big(\sup_{A\in\mathcal{F}_{t_n}} |E(X_{t_n} - Y)^+ I_A - \tilde{\Lambda}_Y^+(A)|\Big) = 0$$

which is in turn equivalent to proving that for any sequence $(A_n, n \in \mathbb{N})$ in $\mathcal{A}$ such that $A_n \in \mathcal{F}_{t_n}$ for each $n \in \mathbb{N}$,

$$\lim_{n\in\mathbb{N}} |E(X_{t_n} - Y)^+ I_{A_n} - \tilde{\Lambda}_Y^+(A_n)| = 0.$$

Therefore, let $(A_n, n \in \mathbb{N})$ be a sequence in $\mathcal{A}$ such that $A_n \in \mathcal{F}_{t_n}$, and let $B_n = \Omega \setminus A_n$. We claim that

$$\overline{\lim_{n\in\mathbb{N}}} \big(E(X_{t_n} - Y)^+ I_{A_n} - \tilde{\Lambda}_Y^+(A_n)\big) \le 0.$$

Suppose the claim is false. Then, without loss of generality, we can assume that there exists $\varepsilon > 0$ such that for every $n \in \mathbb{N}$,

$$E(X_{t_n} - Y)^+ I_{A_n} > \tilde{\Lambda}_Y^+(A_n) + \varepsilon.$$

By an argument similar to the one used in the proof of Lemma 6.1, there exists for every $n \in \mathbb{N}$, $s_n \in T$, $s_n > t_n$, $s_{n+1} \ge s_n$ such that

$$E(X_{s_n} - Y)^+ I_{B_n} + \frac{1}{n} > \tilde{\Lambda}_Y^+(B_n)$$

and

$$|E|X_{s_n} - Y| I_{B_n} - \Lambda_Y(B_n)| < \frac{1}{n}.$$

We define $\tau_n = t_n I_{A_n} + s_n I_{B_n}$, $n \in \mathbb{N}$. Then,

$$E(X_{\tau_n} - Y)^+ = E(X_{t_n} - Y)^+ I_{A_n} + E(X_{s_n} - Y)^+ I_{B_n}$$

$$> \tilde{\Lambda}_Y^+(A_n) + \varepsilon + \tilde{\Lambda}_Y^+(B_n) - \frac{1}{n} = \tilde{\Lambda}_Y^+(\Omega) + \varepsilon - \frac{1}{n}.$$

Therefore, $\overline{\lim}_{n\in\mathbb{N}} E(X_{\tau_n} - Y)^+ \geq \tilde{\Lambda}_Y^+(\Omega) + \varepsilon$. Now by Lemma 6.1,

$$\lim_{n\in\mathbb{N}} |E|X_{t_n} - Y|I_{A_n} - \Lambda_Y(A_n)| = 0, \quad \text{and} \lim_{n\in\mathbb{N}} |E|X_{s_n} - Y|I_{B_n} - \Lambda_Y(B_n)| = 0.$$

Therefore, $\lim_{n\in\mathbb{N}} E|X_{\tau_n} - Y| = \Lambda_Y(\Omega)$. Note that we can always assume that $(\tau_n, n \in \mathbb{N})$ is increasing. If not, we can select a subsequence that is increasing. Hence, we have

$$\overline{\lim}_{n\in\mathbb{N}} E(X_{\tau_n} - Y)^+ \geq \tilde{\Lambda}_Y^+(\Omega) + \varepsilon \quad \text{and} \quad \lim_{n\in\mathbb{N}} E|X_{\tau_n} - Y| = \Lambda_Y(\Omega)$$

which is impossible by definition of $\tilde{\Lambda}_Y^+$. Therefore the claim holds. Now for every $n \in \mathbb{N}$,

$$E(X_{t_n} - Y)^+ I_{A_n} - \tilde{\Lambda}_Y^+(A_n) = \left(E(X_{t_n} - Y)^+ - \tilde{\Lambda}_Y^+(\Omega)\right) - \left(E(X_{t_n} - Y)^+ I_{B_n} - \tilde{\Lambda}_Y^+(B_n)\right).$$

Hence by assumption, and by the claim applied to $(B_n, n \in \mathbb{N})$ we conclude that $\underline{\lim}_{n\in\mathbb{N}} \left(E(X_{t_n} - Y)^+ I_{A_n} - \tilde{\Lambda}_Y^+(A_n)\right) \geq 0$, and thus, $\lim_{n\in\mathbb{N}} \left(E(X_{t_n} - Y)^+ I_{A_n} - \tilde{\Lambda}_Y^+(A_n)\right) = 0$.

Q.E.D.

Let $(t_n, n \in \mathbb{N})$ be an increasing sequence in T. We recall that for a set function $\mu: \mathcal{A} \to \mathbb{R}$ and for $n \in \mathbb{N}$, $R_{t_n}\mu$ is the restriction of μ to $\mathcal{F}_{t_n}$. We also denote by $|\nu|_{t_n}$ the variation of ν as a set function on $\mathcal{F}_{t_n}$.

Lemma 6.5. Let $\mu: \mathcal{A} \to \mathbb{R}$ be a finitely additive measure of finite variation. Let $(t_n, n \in \mathbb{N})$ be an increasing sequence in T. Then,

$$\lim_{n\in\mathbb{N}} |R_{t_n}\mu|_{t_n}(A) = |\mu|(A), \quad A \in \mathcal{A}.$$

Proof: Clearly, for every $A \in \mathcal{A}$, $(|R_{t_n}\mu|_{t_n}(A), n \in \mathbb{N})$ is increasing, and $\lim_{n\in\mathbb{N}} |R_{t_n}\mu|_{t_n}(A) \leq |\mu|(A)$. It is therefore enough to show that $|\mu|(\Omega) \leq \lim_{n\in\mathbb{N}} |R_{t_n}\mu|_{t_n}(\Omega)$. Let $A_1,\ldots,A_n$ be a partition of Ω in $\mathcal{A}$ such that $A_i \in \mathcal{F}_{t_m}$ for every $i \in \{1,\ldots,n\}$. Then,

$$\sum_{i=1}^{n} |\mu(A_i)| = \sum_{i=1}^{n} |R_{t_m}\mu(A_i)| \leq |R_{t_m}\mu|_{t_m}(\Omega)$$

which implies that $|\mu|(\Omega) \leq \lim_{n\in\mathbb{N}} |R_{t_n}\mu|_{t_n}(\Omega)$. Q.E.D.

Proposition 6.1. Let $Y \in L^1(\Omega,\mathcal{F},P)$. Then,

i) There exists an increasing sequence $(t_n, n \in \mathbb{N})$ in T, $t_n \in T(n)$, such that for every $A \in \mathcal{A}$,

$$\tilde{\Lambda}_Y(A) = \lim_{n\in\mathbb{N}} E(X_{t_n} - Y)I_A \quad\text{and}\quad \Lambda_Y(A) = \lim_{n\in\mathbb{N}} E|X_{t_n} - Y|I_A.$$

ii) $|\tilde{\Lambda}_Y|(A) = \Lambda_Y(A)$ for every $A \in \mathcal{A}$.

Proof: By Lemma 6.1, there exists an increasing sequence $(t_n, n \in \mathbb{N})$ in T, $t_n \in T(n)$ such that

$$\tilde{\Lambda}_Y^+(\Omega) = \lim_{n\in\mathbb{N}} E(X_{t_n} - Y)^+ \quad\text{and}\quad \Lambda_Y(\Omega) = \lim_{n\in\mathbb{N}} E|X_{t_n} - Y|.$$

i) Straightforward from Lemma 6.2 and Lemma 6.4.

ii) We know that $|\tilde{\Lambda}_Y| \leq \Lambda_Y$. It is therefore enough to show that $|\tilde{\Lambda}_Y|(\Omega) = \Lambda_Y(\Omega)$. For every $n \in \mathbb{N}$ and $A \in \mathcal{F}_{t_n}$ let $\nu_n(A) = E(X_{t_n} - Y)I_A$. Then, $|\nu|_{t_n}(\Omega) = E|X_{t_n} - E(Y|\mathcal{F}_{t_n})|$. Moreover,

$$|R_{t_n}\tilde{\Lambda}_Y - \nu_n|_{t_n}(\Omega) \leq 2 \sup_{A\in\mathcal{F}_{t_n}} |\tilde{\Lambda}_Y(A) - E(X_{t_n} - Y)I_A|.$$

Therefore, by Lemma 6.4, $\lim_{n\in\mathbb{N}} |R_{t_n}\tilde{\Lambda}_Y - \nu_n|_{t_n}(\Omega) = 0$. Hence, by Lemma 6.5 and Proposition 1.1,

$$|\tilde{\Lambda}_Y|(\Omega) = \lim_{n\in\mathbb{N}} |R_{t_n}\tilde{\Lambda}_Y|_{t_n}(\Omega) = \lim_{n\in\mathbb{N}} |\nu_n|_{t_n}(\Omega) = \Lambda_Y(\Omega). \qquad \text{Q.E.D.}$$

Proposition 6.2. Let $Y \in C$. Then for any $Z \in C$, there exists an increasing $(t_n, n \in \mathbb{N})$ in T, $t_n \in T(n)$ such that X_{t_n} converges to Z in probability and such that for every $A \in \mathcal{A}$, $\lim_{n\in\mathbb{N}} E(X_{t_n} - Z)I_A = \tilde{\Lambda}_Y(A)$ and $\lim_{n\in\mathbb{N}} E|X_{t_n} - Z|I_A = \Phi(A)$.

Proof: First we note that by Corollary 3.1, $\Lambda_Y = \Lambda_Z = \Phi$. Let $(\varepsilon_n, n \in \mathbb{N})$ be a sequence in $\mathbb{R}_+$ decreasing to 0. By Lemma 3.3 Φ is singular with respect to P. Therefore for every $n \in \mathbb{N}$ there exists $A_n \in \mathcal{F}_{k_n}$, $k_{n+1} \geq k_n$, such that $P(A_n) > 1 - \varepsilon_n$ and $\Phi(A_n) < \varepsilon_n$. Hence for every $n \in \mathbb{N}$, there exists $s_n \in T$, $s_n \geq k_n$, $s_{n+1} \geq s_n$, such that $E|X_{s_n} - Z|I_{A_n} < \varepsilon_n$. Let $(\tau_n, n \in \mathbb{N})$ be an increasing sequence in T such that $\lim_{n\in\mathbb{N}} E(X_{\tau_n} - Y)^+ = \tilde{\Lambda}_Y^+(\Omega)$ and $\lim_{n\in\mathbb{N}} E|X_{\tau_n} - Y| = \Phi(\Omega)$. Such a sequence always exists by Lemma 6.1. We may also assume that $\tau_n \geq s_n$ for each $n \in \mathbb{N}$. We define $t_n \in T$ by $t_n = s_n I_{A_n} + \tau_n I_{B_n}$ where $B_n = \Omega\backslash A_n$. We can always assume $(t_n, n \in \mathbb{N})$ increasing. Let $A \in \mathcal{A}$, then $E(X_{t_n} - Z)I_A = E(X_{s_n} - Z)I_{A\cap A_n} + E(Y-Z)I_{A\cap B_n} + E(X_{\tau_n} - Y)I_{A\cap B_n}$. Clearly the first two summands converge to 0. By Lemma 6.4

$$\lim_{n\in\mathbb{N}} |E(X_{\tau_n} - Y)I_{A\cap B_n} - \tilde{\Lambda}_Y(A\cap B_n)| = 0. \quad\text{But,}$$

$$\tilde{\Lambda}_Y(A\cap B_n) = \tilde{\Lambda}_Y(A) - \tilde{\Lambda}_Y(A\cap A_n) \quad\text{and}\quad |\tilde{\Lambda}_Y|(A\cap A_n) \leq \Phi(A_n).$$

Therefore $\lim_{n\in\mathbb{N}} E(X_{\tau_n} - Y)I_{A\cap B_n} = \tilde{\Lambda}_Y(A)$, which implies that

$$\lim_{n\in\mathbb{N}} E(X_{t_n} - Z)I_A = \tilde{\Lambda}_Y(A).$$

Likewise, $\overline{\lim_{n\in\mathbb{N}}} E|X_{t_n} - Z|I_A \leq \overline{\lim_{n\in\mathbb{N}}} E|X_{\tau_n} - Y|I_{A\cap B_n}$, and this time by Lemma 6.1, $\lim_{n\in\mathbb{N}} E|X_{\tau_n} - Y|I_{A\cap B_n} = \Phi(A)$. Hence, $\overline{\lim_{n\in\mathbb{N}}} E|X_{t_n} - Z|I_A \leq \Phi(A) \leq \underline{\lim_{n\in\mathbb{N}}} E|X_{t_n} - Z|I_A$.

Now, for $\varepsilon > 0$, $P(|X_{t_n} - Z| > \varepsilon) \le \frac{1}{\varepsilon} E|X_{t_n} - Z|I_{A_n} + P(B_n) < \frac{\varepsilon_n}{\varepsilon} + \varepsilon_n$. Hence X_{t_n} converges to Z in probability. Q.E.D.

<u>Corollary 6.1</u>. For any $Y, Z \in C$, $\tilde{\Lambda}_Y = \tilde{\Lambda}_Z$, $\tilde{\Lambda}^+_Y = \tilde{\Lambda}^+_Z$ and $\tilde{\Lambda}^-_Y = \tilde{\Lambda}^-_Z$.

<u>Proof</u>: Applying Proposition 6.2 separately to Y and then to Z, we have for any $A \in \mathcal{A}$, $\tilde{\Lambda}^+_Y(A) \le \tilde{\Lambda}^+_Z(A)$ and $\tilde{\Lambda}^+_Z(A) \le \tilde{\Lambda}^+_Y(A)$ and thus $\tilde{\Lambda}^+_Y(A) = \tilde{\Lambda}^+_Z(A)$. Q.E.D.

We will denote by $\tilde{\phi}$ the finitely additive measure such that $\tilde{\phi} = \tilde{\Lambda}_Y$ for any $Y \in C$. Clearly then $|\tilde{\phi}| = \phi$.

<u>Corollary 6.2</u>. For any $Y \in C$, there exists an increasing sequence $(t_n(Y), n \in \mathbb{N})$ in T, such that $X_{t_n(Y)} \to Y$ in probability and such that for every $A \in \mathcal{A}$,

$$\lim_{n\in\mathbb{N}} E(X_{t_n(Y)} - Y)I_A = \tilde{\phi}(A) \quad \text{and} \quad \lim_{n\in\mathbb{N}} E|X_{t_n(Y)} - Y|I_A = \phi(A).$$

<u>Proof</u>: Straightforward from Proposition 6.2 by taking $Y = Z$. Q.E.D.

<u>Note</u>: The sequence $(t_n(Y), n \in \mathbb{N})$ in the above Corollary is different from the one in Proposition 6.1.

We also have two immediate consequences that we state without proofs.

<u>Corollary 6.3</u>. For any $Y \in C$, there exists an increasing sequence $(t_n(Y), n \in \mathbb{N})$ in T such that $X_{t_n(Y)} \to Y$ in probability and such that for any simple, $\mathcal{A}$-measurable function $V = \sum_{i=1}^m \lambda_i I_{A_i}$ we have

$$\lim_{n\in\mathbb{N}} E(X_{t_n(Y)} \cdot V) = E(YV) + \sum_{i=1}^m \lambda_i \tilde{\phi}(A_i).$$

<u>Corollary 6.4</u>. For any $Y, Z \in C$, there exist increasing sequences $(t_n(Y), n \in \mathbb{N})$ and $(t_n(Z), n \in \mathbb{N})$ in T such that $X_{t_n(Y)} \to Y$ in probability, $X_{t_n(Z)} \to Z$ in probability and such that for any simple, $\mathcal{A}$-measurable function V,

$$\lim_{n\in\mathbb{N}} E(X_{t_i(Y)} - X_{t_j(Z)})V = E(Y - Z)V$$

as $i, j \to +\infty$.

<u>Remarks</u>. 6.1. Note that for $Y \in L^1(\Omega, \mathcal{F}, P)$ and $A \in \mathcal{A}$,

$$\tilde{\Lambda}^-_Y(A) = \inf\{\lim_{n\in\mathbb{N}} E(X_{t_n} - Y)^- I_A;\ \lim_{n\in\mathbb{N}} E|X_{t_n} - Y| = \Lambda_Y(A),\ t_{n+1} \ge t_n,\ t_n \in T(n)\}.$$

6.2. Let ℓ^∞ be the set of bounded real-valued sequences, and let L be a Banach limit on ℓ^∞ (see for e.g. [4] p. 73 for the definition and properties of Banach limits). For $Y \in L^1(\Omega, \mathcal{F}, P)$ we define

$$\Gamma^+_Y(A) = \sup\{L(E(X_{t_n} - Y)^+ I_A, n \in \mathbb{N});\ \lim_{n\in\mathbb{N}} E|X_{t_n} - Y|I_A = \Lambda_Y(A),\ t_{n+1} \ge t_n,\ t_n \in T(n)\}$$

for $A \in \mathcal{A}$. We easily deduce from the properties of Banach limits that $\Gamma^+_Y = \tilde{\Lambda}^+_Y$. We therefore cannot hope for a generalization of the definition of $\tilde{\Lambda}_Y$ along the lines of Banach limits.

6.3. There are in general many additive set functions $\tilde{\Lambda}_Y$ usually different from the one we constructed, and satisfying the results of section 6. For example we can replace $\lim_{n\in\mathbb{N}} E(X_{t_n} - Y)^+ I_A$ by $\lim_{n\in\mathbb{N}} E(X_{t_n} - Y)^- I_A$ in Definition 6.1. Alternatively we can replace sup by inf.

6.4. Let $Y \in L^1(\Omega,\mathcal{F},P)$ and let $(M_n, n \in \mathbb{N})$ be the martingale associated with $(X_n, n \in \mathbb{N})$ and Y (see Definition 2.2). Then there exists an L^1-bounded martingale $(\tilde{M}_n, n \in \mathbb{N})$ relative to $(\mathcal{F}_n, n \in \mathbb{N})$ unique up to indistinguishability, such that $|\tilde{M}_n| \le M_n$ a.s., $\tilde{\Lambda}_Y(A) = \int_A \tilde{M}_n\, dP$ for $n \in \mathbb{N}$ and $A \in \mathcal{F}_n$. Moreover, $\Lambda_Y(A) = \lim_{n\in\mathbb{N}} \int_A |\tilde{M}_n| dP$ for $A \in \mathcal{A}$. The proof follows from the fact that $R_n\tilde{\Lambda}_Y$ is absolutely continuous with respect to $R_n\Lambda_Y$ for every $n \in \mathbb{N}$ and the fact that $|\tilde{\Lambda}_Y| = \Lambda_Y$.

6.5. If $Y \in C$ then the martingale $(\tilde{M}_n, n \in \mathbb{N})$ of Remark 6.4 will also satisfy $\lim_{n\in\mathbb{N}} \tilde{M}_n = 0$ a.s. This follows from Proposition 2.3. Therefore, there exists an L^1-bounded martingale $(\tilde{M}_n, n \in \mathbb{N})$ relative to $(\mathcal{F}_n, n \in \mathbb{N})$ such that $\lim_{n\in\mathbb{N}} \tilde{M}_n = 0$ a.s., for $n \in \mathbb{N}$, $A \in \mathcal{F}_n$, $\tilde{\Phi}(A) = \int_A \tilde{M}_n dP$ and such that $\Phi(A) = \lim_{n\in\mathbb{N}} \int_A |\tilde{M}_n| dP$ for $A \in \mathcal{A}$.

7. Extension to the finite-dimensional and infinite-dimensional case

Throughout the section E will be a separable Banach space with norm $\|\cdot\|$ and $\mathcal{B}(E)$ the corresponding σ-field of Borel sets. A mapping $X: \Omega \to E$ is said to be an E-valued random variable if it is measurable as a mapping from $(\Omega,\mathcal{F})$ into $(E,\mathcal{B}(E))$. An E-valued random variable X is said to be (Bochner) integrable if $E\|X\| < \infty$. We will denote by $L^1_E(\Omega,\mathcal{F},P)$ the space of all E-valued, integrable random variables.

It is easy to see that when replacing $|\cdot|$ by $\|\cdot\|$, all the results of sections 2 and 3 remain valid in the Banach space setting. The results of section 4 do not in general extend to the infinite-dimensional case. In the notation of section 4 we have:

Lemma 7.1. The following assertions are equivalent:

i) For any adapted sequence $(X_n, n \in \mathbb{N})$ of E-valued random variables, $\bar{C}[X_n, n \in \mathbb{N}] \neq \emptyset$ if and only if $\bar{C}[\|X_n\|, n \in \mathbb{N}] \neq \emptyset$.

ii) E is of finite-dimension.

Proof: i) => ii). Suppose E is of infinite-dimension. Then, there exist $\varepsilon > 0$ and a sequence $(e_n, n \in \mathbb{N})$ in E such that $\|e_n\| = 1$ for any $n \in \mathbb{N}$ and $\|e_i - e_j\| > \varepsilon$ for any $i,j \in \mathbb{N}$, $i \neq j$. Let $X_n(\omega) = e_n$ for every $n \in \mathbb{N}$ and $\omega \in \Omega$. Then $\bar{C}(\|X_n\|, n \in \mathbb{N}) = \{1\}$, but $\bar{C}[X_n, n \in \mathbb{N}] = \emptyset$ which contradicts i).

ii) => i). We prove the implication only in the case dim E = 2. Let $\{e_1, e_2\}$ be a basis for E. Then for each $n \in \mathbb{N}$, $Z_n = X_n e_1 + Y_n e_2$. Assume $U \in \bar{C}[\|Z_n\|, n \in \mathbb{N}]$. Then there exists $(\tau_n, n \in \mathbb{N})$ a sequence in T, with $\tau_{n+1} \ge \tau_n$ and $\tau_n \in T(n)$ for every $n \in \mathbb{N}$, such that $U = \lim_n \|Z_{\tau_n}\|$ a.s. Let $X'_n = X_{\tau_n}$ and $\mathcal{G}_n = \mathcal{F}_{\tau_n}$ for each $n \in \mathbb{N}$. Then $\overline{\lim}\, |X'_n| < +\infty$ a.s. and we may define $X(\omega) = \overline{\lim}\, X'_n(\omega)$ on the set where this lim sup is finite and 0 otherwise. Clearly then X is a measurable cluster point of $(X'_n, \mathcal{G}_n)_{n \in \mathbb{N}}$. Hence $X = \lim_n X'_{\nu_n}$ a.s., where each ν_n is a stopping time relative to $(\mathcal{G}_k, k \in \mathbb{N})$. Let $\sigma_n = \tau_{\nu_n}$ for each $n \in \mathbb{N}$. Then $(\sigma_n, n \in \mathbb{N})$ is a sequence of stopping times relative to $(\mathcal{F}_k, k \in \mathbb{N})$.

Let now $Y''_n = Y_{\sigma_n}$ and $\mathcal{H}_n = \mathcal{F}_{\sigma_n}$ for each $n \in \mathbb{N}$. Repeating the above argument we find a measurable cluster point Y of $(Y''_n, \mathcal{H}_n)_{n \in \mathbb{N}}$ and a sequence $(\gamma_n, n \in \mathbb{N})$ of stopping times relative to $(\mathcal{H}_k, k \in \mathbb{N})$ such that $Y = \lim_n Y''_{\sigma_n}$ a.s. Then $(\sigma_{\gamma_n}, n \in \mathbb{N})$ is a sequence of stopping times relative to $(\mathcal{F}_k, k \in \mathbb{N})$ and

$$Xe_1 + Ye_2 = \lim_n (X_{\sigma_{\gamma_n}} e_1 + Y_{\sigma_{\gamma_n}} e_2) \quad \text{a.s.}$$

Thus $Z = Xe_1 + Ye_2 \in \bar{C}[Z_n, n \in \mathbb{N}]$. Clearly also $\|Z(\omega)\| = U(\omega)$ a.s. Q.E.D.

Lemma 7.2. If E is of finite-dimension and $(X_n, n \in \mathbb{N})$ is a sequence of E-valued random variables, then for any $U \in \bar{C}[\|X_n\|, n \in \mathbb{N}]$, $U \geq 0$ there exists $Z \in \bar{C}[X_n, n \in \mathbb{N}]$ such that $\|Z\| = U$.

Proof: Follows from the proof of Lemma 7.1.

Therefore all the results of Sections 4 and 5 extend to the finite-dimensional case.

In the infinite-dimensional case we have:

Proposition 7.1. Let E be an infinite-dimensional Banach space and let $(X_n, n \in \mathbb{N})$ be an adapted sequence of E-valued integrable random variables satisfying condition (I). Let Φ be the associated Fatou discrepancy. Then for any $Y \in L^1_E(\Omega, \mathcal{F}, P)$

$$\Phi(\Omega) \leq \Lambda_Y(\Omega) \leq \Phi(\Omega) + \rho(Y,C).$$

The first inequality becomes an equality if and only if $Y \in C$.

Proof: By Corollary 3.1,

$$\Phi(\Omega) = \Phi_Y(\Omega) = \Lambda_Y(\Omega) - E(\underline{\lim}_{n \in \mathbb{N}} \|X_n - Y\|) \leq \Lambda_Y(\Omega)$$

and it is clear that $\Lambda_Y(\Omega) = \Phi(\Omega)$ if and only if $E(\underline{\lim}_{n \in \mathbb{N}} \|X_n - Y\|) = 0$, that is, if and only if $Y \in C$.

If C is empty then $\rho(Y,C) = +\infty$ and the second inequality becomes trivial. If C is nonempty, then for any $X \in C$

$$\|X_n - Y\| \leq \|X_n - X\| + \|X - Y\|$$

whence

$$\underline{\lim}_{n \in \mathbb{N}} \|X_n - Y\| \leq \|X - Y\| \quad \text{a.s.}$$

which implies

$$E(\underline{\lim}_{n \in \mathbb{N}} \|X_n - Y\|) \leq \rho(Y,C)$$

and thus

$$\Lambda_Y(\Omega) = \Phi(\Omega) + E(\underline{\lim}_{n \in \mathbb{N}} \|X_n - Y\|) \leq \Phi(\Omega) + \rho(Y,C)$$

proving the second inequality. Q.E.D.

Proposition 5.1 of section 5 is still valid in the infinite-dimensional case with the same proof. As for section 6, the results carry over in the finite-dimensional case. We do need a few details. Let E be a finite-dimensional Banach space, and $(X_n, n \in \mathbb{N})$ an adapted sequence of E-valued random variables satisfying condition (I). Again for simplicity we assume $\dim E = 2$ and we let (e_1, e_2) be a basis for E. Therefore for each $n \in \mathbb{N}$, $X_n = X_n' e_1 + X_n'' e_2$. Let $Y \in L_E^1(\Omega, \mathcal{F}, P)$, $Y = Y'e_1 + Y'e_2$.

For each $A \in \mathcal{A}$, we define

$$\Lambda_Y'(A) = \sup\{\lim_{n\in\mathbb{N}} E|X_{t_n}' - Y'| I_A : \lim_{n\in\mathbb{N}} E\|X_{t_n} - Y\| I_A = \Lambda_Y(A)\}$$

and

$$\Lambda_Y''(A) = \sup\{\lim_{n\in\mathbb{N}} E|X_{t_n}'' - Y''| I_A : \lim_{n\in\mathbb{N}} E\|X_{t_n} - Y\| I_A = \Lambda_Y(A) \text{ and } \lim_{n\in\mathbb{N}} E|X_{t_n}' - Y'| I_A = \Lambda_Y'(A)\}$$

(both suprema being taken over all possible sequences satisfying also $(t_n, n \in \mathbb{N}) \subset T$, $t_{n+1} \geq t_n$, $t_n \in T(n)$).

We also define

$$(\tilde{\Lambda}_Y')^+(A) = \sup\{\lim_{n\in\mathbb{N}} E(X_{t_n}' - Y')^+ I_A : \lim_{n\in\mathbb{N}} E\|X_{t_n} - Y\| I_A = \Lambda_Y(A), \lim_{n\in\mathbb{N}} E|X_{t_n}' - Y'| I_A = \Lambda_Y'(A)\}$$

and

$$(\tilde{\Lambda}_Y'')^+(A) = \sup\{\lim_{n\in\mathbb{N}} E(X_{t_n}'' - Y'')^+ I_A : \lim_{n\in\mathbb{N}} E\|X_{t_n} - Y\| I_A = \Lambda_Y(A), \lim_{n\in\mathbb{N}} E|X_{t_n}' - Y'| I_A = \Lambda_Y'(A), \lim_{n\in\mathbb{N}} E|X_{t_n}'' - Y''| I_A = \Lambda_Y''(A)\}.$$

Let $\tilde{\Lambda}_Y' = 2(\tilde{\Lambda}_Y')^+ - \Lambda_Y'$, $\tilde{\Lambda}_Y'' = 2(\tilde{\Lambda}_Y'')^+ - \Lambda_Y''$, and $\tilde{\Lambda}_Y = \tilde{\Lambda}_Y' e_1 + \tilde{\Lambda}_Y'' e_2$.

The measure $\tilde{\Lambda}_Y$ thus defined will satisfy all the properties in section 6. The existence of $\tilde{\Phi}$ also follows with all its required properties. The martingale representation of $\tilde{\Lambda}_Y$ and $\tilde{\Phi}$ remains valid since the Radon-Nikodym property is present in finite-dimensional Banach space.

Remarks 7.1. The results of section 6, in particular Proposition 6.2 and Corollary 6.2, fail in the infinitely-dimensional setting. Suppose E is an infinite-dimensional space and let $(e_n, n \in \mathbb{N})$ be a sequence in E such that $\|e_n\| = 1$ and $\|e_i - e_j\| > \varepsilon$, $i \neq j$, for some $\varepsilon > 0$. Let $\Omega = [0,1]$, P the Lebesgue measure, $X_n = 2^n e_n I_{[0,2^{-n}]}$ and $\mathcal{F}_n = \sigma(X_1,\dots,X_n)$ for each $n \in \mathbb{N}$. Then $\lim_{n\in\mathbb{N}} X_n = 0$ a.s. and thus $C \neq \emptyset$. Moreover for each $t \in T$, $EX_t = e_n$ for some n. Therefore $(EX_t, t \in T)$ has no limit points in the norm topology. It is possible to obtain some results from section 6 if one looks at weaker forms of convergence. This will be the object of a future study.

REFERENCES

1. Austin, D. G., Edgar, G. A., and Ionescu Tulcea, A., Pointwise convergence in terms of expectations, *Z. Wahr. verw. Geb.*, *30*, 17-26 (1974).

2. Baxter, J. R., Convergence of stopped random variables, *Adv. Math.*, *21*, 112-115 (1976).

3. Bellow, A., Submartingale characterization of measurable cluster points. Probability in Banach spaces. Advances in Probability and Related Topics *4*, 69-80 (1978).

4. Dunford, N., Schwartz, J. T., *Linear operators, Part I.* New York: Interscience (1958).

5. Dvoretzky, A., On the Fatou Inequality, Preprint (1983).

6. Meyer, P. A., Dellacherie, C., *Théorie des martingales*, Hermann, Paris (1980).

7. Neveu, J., *Martingales à temps discret*, Masson, Paris (1972).

D. G. Austin
A. Bellow[1)]
Department of Mathematics
Northwestern University
Evanston, Illinois 60201

N. Bouzar[2)]
Department of Industrial Engineering
Northwestern University
Evanston, Illinois 60201

[1)] Research supported by the National Science Foundation.

[2)] Research supported by the Air Force Office of Scientific Research under Grant AFOSR-82-0189.

Limit Theorems for Sojourns
of Stochastic Processes

by

Simeon M. Berman
Courant Institute of Mathematical Sciences
New York University
New York, N. Y. 10012

1. Introduction and Summary.

Let $X(t)$, $t \in B$, be a stochastic process assuming values in a measure space $\mathcal{X}$, and let (A_u), $u > 0$, be a family of measurable subsets of $\mathcal{X}$ such that $P(X(t) \in A_n) > 0$ and $\lim_{u\to\infty} P(X(t) \in A_u) = 0$ for every $t \in B$. Define the random variable $L_u = \text{mes}(t: t \in B, X(t) \in A_u)$. The subject of this paper is the asymptotic behavior of the ratio

$$(1.1) \qquad \frac{\int_0^x y \, dP(vL_u \leq y)}{E(vL_u)}, \quad x > 0,$$

for a suitable function $v = v(u)$, and $u \to \infty$. Results about (1.1) are shown to imply corresponding results about the ratio

$$(1.2) \qquad \frac{P(vL_u > x)}{E(vL_u)}, \quad \text{for } x > 0 .$$

This problem was first considered in the special case where $X(t)$, $0 \leq t \leq 1$, is a real separable measurable, stationary Gaussian process, and $A_u = (u,\infty)$, so that L_u is the sojourn time above the level u [2]. Then the hypothesis of stationarity was

This paper represents results obtained at the Courant Institue of Mathematical Sciences, New York University, under the sponsorship of the National Science Foundation, Grant MCS-82-01119.

weakened to include Gaussian processes which were locally stationary in a specific sense; this allowed for the consideration of Gaussian processes with stationary increments [4]. As a next step in the extension of the methods, the assumption of a Gaussian property was replaced by more general conditions on the finite dimensional distributions of the process, but the assumption of stationarity was kept [5]. Finally the assumption that the process was real valued was dropped, and the family of sets (u,∞), $u > 0$, was replaced by a more general family of "rare" sets [6]. But the most general results up to this point require stationarity (or, in the Gaussian case, "local stationarity") and also are restricted to the case of a one dimensional time parameter. Indeed, the primary identity used in all the proofs depends crucially on the one dimensional nature of the time set.

The main contributions of this paper are

i) The extension of previous results from linear time sets to more general sets such as measurable subsets R^N.

ii) The elimination of the assumption of stationarity and its replacement by a relatively weak uniformity condition on the marginal probabilities $P(X(t)\in A_u)$.

The first extension is carried out by means of a new and simple identity for the distribution of the sojourn time of a general stochastic process: For sets B and A in the time and space variables, respectively,

$$(1.3) \qquad \int_0^x y\,dP(vL_u \leq y) = v\int_B P(vL_u \leq x,\ X(t)\in A)dt.$$

Note that

$$EL_u = \int_B P(X(t)\in A_u)dt ,$$

and define

$$(1.4) \qquad g_u(t) = \frac{P(X(t)\in A_u)}{\int_B P(X(s)\in A_u)ds}, \qquad t\in B.$$

Then, by (1.3) and (1.4), the ratio (1.1) is equal to

$$(1.5) \qquad \int_B P(vL_u) \leq x \mid X(t)\in A_u)g_u(t)dt .$$

It is clear from (1.4) that g_u is a density function on B. The assumption of stationarity is now replaced by an hypothesis called "g-stationarity": There exists a density function g on B such that $g_u \to g$, a.e. on B, for $u \to \infty$. The limit of (1.5) is then shown to be a mixture with respect to g(t) of the limit of the conditional distribution of vL_u, given $X(t) \in A_u$.

These extensions permit the application of our results to familiar processes which are not stationary, such a stable processes with independent increments. Our main theorems are applied to results about the sojourn times of stable processes above high levels, the sojourns of stationary Gaussian random fields above high moving barriers of the form $\sqrt{(u^2+f(t))}$, for $u \to \infty$, and the sojourns of a stable process in R^d in a neighborhood of the origin.

2. Elementary results on the sojourns of stochastic processes.

Let $X(t)$, $t \in R^N$, be a stochastic process assuming values in some measure space $\mathcal{X}$, where $N \geq 1$. For an arbitrary measurable subset A of $\mathcal{X}$, let $\chi_A(x)$, $x \in \mathcal{X}$, be the indicator function of A, and consider the family of random variables $\{\chi_A(X(t)), t \in R^N\}$. Throughout this work, we will always assume that the set A satisfies the following condition relative to the distributions of X(t): The functions $P(X(t) \in A)$ and $P(X(s) \in A, X(t) \in A)$ are continuous on R^N and R^{2N}, respectively. This implies that the real valued process $\chi_A(X(t))$, $t \in R^N$, is continuous in mean square, and so, by Doob's fundamental theorem [9], p. 61, has a separable, measurable version.

For any measurable subset B of R^N of finite Lebesgue measure, we may define

$$(2.1) \qquad L = \int_B \chi_A(X(t))dt ,$$

where dt is Lebesgue measure, for the separable, measurable version of the integrand. L is the sojourn time of X(t), $t \in B$, in A.

The fact that (2.1) is defined only for the particular version of the integrand is merely a technical point of the general theory. In most applications of interest, $\mathcal{X}$ has a topology, and the separability-measurability hypothesis is more naturally stated in terms of the process X itself.

LEMMA 2.1. <u>For every measurable subset J of $(0,\infty)$ we have</u>

$$(2.2) \qquad \int_J y\ dP(L \leq y) = \int_B P(L \in J | X(t) \in A) P(X(t) \in A)dt.$$

PROOF. By (2.1) and Fubini's theorem, we have

$$\int_J y\ dP(L \leq y) = EL\chi_J(L)$$

$$= E \int_B \chi_A(X(t))\chi_J(L)dt = \int_B P(L\in J,\ X(t)\in A)dt \ .$$

By the definition of conditional probability, the last integral is equal to the right hand member of (2.2), and the latter equation is established.

The relation (2.2) yields an explicit form for the distribution of L:

$$P(L\in J) = \int_B \int_J y^{-1}\, dP(L\leq y|X(t)\in A)P(X(t)\in A)dt \ . \tag{2.3}$$

Indeed, this follows from (2.2) by writing it formally as

$$y\ dP(L\leq y) = \int_B dP(L\leq y|X(t)\in A)P(X(t)\in A)dt \ ,$$

and then dividing each side by y, and integrating over J.

The primary concern of this work is the limiting behavior of the left hand member of (2.2) after appropriate normalization. The interest in such results is that they provide corresponding limit theorems for the left hand member of (2.3). The continuity of the correspondence between (2.2) and (2.3) is based on the following elementary result:

LEMMA 2.2. _Let $F_n(x)$, $x \geq 0$, be a sequence of distribution functions with finite first moments. Suppose that there is a distribution function $G(x)$, $x \geq 0$, such that for $n \to \infty$_

$$\frac{\int_0^x y\ dF_n(y)}{\int_0^\infty y\ dF_n(y)} \to G(x) \ , \tag{2.4}$$

at all points of continuity $x \geq 0$. Then

$$\frac{1 - F_n(x)}{\int_0^\infty y\ dF_n(y)} \to \int_x^\infty y^{-1}dG(y) \tag{2.5}$$

at all continuity points $x \geq 0$.

PROOF. The proof is similar to that for (2.3). The relation (2.4) implies that

$$\frac{\int_a^b y\, dF_n(y)}{\int_0^\infty y\, dF_n(y)} \to G(b) - G(a) .$$

for all continuity points $0 < a < b$. Hence, by considering small intervals, and then using them to approximate over large intervals of continuity, we obtain,

$$\frac{F_n(b) - F_n(a)}{\int_0^\infty y\, dF_n(y)} \to \int_a^b y^{-1} d\, G(y) . \tag{2.6}$$

Passage to the limit $b \to \infty$ is permitted on each side of the limit relation above, and this proves the lemma. Indeed, for arbitrary $b > 0$, Markov's inequality implies

$$\frac{1 - F_n(b)}{\int_0^\infty y\, dF_n(y)} \leq 1/b ,$$

and also,

$$\int_b^\infty y^{-1} dG(y) \leq [1 - G(b)]/b .$$

Each of the right hand members above tends to 0 for $b \to \infty$, and so the contribution of the interval (b,∞) to (2.6) is negligible.

We note that (2.2) has some similarity to an identity which have repeatedly used in the case $N = 1$. Define, for $t > 0$,

$$L_t = \int_0^t \chi_A(X(s))ds ;$$

then the distribution of L_t satisfies

$$\int_0^x P(L_t > y)dy = \int_0^t P(L_s \leq x | X(s) \in A)P(X(s) \in A)ds . \tag{2.7}$$

(See, for example, [6] and the references to earlier work.) The most important difference between (2.2) and (2.7) is that the random variable L_s on the right hand side is a partial sojourn up to time s, while L_t on the left hand side is the complete sojourn. The proof of (2.7) depends heavily on the complete order relation of the real line, and does not generalize to higher dimensions. For this reason we have shifted to (2.2) which is more generally valid and which, in the context it arises, is just as useful.

In the following, we take A_u, $u > 0$, to be a family of measurable subsets of $\mathcal{X}$; and then we define L_u in accordance with (2.1) for $A = A_u$:

$$L_u = \int_B \chi_{A_u}(X(t))dt \ . \tag{2.8}$$

(Thus, L_u is not analogous to the partial sojourn in (2.7).) Let $v > 0$ be arbitrary, and substitute vL_u for L in the left hand member of (2.2); then, by a change of variable on the right hand side, it is seen that

$$\int_0^x y \, dP(vL \le y) = v \int_B P(vL \le x | X(t) \in A) P(X(t) \in A) dt \ .$$

Note also that

$$E(vL_u) = v \int_B P(X(t) \in A_u) dt \ .$$

Then the two preceding equations imply

$$\frac{\int_0^x y \, dP(vL_u \le y)}{E(vL_u)} \tag{2.9}$$

$$= \frac{\int_B P(vL_u \le x | X(t) \in A_u) P(X(t) \in A_u) dt}{\int_B P(X(t) \in A_u) dt} \ .$$

Put

$$g_u(t) = \frac{P(X(t) \in A_u)}{\int_B P(X(s) \in A_u) ds} \ , \quad t \in B; \tag{2.10}$$

then g_u is a density function on B, and the right hand member of (2.9) is expressible as

$$\int_B P(vL_u \leq x | X(t) \in A_u) g_u(t) dt \ . \tag{2.11}$$

Formula (2.11) above is the starting point of our derivation of the limit of the ratio (2.9) for $u \to \infty$. Here v is taken to be a positive function $v = v(u)$, $u > 0$. A brief sketch of the analysis is now given. Define

$$G_u(x;t) = P(vL_u \leq x | X(t) \in A_u) \ ; \tag{2.12}$$

and define the probability measure

$$\Gamma_u(C) = \int_C g_u(t) dt \ , \quad u > 0 \ , \tag{2.13}$$

for Borel subsets C of B; then (2.11) is identical with

$$\int_B G_u(x;t) \Gamma_u(dt) \ . \tag{2.14}$$

If there is a probability measure Γ on B such that $\Gamma_u \to \Gamma$ weakly for $u \to \infty$, and, for each t, a distribution function $G(x;t)$, $x > 0$, such that $G_u(\cdot\,;t) \to G(\cdot\,;t)$ weakly, then under appropriate conditions, we expect (2.14) to converge to the limit

$$\int_B G(x;t) \Gamma(dt) \ . \tag{2.15}$$

The passage from (2.11) to (2.15) depends on the form of Γ. In this paper we consider the case where Γ is absolutely continuous. In a forthcoming paper, we will consider the case where Γ consists of a single atom. These two cases include all of our applications.

3. An hypothesis on the marginal distributions: g-stationarity.

In this section we formulate a condition on the marginal distributions of the process which determines the limiting behavior of Γ_u for $u \to \infty$.

DEFINITION 3.1. *$X(t)$, $t \in B$, is said to be marginally g-stationary with respect to the family (A_u) if*

(3.1) $$g(t) = \lim_{u\to\infty} \frac{P(X(t)\in A_u)}{\int_B P(X(s)\in A_u)ds}$$

exists for almost all $t \in B$, and

(3.2) $$\int_g g(t)dt = 1 .$$

It is obvious that a process with identical marginal distributions is marginally g-stationary with $g \equiv (\text{mes } B)^{-1}$. The assumptions (3.1) and (3.2) imply that the function g_u in (2.10) satisfies

(3.3) $$\lim_{u\to\infty} \int_B |g_u(t) - g(t)|dt = 0 .$$

Indeed, this follows by an application of a theorem of Scheffe [12].

Under the hypothesis of g-stationarity the integral (2.11) is asymptotically equal to

(3.4) $$\int_B P(vL_u < x|X(t)\in A_u)g(t)dt .$$

Indeed-(3.4) implies that the limit of $g_u(t)$ may be taken under the integral sign.

Let $G_u(x;t)$ represent the conditional probability appearing in (3.5) and let $G(x;t)$ be a distribution function in x for each fixed t.

LEMMA 3.1. If for each $t\in B$,

(3.5) $$\lim_{u\to\infty} G_u(x;t) = G(x;t)$$

weakly in $x > 0$, then in the same way,

(3.6) $$\lim_{u\to\infty} \int_B G_u(x;t)g(t)dt = \int_{R^N} G(x;t)g(t)dt .$$

PROOF. Let x_0 be a point of continuity of the right hand member of (3.7). Then, by a monotone convergence argument, x_0 is also a point of continuity of the function $G(x;t)g(t)$ for almost all t. Now (3.5) implies that $G_u(x_0;t)g(t) \to G(x_0;t)g(t)$ for almost all t, so that, by dominated convergence, (3.6) holds for $x = x_0$.

As a consequence of this lemma and Lemma 2.1, and the asymptotic equivalence of (2.11) and (3.5), we have:

LEMMA 3.2. *Under the hypothesis of marginal g-stationarity, if for each t,*

$$\lim_{u\to\infty} P(vL_u \leq x \mid X(t)\in A_u) = G(x;t), \tag{3.7}$$

weakly in x, then in the same way,

$$\lim_{u\to\infty} \frac{\int_0^x y\,dP(vL_u \leq y)}{E(vL_u)} = \int_B G(x;t)g(t)dt\ . \tag{3.8}$$

4. *A convergence lemma for a family of zero-one valued stochastic processes.*

We prove the following general result to be used in the proofs of our main theorems:

LEMMA 4.1. *Let $\{\xi_u(t),\ t\in R^N\}$, $u > 0$, be a family of stochastic processes such that $P(\xi_u(t) = 0) + P(\xi_u(t) = 1) = 1$ for all u and t. Suppose that for each $m \geq 1$, and for each finite set $t_1,\dots,t_m$ of points in R^N, the limit*

$$q_m(t_1,\dots,t_m) = \lim_{u\to\infty} E\xi_u(t_1)\dots\xi_u(t_m) \tag{4.1}$$

exists, and that the functions q_1 and q_2 are continuous on R^N and R^{2N}, respectively. Then there exists a separable measurable process $\eta(t)$, $t\in R^N$ such that $P(\eta(t) = 0) + P(\eta(t) = 1) = 1$ for all t, and

$$q_m(t_1,\dots,t_m) = E\eta(t_1)\dots\eta(t_m)\ , \tag{4.2}$$

for all $t_1,\dots,t_m$ and $m \geq 1$.

PROOF. Since $\xi_u(t)$ assumes only the values 0 and 1, it follows that the finite dimensional distributions of the process are completely specified by functions

$$E\xi_u(t_1)\dots\xi_u(t_m) = P(\xi_u(t_1) = \dots = \xi_u(t_m) = 1)\ .$$

By (4.1), the finite dimensional distributions of the process converge to limits which are finite dimensional distributions on product sets of the form $\{0,1\}^m$ for integer $m \geq 1$. The consistency of the system of limiting finite dimensional distributions follows from the consistency of the system of the distributions of the original process. Hence, by the fundamental Kolmogorov consistency theorem for stochastic processes, there exists a process $\eta(t)$ having the finite dimensional distributions obtained as limits; furthermore, $\eta(t)$ necessarily assumes only the values 0 and 1 by virtue of the nature of the distributions.

The assumed continuity of q_1 and q_2 implies that, as a second order process, $\eta(t)$ is mean square continuous because its mean and covariance function are q_1 and $q_2 - q_1 \cdot q_1$, respectively. Doob's fundamental result [9], p.61, now implies that η has a separable measurable version, and (4.2) is valid also for this version. The proof is complete.

In our applications the process

$$\chi_{A_u}(X(t))$$

appearing in the integrand of (2.8) plays the role of a process closely related to $\xi_u(t)$. In our previous work, the hypotheses of the theorems required not only the convergence of the finite dimensional distributions of the process, but also the identification of the limiting process (see [6], Section 2). The current lemma permits us to drop the latter requirement.

5. The Sojourn Limit Theorem with $v(u) = 1$.

In all of our previous work on the sojourn L_u, the latter random variable was first multiplied by the scaling function $v(u) \to \infty$ before passing to the limit in (1.1). In this section we present a new and simpler version of the theorem in which v is taken to be constant.

THEOREM 5.1. For a measurable set $B \subset R^N$ of finite measure, let $X(t)$, $t \in B$, be marginally g-stationary with respect to (A_u) for some density function g. Assume that for every $m \geq 1$, $t \in B$ and $s_1, \ldots, s_m \in B$, the limit

$$q_m(s_1,\ldots,s_m;t) = \lim_{u\to\infty} P(X(s_i)\in A_u, i = 1,\ldots,m \mid X(t)\in A_u) \tag{5.1}$$

exists, and that $q_1(s;t)$ and $q_2(s_1,s_2;t)$ are continuous in s and (s_1,s_2), respectively. Then for such t there exists a separable, measurable stochastic process $\eta_t(s)$, $s \in B$, assuming the values 0 and 1, such that

(5.2) $$E\eta_t(s_1)\ldots\eta_t(s_m) = q_m(s_1,\ldots,s_m;t)$$

and

(5.3) $$\lim_{u\to\infty} \frac{\int_0^x y\, dP(L_u < y)}{EL_u} = \int_B G(x;t)g(t)dt ,$$

<u>at all continuity points of the latter function, where</u>

(5.4) $$G(x;t) = P(\int_B \eta_t(s)ds < x) .$$

PROOF. For fixed $t \in B$, define the family of processes

(5.5) $$\xi_{u,t}(s) - \chi_{A_u}(X(s)),\ s \in B,$$

conditioned by $X(t) \in A_u$. Then (5.1) implies

(5.6) $$\lim_{u\to\infty} E[\xi_{u,t}(s_1)\ldots\xi_{u,t}(s_m)|X(t)\in A_u] = q_m(s_1,\ldots,s_m;t)$$

Thus, by Lemma 4.1, there is a process $\eta_t(s)$ of the form stated in the theorem such that (5.2) holds.

Our next step is to verify the condition (3.7). By Fubini's theorem, we have

(5.7) $$E(L_u^m|X(t)\in A_u) = \int_B\cdots\int_B P(X(s_i)\in A_u,\ i=1,\ldots,m|X(t)\in A_u)ds_1\ldots ds_m .$$

By (5.6), the latter integrand converges to $q_m(s_1,\ldots,s_m;t)$. Therefore, by (5.2), the limit of the integrand in (5.7) is representable as $E\eta_t(s_1)\ldots\eta_t(s_m)$. By bounded convergence, the limit of the integral is

$$\int_B\cdots\int_B E\,\eta_t(s_1)\ldots\eta_t(s_m)ds_1\ldots ds_m ,$$

which by Fubini's theorem, is equal to

(5.8) $$E(\int_B \eta_t(s)ds)^m .$$

Thus, we have shown that the conditional moment (5.7) converges to the corresponding moment (5.8). The sequence of moments (5.8) for $m \geq 1$ determines a unique distribution because B is of finite measure, and so the moment (5.8) is bounded by $(\text{mes } B)^m$. The moment convergence theorem now implies the convergence of the conditional distribution of L_u to the limit defined by (5.4). Thus, the condition (3.7) holds, and so the conclusion (5.3) now follows by application of Lemma 3.2.

REMARK. The assumption that (5.1) holds for all $t \in B$ can be weakend to the condition that it holds for all t such that $g(t) > 0$. Indeed, the expression on the right hand side of (5.3) does not depend on the nature of $G(x;t)$ for t-values outside the support of g.

As an application of this theorem we have the following corollary in the case $N = 1$.

COROLLARY 5.1. *Let $X(t)$, $0 \leq t \leq 1$, be marginally g-stationary for some continuous function g. If*

$$(5.9) \qquad \lim_{u\to\infty} P(X(t) \in A_u | X(s) \in A_u) = 1, \quad \text{for } s \leq t ,$$

then g is necessarily nondecreasing, and

$$(5.10) \qquad \lim_{u\to\infty} \frac{\int_0^x y \, dP(L_u \leq y)}{EL_u} = \int_{1-x}^{1} g(t)dt - xg(1-x) ,$$

for $0 < x < 1$.

PROOF. According to the remark following the proof of Theorem 5.1, it suffices to prove our corollary under the assumption $g(t) > 0$ for all $0 < t \leq 1$.

We note first that g is nondecreasing. Indeed, if $s < t$, then,

$$P(X(s) \in A_u | X(t) \in A_u) = P(X(t) \in A_u | X(s) \in A_u) \frac{P(X(s) \in A_u)}{P(X(t) \in A_u)} .$$

Since the left hand member above is a probability, it follows that

$$\frac{P(X(s) \in A_u)}{P(X(t) \in A_u)} \leq \frac{1}{P(X(t) \in A_u | X(s) \in A_u)} ,$$

so that, by (5.9),

$$\limsup_{u\to\infty} \frac{P(X(s)\in A_u)/\int_0^1 P(X(s')\in A_u)ds'}{P(X(t)\in A_u)/\int_0^1 P(X(t')\in A_u)dt'} \le 1 .$$

Then (3.1) implies that $g(s) \le g(t)$.

We claim that

(5.11) $$\lim_{u\to\infty} P(X(s_i)\in A_u,\ i=1,..,m \mid X(t)\in A_u) = \min\left[1, \frac{g(s_i)}{g(t)},\ i=1,...,m\right] .$$

Indeed, if $t \le \min s_i$, then (5.9) implies that the limit above is equal to 1. If $t > \min s_i$, then (5.9) implies that the conditional probability in (5.11) is asymptotically equal to

$$P(X(\min s_i)\in A_u)/P(X(t)\in A_u) .$$

Upon division of the numerator and denominator by $\int_0^1 P(X(s)\in A_u)ds$, and application of (3.1), the ratio above converges to $\min g(s_i)/g(t)$. This completes the proof of (5.11). Thus (5.1) holds with

(5.12) $$q_m(s_1,...,s_m;t) = \min[1, g(s_i)/g(t), i=1,...,m] ,$$

and q_1 and q_2 are continuous because g is.

Now we identify the process $\eta_t(s)$. If $s > t$, then $\eta_t(s) = 1$ almost surely; indeed, by (5.12), $E\,\eta_t(s) = 1$, and $\eta_t(s)$ assumes only the values 0 and 1. Next let T_t be a nonnegative random variable with the distribution function equal to

(5.13) $$\begin{cases} 0 & , \text{ for } s \le 0 \\ g(s)/g(t) , & \text{for } 0 < s \le t \\ 1 & , \text{ for } s > t \end{cases}$$

Then

(5.14) $\quad \eta_t(s) = \chi_{(0,s]}(T_t) \quad , \quad 0 < s \leq t ,$

in the sense of equivalence of distributions; indeed, it is simple to see that $E\,\eta_t(s_1)\ldots\eta_t(s_m)$ is equal to (5.12) for $\max s_i \leq t$.

We infer from the last paragraph that

$$\int_0^1 \eta_t(s)ds = \int_0^t \chi_{(0,s]}(T_t)ds + 1-t = 1-T_t \ .$$

Hence, by (5.13) the function $G(x,t)$ in (5.4) takes the form, for $0 \leq x \leq 1$,

$$\begin{aligned} G(x;t) &= 0 \ , \ t \leq 1-x \\ &= 1 - \frac{g(1-x)}{g(t)} \ , \ t > 1-x \ . \end{aligned}$$

Therefore, the right hand member of (5.3) is

$$\int_{1-x}^{1} [g(t) - g(1-x)]dt \ ,$$

which agrees with (5.10).

EXAMPLE 5.1. We apply Corollary 5.1 to the sojourns above a high level for a process with stable stationary independent increments. As a preliminary, we prove,

LEMMA 5.1. _Let X and Y be independent random variables. If_

(5.15) $$\lim_{\varepsilon\to 0} \liminf_{u\to\infty} \frac{P(X > u(1+\varepsilon))}{P(X > u)} = 1 \ ,$$

then

(5.16) $$\lim_{u\to\infty} P(X + Y > u | X > u) = 1 \ .$$

PROOF. The conditional probability in (5.16) is, by definition, equal to $P(X+Y > u, X > u)/P(X > u)$, which, for arbitrary $\varepsilon > 0$, is at least equal to

$$P(X > u(1+\epsilon),\ Y > -u\epsilon)/P(X > u)\ .$$

By independence, the latter is equal to

$$\frac{P(X > u(1+\epsilon))}{P(X > u)}\, P(Y > -u\epsilon)\ .$$

This converges to 1 under the limiting operation in (5.15).

Let $X(t)$, $0 \leq t \leq 1$, have stationary independent increments which are symmetric and stable of index α, $0 < \alpha < 2$. Put $A_u = (u,\infty)$, so that L_u is the sojourn time above u. It is known that $P(X(t) > u) \sim$ constant $t\, u^{-\alpha}$, for $u\to\infty$. This implies that $X(t)$ is marginally g-stationary with

(5.17) $$g(t) = 2t\ ,\quad 0 \leq t \leq 1\ ;$$

and that $X = X(s)$ satisfies (5.15) for every $0 < s < 1$. Therefore, by applying Lemma 5.1 with $X = X(s)$ and $Y = X(t) - X(s)$ for $s < t$, and noting the independence of the increments, we see that the hypothesis (5.9) holds. Hence, the conclusion (5.10) follows with g as in (5.17):

(5.18) $$\lim_{u\to\infty} \frac{\int_0^x y\, dP(L_u \leq y)}{EL_u} = x^2\ ,\quad 0 \leq x \leq 1\ .$$

Then Lemma 2.2 implies

(5.19) $$\lim_{u\to\infty} \frac{P(L_u > x)}{EL_u} = 2(1-x)\ ,\quad 0 \leq x \leq 1\ .$$

6. A condition for the separation of the times of sojourn.

As demonstrated in Lemma 3.2, the limit of the ratio (2.9) is determined, under g-stationarity, by the limit of the probability in (3.7), namely, the probability,

(6.1) $$P(vL_u \leq x \mid X(t)\in A_u)\ .$$

Thus we are led to analyze the conditional distribution of vL_u given $X(t) \in A_u$. In many applications the process X and the family (A_u) exhibit the following kind of asymptotic behavior: If at some point t, X(t) is in the set A_u, then the contributions to the integral L in (2.1) come almost entirely from the portion of the time domain consisting of a ball centered at t and of radius of the order $v^{-1/N}$ if $v(u) \to \infty$. Thus the sample function tends to leave A_u quite soon after making a visit, and so the sojourns are relatively brief, and are locally separated. This behavior is just the opposite of that which is assumed in Corollary 5.1, where the condition (5.9) states that the sample function tends to stay in A_u after it steps in.

For the purpose of formulating the separation condition, we define

$$(6.2) \qquad L_u(t;r) = \int_{B \cap \{s:|s-t|<r\}} \chi_{A_u}(X(s))ds \; .$$

The separation condition is that the conditional probability (6.1) is asymptotically unchanged if vL_u is replaced by (6.2) for $r \to 0$ at a prescribed rate.

LEMMA 6.1. <u>Assume $v(u) \to \infty$ for $u \to 0$ and the condition</u>

$$(6.3) \qquad \lim_{r\to\infty} \limsup_{u\to\infty} \frac{v \iint_{\{(s,t):s,t \in B,\ |s-t|>rv^{-1/N}\}} P(X(s)\in A_u, X(t)\in A_u)ds\,dt}{\int_B P(X(t)\in A_u)dt} = 0$$

<u>Then the limit of the ratio on the right hand side of (2.9) may be determined to exist and its value computed on the basis the limit of the expression obtained by substituting $L_u(t;rv^{-1/N})$ for L_u,</u>

$$(6.4) \qquad \frac{\int_B P(vL_u(t;rv^{-1/N}) \le x | X(t)\in A_u)P(X(t)\in A_u)dt}{\int_B P(X(t) \in A_u)dt},$$

<u>and letting $u \to \infty$ and $r \to \infty$.</u>

PROOF. Since $L_u(t;r) \le L_u$ for every $r > 0$, it is obvious that the ratio in (2.9) is not greater than (6.4), so that

(6.5) $\limsup_{u\to\infty}$ (ratio (2.9)) $\leq \lim_{r\to\infty} \limsup_{u\to\infty}$ (ratio (6.4)).

Now we derive the reverse inequality for the liminf. For arbitrary $r > 0$, write

$$L_u = L_u(t;rv^{-1/N}) + (L_u - L_u(t;rv^{-1/N})) ,$$

the sum of two nonnegative terms. It is elementary that for any set of nonnegative numbers ξ, η, x and y,

$\xi \leq x$ <u>implies</u> : <u>Either</u> $\xi + \eta \leq x + y$ <u>or</u> $\eta \geq y$.

It follows by an application of this remark that for each $x > 0$ and $0 < \varepsilon < 1$,

(6.6) $$P(vL_u(t;rv^{-1/N}) \leq x(1-\varepsilon)|X(t)\in A_u)$$
$$\leq P(vL_u \leq x|X(t)\in A_u)$$
$$+ P(v[L_u - L_u(t;rv^{-1/N})] > x\varepsilon\,|X(t)\in A_u) .$$

By an application of Markov's inequality and Fubini's theorem, the last member of (6.6) is at most equal to

$$\frac{v}{x\varepsilon} \int_{\{s:s\in B, |s-t|>rv^{-1/N}\}} P(X(s)\in A_u|X(t) \in A_u)ds .$$

Thus, if we multiply the terms in (6.6) by

$$\frac{P(X(t)\in A_u)}{\int_B P(X(t')\in A_u)dt'} ,$$

and integrate over B, and then pass to the limit, we obtain, by means of the elementary relation $\inf(a_n+b_n) \leq \inf a_n + \sup b_n$,

(6.7) $\lim_{r\to\infty} \liminf_{u\to\infty}$

$$\frac{\int_B P(vL_u(t;rv^{-1/N}) \le x(1-\epsilon)|X(t)\in A_u)P(X(t)\in A_u)dt}{\int_B P(X(t)\in A_u)dt}$$

$$\le \liminf_{u\to\infty} \frac{\int_B P(vL_u \le x|X(t)\in A_u)P(X(t)\in A_u)dt}{\int_B P(X(t)\in A_u)dt}$$

$$+ \lim_{r\to\infty} \limsup_{u\to\infty}$$

$$\frac{v}{x\epsilon} \frac{\iint_{\{(s,t):s,t\in B,|s-t|>rv^{-1/N}\}} P(X(s)\in A_u, X(t)\in A_u)ds\, dt}{\int_B P(X(t)\in A_u)dt} .$$

By the assumed condition (6.3), the last expression above is equal to 0; thus, the first term in (6.7) is at most equal to the term following the sign of inequality. Since $\epsilon > 0$ is arbitrary, a standard argument concerning the countability of the set of points of discontinuity of a bounded monotonic function may be used to show that (6.7) implies

(6.8) $$\lim_{r\to\infty} \liminf_{u\to\infty} (\text{ratio (6.4)}) \le \liminf_{u\to\infty} (\text{ratio (2.9)}) ,$$

on a dense subset $x > 0$. The conclusion of the lemma is now inferred from (6.5) and (6.8).

By a change of variables of integration, we find the following upper bound for the numerator in (6.3):

(6.9) $$v \int_B \int_{(B-B)\cap\{t:|t|>rv^{-1/N}\}} P(X(s)\in A_u,\ X(s+t)\in A_u)dt\, ds ,$$

where B-B is the set of differences $\{t-s:s,t\in B\}$. In the case where X is stationary, this yields a condition which is sufficient for (6.3),

(6.10) $$\lim_{r\to\infty} \limsup_{u\to\infty} \frac{v \int_{(B-B)\cap\{t:|t|>rv^{-1/N}\}} P(X(0)\in A_u, X(t)\in A_u)dt}{P(X(0)\ \ A_u)} ,$$

which reduces to the known condition in [6] for N=1.

7. The Sojourn Limit Theorem with $v(u) \to \infty$.

THEOREM 7.1. Let $X(t)$, $t \in B$, where $B \subset R^N$, be marginally g-stationary with respect to (A_u) for some density function g. Suppose that there is a function $v(u)$ such that $v(u) \to \infty$ for $u \to \infty$ such that (6.3) is satisfied; and, for every $m \geq 1$, $t \subset B$ and $s_1, \dots, x_m \in R^N$, the limit

$$(7.1) \qquad q_m(s_1,\dots,s_m;t) = \lim_{u\to\infty} P(X(t+s_i v^{-1/N}) \in A_u, i=1,\dots,m \mid X(t) \in A_u)$$

exists, and $q_1(s;t)$ and $q_2(s_1,s_2;t)$ are continous in s and (s_1,s_2), respectively. Assume also that the boundary of B has Lebesgue measure 0.

Then, for each t in the interior of B, there exists a separable, measurable stochastic process $\eta_t(s)$, $s \in R^N$, assuming only the values 0 and 1, such that

$$(7.2) \qquad E\eta_t(s_1)\dots\eta_t(s_m) = q_m(s_1,\dots,s_m;t)$$

and

$$(7.3) \qquad \lim_{u\to\infty} \frac{\int_0^x y \, dP(vL_u \leq y)}{E(vL_u)} = \int_B G(x;t)g(t)dt ,$$

at all continuity points $x > 0$, where

$$(7.4) \qquad G(x;t) = P(\int_{R^N} \eta_t(s)ds \leq x), \; x > 0 .$$

PROOF. For fixed t, define the family of stochastic processes

$$\xi_{u,t}(s) = \chi_{A_u}(X(t+sv^{-1/N})) , \quad s \in R^N ,$$

conditioned by $X(t) \in A_u$, for $u > 0$. Then (7.1) implies

$$(7.5) \qquad \lim_{u\to\infty} E[\xi_{u,t}(s_1)\dots\xi_{u,t}(s_m) \mid X(t) \in A_u] = q_m(s_1,\dots,s_m;t).$$

Thus, by Lemma 4.1, there is a process $\eta_t(s)$, $s \in R^N$, of the form stated in the theorem, such that (7.2) holds.

Let t be a fixed interior point of B; boundary points form a null set, and may be ignored. Then for every $r > 0$, there exists u_0 sufficiently large, and thus v_0 sufficiently large, such that the distance of t to the boundary of B is at least equal to $rv_0^{-1/N}$, for all $u > u_0$. According to Lemma 6.1, under the hypothesis (6.3), it suffices, for the determination of (7.3), to consider $L_u(t;rv^{-1/N})$ in the place of L_u in formula (6.4). Then for all $u > u_0$, $vL_u(t;rv^{-1/N})$ is equal to

$$v \int_{\{s:|s-t|<rv^{-1/N}\}} \chi_{A_u}(X(s))ds ,$$

which, by a change of variable of integration, is equal to

$$\int_{\{s:|s|<r\}} \xi_{u,t}(s)ds .$$

Then, by Fubini's theorem, we have

$$E[(\int_{\{s:|s|<r\}} \xi_{u,t}(s)ds)^m \,|X(t)\in A_u]$$

$$= \int_{|s_1|<r} \cdots \int_{|s_m|<r} E[\xi_{u,t}(s_1)\cdots\xi_{u,t}(s_m)|X(t)\in A_u]\, ds_1\cdots ds_m .$$

By (7.5), the latter converges to

$$\int_{|s_1|<r} \cdots \int_{|s_m|<r} q_m(s_1,\ldots,s_m;t)ds_1\cdots ds_m ,$$

which, by (7.2) and Fubini's theorem, is equal to

$$(7.6) \qquad E[\int_{|s|<r} \eta_t(s)ds]^m .$$

Thus, as in the proof of Theorem 5.1, it follows from the moment convergence theorem, that

$$(7.7) \qquad P(vL_u(t;rv^{-1/N}) \le x|X(t)\in A_u) \to P(\int_{|s|<r} \eta_t(s)ds \le x)$$

at all continuity points $x > 0$.

By Lemma 3.2, the relation (7.7) implies

$$(7.8)\qquad \lim_{u\to\infty} \int_{R^N} P(vL_u(t;rv^{-1/N}) < x|X(t)\in A_u)g_u(t)dt$$

$$= \int_{R^N} P(\int_{|s|<r} \eta_t(s)ds < x)g(t)dt \ .$$

Letting $r \to \infty$ on each side of (7.8), and applying Lemma 6.1, we arrive at (7.3).

8. High level sojourns of Gaussian random fields.

Let $X(t)$, $t\in R^N$, be a real separable measurable stationary Gaussian process over R^N, $N \geq 1$. Suppose that $EX(t) \equiv 0$ and $EX^2(t) \equiv 1$, and put $EX(s)X(t) = r(s-t)$, where r is continuous. In earlier work we had considered the limit of the ratio (1.2) for the sojourn time above the level u, for $u \to \infty$, and in the particular case $N = 1$ [5]. In this section, we apply Theorem 7.1 to extend the result above in two directions, to the case $N > 1$, and to the case of a variable barrier of the form

$$(8.1)\qquad (u^2 + f(t))^{1/2} ,$$

where $f(t)$ is a real valued continuous function on B. First we consider the special case of (8.1) where $f(t) \equiv 0$; here we again take $A_u = (u,\infty)$.

THEOREM 8.1. *Suppose that there is a positive function $v = v(u)$, $u > 0$, and a nonnegative, continuous function $\psi(t)$, $t\in R^N$, not identically equal to 0, such that*

$$(8.2)\qquad \psi(t) = \lim_{u\to\infty} u^2(1 - r(tv^{-1/N})) ,$$

and

$$(8.3)\qquad \lim_{d\to\infty} \limsup_{u\to\infty} \int_{d<|s|<t_0v^{1/N}} \exp(-\frac{u^2}{4}(1-r(sv^{-1/N})))ds = 0 ,$$

for some $t_0 > 0$. Then $\psi(t)$ is necessarily the incremental variance function of a Gaussian field over R^N with stationary increments, and (7.3) holds with g a constant density on B, and with the process $\eta_t(s)$ in (7.4) independent of t and defined by

$$(8.4)\qquad \eta(s) = \chi_{(0,\infty)}(W(s) - \psi(s) + \zeta),\ s\in R^N ,$$

where $W(s)$ is a Gaussian field over R^N with mean 0, $W(0) = 0$, and with stationary incremental variance function

(8.5) $$E(W(t)-W(s))^2 = 2\psi(t-s) ,$$

and where ζ is exponentially distributed, independently of $W(\cdot)$.

PROOF. This is a generalization of the proof of the case N=1 considered in [5]. Conditions (8.2) and (8.3) are equivalent in the latter case to the assumption that $1-r(t)$ is of regular variation for $t \to 0$, and they are implied by the condition of local isotropy of Qualls and Watanabe in [11] and the condition of Bickel and Rosenblatt in [7].

It follows from the definitions that $u^2(1-r(tv^{-1/N}))$ is the incremental variance function of the process $2^{-1/2}(X(tv^{-1/N})-X(0))$. Therefore, as the continuous limit of such a function, $\psi(t)$ is also an incremental variance function. (In fact, it follows from the theory of infinitely divisible laws that $\psi(t)$ is necessarily the negative of the logarithm of a stable distribution in R^N (see Lévy [10], Chapter 7).

The estimates needed for the verification of assumption (7.1) of Theorem 7.1 are similar to those for N=1, and are based on (8.2) and the relations,

$$E[X(sv^{-1/N})|X(0) = u + \frac{y}{u}] = (u + \frac{y}{u})r(sv^{-1/N}) ,$$

and

$$\text{var}\{X(tv^{-1/N}) - X(sv^{-1/N})|X(0)\}$$

$$= 2[1-r((t-s)v^{-1/N})] - [r(tv^{-1/N}) - r(sv^{-1/N})]^2 .$$

(See [5], formula (7.4) and (7.5).)

For the verification of the assumption (6.3), we use the same kind of estimates as those leading to [5], formula (7.8): It suffices to show that

$$\lim_{r\to\infty} \limsup_{u\to\infty} \int_{\{s: r<|s|<t_0 v^{1/N}\}} [1 - \Phi(\frac{u}{\sqrt{2}} (1-r(sv^{-1/N}))^{1/2})]ds = 0 ,$$

where Φ is the standard normal distribution function. But the latter follows from the assumption (8.3) and the well known inequality

$$1 - \Phi(x) \leq e^{-\frac{1}{2}x^2} , \quad x > 0.$$

Thus (6.3) is verified, and the proof of the theorem is complete.

Now we extend Theorem 8.1 to the sojourn above the variable barrier (8.1). Our next result states that the sojourn times above (8.1) are asymptotically (in the sense of the value of the ratio

$$\int_0^x y \, dP(vL_u \leq y)/E(vL_u))$$

equal to those above the variable barrier

(8.6) $$u + f(t)/2u \, .$$

LEMMA 8.1. *Let L_u' and L_u'' be defined as*

$$L_u' = \int_B \chi_{(u + f(t)/2u, \infty)}(X(t))dt$$

and

$$L_u'' = \int_B \chi_{((u^2+f(t))^{1/2}, \infty)}(X(t))dt \, .$$

Then, for $u \to \infty$,

(8.7) $$EL_u' \sim EL_u'' \, .$$

If there is a distribution function $G(x)$ such that

(8.8) $$\lim_{u\to\infty} \frac{\int_0^x y \, dP(vL_u' \leq y)}{E(vL_u')} = G(x)$$

weakly for $x > 0$, then (8.8) remains true if L' is replaced by L''.

PROOF. The relation (8.7) is a consequence of a standard computation based on the stationarity of X, Fubini's theorem, and the well known relation for the tail of the standard normal distribution,

(8.9) $$\Phi'(x)(x^{-1}-x^{-3}) \le 1 - \Phi(x) \le \Phi'(x)x^{-1}, \quad x > 0 .$$

Next we observe that

(8.10) $$L''_u \ge L'_u$$

indeed, as is seen by squaring the functions in (8.1) and (8.6), the latter is at most equal to the former.

Under the hypothesis (8.8), Lemma 2.2 implies

(8.11) $$\lim_{u\to\infty} \frac{P(vL'_u>x)}{E(vL'_u)} = \int_x^\infty y^{-1}\, dG(y) ,$$

weakly for $x > 0$. Furthermore, the argument used above to establish (8.7) also shows that

$$\lim_{u\to\infty} \frac{E(L''_u-L'_u)}{EL'_u} = 0 .$$

Thus, if we write $L''_u = L'_u + (L''_u-L'_u)$, then by (8.10) and (8.11), it follows that

$$\lim_{u\to\infty} \frac{P(vL''_u>x)}{E(vL'_u)} - \frac{P(vL'_u>x)}{E(vL'_u)} = 0$$

at all continuity points $x > 0$ in (8.11). The last statement of the lemma now follows from the relation above, from (8.7), and the bound obtained by integration by parts,

$$\left| \int_0^x y\, dP(vL''_u\le y) - \int_0^x y\, dP(vL'_u\le y)\right|$$

$$\le x(P(vL''_u>x) - P(vL'_u>x))$$

$$+ \int_0^x (P(vL''_u>y) - P(vL'_u>y))dy .$$

Now we extend Theorem 8.1 to sojourns above the barrier (8.1), or equivalently, the barrier (8.6).

THEOREM 8.2. *Let $X(t)$ satisfy the conditions of Theorem 8.1 and let $f(t)$, $t \in R^N$, be a continuous real valued function, and*

$$L'_u = \int_B \chi_{(u+f(t)/2u,\infty)}(X(t))dt \ .$$

Then the conclusion of Theorem 8.1 holds for L'.

PROOF. In the place of the single process, we consider the family of processes $\{X(t) - f(t)/2u,\ t \in R^N\}$, for $u > 0$. Although the hypothesis of g-stationarity was defined for a single process relative to the family (A_u), it extends directly to a family $\{X_u(t)\}$ of processes by attaching the index u in (3.1)). Here the family is g-stationary with

$$g(t) = \frac{e^{-f(t)/2}}{\int_B e^{-f(s)/2} ds}\ , \quad t \in B. \tag{8.12}$$

Indeed, this follows by Fubini's theorem, the stationarity of X, and the formula (8.9) by noting that $P(X(t) - f(t)/2u > u) = 1 - \Phi(u+f(t)/2u)$.

To verify the assumption (7.1) of Theorem 7.1, we extend the method used in the proof of Theorem 8.1 to the process $X(t+sv^{-1/N}) - f(t+sv^{-1/N})/2u$, conditioned by $X(t) - f(t)/2u = u + y/u$. By the assumed stationarity of X, the latter process is identical to the process $X(sv^{-1/N}) - f(t+sv^{-1/N})/2u$, $s \in R^N$, conditioned by $X(0) = u + (y+f(t)/2)/u$, for fixed t. By a standard computation we find

$$E[X(sv^{-1/N}) - f(t+sv^{-1/N})/2u \,|\, X(0) = u + \frac{y+\frac{1}{2}f(t)}{u}]$$

$$= (u+\frac{y}{u})r(sv^{-1/N}) + o(u^{-1})\ ,$$

$$\mathrm{var}\{X(sv^{-1/N}) - \frac{1}{2u}f(t+sv^{-1/N}) - X(s'v^{-1/N}) + \frac{1}{2u}f(t+s'v^{-1/N}) \,|\, X(0)\}$$

$$= \mathrm{var}\{X(sv^{-1/N}) - X(s'v^{-1/N}) \,|\, X(0)\}\ .$$

Hence, the first and second order moments of the conditional process are asymptotically identical to those obtained in the case $f=0$ in the proof of Theorem

8.1. Therefore, according to the reasoning employed there, assumption (7.1) of Theorem 7.1 holds, and the process $\eta_t(s)$ is the same as is (8.4), and is independent of t.

Finally we indicate the confirmation of (6.3). The bivariate probability in the numerator in (6.3) takes the form

$$P(X(s) > u + f(s)/2u,\ X(t) > u + f(t)/2u) ,$$

which is at most equal to $P(X(s) > u', X(t) > u')$, where $u' = u + (2u)^{-1} \min_B f$. With the assistance of (8.9) and the fact that $v(u) \sim v(u')$ for $u \to \infty$ (see [5]), it follows that the estimate of the ratio (6.3) is of the same order as that in the case f=0 considered in the proof of Theorem 8.1. It follows that the conditions of Theorem 7.1 are satisfied, and the conclusion (7.3) holds with $\eta_t(s)$ given by (8.4) and g(t) by (8.12). Since the former is independent of t, the limit (7.3) is the same as in the case f=0 in Theorem 8.1.

By Lemma 8.1, Theorem 8.2 also covers the case of the barrier (8.1).

In [3] and [4] we considered sojourns above barriers having a "spike". The barrier (8.6) here is different in that it flattens for $u \to \infty$.

There has been interest in the related problem of determining the limit of the probability that a stationary Gaussian process X(t), $t > 0$, will exceed the barrier f(t) somewhere on a long time interval. Cuzick [8] has studied a class of rising barriers $f_n(t)$, where $f_n(t) \to \infty$, for each t, as $n \to \infty$. The barrier sequence $f_n(t) = n + f(t)/2n$ in (8.6) satisfies the local conditions required in [8], and it can be easily shown that the function g considered there is necessarily identically equal to 0. Bickel and Rosenblatt [7] have considered the more general problem of a several dimensional time parameter. Finally, we mention that Adler [1] has given a version of Theorem 8.1 with $v \equiv 1$.

9. Sojourns in small sets for stable processes

Let X(t), $t \in R^N$, assume values in R^d, $d > 1$, and let A be a compact subset of R^d of positive measure. Define the family

$$(9.1) \qquad A_u = \{x\colon x \in R^d,\ ux \in A\} = Au^{-1} .$$

We will discuss various conditions under which Theorem 7.1 can be applied.

LEMMA 9.1. If X(t) has a density function $f_t(x)$ which is positive at x=0 and continuous there, uniformly for $t \in B$, then X(t) is marginally g-stationary relative to the family (9.1) with

$$(9.2)\qquad g(t) = \frac{f_t(0)}{\int_B f_s(0)ds}, \quad t \in B .$$

PROOF. Write

$$g_u(t) = \frac{\int_{Au^{-1}} f_t(x)dx}{\int_B \int_{Au^{-1}} f_s(x)dx\, ds},$$

divide the numerator and denominator by $mes(Au^{-1})$, let $u \to -\infty$, and then use the uniform continuity and positivity of f at x = 0. It follows that $g_u(t) \to g(t)$ for g in (9.2).

LEMMA 9.2. _If $f_t(x)$ is continuous and positive at x=0, then the conditional distribution of uX(t), given uX(t)$\in$A, converges to the uniform distribution on A._

PROOF. For any Borel set B,

$$P(uX(t) \in B | uX(t) \in A) = \frac{\int_{(A \cap B)u^{-1}} f_t(x)dx}{\int_{Au^{-1}} f_t(x)dx}$$

$$\to \frac{mes(A \cap B)}{mes\ A} .$$

In [6] we considered the sojourns of a stationary Gaussian process in a cube centered at the origin. Using the concept of g-stationarity, we extend the method to stable processes with independent increments. Let X(t) be a real valued separable measurable process with stationary independent increments which are symmetric and stable of index α, $0 < \alpha < 1$. Assume that X(0) = 0, a.s. As is well known, we have

$$(9.3)\qquad Ee^{iw(X(t)-X(s))} = e^{-(t-s)|w|^\alpha}, \quad 0 < s < t ,$$

where, for convenience, the scale factor in the exponent is taken to be 1. The density function is

$$(9.4)\qquad f_t(x) = \frac{1}{2\pi}\int_{-\infty}^{\infty} e^{-ixw-t|w|^{\alpha}}\,dw\ .$$

Take B as the interval $[\delta,1]$, for $0 < \delta < 1$. By Lemma 9.1, the process is marginally g-stationary with

$$(9.5)\qquad g(t) = \frac{t^{-1/\alpha}}{\int_{\delta}^{1} s^{-1/\alpha}\,ds}\ ,\ \delta \leq t \leq 1\ .$$

THEOREM 9.1. <u>For the symmetric stable process of index $0 < \alpha < 1$, and the family of sets (A_u) in (9.1) with A a bounded interval, the conclusion (7.3) holds with $g(t)$ as in (9.5), and $v(u) = u^{1/\alpha}$, and the process $\eta_t(s)$ in (7.4) defined as</u>

$$(9.6)\qquad \eta_t(s) = \chi_A(X_1(s) + \xi)\ ,\ s > 0$$

$$= \chi_A(X_2(-s) + \xi)\ ,\ s < 0\ ,$$

<u>where $X_1(s)$ and $X_2(s)$ are independent processes with the same distributions as the process $X(s)$, and where ξ is a random variable which is uniformly distributed over A, and is independent of $X_i(s)$, $i=1,2$.</u>

PROOF. Let us verify condition (7.1) of Theorem 7.1. For simplicity let the points $s_1,\ldots,s_m$ be arranged so that $s_1 <\ldots< s_m$, with $s_1 <\ldots< s_k < 0 < s_{k+1} < \ldots < s_m$. Since X is a Markov process, the processes $X(t + su^{-1/\alpha})$, $s > 0$ and $X(t-su^{-1/\alpha})$, $s > 0$, are conditionally independent given $X(t)$. Therefore the conditional probability in (7.1) has the representation

$$(9.7)\qquad (P(uX(t)\in A))^{-1} \int_A f_t(\tfrac{x}{u})\ P(uX(t + s_i u^{-1/\alpha})\in A,\ i=1,\ldots,k\,|\,uX(t) = x)$$

$$P(uX(t + s_i u^{-1/\alpha})\in A,\ i=k+1,\ldots,m\,|\,uX(t) = x)dx\ .$$

For the second conditional probability in the integrand in (9.7), we have

$$(9.8)\qquad P(uX(t + s_i u^{-1/2})\in A,\ i=k+1,\ldots,m\,|\,uX(t) = x)$$

$= P(u[X(t + s_i u^{-1/\alpha}) - X(t)] + x \in A,\ i=k+1,\dots,m)$

(by independence of increments)

$= P(uX(s_i u^{-1/\alpha}) + x \in A,\ i=k+1,\dots,m)$

(by stationarity of increments and $X(0) = 0$),

$= P(X(s_i) + x \in A,\ i=k+1,\dots,m)$

(by self-similarity).

Next we consider the first conditional probability in the integrand in (9.7). It may be written as

(9.9) $$P(u[X(t + s_i u^{-1/\alpha}) - X(t)] + x \in A,\ i=1,\dots,k \mid uX(t) = x)$$

Since $s_i < 0$ in (9.9), the increments of the conditioned process are neither stationary nor independent, so that the arguments in (9.8) cannot be used here. However, as we will indicate, the conditioned process $u[X(t + su^{-1/\alpha}) - X(t)]$ $s < 0$, given $uX(t) = x$, is asymptotically independent of t and x, so that the conditional probability (9.9) has the limit

(9.10) $$P(X(s_i) + x \in A, i=1,\dots,k) .$$

For the proof, let us show that for $s > 0$ the conditional density of $U = u[X(t - su^{-1/\alpha}) - X(t)]$ at z, given $X = uX(t) = x$, converges to $f_s(z)$, which is independent of x. (This is sufficient because the proof extends to a finite set of increments.) Put $Y = U + X = uX(t-su^{-1/\alpha})$; then, by the independence of the increments, the joint density of (U,Y) at the point (z,y) is

$$f_s(z)u^{-1}f_{t-su^{-1/\alpha}}(y/u) .$$

By an elementary formula, it follows that the conditional density of U at the point z, given $X = Y-U = x$ is

$$\frac{f_s(z)u^{-1}f_{t-su^{-1/\alpha}}\left(\frac{x+z}{u}\right)}{\int_{-\infty}^{\infty} f_s(w)u^{-1}f_{t-su^{-1/\alpha}}\left(\frac{x+w}{u}\right) dw},$$

which converges, for $u \to \infty$, to $f_s(z)$.

By Lemma 9.2, and by (9.8) and (9.10), it follows that (9.7) has the limit

$$(\text{mes } A)^{-1} \int_A P(X(s_1)+x\in A,\ldots,X(s_k)+x\in A)$$
$$P(X(s_{k+1})+x \in A,\ldots,X(s_m)+x\in A)dx,$$

which verifies the form (9.6) of the process $\eta_t(s)$.

Next we confirm that (6.3) holds. By (9.3) we have, for $s < t$,

$$E\, e^{ivX(s) + iwX(t)} = E\, e^{i(v+w)X(s) + iw(X(t)-X(s))}$$
$$= \exp[-s|v+w|^\alpha - (t-s)|w|^\alpha] .$$

Thus, by the inversion formula, we have

$$P(uX(s)\in A,\ uX(t)\in A)$$

$$\leq (2\pi)^{-2}(\text{mes } A)^2 \int_{-\infty}^{\infty}\int_{-\infty}^{\infty} \exp[-s|v+w|^\alpha - (t-s)|w|^\alpha]\, dv\, dw$$

$$= (2\pi)^{-2}(\text{mes } A)^2(s(t-s))^{-1/\alpha}\ 2\Gamma(\frac{1}{\alpha}+1) .$$

It follows that the ratio in (6.3) with N=d=1 and $v(u) = u^{1/\alpha}$ is most equal to

$$\underline{\text{constant}}\ u^{1/\alpha-1} \iint_{\delta\leq s,t\leq 1,\, ru^{-1/\alpha}<t-s} [s(t-s)]^{-1/\alpha} ds\, dt ,$$

which, after a change of variable of integration $x=s$, $y=t-s$, and a standard computation, is seen to be at most equal to <u>constant</u> $r^{1-1/\alpha}$. The latter tends to 0 for $r\to\infty$ because $\alpha < 1$. This completes the proof of the theorem.

We have the following extension to stable processes in R^d.

THEOREM 9.2. <u>For $d > 1$, let $X(t) = (X_1(t),\ldots,X_d(t))$ be a vector process whose components are independent copies of a process with independent symmetric stable increments of index α, $0 < \alpha \leq 2$; and let A be a compact subset of R^d of positive measure and with a boundary of measure 0. If $d > \alpha$, then the conclusion (7.3) holds for the family (A_u) in (9.1) with $v(u) = u^{1/\alpha}$, and</u>

$$(9.11)\qquad g(t) = t^{-d/\alpha}\left(\int_\delta^1 s^{-d/\alpha}ds\right)^{-1},\ \delta < t < 1,$$

and the process $\eta_t(s)$ in (7.4) equal to the vector version of (9.6) namely, $\chi_A(X(\pm s) + \xi)$, where ξ is uniformly distributed on A.

PROOF. The formula (9.11) for g(t) in d dimensions follows from (9.2) and the fact that the d-dimensional density is the product of the one-dimensional densities.

The validity of the condition (7.1) in the d-dimensional case follows from its validity for each of the independent component processes as established in the proof of Theorem 9.1.

For the verification of condition (6.3) we first note that since A is bounded, it can be enclosed in a sufficiently large cube $[-c,c]^d$, for some $c > 0$. Hence the 2d-variate probability in (6.3), $P(uX(s)\in A,\ uX(t)\in A)$, is at most equal to $[P(|uX_1(s)| < c,\ |uX_1(t)| < c)]^d$. Hence, by the same calculation as at the end of the proof of Theorem 9.1, we find that the ratio in (6.3) is at most equal to

$$\underline{\text{constant}}\ u^{\frac{1}{\alpha}-d} \iint_{\delta<s,t<1,\ ru^{-1/\alpha}<t-s} [s(t-s)]^{-d/\alpha}\,ds\,dt\ ,$$

which, by the corresponding method, is at most equal to constant $r^{1-d/\alpha}$. The latter tends to 0 for $r\to\infty$ because, by hypothesis, $\alpha < d$. Thus (6.3) is verified, and the proof is complete.

We note that in the case $\alpha=2$, the process is Brownian motion, and the theorem holds in dimension $d\geq 3$. This is analogous to the results in [6] for stationary Gaussian processes, and, in particular, the Ornstein-Uhlenbeck process.

We also note that the left endpoint δ of the time interval must be positive. Otherwise the assumption of g-stationarity would not be valid (see (9.5)).

References

1. Adler, R.J.: Distribution results for the occupation measures of continuous Gaussian fields. Stochastic Processes Appl. 7 (1978) 299-310.

2. Berman, S.M.: Excursions above high levels for stationary Gaussian processes. Pacific J. Math. 36 (1971) 63-79.

3. Berman, S.M.: Excursions of stationary Gaussian processes above high moving barriers. Ann. Probability 1 (1973) 365-387.

4. Berman, S.M.: Sojourns and extremes of Gaussian processes. Ann. Probability 2 (1974) 999-1026, Correction 8 (1980) 999, Correction 12 (1984) 281.

5. Berman, S.M.: Sojourns and extremes of stationary processes. Ann. Probability 10 (1982) 1-46.

6. Berman, S.M.: Sojourns of stationary processes in rare sets. Ann. Probability 11 (1983) 847-866.

7. Bickel, P. and Rosenblatt, M.: Two dimensional random fields. Multivariate Analysis III, P.R. Krishnaiah, Editor, Academic Press, New York, 1973, p. 3-15.

8. Cuzick, J.M.: Boundary crossing probabilities for stationary Gaussian processes and Brownian motion. Trans. Amer. Math. Soc. 263 (1981) 469-492.

9. Doob, J.L.: Stochastic Processes. John Wiley, New York, 1953.

10. Levy, P.: Theorie de l'Addition des Variables Aleatoires, Gauthier-Villars, Paris 1937.

11. Qualls, C.R. and Watanabe, H.: Asymptotic properties of Gaussian random fields. Trans. Amer. Math. Soc. 177 (1973) 155-171.

12. Scheffe, H.: A useful convergence theorem for probability distributions. Ann. Math. Statist. 18 (1947) 434-438.

INTRINSIC BOUNDS ON SOME REAL-VALUED STATIONARY RANDOM FUNCTIONS

By

Christer Borell

1. Introduction

Let $X = (X(t);\ t \in T)$ be a real-valued stochastic process such that each individual $X(t)$ has a continuous distribution function. We shall write $X \in \mathfrak{M}_-$ if the following is true.

For any $p_0, \dots, p_n \in]0,1[, t_0, \dots, t_n \in T$, and $n \in \mathbb{N}$ and any Borel sets $A_0, \dots, A_n \subseteq \mathbb{R}$ satisfying $\mathbb{P}(X(t_0) \in A_0) = p_0, \dots, \mathbb{P}(X(t_n) \in A_n) = p_n$, the joint probability

$$\mathbb{P}(X(t_0) \in A_0, \dots, X(t_n) \in A_n)$$

is maximal provided $A_0 \times \dots \times A_n$ is of the form

$$]-\infty, a_0] \times \dots \times]-\infty, a_n] \quad (a_0, \dots, a_n \in \mathbb{R}).$$

From the author's paper [1] we know that $X \in \mathfrak{M}_-$ if

a) $X = (X_0, X_1)$ is mean zero Gaussian with $\mathbb{E}\, X_0 X_1 \geq 0$.

b) $X = (X(t); t \geq 0)$ is mean zero Gaussian with $\mathbb{E}\, X(s)X(t) = \exp(-\frac{1}{2}|s-t|)$ (the one-dimensional Ornstein-Uhlenbeck [velocity] process [in equilibrium]).

By simple means one may verify that these results are, in fact, equivalent. The purpose of this paper is to exhibit an interesting and fairly large subclass of the class $\mathfrak{M}_-$.

Let $\mu(dx) = \rho(x)dx$ be a probability measure on $\mathbb{R}$ and suppose $\rho > 0$ has a locally Lipschitz continuous derivative. Set $R = \frac{1}{2} \ln \rho$ and denote by ξ any solution of the following stochastic differential equation

(1.1) $\quad d\xi(t) = R'(\xi(t))dt + d\omega(t), \ t \geq 0$

where ω is the Wiener process in $\mathbb{R}$ ($\mathbb{E}\omega^2(t) = t$). Here a.s. no explosions occur by Feller's test. Furthermore, we define a stationary random function ζ in $C([0,+\infty[)$ by assigning ξ the initial distribution μ i.e.

$$\mathbb{P}(\zeta \in \cdot) = \int \mathbb{P}_x(\xi \in \cdot)d\mu(x).$$

Since μ is the unique invariant probability measure of ξ, ζ is simply called the equilibrium solution of (1.1). The random function ζ is symmetric, that is

$$\mathbb{E}\, f(\zeta(s))g(\zeta(t)) = \mathbb{E}\, f(\zeta(t))g(\zeta(s)), \ s, \ t \geq 0$$

for all $f,g \in C_b([0,+\infty[)$ (b refers to bounded). If $\mu = \mathcal{L}(\omega(1))$, then ζ reduces to the one-dimensional Ornstein-Uhlenbeck process.

For a very direct account on the above topic, the reader may consult Priouret and Yor [13]. To line up with our premises the basic books of McKean [11] and Stroock and Varadhan [16] are helpful. In mathematical physics the random function ζ is often called a $P(\phi)_1$-process (see e.g. Courrège and Renouard [6] and Simon [15]). For connections with Dirichlet semigroups, see Carmona [5].

The main result in this paper is

Theorem 1.1. *Let* (μ,ρ,R,ζ) *be as above and suppose*

$$\int_{-\infty}^{G(p)} d\mu = p, \ 0 < p < 1.$$

Set $\phi = \rho \circ G = 1/G'$. *The random function* $\zeta \in \mathfrak{M}_-$ *if and only if*

(1.2) $\quad |\phi(p) - \phi(q)| \leq \phi(p+q), \ 0 < p,q,p+q < 1.$

A measure μ as above satisfying (1.2) has a point of symmetry and its tails fall off exponentially. The converse implication is wrong. However, if μ possesses a point of symmetry and if the distribution function $F = G^{-1}$ is log concave, then (1.2) holds. Here recall that F is log concave if ρ is log concave (see e.g. Borell [2]).

It is a good tradition in papers on isoperimetric inequalities to give physical interpretations of the mathematics and we will do so, too. However, because of some extra difficulties our prototype example (the electron in the potential well) will not appear until Section 5.

2. Isoperimetric measures on $\mathbb{R}$

Below Q denotes the class of all absolutely continuous probability measures $\mu(dx) = \rho_\mu(x)dx$ on $\mathbb{R}$ possessing the following properties:

(i) $\operatorname{supp}\mu$ is an interval $I_\mu = (\alpha_\mu, \beta_\mu)$ such that

$-\infty < \alpha_\mu < \beta_\mu < +\infty$ or $\alpha_\mu = -\beta_\mu = -\infty$

and

(ii) $\rho_\mu(x), \alpha_\mu < x < \beta_\mu$, is continuous and strictly positive.

For any $\mu \in Q$, let $F_\mu(x) = \mu(]-\infty, x])$, $x \in \overset{\circ}{I}_\mu =]\alpha_\mu, \beta_\mu[$.

Throughout, the Borel field in $\mathbb{R}$ is denoted by $\mathcal{B}$. If $A \subseteq \mathbb{R}$ and $r > 0$, set $A_r = \{x+y;\ x \in A \text{ and } |y| < r\}$. A measure $\mu \in Q$ is said to be isoperimetric if, given $A \in \mathcal{B}$

$$\mu(A) = F_\mu(a) \Rightarrow \mu(A_r) \geq F_\mu(a+r),\ 0 < r < \beta_\mu - a.$$

If this implication is only postulated for subintervals A of $\mathbb{R}$, then μ is called weakly isoperimetric. It is immediate that the measure $1_{[0,1]}dx$ is isoperimetric. Furthermore, any non-degenerated Gaussian measure on $\mathbb{R}$ is isoperimetric (Borell [3]). Recently, Ehrhard [7] treats the Gaussian case by new methods; the crucical Theorem 2.2 below is an abstraction from his paper. Theorem 2.2 simply says that a weakly isoperimetric measure is, in fact, isoperimetric.

In what follows, we very often drop μ as a subindex if there is no ambiguity.

<u>Theorem 2.1.</u> <u>Let</u> μ <u>be weakly isoperimetric. Then there exists a unique</u> $c = c_\mu \in \mathbb{R}$ <u>such that</u> $\mu(c-A) = \mu(c+A)$, $A \in \mathcal{B}$. <u>Here</u> $c = (\alpha_\mu + \beta_\mu)/2$ <u>if</u> I_μ <u>is compact</u>.

Proof. First suppose $I_\mu = [\alpha,\beta]$ is compact and choose $0 < \delta < \beta-\alpha$. Let

(2.1) $\mu([\beta-\delta,\beta]) = F(\kappa(\delta))$.

Then

(2.2) $\mu([\beta-\delta-r,\beta]) \geq F(\kappa(\delta)+r),\ 0 < r < (\beta-\kappa(\delta))\wedge(\beta-\alpha-\delta)$.

Here, if strict inequality occurs for some r,

$$F(\beta-\delta-r) < \mu([\kappa(\delta)+r,\beta])$$

thereby forcing

$$F(\beta-\delta) < \mu([\kappa(\delta),\beta])$$

which contradicts (2.1). Accordingly, (2.2) is, in fact, an identity and by letting $\delta \downarrow 0$, we have

$$\mu([\beta-r,\beta]) = \mu([\alpha,\alpha+r]),\ 0 < r < \beta-\alpha.$$

It is now evident that μ is symmetric with respect to the point $(\alpha+\beta)/2$. Noting that a finite measure on $\mathbb{R}$ has at most one point of symmetry, the case I_μ compact is completely proved. The case $I_\mu = \mathbb{R}$ is very similar and the proof is excluded here. □

Corollary 2.1. Let $f:\mathbb{R} \to \mathbb{R}$ be affine with $f' \neq 0$. If μ is isoperimetric then the image measure $f(\mu) = \mu(f^{-1}(\ ))$ is isoperimetric.

Now suppose $\mathcal{G}(\mathcal{F})$ denotes the class of all open (closed) subsets of $\mathbb{R}$. Given $\mu \in Q$ and $A \in \mathcal{G}$, $L_\mu(A) \subseteq \mathbb{R}$ is empty if $\mu(A) = 0$, $L_\mu(A) = \mathbb{R}$ if $\mu(A) = 1$, and, otherwise, $L_\mu(A)$ is an open interval $]-\infty,a[$ satisfying $F_\mu(a) = \mu(A)$. The set $A^* = L_\mu(A)$ is called the left-hand side μ-rearrangement of A. The right-hand side μ-rearrangement of A, $R_\mu(A) = A_*$, is defined in a similar way. Finally, the maps L_μ and R_μ are extended to mappings of $\mathcal{G} \cup \mathcal{F}$ into $\mathcal{G} \cup \mathcal{F}$ satisfying

$$L_\mu = \sim R_\mu \sim \quad (\sim \equiv \complement_{\mathbb{R}}).$$

<u>Theorem 2.2. Suppose</u> μ <u>is weakly isoperimetric. Then for any</u> $A \in \mathcal{J}$,

(2.3) $$L_\mu(A_r) \supseteq (L_\mu(A))_r, \quad r > 0.$$

<u>In particular,</u> μ <u>is isoperimetric</u>.

<u>Proof</u>. Let $A, B \in \mathcal{J}$, where B is an interval. Furthermore, suppose $A_{r_0} \cap B_{r_0} = \phi$ $(r_0 > 0)$ and that A satisfies (2.3). We next show that

$$L((A \cup B)_r) \supseteq (L(A \cup B))_r, \quad 0 < r < r_0.$$

To this end, let $0 < r < r_0$ be fixed and note that

$$L((A \cup B)_r) = L(A_r \cup B_r) = L(L(A_r) \cup R(B_r)).$$

Hence,

$$L((A \cup B)_r) \supseteq L((L(A))_r \cup (R(B))_r)$$

by Theorem 2.1. Set $C = L(A) \cup R(B)$. Then $\sim C_r = \mathbb{R} \setminus (C_r)$ is an interval and we get

$$R((\sim C_r)_r) \supseteq (R(\sim C_r))_r.$$

Usin ,

$$\sim (\ ) \supseteq (\sim (\ )_r)_r$$

it now follows that

$$L((A \cup B)_r) \supseteq L(C_r) = \sim R(\sim C_r) \supseteq (\sim (R(\sim C_r))_r)_r \supseteq (\sim R((\sim C_r)_r))_r \supseteq$$

$$\supseteq (\sim R(\sim C))_r = (L(C))_r = (L(A \cup B))_r.$$

From this we conclude that (2.3) is valid if A is a finite union of open intervals. The general case is now evident. □

If $\mu \in Q$ we let $G_\mu = F_\mu^{-1}$ be the inverse function of F_μ and we introduce

$$\phi_\mu = \rho_\mu \circ G_\mu = 1/G'_\mu.$$

Note that

$$F_\mu = (G_\mu(1/2) + \int_{\frac{1}{2}}^{\bullet} \frac{dp}{\phi_\mu})^{-1}.$$

Set $Q_c = \{\mu \in Q; \mu(c-A) = \mu(c+A),\ A \in \mathcal{B}\}$, $c \in \mathbb{R}$. If $\mu \in Q_c$, then $G_\mu(1/2) = c$ and $\phi_\mu(1-p) = \phi_\mu(p)$, $0 < p < 1$. Conversely, if $\phi:]0,1[\to]0,+\infty[$ is continuous and symmetric with respect to the point 1/2, then to each $c \in \mathbb{R}$ there exists a unique $\mu \in Q_c$ with $\phi_\mu = \phi$.

Theorem 2.3. Let $\mu \in Q_c \cup (Q \cap \{I_\mu = \mathbb{R}\})$. The following conditions are equivalent:

(i) μ is isoperimetric

(ii) for any $\alpha_\mu < x,y,z < \beta_\mu$,

$F_\mu(y) - F_\mu(x) = F_\mu(z) \Rightarrow \rho_\mu(y) + \rho_\mu(x) \geq \rho_\mu(z)$

(iii) $|\phi_\mu(p) - \phi_\mu(q)| \leq \phi_\mu(p+q)$, $0 < p,q,p+q < 1$.

Proof. (i) $\Leftrightarrow$ (ii): Let $z = z(x,y)$ be as in (ii). The premise on μ implies that μ is weakly isoperimetric if and only if

$$z(x-r,y+r) \geq z(x,y)+r, \quad \alpha < x-r < y+r < \beta,\ r > 0,$$

which equivalently means that $z'_y - z'_x \geq 1$. But

$$\rho(y) + \rho(x) = \rho(z)(z'_y - z'_x)$$

and we are done in view of Theorem 2.2.

(ii) $\Leftrightarrow$ (iii): Obvious from the very definition of ϕ. □

Example 2.1. A measure $\mu \in Q_c$ is isoperimetric if

$$2 \inf_{\alpha_\mu < x < \beta_\mu} \rho_\mu(x) \geq \sup_{\alpha_\mu < x < \beta_\mu} \rho_\mu(x).$$ □

Example 2.2. If μ is isoperimetric, then

$$2 \inf_{\alpha_\mu < x < \beta_\mu} \rho_\mu(x) \geq \overline{\lim_{x \downarrow \alpha_\mu}} \rho_\mu(x).$$

In particular, when $I_\mu = \mathbb{R}$, $\rho_\mu(-\infty) = 0$ and $\phi_\mu(0+) = 0$. Indeed, suppose $x_n \downarrow \alpha$ and $\lim_{n\to+\infty} \rho(x_n) = \overline{\lim_{x\downarrow\alpha}} \rho(x)$. For any $\alpha < y < \beta$ and n large enough, there exists a $y_n > y$ such that $F(y_n) - F(y) = F(x_n)$. Hence $\rho(y_n) + \rho(y) \geq \rho(x_n)$ by Theorem 2.3 and the claim above follows since $y_n \downarrow y$. □

A function $\phi:]0,1[\to]0,+\infty[$ is called a dome function if

$$\phi(\theta p+(1-\theta)q) \geq \theta\phi(p), \; 0 < \theta,p < 1, \; q = 0,1.$$

Stated otherwise, a strictly positive function on $]0,1[$ is a dome function if and only if it may be represented as the supremum of a uniformly bounded family of strictly positive and concave functions on $]0,1[$.

<u>Theorem 2.4.</u> <u>Let</u> $c \in \mathbb{R}$. <u>Then</u> $\{\phi_\mu; \mu \in Q_c$ <u>and</u> F_μ <u>log concave</u>$\}$ = $\{$<u>dome functions symmetric with respect to the point</u> $1/2\}$.

<u>Proof.</u> Suppose $\mu \in Q_c$. The function F is log concave if and only if the function $G(e^x)$, $x < 0$, is convex. Equivalently this means that the function $G'(e^x)e^x$, $x < 0$, increases, that is $\phi(\theta p) \geq \theta\phi(p)$, $0 < \theta,p < 1$. □

<u>Theorem 2.5.</u> <u>A measure</u> $\mu \in Q_c$ <u>is isoperimetric if</u> F_μ <u>is log concave.</u>

<u>Proof.</u> Suppose $0 < p,q,p+q < 1$. From $q < 1-p$ we have

$$\phi(q) \geq \frac{q}{1-p}\,\phi(1-p) = \frac{q}{1-p}\,\phi(p).$$

Moreover, as $1-p-q < 1-p$,

$$\phi(p+q) = \phi(1-p-q) \geq \frac{1-p-q}{1-p}\,\phi(p).$$

By adding these inequalities, we have $\phi(p)-\phi(q) \leq \phi(p+q)$ and the result follows from Theorem 2.3. □

Corollary 2.2. A measure $\mu \in Q_c$ is isoperimetric if ρ_μ is log concave.

Here recall that F_μ is log concave if ρ_μ is so (see e.g. [2]). It is also simple to give a direct proof of Corollary 2.2 using Theorem 2.3 (ii).

The next theorem and the subsequent example have no applications in this paper but we think they are of independent interest. Henceforth, we spell some time on them.

Theorem 2.6. If μ is isoperimetric, $F_\mu \leq \exp f$ for an appropriate affine f with strictly positive slope.

Proof. We claim that $\underline{\lim}_{p \downarrow 0} \phi(p)/p > 0$. Indeed, if $p_n \downarrow 0$ and $\phi(p_n)/p_n \to 0$ as $n \to +\infty$, then

$$-\frac{\phi(p_n)}{p_n} \leq \frac{\phi(q+p_n)-\phi(q)}{p_n}, \quad n \text{ large},$$

uniformly in q on compacts of $]0,1[$. Hence $\phi' \geq 0$ in the sense of distribution theory. Since ϕ is symmetric with respect to the point 1/2 it follows that ϕ is constant > 0, which is absurd. Finally, if $\phi(p) \geq Kp$ on $]0,1/2[$ $(K>0)$, a simple argument leads to $F(x) \leq \exp(Kx-Kc-\ln 2)$, $x \leq c$. □

Example 2.3. (Estimate of the Landau and Shepp type [10]). Suppose $\mu \in Q_0$ is isoperimetric and let J be an interval such that $\mu(J) > 1/2$. Then

$$\mu(J) = F_\mu(a) \Rightarrow \mu(\gamma J) \geq F_\mu(\gamma a), \quad 1 < \gamma < \beta_\mu/a .$$

To see this set $J = (x,y)$ and choose $r = (\gamma-1)(|x| \wedge y)$ so that $\gamma J \supseteq J_r$ and $a+r \geq \gamma a$. The implication now follows at once. □

Next suppose $\mu \in Q$ and write $L_\mu = (\ )^*$ and $R_\mu = (\ )_*$ as above. For any $f \in C(\mathbb{R})$, set $M_f^* = \{\{f>y\}^* \times \{y\};\ y \in \mathbb{R}\}$ and

$$f^*(x) = \sup\{y \in \mathbb{R};\ (x,y) \in M_f^*\}, \quad x \in \mathbb{R} .$$

By simple means one may verify that $\{f^* > y\} = \{f > y\}^*$ and $\{f^* \geq y\} \cap \mathring{I}_\mu = \{f \geq y\}^* \cap \mathring{I}_\mu$ for every $y \in \mathbb{R}$. Hence f^* is continuous. The function $f^* = L_\mu f$

is called the decreasing μ-rearrangement of f. Recall that μ is inner regular with respect to compacts. Therefore, given any metric space T, L_μ extends to a mapping of $C(T \times \mathbb{R})$ into $C(T \times \mathbb{R})$ by setting

$$(L_\mu f)(t,x) = (L_\mu f(t,\cdot))(x).$$

Below

$$\mathrm{Lip}\,K = \{f \in C(\mathbb{R});\ \sup_{x \neq y} |f(x)-f(y)|/|x-y| \leq K\},\ K > 0.$$

Theorem 2.7. Let $\mu \in Q$. The follwing conditions are equivalent:

(i) μ is isoperimetric

(ii) $L_\mu \mathrm{Lip}\,K \subseteq \mathrm{Lip}\,K$

(iii) for any bounded interval J and $r > 0$,

$$L_\mu(r \wedge d(\cdot, \sim J_r)) \in \mathrm{Lip}\,1.$$

Here $d(x,A) = \inf\{|x-a|;\ a \in A\}$, $x \in \mathbb{R}$, $A \subseteq \mathbb{R}$.

Proof. (i) $\Rightarrow$ (ii): Suppose $f \in \mathrm{Lip}\,K$ and note that f^* is constant off I_μ. If $f^* \notin \mathrm{Lip}\,K$ there exist $\alpha_\mu < x,y < \beta_\mu$ such that $|x-y| < r$ and $f^*(x) \leq f^*(y) - Kr$. From

$$\{f > f^*(x)\} \supseteq \{f \geq f^*(y)\}_r$$

we now get

$$\mu(]-\infty,x[) \geq \mu(]-\infty,y]_r) > \mu(]-\infty,x])$$

which is a contradiction.

(ii) $\Rightarrow$ (iii): L_μ is positive homogeneous.

(iii) $\Rightarrow$ (i): Suppose J is an interval such that $0 < \mu(J) < 1$. Set $a = G_\mu(\mu(J))$. Let $r > 0$ satisfy $\mu(J_r) < 1$. Since a weakly isoperimetric measure is isoperimetric it is enough to show that $G_\mu(\mu(J_r)) \geq a + r$. To this end, set $f = r \wedge d(\cdot, \sim J_r)$. Then $f^*(a) = r$ and $f^*(G_\mu(J_r)) = 0$ so that

$$r = f^*(a) - f^*(G_\mu(J_r)) \leq 1 \cdot (G_\mu(J_r) - a)$$

which gives the desired estimate at once. □

Before stating the last theorem in this section we introduce the following convention. Suppose $\mu \in Q$ with I_μ compact. If $f, g \in C(\mathbb{R})$ and $f = g$ on I_μ, then $L_\mu f = L_\mu g$. Therefore, given $f \in C(I_\mu)$ we may define $L_\mu f = L_\mu f_1$, where $f_1 \in C(\mathbb{R})$ is any extension of f. The map L_μ is defined in a similar way on $C(T \times I_\mu)$ for any metric space T.

Theorem 2.8. *Let* $\mu(dx) = \rho(x)dx$ *be isoperimetric and suppose* $R = \frac{1}{2} \ln \rho \in C^1(\mathbb{R})$. *There exists a sequence* $\mu_n(dx) = \rho_n(x)dx$, $n \in \mathbb{N}$, *of isoperimetric measures possessing the following properties:*

(i) $c_{\mu_n} = c_\mu$

(ii) I_{μ_n} *is compact and* $\lim_{n\to\infty} \alpha_{\mu_n} = -\infty$

(iii) *the function* $R_n = \frac{1}{2} \ln \rho_n |_{\overset{\circ}{I}_{\mu_n}}$ *is real analytic in* I_{μ_n} *(i.e. in a neighbourhood of* I_{μ_n}*)*

(iv) $R_n' \to R'$ *uniformly on compacts as* $n \to +\infty$

(v) *if* $f_n \in C(I_{\mu_n})$, $n \in \mathbb{N}$, $f \in C(\mathbb{R})$, *and* $f_n \to f$ *uniformly on compacts as* $n \to +\infty$, *then* $L_{\mu_n} f_n \to L_\mu f$ *uniformly on compacts as* $n \to +\infty$.

Proof. Set $\kappa = \phi = \phi_\mu$ in $]0,1[$ and $= 0$ elsewhere on $\mathbb{R}$. Note that κ is continuous by Example 2.2. We define

$$\kappa_k = \mathbb{E}_{\bullet}\, \kappa(\omega(2^{-k})),\ k \in \mathbb{N},$$

so that each κ_k is real analytic and symmetric with respect to the point 1/2. Since $\kappa_k \to \kappa$ uniformly on $\mathbb{R}$ as $k \to +\infty$ there exists for each $n \in \mathbb{N}$ a $k_n \in \mathbb{N}$ such that

$$|\kappa_{k_n}(p) - \kappa_{k_n}(q)| \leq \kappa_{k_n}(p+q) + 2^{-n},\ 0 < p,q,p+q < 1.$$

Set $\phi_n = \kappa_{k_n|]0,1[} + 2^{-n}$ and

$$F_n = (c_\mu + \int_{\frac{1}{2}}^{\cdot} \frac{dp}{\phi_n})^{-1}.$$

The function F_n is real analytic in the closure of its domain of definition. Furthermore, by Theorem 2.3 there exists a unique isoperimetric measure μ_n with $c_{\mu_n} = c_\mu$ and $F_{\mu_n} = F_n$. Since $0 < \phi \in C(]0,1[)$ it follows that $F_n \to F_\mu$ uniformly on compacts as $n \to +\infty$. Moreover, $\phi_n' = \kappa'_{k_n|]0,1[} \to \phi'$ uniformly on compacts as $n \to +\infty$. Noting that

$$\begin{cases} \phi_n'(F_n) = 2R_n' & \text{in } \overset{\circ}{I}_{\mu_n} \\ \phi'(F) = 2R' \end{cases}$$

the properties (i) - (iv) are thereby completely proved.

To prove (v) suppose first that $x_n, x \in \mathbb{R}$, $n \in \mathbb{N}$, $\lim_{n\to+\infty} x_n = x$, and

$$f_n^{*n}(x_n) \geq f^*(x) + \delta, \; n \in \mathbb{N},$$

for an appropriate $\delta > 0$. Here $f_n^{*n} = L_{\mu_n} f_n$ and $f^* = L_\mu f$. Then

$$\mu_n(f_n \geq f^*(x)+\delta) \geq \mu_n(f_n \geq f_n^{*n}(x_n)) \geq \mu_n(]-\infty, x_n]).$$

However, from (iv), $\rho_n \to \rho$ uniformly on compacts as $n \to +\infty$ and we get

$$\mu(f \geq f^*(x)+\delta) \geq \mu(]-\infty, x]).$$

But

$$\mu(f \geq f^*(x)+\delta) < \mu(]-\infty, x])$$

and we have got a contradiction. On the other hand, if

$$f_n^{*n}(x_n) \leq f^*(x)-\delta, \; n \in \mathbb{N},$$

where $x_n \to x$ as $n \to +\infty$ and $\delta > 0$, then from

$$\mu_n(f_n > f^*(x)-\delta) \leq \mu_n(f_n > f_n^{*n}(x_n)) \leq \mu_n(]-\infty, x_n])$$

we again get a contradiction. Thus $f_n^{*n} \to f^*$ uniformly on compacts as $n \to +\infty$ and (v) is proved, too. □

3. A key lemma

The main point in the proof of Theorem 1.1 is the following

Lemma 3.1. *Suppose* $\mu(dx) = \rho(x)dx$ *is isoperimetric with* $I_\mu = [\alpha,\beta]$ *compact. Moreover, assume the function* $R = \frac{1}{2}\ln \rho_{|]\alpha,\beta[}$ *is real analytic in* $[\alpha,\beta]$. *Set*

$$N = -\tfrac{1}{2}(d/dx)^2 - R'(x)d/dx$$

and

$$M = -\tfrac{1}{2}(d/dx)^2 + R'(x)d/dx.$$

Finally, let $f \in C^+(\mathbb{R})$ *satisfy* $\operatorname{supp} f \subseteq [\alpha,\beta]$.

a) *Suppose*

$$\begin{cases} (\partial_t + N)u = 0, \ \alpha < x < \beta, \ t > 0 \\ u(t,\alpha) = u(t,\beta) = 0, \ t > 0 \\ u(0,x) = f(x), \alpha \leq x \leq \beta, \ u \in C([0,+\infty[\times [\alpha,\beta]), \end{cases}$$

and introduce

$$\tilde{u}(t,x) = \int_\alpha^x (L_\mu u)(t,y)d\mu(y), \ t \geq 0, \ \alpha \leq x \leq \beta.$$

Then

$$(\partial_t + M)\tilde{u} \leq 0, \ t > 0, \ \alpha < x < \beta.$$

b) *Suppose*

$$\begin{cases} (\partial_t + N)v = 0, \ \alpha < x < \beta, \ t > 0 \\ v(t,\alpha) = (L_\mu f)(\alpha), \ t > 0 \\ v(t,\beta) = 0, \ t > 0 \\ v(0,x) = (L_\mu f)(x), \ \alpha \leq x \leq \beta, \ v \in C([0,+\infty[\times [\alpha,\beta]). \end{cases}$$

Then the function $v(t,\cdot)$ *decreases for each* $t \geq 0$. *Furthermore, the function*

$$\tilde{v}(t,x) = \int_\alpha^x v(t,y)d\mu(y), \ t \geq 0, \alpha \leq x \leq \beta,$$

satisfies

$$(\partial_t + M)\tilde{v} \geq 0, \ t > 0, \alpha < x < \beta.$$

c) $\tilde{u} \leq \tilde{v}$.

Here and from now on all derivatives shall be interpreted in the weak sense with C_0^∞ as the underlying class of test functions. Recall that a positive distribution is a positive Radon measure.

The following proof of Lemma 3.1 is very similar to the line of reasoning in [1]. Nevertheless, we include most details here.

Proof. a) Throughout, $t > 0$ and $\alpha < x < \beta$. Set $u^*(t,x) = (L_\mu u)(t,x)$ and suppose $f \not\equiv 0$.

Since the function $u(t,x)$, $\alpha < x < \beta$, is real analytic (see e.g. Friedman [8]) the function $u^*(t,y)$, $\alpha \leq y \leq \beta$, is strictly decreasing and

$$\tilde{u}(t,x) = \int_{C(t,x)} u(t,y)d\mu(y)$$

where $C(t,x) = \{u(t,\cdot) \geq u^*(t,x)\}$ is a compact subset of $]\alpha,\beta[$. Therefore, if $K \subseteq]0,+\infty[\times]\alpha,\beta[$ is compact, the continuity of u^* implies that the union

$$\bigcup_{(t,x)\in K} C(t,x)$$

is a compact subset of $]\alpha,\beta[$.

In the following r symbolizes a real number so close to zero that the actual formulas where it enters become meaningful. From

$$\tilde{u}(t+r,x) \geq \int_{C(t,x)} u(t+r,y)d\mu(y)$$

we thus have

$$\partial_t \tilde{u}(t,x) = \int_{C(t,x)} \partial_t u(t,y)d\mu(y) \; . \tag{3.1}$$

Moreover, depending on real analyticity we have

$$C(t,x) = \bigcup_{k=0}^{n} [a_k,b_k] \quad (\alpha < a_0 < b_0 < \ldots < a_n < b_n < \beta)$$

and we may define a function δ by

$$\sum_{k=0}^{n} \int_{a_k-\delta(r)}^{b_k+\delta(r)} d\mu = F(x+r) \; .$$

Then

$$\delta'(r) \sum_{k=0}^{n} (\rho(b_k+\delta(r)) + \rho(a_k-\delta(r)) = \rho(x+r),$$

so that

$$\delta'(0) \sum_{k=0}^{n} (\rho(b_k) + \rho(a_k)) = \rho(x) \tag{3.2}$$

and

$$\begin{aligned} &2\delta'(0)^2 \sum_{k=0}^{n} (R'(b_k)\rho(b_k) - R'(a_k)\rho(a_k)) + \\ &+ \delta''(0) \sum_{k=0}^{n} (\rho(b_k) + \rho(a_k)) = 2R'(x)\rho(x). \end{aligned} \tag{3.3}$$

Set $u_0 = u(t,a_0) = \ldots = u(t,b_n)$. We claim that

$$\partial_x \tilde{u}(t,x) = u_0\rho(x). \tag{3.4}$$

Indeed,

$$\Big(\int_{C(t,x+r)} - \int_{C(t,x)}\Big)\, u(t,y)d\mu(y) \geq \sum_{k=0}^{n} \Big(\int_{a_k-\delta(r)}^{b_k+\delta(r)} - \int_{a_k}^{b_k}\Big)\, u(t,y)d\mu(y)$$

and (3.4) follows from (3.2) We next show that

$$\partial_x^2 \tilde{u}(t,x) \geq \delta'(0)^2 \sum_{k=0}^{n} (u_x'(t,b_k)\rho(b_k) - u_x'(t,a_k)\rho(a_k)) + 2u_0 R'(x)\rho(x). \tag{3.5}$$

To see this, we use

$$\Big(\int_{C(t,x+r)} + \int_{C(t,x-r)} - 2\int_{C(t,x)}\Big)\, u(t,y)d\mu(y) \geq$$

$$\geq \sum_{k=0}^{n} \Big(\int_{a_k-\delta(r)}^{b_k+\delta(r)} + \int_{a_k-\delta(-r)}^{b_k+\delta(-r)} - 2\int_{a_k}^{b_k}\Big)\, u(t,y)d\mu(y)$$

and we have

$$\partial_x^2 \tilde{u}(t,x) \geq \delta'(0)^2 \sum_{k=0}^{n} (\partial_y (u(t,y)\rho(y))\big|_{y=b_k} - (\partial_y (u(t,y)\rho(y))\big|_{y=a_k}) +$$

$$+ \delta''(0) \sum_{k=0}^{n} (u(t,b_k)\rho(b_k) + u(t,a_k)\rho(a_k)).$$

The inequality (3.5) is now immediate remembering (3.3). However, from the very definition of isoperimetric measures, $0 \leq \delta'(0) \leq 1$, and (3.5) yields

$$\partial_x^2 \tilde{u}(t,x) \geq \sum_{k=0}^{n} (u_x'(t,b_k)\rho(b_k) - u_x'(t,a_k)\rho(a_k)) + 2u_0 R'(x)\rho(x). \tag{3.6}$$

But by (3.1),

$$\partial_t \tilde{u}(t,x) = \sum_{k=0}^{n} \int_{a_k}^{b_k} u_t'(t,y)d\mu(y) = \sum_{k=0}^{n} \int_{a_k}^{b_k} (\tfrac{1}{2}u_{xx}''(t,y) + R'(y)u_x'(t,y))d\mu(y) =$$

$$= \tfrac{1}{2} \sum_{k=0}^{n} (u_x'(t,b_k)\rho(b_k) - u_x'(t,a_k)\rho(a_k))$$

and Part a) now follows at once from (3.6) and (3.4).

b) Let $R' \in \mathrm{Lip}\,K$ be any extension of the function R' in the formulation of Lemma 3.1. Furthermore, denote by ξ_x the solution of (1.1) satisfying $\xi_x(0) = x$. Then, with τ_x the first point of time the random function $(t-s,\xi_x(s)), 0 < s \leq t$, hits the boundary of $]0,+\infty[\times]\alpha,\beta[$, we have

$$v(t,x) = \mathbb{E}\, f^*(t-\tau_x,\xi_x(\tau_x)),\ t > 0,\ \alpha < x < \beta$$

where $f^*(\cdot,\cdot)$ corresponds to the boundary data of the Cauchy problem in Part b) (see e.g. Friedman [9, Th. 5.2]). However, if $\alpha < x < y < \beta$, $\mathbb{P}(f^*(t-\tau_x,\xi_x(\tau_x)) \geq f^*(t-\tau_y,\xi_y(\tau_y))) = 1$ because ξ_y starts afresh at the stopping time $\inf\{s > 0;\ \xi_y(s) = \xi_x(s)\}$ (see e.g. [11, Problem 4, p. 58]). Consequently, $v(t,x) \geq v(t,y)$.

Now let $\alpha < a < \beta$ and define

$$\tilde{v}_a(t,x) = \int_a^x v(t,y)d\mu(y),\ t > 0,\ \alpha < x < \beta.$$

By a straightforward calculation we have

$$(\partial_t + M)\tilde{v}_a = -\tfrac{1}{2}v_x'(\cdot,a)\rho(a) \geq 0$$

and since $\tilde{v}_a \to \tilde{v}$ in the distribution sense as $a \downarrow \alpha$, Part b) is proved.

c) Let $\varphi \in C([\alpha,\beta])$ be equal to $1 - F_\mu$ in $]\alpha,\beta[$ and note that $M\varphi = 0$. Moreover, let $\delta > 0$ be arbitrary but fixed and define $w = \tilde{v} - \tilde{u} - \delta\varphi$. Then $(\partial_t + M)w \geq 0$ by Parts a) and b) and, in addition, w is continuous. Therefore, for each fixed $T > 0$,

$$\min_{t \leq T} w \geq \min\{w;\ t \leq T \text{ and } t = 0 \text{ or } x \in \{\alpha,\beta\}\}. \tag{3.7}$$

This follows from the standard minimum principle if w is smooth enough. The author has had troubles to find an adequate reference in the general case. However, one may proceed exactly as in [1] with some minor changes and we do not go into details here.

We next show that, if $t > 0$, then $w(t,\cdot)$ cannot have a minimum at the point β. In fact, assuming the converse,

$$0 \geq \int_x^\beta v(t,y)d\mu(y) - \int_x^\beta u^*(t,y)d\mu(y) + \delta\varphi(x), \quad \alpha < x < \beta.$$

By dividing this inequality by $\beta - x$ and letting $x \uparrow \beta$, we get $\delta\rho(\beta-) \leq 0$, which is a contradiction. In view of (3.7) and the definitions of $\tilde{u}$ and $\tilde{v}$ it now follows that $w \geq -\delta$. Finally, as $\delta \downarrow 0$, we have $\tilde{v} \geq \tilde{u}$. □

4. The main result

Throughout this section we abide by the notation and conditions stated in the Introduction. To repeat these $\mu(dx) = \rho(x)dx$ is a probability measure on $\mathbb{R}$ such that $0 < \rho \in C^1(\mathbb{R})$ with ρ' locally Lipschitz continuous, $R = \frac{1}{2}\ln\rho$, and ζ denotes the equilibrium solution of (1.1).

Below $(\ )^* = L_\mu$.

<u>Theorem 4.1.</u> <u>The following assertions are equivalent:</u>

(i) $\mathbb{E}[f(\zeta(s))g(\zeta(t))] \leq \mathbb{E}[f^*(\zeta(s))g^*(\zeta(t))]$, $f,g \in C_b(\mathbb{R})$, $s,t \geq 0$

(ii) μ <u>is isoperimetric</u>.

To prove Theorem 4.1 we need

<u>Lemma 4.1.</u> <u>Suppose</u> $h \in C^2(\mathbb{R})$, <u>where</u> $\operatorname{supp} h'$ <u>is compact</u>. Then

$$\lim_{t\to 0+} t^{-1}\,\mathbb{E}\,[h(\zeta(0))(h(\zeta(0))-h(\zeta(t))] = \tfrac{1}{2}\,\|h'\|^2_{2,\mu}. \tag{4.1}$$

<u>Proof.</u> We first use the Itô lemma to get

$$dh(\xi(t)) = h'(\xi(t))d\xi(t) + \tfrac{1}{2}h''(\xi(t))dt$$

so that

$$\mathbb{E}_x h(\xi(t)) - h(x) = \int_0^t \mathbb{E}_x(h'(\xi(s))R'(\xi(s)) + \tfrac{1}{2}h''(\xi(s)))ds.$$

By dominated convergence the limit in (4.1) now equals

$$-\int h(h'R' + \tfrac{1}{2}h'')d\mu = \tfrac{1}{2}\,\|h'\|^2_{2,\mu}.$$ □

Proof of Theorem 4.1. (i) ⇒ (ii): Let $J = [a,b]$ be a compact interval and suppose $r > 0$. Set $f = r \wedge d(\cdot, \sim J_r)$. By Theorem 2.7 it is enough to prove that $f^* \in \mathrm{Lip}\,1$. To this end, suppose $\varphi \in C_0^\infty(\mathbb{R})$ is an even probability density with support equal to $[-1,1]$. For arbitrary $0 < A < B < r$ and $0 < \delta < \frac{1}{2}((B-A) \wedge (b-a))$ we introduce $\varphi_\delta = \varphi(\cdot/\delta)/\delta$, $g = A \vee (f \wedge B)$ and $g_\delta = \varphi_\delta * g$. Note that $g'_\delta = \varphi_\delta * g'$ so that g_δ is unimodal and $g'_\delta \neq 0$ in $\{A < g_\delta < B\} =]e_1,e_2[\cup]e_3,e_4[$ $(e_1 < e_2 < e_3 < e_4)$. Consequently, the maps $g_{\delta}|]e_1,e_2[$ and $g_{\delta}|]e_3,e_4[$ both possess C^∞ inverses. Furthermore, since $g_\delta^{(i)}(e_k) = 0$, $i = 1,2$, $k = 1,2,3,4$, it now follows that $g_\delta^* \in C^2(\mathbb{R})$. We next use

$$\mathbb{E}[g_\delta(\zeta(0))(g_\delta(\zeta(0)) - g_\delta(\zeta(t))] \geq \mathbb{E}[g_\delta^*(\zeta(0))(g_\delta^*(\zeta(0)) - g_\delta^*(\zeta(t)))]$$

and Lemma 4.1 to get

$$\|g'_\delta\|_{2,\mu} \geq \|g_\delta^{*\prime}\|_{2,\mu} .$$

Noting that $g_\delta = g$ off $]e_1, e_1 + 2\delta[\cup]\, e_2 - 2\delta, e_2[\cup]e_3, e_3 + 2\delta[\cup]e_4 - 2\delta, e_4[$ we have in the limit as $\delta \downarrow 0$ that

$$\int_{A<f<B} d\mu \geq \int_{A<f^*<B} (f^{*\prime})^2 d\mu.$$

But then $|f^{*\prime}| \leq 1$ in $0 < f^* < r$ so that $f^* \in \mathrm{Lip}\,1$.

(ii) ⇒ (i): Suppose $f \in C_b^+(\mathbb{R})$ has compact support and set for all $t \geq 0$ and $x \in \mathbb{R}$,

$$u(t,x) = \mathbb{E}_x f(\xi(t))$$

$$v(t,x) = \mathbb{E}_x f^*(\xi(t))$$

$$\tilde{u}(t,x) = \int_{-\infty}^{x} u^*(t,y)\,d\mu(y)$$

and

$$\tilde{v}(t,x) = \int_{-\infty}^{x} v(t,y)\,d\mu(y).$$

Since $\langle h_0,h_1\rangle_\mu \leq \langle h_0^*,h_1^*\rangle_\mu$, $h_0,h_1 \in C_b(\mathbb{R})$, it is enough to prove that $\tilde{u} \leq \tilde{v}$. To this end, let $\mu_n,\rho_n,R_n,\ldots$ be as in Theorem 2.8. Furthermore, let $R_n' \in \mathrm{Lip}\,K$ denote any extension of R_n' and suppose

$$d\xi_n(t) = R_n'(\xi_n(t))dt + d\omega(t),\ t \geq 0.$$

We next introduce $\tau_n = \inf\{t>0;\xi_n(t) \notin I_{\mu_n}\}$ and set for each $t \geq 0$ and each $x \in I_{\mu_n}$,

$$u_n(t,x) = \mathbb{E}_x(f(\xi_n(t));\tau_n > t)$$

and

$$v_n(t,x) = \mathbb{E}_x(f^{*n}(\xi_n(t));\tau_n > t) + (\max f)\mathbb{P}_x(\tau_n \leq t,\xi_n(\tau_n) = \alpha_n),$$

where $(\ )^{*n} = L_{\mu_n}$. In the following, assume n is so large that $\mathrm{supp}\,f \subseteq I_{\mu_n}$. The functions u_n and v_n solve the Cauchy problems in Lemma 3.1 a) and b), respectively, with μ replaced by μ_n. Hence, setting

$$\begin{cases} \tilde{u}_n(t,x) = \int_{\alpha_n}^{x} u_n^{*n}(t,y)d\mu_n(y) \\ \tilde{v}_n(t,x) = \int_{\alpha_n}^{x} v_n(t,y)d\mu_n(y),\ x \in I_{\mu_n}, \end{cases}$$

we have that $\tilde{u}_n \leq \tilde{v}_n$. Therefore, if $u_n(t,\cdot) \to u(t,\cdot)$ and $v_n(t,\cdot) \to v(t,\cdot)$ uniformly on compacts as $n \to +\infty$, Theorem 2.8 (v) yields $\tilde{u} \leq \tilde{v}$ and we are done. However, if $x,x_n \in \mathbb{R}$, $n \in \mathbb{N}$, and $\lim_{n\to+\infty} x_n = x$ then the sequence $(\mathbb{P}_{x_n}(\xi_n \in \cdot))_{n\in\mathbb{N}}$ converges weakly to the measure $\mathbb{P}_x(\xi \in \cdot)$ (see e.g. [16, Theorems 6.3.4, 10.1.2, and 11.1.4]). This proves (i). □

<u>Proof of Theorem 1.1</u>. Let $0 \leq t_0 \leq \ldots \leq t_n$ and $f_0,\ldots,f_n \in C_b^+(\mathbb{R})$. Set $\mathbb{E}(f_0,\ldots,f_n) = \mathbb{E}[\prod_0^n f_k(\zeta(t_k))]$. In view of Theorems 2.3 and 4.1 it is enough to prove that

(4.2) $$\mathbb{E}(f_0,\ldots,f_n) \leq \mathbb{E}(f_0^*,\ldots,f_n^*)$$

for $n \geq 2$ if μ is isoperimetric. To this end, we write

$$\mathbb{E}(f_0,\dots,f_n) = \langle f_0, \mathbb{E}_\bullet[\prod_1^n f_k(\xi(t_k-t_0))]\rangle_\mu =$$

$$= \langle f_0, \mathbb{E}_\bullet\{f_1(\xi(t_1-t_0))\mathbb{E}_{\xi(t_1-t_0)}[\prod_2^n f_k(\xi(t_k-t_1))]\}\rangle_\mu =$$

$$= \langle \mathbb{E}_\bullet[f_0(\xi(t_1-t_0))], f_1\mathbb{E}_\bullet[\prod_2^n f_k(\xi(t_k-t_1))]\rangle_\mu$$

where we used the symmetry of ζ in the last step. Consequently,

$$\mathbb{E}(f_0,\dots,f_n) \le \langle f_1^*(\mathbb{E}_\bullet[f_0(\xi(t_1-t_0))])^*, (\mathbb{E}_\bullet[\prod_2^n f_k(\xi(t_k-t_1))])^*\rangle_\mu .$$

Now assuming that (4.2) is true with n replaced by $n-1$,

$$\mathbb{E}(f_0,\dots,f_n) \le \langle f_1^*(\mathbb{E}_\bullet[f_0(\xi(t_1-t_0))])^*, h\rangle_\mu$$

where $h = \mathbb{E}_\bullet[\prod_2^n f_k^*(\xi(t_k-t_1))]$ is decreasing. Indeed, as $\langle g,h\rangle_\mu \le \langle g^*,h\rangle_\mu$ for all $g \in C_b^+(\mathbb{R})$, the function h decreases. Finally, since the function f_1^*h decreases we may apply Theorem 4.1 to get $\mathbb{E}(f_0,\dots,f_n) \le \mathbb{E}(f_0^*,\dots,f_n^*)$. □

5. Applications to one-dimensional stochastic mechanics

Throughout this section we assume $V:\mathbb{R} \to \mathbb{R}$ satisfies the following conditions:

(i) $V(x) \in L_0^+(dx)$

(ii) V is non-constant

(iii) V is quasi-convex i.e. all the level sets $\{V \le a\}$, $a \in \mathbb{R}$, are convex

and

(iv) $V(-\infty) = V(+\infty)$.

It is well-known that the operator $H = -\frac{1}{2}(d/dx)^2 + V(x)$ in $L_2(dx)$ is essentially self-adjoint on $C_0^\infty(\mathbb{R})$ (see e.g. Reed and Simon [14, Th. X.28]). Furthermore, $E_0 = \inf \operatorname{spec}(H)$ gives a one-dimensional eigenspace of H and there exists a unique $\Psi_0 \in L_2(dx)$ such that $H\Psi_0 = E_0\Psi_0$, $\Psi_0 > 0$, and $\int \Psi_0^2 dx = 1$. The eigenfunction Ψ_0 is called the ground state of H (for existence, see below).

Theorem 5.1. Ψ_0 is log concave.

Theorem 5.1 is well-known if the quasi-convexity of V is replaced by convexity (Brascamp and Lieb [4]). However, the general case seems to be new.

Our proof of Theorem 5.1 is partly sketchy; first from [15, Theorems 25.11 and 25.15] the existence of Ψ_0 is rather obvious. Furthermore, by applying the same theorems and the following lemma it is simple to complete the proof of Theorem 5.1.

Lemma 5.1. Let $0 < b-a < +\infty$ and suppose $U:]a,b[\to \mathbb{R}$ is quasi-convex with finite range. If

$$\begin{cases} -\frac{1}{2}\Psi'' + U\Psi = 0 \text{ in }]a,b[\quad \text{(weak sense)} \\ \Psi(a) = \Psi(b) = 0 \\ \Psi > 0 \text{ in }]a,b[, \; \Psi \in C([a,b]) \end{cases}$$

then Ψ is log concave.

Proof. Set $\{U \leq 0\} = (A,B)$. It is obvious that $a \leq A < B \leq b$. Since Ψ is concave in (A,B) and C^1 in $]a,b[$ it is enough to prove that Ψ is locally log concave off (A,B). By symmetry, we may restrict ourselves to the interval $]a,A[$.

Suppose $a = a_0 < a_1 < \ldots < a_n < a_{n+1} = A$ and that $U_{|]a,A[}$ is continuous off $\{a_1,\ldots,a_n\}$. Set $\theta_k = (2U(a_k+))^{\frac{1}{2}}$, $k = 0,\ldots,n$. Then, if $a_k < x < a_{k+1}$,

$$\Psi(x) = M_k e^{\theta_k x} + N_k e^{-\theta_k x} \quad (M_k, N_k \in \mathbb{R})$$

so that

$$\Psi(x)\Psi''(x) - (\Psi'(x))^2 = 4\theta_k^2 M_k N_k .$$

Accordingly, Lemma 5.1 follows if we show that

(5.1) $\qquad M_k > 0, \; N_k < 0.$

However,

$$\begin{cases} M_k e^{\theta_k a_k} + N_k e^{-\theta_k a_k} = \Psi(a_k) \\ M_k e^{\theta_k a_k} - N_k e^{-\theta_k a_k} = \Psi'(a_k)/\theta_k \end{cases}$$

and hence (5.1) equivalently means that

$$\Psi(a_k) - \Psi'(a_k)/\theta_k < 0.$$

Furthermore, we have

$$\Psi(a_{k+1}) - \Psi'(a_{k+1})/\theta_{k+1} = M_k e^{\theta_k a_{k+1}} + N_k e^{-\theta_k a_{k+1}} -$$

$$- \frac{\theta_k}{\theta_{k+1}} (M_k e^{\theta_k a_{k+1}} - N_k e^{-\theta_k a_{k+1}}) = (1 - \frac{\theta_k}{\theta_{k+1}}) M_k e^{\theta_k a_{k+1}} + (1 + \frac{\theta_k}{\theta_{k+1}}) N_k e^{-\theta_k a_{k+1}}.$$

Finally, since $\theta_k \geq \theta_{k+1}$ and

$$\Psi(x) = M_0 (e^{\theta_0 x} - e^{\theta_0 (2a-x)}), \quad a < x < a_1,$$

where $M_0 > 0$ as $\Psi > 0$, (5.1) follows by the principle of mathematical induction. □

Corollary 5.1. *If (i) - (iv) hold and*

(v) *V is even*

then the ground state distribution

$$\mu(dx) = \Psi_0^2(x)dx$$

is isoperimetric.

Proof. The measure $\mu \in Q_0$ and Ψ_0^2 is log concave. The result is now a consequence of Corollary 2.2. □

In the sequel we assume (i) - (v) hold.

Example 5.1. Suppose we think of an electron in an one-dimensional situation in which its potential energy equals $V(x)$ at the point $x \in \mathbb{R}$. Choosing Coulomb units

($\hbar$=m=1) the Hamiltonian is given by $H = -\frac{1}{2}(d/dx)^2 + V(x)$. Below we suppose the electron is in the ground state. According to stochastic mechanics (Nelson [12]) the electron then performs a Markov process ζ, which is simply the equilibrium solution of (1.1) with $R = \ln \psi_0$. From the above it follows that $\zeta \in \mathcal{M}_-$. Especially, whenever the (quantum) probabilities $\mu(A)$ and $\mu(B)$ are prescribed, the joint event $[\zeta(s) \in A, \zeta(t) \in B]$ $(s \neq t)$ has maximal probability provided A and B are unbounded and similarly ordered intervals. □

6. Invitation to further research

Let $X = (X(t);\ t \in T)$ be an $\mathbb{R}^d$-valued stochastic process and suppose $\nu \in (\mathbb{R}^d)' = \mathbb{R}^d$ is a fixed unit vector such that each individual $\langle X(t),\nu\rangle$ has a continuous distribution function. We shall write $X \in \mathcal{M}_\nu$ if the following is true.

<u>For any</u> $p_0,\dots,p_n \in]0,1[$ $,t_0,\dots,t_n \in T$, <u>and</u> $n \in \mathbb{N}$ <u>and any Borel sets</u> $A_0,\dots,A_n \subseteq \mathbb{R}^d$ <u>satisfying</u> $\mathbb{P}(X(t_0) \in A_0) = p_0,\dots,\mathbb{P}(X(t_n) \in A_n) = p_n$, <u>the joint probability</u>

$$\mathbb{P}(X(t_0) \in A_0,\dots,X(t_n) \in A_n)$$

<u>is maximal provided</u> $A_0 x \dots x A_n$ <u>is of the form</u>

$$\{\langle\cdot,\nu\rangle \geq a_0\} x \dots x \{\langle\cdot,\nu\rangle \geq a_n\}.$$

In [1] we proved that the d-dimensional Ornstein-Uhlenbeck process belongs to $\mathcal{M}_\nu$ for each $\nu \in S^{d-1}$. It should be very interesting to bring forth further non-trivial examples in $\mathbb{R}^d$. From the point of view of stochastic mechanics the electron in a 3-dimensional convex potential well $V(|x|)$ must be interesting to examine. Needless to say, the same is true for the hydrogen electron.

REFERENCES

1. Borell, C: Geometric bounds on the Ornstein-Uhlenbeck velocity process in equilibrium (to appear).

2. Borell, C: Convex measures on locally convex spaces. Ark. Mat. 12(1974), 239-252.

3. Borell, C: The Brunn-Minkowski inequality in Gauss spaces. Inventiones math. 30(1975), 207-216.

4. Brascamp, H.J. and E.H. Lieb: On the extension of the Brunn-Minkowski and Prékopa-Leindler theorems, including inequalities for log concave functions, and with an application to the diffusion equation. J. Functional Analysis 22(1976), 366-389.

5. Carmona, R: Regularity properties of Schrödinger and Dirichlet semigroups. J. Functional Analysis 33(1979), 259-296.

6. Courrège, P. and P. Renouard: Oscillateur anharmonique, measures quasi-invariantes dans $C(\mathbb{R}, \mathbb{R})$ et théorie quantique des champs en dimension $d = 1$, Astérisque 22-23(1975), 1-245.

7. Ehrhard, A: Symetrisation dans l'espace de Gauss. Math. Scand.53(1983), 281-301.

8. Friedman, A: Classes of solutions of linear systems of partial differential equations of parabolic type. Duke Math. J. 24(1957), 433-442.

9. Friedman, A: Stochastic Differential Equations and Applications. Vol. 1. Academic Press, New York-San Francisco-London 1975.

10. Landau, H.J. and L.A. Shepp: On the supremum of a Gaussian process. Sankyā Ser. A. 32(1971), 369-378.

11. McKean, H.P: Stochastic Integrals. Academic Press, New York-London 1969.

12. Nelson, E: Dynamical Theories of Brownian Motion. Math. Notes, Princeton Univ. Press, Princeton 1967.

13. Priouret, P. and M. Yor: Processus de diffusion a valeur dans $\mathbb{R}$ et mesures quasi-invariantes sur $C(\mathbb{R}, \mathbb{R})$. Astérisque 22-23(1975), 248-290.

14. Reed, M. and B. Simon: Methods of Modern Mathematical Physics II: Fourier Analysis, Self-Adjointness. Academic Press, New York-San Francisco-London 1975.

15. Simon, B: Functional Integration and Quantum Physics. Academic Press, New York-San Francisco-London 1979.

16. Stroock, D.W., and S.R.S. Varadhan: Multidimensional Diffusion Processes. Springer Verlag, Berlin-Heidelberg-New York 1979.

Department of Mathematics
Case Western Reserve University
Cleveland, Ohio 44106, U.S.A.
and
Chalmers University of Technology
S-412 96 Göteborg, Sweden

SUBSPACES OF L_N^∞, ARITHMETICAL DIAMETER AND SIDON SETS

J. Bourgain

Vrije Universiteit Brussel
Pleinlaan 2-F7
1050 Brussels BELGIUM

and

IHES
Bures-sur-Yvette
FRANCE

INTRODUCTION

Let G be a compact Abelian group, $\Gamma = \hat{G}$ the dual group and Λ a subset of Γ. Denote C_Λ the subspace of the Banach space $C(G)$ of continuous functions on G of those functions f such that

$$\hat{f}(\gamma) = \int_G f(x)\, \overline{\gamma(x)}\, dx = 0 \qquad \text{if } \gamma \notin \Lambda$$

thus with Fourier transform $\hat{f}$ supported by the set Λ. Let us recall also that the norm on $C(G)$, hence on C_Λ, is the sup-norm

$$\|f\|_{C(G)} = \|f\|_\infty = \sup_{x \in G} |f(x)| .$$

If Λ is a finite set, then $\dim C_\Lambda = |\Lambda|$ ($\overset{\text{def}}{=}$ cardinal Λ) obviously and C_Λ consists of the linear combinations of the characters (seen here as functions on the group G) belong=ng to Λ.

We next formulate our key notion (cf. [11], p. 403) which in turn is a concept from local theory of Banach spaces.

DEFINITION 1: If $\Lambda \subset \Gamma$, $|\Lambda| < \infty$, we call the arithmetical diameter $d(\Lambda)$ of Λ the smallest positive integer N such that C_Λ is isomorphic up to a factor 2 say, to a subspace of the N-dimensional space ℓ_N^∞.

Explicitly we require the existence of a linear operator $T\colon C_\Lambda \to \ell_N^\infty$ such that

T is one-to-one and $\|T\|\|T^{-1}\| \leq 2$. For our purpose, any other fixed bound instead of 2 would do as well. Let us give some examples.

If $G = \mathbb{T}$ = circle group and $\Lambda = \{1,2,\ldots,n\} \subset \mathbb{Z}$ then $d(\Lambda) \sim n$. Letting now $G = \{1,-1\}^{\mathbb{N}}$ = the Cantor group and $\Lambda = \{\varepsilon_1,\ldots,\varepsilon_n\}$ the set of the first n coordinate projections (= Rademacher functions), we have the extreme case $\log (d\Lambda) \sim n$. This is due to the fact that if T is a norm-1 operator from C_Λ into ℓ^∞_N, then

$$\|T^{-1}\|^{-1} n \leq \int_{\{1,-1\}^n} \| \sum_{1\leq j\leq n} \delta_j (T(e_j)\|_{\ell^\infty_N} d\delta \leq c(\log N)^{1/2} (\sum \|T(e_j)\|^2)^{1/2}$$

$$\leq cn^{1/2} (\log N)^{1/2}$$

More conceptually, this may be reformulted by saying that the space ℓ^1_n can not be embedded in ℓ^∞_n (up to some fixed isomorphism-factor) unless N is exponential in n. This principle could as well be applied in case $G = \Pi$ and Λ a finite Hadamard-lacunary set (see [13] for more details).

DEFINITION 2: A subset Λ of Γ is called a Sidon set provided there is a constant C for which the inequality

$$\sum_\Lambda |a_\gamma| \leq C \| \sum_\Lambda a_\gamma \gamma \|_{C(G)} \tag{1}$$

holds, whenever $\{a_\gamma\}_{\gamma\in\Lambda}$ is a finitely supported scalar sequence (Again characters are seen here as functions on the group). The smallest constant C satisfying (1) is denoted $S(\Lambda)$, the Sidon constant of Λ.

If we consider the space $A(G)$ of absolutely convergent Fourier series, i.e. $\|f\|_{A(G)} \overset{\text{def}}{=} \sum |\hat{f}(\gamma)|$, then clearly

$$S(\Lambda) = \sup_{f\in C_\Lambda} \frac{\|f\|_{A(G)}}{\|f\|_\infty}$$

If Λ is a Sidon set, then C_Λ is ismorphic to the space ℓ^1_Λ (up to isomorphism factor $S(\Lambda)$) and the characters in Λ provide the natural ℓ^1-basis. Previous considerations show therefore that $\log d(\Lambda_1) \sim \Lambda_1$ for any finite subset Λ_1 of

Λ. Our first main result, which proof is much less trivial, is the converse property (see [1]).

THEOREM 1: Let Λ be a set of characters and assume the existence of a constant $\delta > 0$ satisfying

$$\log (d(\Lambda_1) \geq \delta |\Lambda_1| \quad \text{whenever } \Lambda_1 \text{ is a finite subset of } \Lambda.$$

Then Λ is a Sidon set.

The theorem is a characterization of Sidonicity in terms of arithmetical diameter. In case of bounded groups, say $G = \prod_{j=1}^{\infty} Z_{p_j}$ with $\sup p_j < \infty$, the proof may be derived from combinatorial principles. The general case, say $G = \Pi$, is done by different arguments however. Details will be presented in section 3 of this paper.

It was shown by N. Varapoulos (see [81]) that if the space C_Λ is isomorphic to $\ell^1_{|\Lambda|}$ as a Banach space, or more generally if C_Λ is a $\mathcal{L}^1$-space in the sense of [33], then the isomorphism can be taken naturally, i.e. Λ is a Sidon set. This result was later improved by G. Pisier (see [26]) who showed that Λ is Sidon iff C_Λ has cotype 2.

DEFINITION 3: Let X be a Banach space and $2 \leq q < \infty$. Say that X has cotype q provided for some constant C, the inequality

$$\left(\sum_i \|x_i\|^q \right)^{1/q} \leq C\left\{\int \| \sum_i \varepsilon_i x_i \|^2 \, d\varepsilon\right\}^{1/2} \tag{2}$$

holds, whenever $\{x_i\}$ is a finite sequence of vectors in X. The smallest constant C in (2) is denoted $C_q(X)$, the cotype q constant of X.

Next theorem is derived from Theorem 1 and methods from geometry of Banach spaces to be explained later. It solves affirmatively a problem raised by S. Kwapien and A. Pelczynski in [14] (see problem 5) and G. Pisier (see [25]).

THEOREM 2: For each space C_Λ the following dichotomy principle holds. Either C_Λ has cotype 2, thus Λ is a Sidon set, or C_Λ has no finite cotype, which means that it contains ℓ^∞_n-spaces ($1+\varepsilon$-isometrically) of arbitrary large dimension n.

In order to derive Theorem 2 from Theorem 1, it should clearly suffice to show that a finite dimensional normed space X with bounded cotype q constant $C_q(x)$ (for a fixed $q < \infty$) only well embeds in ℓ_N^∞ when $\log N \sim \dim X$. At this point we are unable to prove this and are obliged to use a more subtle approach. This approach uses partial solutions to the more general conjecture formulated above. Results on subspaces of ℓ_N^∞ of "large" dimension will be presented in a separate section.

Theorem 2 expresses a dichotomy which in spirit is closely related to the dichotomy problem for restriction algebras.

Denote $M(G)$ (resp. $M_d(G)$) the convolution algebra of bounded Borel measures (resp. discrete measures) on G. For $\Lambda \subset \Gamma$, denote $B(\Lambda) = \{\hat{\mu}|_\Lambda;\ \mu \in M(G)\}$ and $B_d(\Lambda) = \{\hat{\mu}|_\Lambda;\ \mu \in M_d(G)\}$ the restriction algebras equipped with the natural quotient norm. Notice that $B(\Lambda)$ identifies with the Banach space dual of C_Λ.

The <u>dichotomy conjecture</u> states that either Λ is a Sidon set or the only functions operating on the algebra $B(\Lambda)$ are restrictions of entire functions (see [12] and also [11] for more details).

In some sense, our Theorem 2 solves the Banach space version of this conjecture. It was observed by C. Graham (see [12]) that a positive solution to the dichotomy conjecture implies following fact, which we state as a theorem

<u>THEOREM 3</u>: Let Λ be a set of characters on G satisfying the equality $B(\Lambda) = B_d(\Lambda)$. Then Λ is a Sidon set.

Consequently, if $B(\Lambda) = B_d(\Lambda)$, then any bounded scalar sequence $(a_\gamma)_{\gamma\varepsilon\Lambda}$ is the restriction of the Fourier transform of a discrete measure on G, i.e. Λ is a so-called <u>I-set</u> (in contrast to Sidonicity, this notion is not stable for finite union).

The proof of Theorem 3 again makes crucial use of Theorem 1.

In the last section of the paper, some partial results to the $A(\Lambda) = B_0(\Lambda)$ problem are presented. Here the restriction algebras

$A(\Lambda) = \{\hat{f}|_\Lambda;\ f \in L^1(G)\}$ and $B_0(\Lambda) = \{\hat{\mu}|_\Lambda;\ \mu \in M(G),\ \hat{\mu} \to 0\}$ are involved. This problem may be seen as a strong version of the "tilde problem" (see [11] again).

These notes aim to give a survey of some recent developments in the area of thin sets mainly corrected with local theory and geometry of Banach spaces. Beside some new results on subspaces of ℓ_N^∞, the content of this paper appears in separate articles, mainly [1], [3], [5], [4], [21]. We will often refer to them for more complete proofs.

1. STRUCTURAL THEORY OF SIDON SETS

In this section, we give a more analytical description of a Sidon set. The structure of a general Sidon set is explained in terms of explicit and elementary objects, such as Riesz products and dissociated sets.

DEFINITION 4: A subset Λ of Γ is called dissociated or quasi-independent provided (using the additive notation for the group operation)

$$\Sigma'_\Lambda \ \varepsilon_\gamma \gamma = 0 \ \text{ in } \Gamma \ (\varepsilon_\gamma = 0, 1, -1) \Rightarrow \varepsilon_\gamma = 0 \quad \text{if} \quad \gamma \neq 0$$

It is well known and easily seen that if Λ is dissociated and $(a_\gamma)_{\gamma\varepsilon\Lambda}$ complex numbers of modulus ≤ 1, the product

$$\prod_{\gamma\varepsilon\Lambda} \left(1 + \frac{1}{2} a_\gamma \gamma + \frac{1}{2} \overline{a}_\gamma \overline{\gamma}\right)$$

defines a positive measure of mean 1, which we call a Riesz-product. For a measure μ on G, it will be convenient to denote $\|\mu\|_{PM} = \sup_{\gamma\varepsilon\Gamma} |\hat{\mu}(\gamma)|$. The norm $\| \ \|_{PM}$ is dual to the A(G)-norm defined earlier.

PROPOSITION 1: If Λ is dissociated, then Λ is a Sidon set and $S(\Lambda) \leq C$ (= numerical constant)

Proof: We use Riesz products to interpolate bounded sequences indexed by Λ. Without restriction, we assume $0 \notin \Lambda$. Let $f = \sum'_{\gamma\varepsilon\Lambda} a_\gamma \gamma$ be a polynomial and define for $0 < \delta \leq 1/2$

$$\mu_\delta = \prod_\Lambda \left[1 + \delta \frac{a_\gamma}{|a_\gamma|} \gamma + \delta \frac{\overline{a}_\gamma}{|a_\gamma|} \overline{\gamma}\right] + i \prod_\Lambda \left[1 + \delta \frac{ia_\gamma}{|a_\gamma|} \gamma - \delta \frac{i\overline{a}_\gamma}{|a_\gamma|} \overline{\gamma}\right]$$

$$= 1 + i + 2\delta \sum_{\gamma\varepsilon\Lambda} \frac{\bar{a}_\gamma}{|a_\gamma|} \bar{\gamma} + \delta^2\mu^{(2)} + \delta^3\mu^{(3)} + \dots$$

with thus $\|\mu_\delta\|_{M(G)} \leq 2$ (the products appearing above are finite). The main point is the fact that

$$\|\mu^{(2)}\|_{PM}, \ \|\mu^{(3)}\|_{PM}, \dots \leq 2 \ PM\big(\prod_\Lambda{}' (1 + \tfrac{1}{2}\gamma + \tfrac{1}{2}\bar{\gamma})\big) = 2$$

as follows from a closer analysis. Using μ_δ as test measure, we get

$$2\|f\|_{L^\infty(G)} \geq \left|\int f\, d\mu_\delta\right| = \left|2\delta \sum_{\gamma\varepsilon\Lambda} |a_\gamma| + \delta^2 < \mu^{(2)},f> + \delta^3<\mu^{(3)},f> + \dots\right|$$

It suffices to take $\delta > 0$ small enough to get the result.

It turns out that Riesz-products are the only measures needed to develop Sidon set theory in general. This is a consequence of the next result.

<u>THEOREM 4</u>: For a set of characters $\Lambda \subset \Gamma$, following conditions are equivalent

(1) Λ is a Sidon set

(2) There is a constant C such that for all finite scalar sequences $(a_\gamma)_{\gamma\varepsilon\Gamma}$ and $p \geq 1$ the inequality

$$\|\sum_\Lambda a_\gamma \gamma\|_{L^p(G)} \leq c\sqrt{p}(\sum |a_\gamma|^2)^{1/2}$$

holds, or equivalently

$$\|f\|_{L^{\Phi_2}(G)} \leq c\|f\|_{L^2(G)} \quad \text{if} \quad \text{supp}\, \hat{f} \subset \Lambda.$$

(3) There is $\delta > 0$ such that given any finite subset A of Λ, there is a subset B of A, B dissociated and $|B| > \delta|A|$.

(4) There is $\rho > 0$ such that if $(a_\gamma)_{\gamma\varepsilon\Lambda}$ is a (finite) sequence of "weights", there is a dissociated subset Λ_0 of Λ such that $\sum_{\gamma\varepsilon\Lambda_0} a_\gamma \geq \rho \sum |a_\gamma|$.

The proof yields following interdependence of constants:

$$S(\Lambda)^{-1/2} > \rho_{(4)} \sim \delta_{(3)} \gtrsim C_{(2)}^{-2} \gtrsim S(\Lambda)^{-2}$$

The theorem implies in particular that a finite union of Sidon sets is a Sidon set, a result due to S. Drury [8].

Implication (1) => (2) is due to W. Rudin. The equivalence of (1), (2), (3) was obtained by G. Pisier in [20]. The complete proof of Theorem 4 including (4) appears in [3], based on an argument substantially different from [20]. The approach in [3], sketched below, does not involve techniques of random measures. Also, by dualization (4) implies (together with the proof of Prop. 1)

COROLLARY 5: Interpolating measures of Sidon sets are obtained by averaging Riesz products built on dissociated subsets.

Assume $S(\Lambda) < \infty$ and $\mu_\varepsilon \in M(G)$ satisfy $\|\mu_\varepsilon\| \leq S(\Lambda)$ and $\hat{\mu}_\varepsilon(\gamma) = \varepsilon_\gamma = \pm 1$ for $\gamma \in \Lambda$. For $f = \sum'_{\Lambda} a_\gamma \gamma$, it follows for each sign-combination ε

$$f = (\sum \varepsilon_\gamma a_\gamma \gamma) * \mu_\varepsilon \Rightarrow \|f\|_{L^p(G)} \leq S(\Lambda) \, \|\sum \varepsilon_\gamma a_\gamma \gamma\|_{L^p(G)}$$

Integrating in ε, Khintchine's inequality and Fubini's theorem give the implication (1) => (2).

The interesting feature of the implication (2) => (3) is the fact that the extraction of a dissociated subset is partly probabilistic, partly deterministic. Denote $C = C_{(2)}$. We first use a random argument to obtain $A_1 \subset A$, $|A_1| \sim C^{-2}|A|$ without too long relations, in the sense that

$$\sum_{A_1} \varepsilon_\gamma \gamma = 0, \quad \varepsilon_\gamma = 0,1,-1 \Rightarrow \sum |\varepsilon_\gamma| \leq \tfrac{1}{2}|A_1|$$

If then $A_2 \subset A_1$ is the support of a (1,-1)-relation of maximal length, $B = A\,A_1$ has to be dissociated and $|B| \geq \frac{1}{2}|A_1|$.

To construct A_1, define $\tau = \frac{1}{C_1} C^{-2}$ and $\ell = \frac{1}{4}\tau|A|$ (C_1 is a numerical constant).

Consider a system $(\xi_\gamma)_{\gamma\in A}$ of independent (0,1)-valued mean-τ random variables (= selectors) in ω and define for $x \in G$

$$F_\omega(x) = \sum_{m=\ell}^{|A|} \sum_{\substack{S \subset A \\ |S|=m}} \prod_{\gamma \in S} \xi_\gamma(\omega)[\gamma(x) + \overline{\gamma(x)}]$$

Let us explain interest of this function. Clearly, $\forall \omega$, $\int_G F_\omega$ is a poitive integer and hence $\int_G F_\omega = 0$ as soon as $\int_G F_\omega < 1$. It is clear that this will be the case iff the random set $A_\omega = \{\gamma \in A \mid \xi_\gamma(\omega) = 1\}$ does not admit relations of length $\geq \ell$. Hence the existence of some ω has to be proved for which $|\int_\Gamma F_\omega| < 1$ and $|A_\omega| > 2\ell$. By integration in ω and hypothesis (2) of Theorem 4

$$\iint_G F_\omega(x)dxd\omega \leq \sum_{m=\ell}^{A} \tau^m \frac{1}{m!} \int_G |\sum_A \gamma(x) + \sum_A \overline{\gamma(x)}|^m dx$$

$$\leq \sum_{\ell \leq m \leq |A|} \tau^m \frac{C^m m^{m/2}(2|A|)^{m/2}}{m!} = \sum_{\ell \leq m \leq |A|} \left[\frac{\text{const. } C^2\tau^2|A|}{n} \right]^{m/2}$$

which is dominated by $2^{\ell/2}$ by the choice of τ. Therefore

$$\frac{\tau|A|}{2} + 2^{\ell/2} \iint_G F_\omega(x)\, dd\omega \leq \int \sum_{\gamma \in A} \xi_\gamma(\omega)d\omega$$

implying the existence of some ω for which

$$\frac{\tau|A|}{2} + 2^{\ell/2} \int_G F_\omega \leq |A_\omega|$$

where $\frac{\tau|A|}{2} \geq 2\ell$. This proves (2) => (3).

As a consequence of the proof of Proposition 1, the implication (4) => (1) is obtained. The central point is (3) => (4). Fixing a scalar sequence $(a_\gamma)_{\gamma\in\Lambda}$, we consider the level sets $\Lambda_k = \{\gamma \in \Lambda;\ R_1^{-k} \geq |a_\gamma| > R_1^{-k-1}\}$ for $k = 0,1,2,\ldots,$ assuming $\sum |a_\gamma| = 1$. Here R_1 stands for a certain numerical constant. Using hypothesis (3) in Th. 4, replace each Λ_k by a dissociated proportional subset Λ'_k. By elementary considerations, we may come to the situation where

$$\frac{|\Lambda'_{k+1}|}{\Lambda'_k} > 100$$

We then apply the following lemma, with a proof in the same spirit as (2) => (3)

(but a bit more complicated)

LEMMA 1: (cf. [3]) Assume $\Lambda_1, \ldots, \Lambda_J$ finite disjoint, dissociated subsets of Γ and $|\Lambda_{j+1}| > 100|\Lambda_j|$ for $j = 1, \ldots, J-1$. Then we may extract subsets $\bar{\Lambda}_j \subset \Lambda_j$ such that

$$|\bar{\Lambda}_j| > \frac{1}{10}|\Lambda_j| \quad \text{for each } j \text{ and } \bigcup_{j=1}^{J} \bar{\Lambda}_j \text{ is still dissociated.}$$

The argument is again probabilistic and deterministic.

The main question arising from Theorem 4 is whether or not any Sidon set is a finite union of dissociated sets. For some cases, the problem was settled (in the affirmative). A significant result in this direction was obtained by Malliavin [28] as a corollary to the ingenious combinatorial lemma of Rado and Horn [15].

PROPOSITION 2: Let p be a prime number and consider the product group $G = \Pi Z(p)$ where $\mathbb{Z}(p) = \mathbb{Z}/p\mathbb{Z}$. Any Sidon subset of $\Gamma = \hat{G}$ is a finite union of sets which are independent in Γ as a vector space over the field $Z(p)$, hence of dissociated sets.

Already for the group $\Pi\mathbb{Z}(4)$, the previous question is unsettled at the time of this writing. In [2], the following fact is proved on the structure of Sidon sets in product groups.

PROPOSITION 3: Let $\Lambda \subset (G_1 \times G_2)^\wedge$ be a Sidon set. Then Λ decomposes as a finite union of sets Λ_α such that for each α for either $i = 1$ or 2, $\pi_i|_{\Lambda_\alpha}$ is one-to-one and $\pi_i(\Lambda_\alpha)$ is a Sidon set in G_i (π_i denotes here the natural projection of $(G_1 \times G_2)^\wedge$ onto $\hat{G}_i$).

It is well known that if G is an Abelian group with elements of bounded order (we may for simplicity call G a bounded group), G is obtained as a direct product of its p-groups $G(p)$, each $G(p)$ being a finite product of groups $\Pi\, \mathbb{Z}(p^r)$. Proposition 3 clearly reduces the decomposition problem for bounded groups to the case $G = \Pi\, \mathbb{Z}(p^r)$, with p^r a fixed power of the prime p. As a corollary to Prop. 2, we get

PROPOSITION 4: If the integer n is a distinct product of prime numbers, then any Sidon set in $\hat{G}$, $G = \Pi\, \mathbb{Z}(n)$, is a finite union of dissociated sets.

Let us mention one more partial result around decomposition.

PROPOSITION 5: (see also [2]): Any Sidon set admits for any positive integer ℓ a decomposition in finitely many subsets admitting no non-trivial relation of length at most ℓ.

2. CHARACTERIZATION OF SIDON SETS BY ENTROPY PROPERTIES

In [20], [21], G. Pisier proved the equivalence of (1), (2), (3) in Theorem 4 by methods of random Fourier series, relying in particular on variants of the Dudley-Fernique theorem for stationary Gaussian processes. We will make direct use of the latter theorem in order to obtain some of the results of this section. Since interdependence of constants will be of some importance later on, we state explicit estimations (which may not be sharp).

LEMMA 2: If $\Lambda \subset \Gamma = \hat{G}$ satisfies the condition

$$\int \| \sum_{A} \varepsilon_\gamma \gamma \|_{C(G)}\, d\varepsilon \geq \delta |A|$$

whenever A is a finite subset of Λ and where $\delta > 0$ is a fixed constant, then Λ is a Sidon set and $S(\Lambda) \leq c\delta^{-4}$

Proof: Fix a finite set A. By hypothesis, there is a measuare μ_ε such that

$$\|\mu_\varepsilon\|_{M(\varepsilon)} \leq 1 \quad \text{for each } \varepsilon \text{ in } \{1,-1\}^{|A|}$$

$$\sum |\langle \varepsilon_\gamma \otimes \gamma, \mu_\varepsilon \rangle| \geq \delta |A|$$

The latter fact easily implies the existence of $\delta < \delta_1 < 1$ and $B \subset A$ for which

$$|\langle \mu_\varepsilon, \varepsilon_\gamma \otimes \gamma \rangle| > \delta_1 \quad \text{if } \gamma \in B, \qquad |B| > \left(\frac{\delta}{\delta_1}\right)^2 |A|.$$

Hence, by Khintchine's inequality

$$cp^{1/2}\ B \geq \int\{\int \|\sum_B \varepsilon_\gamma \overline{\gamma(x)}\gamma\|_p\ \mu_\varepsilon(dx)\}d\varepsilon \geq \delta_1 \|\sum_B \gamma\|_p$$

From theorem 4, it now follows that B has a dissociated subset B with

$$B_1 \geq c\delta_1^2|B| \geq c\delta^2|A|$$

Again by Theorem 4, we may therefore conclude that Λ is a Sidon set with $S(\Lambda) \leq c\delta^{-4}$.

The next corollary is known as Rider's theorem (the estimation can be improved)

COROLLARY 6: If the set Λ is a Sidon set in the "Rademacher sense", i.e.

$$C \int \|\sum_\Lambda{}' \ \varepsilon_\gamma a_\gamma \gamma\|_{C(G)} d\varepsilon \geq \sum_\Lambda{}' \ |a_\gamma|$$

then Λ is a Sidon set with $S(\Lambda) \lesssim C^4$.

Suppose $\Lambda \subset \Gamma = \hat{G}$ and $(a_\gamma)_{\gamma\varepsilon\Lambda}$ a finite scalar sequence. Consider the pseudo-distance d_2 on G defined by

$$d_2(x,y) = (\sum_\Lambda |a_\gamma|^2 |\gamma(x) - \gamma(y)|^2)^{1/2}$$

and let $N_{d_2}(\varepsilon)$ stand for the corresponding entropy-numbers, i.e. the minimum number of ε-balls for the d_2-metric needed to cover G. If $(g_\gamma)_{\gamma\varepsilon\Lambda}$ denotes a sequence of independent Gaussian variables, then d_2 is the metric associated to the stationary process $\sum g_\gamma(\omega)a_\gamma\gamma$. The following result is a consequence of the Dudley-Fernique theorem [9]. The equivalence (2) is due to Marcus and Pisier [16].

PROPOSITION 6: Using the previous notations, following quantities are equivalent

(1) $\int \|\sum g_\gamma(\omega)a_\gamma\gamma\|_{C(G)} d\omega$

(2) $\int \|\sum \varepsilon_\gamma a_\gamma \gamma\|_{C(G)}\ d\varepsilon$

(3) $\int_0^\infty [\log N_{d_2}(\varepsilon)]^{1/2}\ d\varepsilon$

As a consequence of the latter fact and Lemma 2.

COROLLARY 7: A subset Λ of $\Gamma = \hat{G}$ is Sidon provided for some constant $\rho > 0$

$$\int_0^\infty [\log N_{d_{2,A}}(\varepsilon)]^{1/2}\, d\varepsilon \geq \rho\, A$$

wherever A is a finite subset of Λ and denoting

$d_{2,A}(x,y) = (\sum_A |\gamma(x) - \gamma(y)|^2)^{1/2}$. Moreover $S(\Lambda) \lesssim \rho^{-4}$.

For later use, we will need the analogue of Corollary 7 with respect to ℓ^∞-entropy instead of the ℓ^2-entropy. This is obtained in the next (see [21])

PROPOSITION 7: Assume that Λ satisfies the following condition: For each finite subset A of Λ, there is a set of points in G, say $\mathcal{E}$, such that

$${}^2\log |\mathcal{E}| > \rho |A|$$

$$d_{\infty,A}(x,y) \overset{\text{def}}{=} \sup_{\gamma\varepsilon A} |\gamma(x) - \gamma(y)| > \rho \quad \text{if } x \neq y \text{ in } \mathcal{E}$$

Then Λ is Sidon and $S(\Lambda) \leq C\rho^{-10}$.

Proof: Fix a finite subset A of Λ and let $\mathcal{E}$ be as above. Let $\tau > 0$ and $\mathcal{E}_1$ a maximal subset of $\mathcal{E}$ for which $d_{2,A}(x,y) \geq \tau |A|^{1/2}$ if $x \neq y$ in $\mathcal{E}_1$. Thus

$$\int_0^\infty [\log N_{d_2}(\varepsilon)]^{1/2} \geq \tau(\log |\mathcal{E}_1|)^{1/2}\, |A|^{1/2}$$

To each point $x \in \mathcal{E}$ corresponds $x_1 \in \mathcal{E}_1$ such that $d_2(x,x_1) < \tau|A|^{1/2}$.
Consequently we may fix some $\bar{x} \in G$ so that

$$\#\, \mathcal{E}_2 > \frac{\#\mathcal{E}}{\#\mathcal{E}_1}$$

defining $\mathcal{E}_2 = \{x \in \mathcal{E};\ d_2(x,\bar{x}) < \tau |A|^{1/2}\}$.

For each $x \in \mathcal{E}_2$, consider the set $A_x = \{\gamma \in A;\ \gamma(\alpha) - \gamma(\bar{\alpha}) > \rho\}$. Thus, by definition

$$|A_x| < \frac{\tau^2}{\rho^2}|A| \equiv m$$

Elementary entropy considerations show therefore that

$$|\mathcal{E}_2| \leq \binom{|A|}{m}\left(\frac{2}{\rho^2}\right)^m$$

$$\log |\mathcal{E}| \leq \log |\mathcal{E}_1| + cm \log \frac{|A|}{m} + cm \log \frac{2}{\rho}$$

$$\rho|A| \leq \log |\mathcal{E}_1| + c\left(\frac{\tau^2}{\rho^2} \log \frac{\rho}{\tau}\right)|A| + c\left[\frac{\tau^2}{\rho^2} \log \frac{1}{\rho}\right]|A|$$

Choose $\tau \sim \rho^2$ in which case $\log |\mathcal{E}_1| > \frac{1}{2}\rho|A|$. Consequently

$$\int_0^\infty [\log N_{d_2}(\varepsilon)]^{1/2}\, d\varepsilon \geq \rho^{5/2}|A|$$

and Corollary 7 implies the desired result.

The use of the Dudley-Fernique theorem in the context of Sidon sets relies on the structure of homogeneous space of the group G. In order to prove Theorem 1 in the general case (in particular for the group T), an additional lemma on comparision of entropy numbers is needed. In this lemma, the full group structure of G is exploited.

LEMMA 3: Denote for simplicity $N(A,\varepsilon) = N_{d_{\infty,A}}(\varepsilon)$. The following general inequality then holds

$$\log N\left(A, \frac{1}{20}\right) \geq \frac{\log N(A,\varepsilon)}{\log \frac{2}{\varepsilon}}$$

Proof: Since clearly $N(A,\varepsilon) = \prod_{j=1,2,\ldots,\log \frac{2}{\varepsilon}} \frac{N(A,2^{j-1}\varepsilon)}{N(A,2^j\varepsilon)}$ we may find some $\varepsilon \leq \delta \leq 1$ satisfying

$$\log \frac{N(A,\delta)}{N(A,2\delta)} \geq A = \frac{\log N(A,\varepsilon)}{\log \frac{2}{\varepsilon}}$$

From this results the existence of a subset P of G such that for $x \neq y$ in P

$$\delta \leq d_\infty(x,y) \leq 4\delta \tag{3}$$

where

$$\log |P| \geq A.$$

If $\gamma \in \Gamma$, $x, y \in G$ and k is a positive integer, we have

$$|\gamma(kx) - \gamma(ky)| \geq k|\gamma(x) - \gamma(y)| \left(1 - \frac{k-1}{2}|\gamma(x) - \gamma(y)|\right) \tag{4}$$

since γ is a character. The reader will easily check details.

Suppose $\delta < 1/10$ and let $k = [1/4\delta]$. Replace P by $P_k = \{k.x;\ x \in P\}$. It results from (3), (4) that $d_\infty(kx,ky) > 1/10$ for any pair of distinct points x, y in P. Thus $N(A, 1/20) \geq |P_k| = |P|$, which proves the lemma. Combining Proposition 7 and Lemma 3 leads to

COROLLARY 8: Let $\Lambda \subset \hat{G}$ satisfy the following condition: For each finite subset A of Λ, there is a net $\mathcal{E}$ of points in G such that

$$^2\log |\mathcal{E}| > \rho|A|$$

$$d_{\infty,A}(x,y) > \tau \quad \text{whenever} \quad x \neq y \quad \text{in } \mathcal{E}.$$

Then Λ has Sidon constant $S(\Lambda)$ at most $C\rho^{-10}\left(\log \frac{2}{\tau}\right)^{10}$.

The interesting feature in this estimation of $S(\Lambda)$ is the mild dependence on the separation constant τ.

Proof of Theorem 1: Let $\Lambda \subset \Gamma$ satisfy the condition on arithmetic diameters. We prove an a priori inequality on the Sidon constant of finite subsets Λ_1 of Λ. Let for a finite subset A of Λ

$$\bar{d}_A(x,y) = \sup_{f \in C_A,\ \|f\| \leq 1} |f(x) - f(y)|$$

and $\bar{N}(A,\tau)$ refer to the corresponding entropy numbers. Obviously $d_{\infty,A} \leq \bar{d}_A$. On the other hand, for $A \subset \Lambda_1$

$$|f(x) - f(y)| \leq \|f\|_{A(G)}\, d_A(x,y)$$

implying

$$\bar{d}_A(x,y) \leq S(\Lambda_1)\, d_A(x,y) \tag{5}$$

It is easily seen from the definition of the arithmetical diameter $d(A)$ of the set

A and the pseudo-distance $\overline{d}_A$, that $d(A) \leq \overline{N}(A, \frac{1}{3})$. By hypothesis on Λ,

$$\log d(A) \geq \delta|A| \quad \text{for any finite subset } A \text{ of } \Lambda.$$

It results from (5) that $N_{d_{\infty,A}}(\frac{1}{3S(\Lambda_1)}) \geq d(A)$ if A is a subset of Λ_1. Therefore, Corollary 8 yields the estimation $S(\Lambda_1) \leq c\delta^{-10}\,{}^2\log(6S(\Lambda_1))$ if Λ is a finite subset of Λ. This means that Λ is a Sidon set with $S(\Lambda) \leq C\,\delta^{-11}$, proving the theorem.

Remark: The polynomial dependence of $S(\Lambda)$ with respect to δ in Theorem 1 will play a role in later arguments.

3. "LARGE" SUBSPACES OF L_N^∞

The meaning of "large", which we do not specify, does not mean here "proportional dimensional" of (cf [10]). In this section, we propose two different approaches to generate ℓ_k^∞-subspaces in "large" dimensional subspaces E of ℓ_N^∞. The first method, improving on some of the work in [10], gives essentially best possible results when $\log \dim E \sim \log N$. The second method, which will lead to Theorem 2, deals with the situation $\dim E >> \log N$ in which case we only try to show that cotype constant $c_q(E)$ can not remain bounded, given a fixed $q < \infty$.

LEMMA 4: Let $\{\alpha_i\}_{i=1,\dots,n}$ be positive numbers satisfying $\alpha_i < \varepsilon$ and $\sum \alpha_i^2 \leq 1$. Define, for $c > 0$ to be fixed later and for $\gamma > 0$

$$\mathcal{S} = \{S \subset \{1,\dots,n\};\ |S| = c\gamma\sqrt{n}\}$$

If P denotes the normalized counting measure on , then for n large enough

$$\mathbb{P}[S;\ \sum_{i\in S} \alpha_i > \gamma] < n^{-c\gamma\varepsilon^{-1}}$$

Proof: Define $I_k = \{2^{-k} \leq \alpha_i < 2^{-k+1}\}$ for $k = 0,1,2,\dots$ and $c_k = \sum_{i\in I_k} \alpha_i^2$. Then $|I_k| \leq 4^k c_k$ and $\sum c_k \leq 1$. Consider a sequence of positive numbers ε_k such that

$\sum \varepsilon_k \leq 2$. Define k_0, k_1 by $2^{k_0} \sim \frac{1}{\varepsilon}$ and $2^{k_1} \sim c\sqrt{n}$. Then clearly

$$\mathbb{P}[S \sum_{i\varepsilon S} \alpha_i > \gamma] \leq \sum_{k_0<k<k_1} \mathbb{P}[S| \sum_{i\varepsilon S\cap I_k} \alpha_i > \frac{1}{3}\gamma\varepsilon_k]$$

$$\leq \sum_{k_0<k<k_1} \mathbb{P}[S;\ |S \cap I_k| > \frac{1}{6}\gamma\varepsilon_k 2^k]$$

where

$$\mathbb{P}[|S \cap I| > T] \leq \binom{|I|}{T} \cdot (cC_2\gamma n^{-1/2})^T \leq (cC_2\gamma \frac{|I|}{Tn^{1/2}})^T$$

Define $\varepsilon_k = c_k + (k - k_0 + 1)^{-2}$ for $k > k_0$. Thus $\sum \varepsilon_k < 2$. For c small enough, it follows

$$P[S; \sum_S \alpha_i > \gamma] \leq \sum_{k_0<k<k_1} (6cC_2\, n^{-1/2}\, 2^k)^{\frac{1}{6}2^k\gamma\varepsilon_k} < n^{-c\gamma/\varepsilon}.$$

<u>Proposition 8</u>: Let E be a subspace of ℓ_N^∞, $\dim E = n = N^\delta$. For each $\tau > 0$, there is a subspace F of E, $\dim F = m$ where

$$d(F, \ell_n^\infty) < 1 + \tau \quad \text{and} \quad m > c\tau^5 \frac{\delta^2}{\log\frac{1}{\delta}} \cdot \sqrt{n}$$

<u>Proof</u>: By N. Tomczak's theorem on computing 2-summing norm (see [30])

$$\pi_2^{(n)}(E) > \frac{1}{\sqrt{2}}\pi_2(E) = \frac{\sqrt{n}}{\sqrt{2}}$$

This easily implies the existence of norm-1 vectors $x_1,\dots,x_{n_1}$; n n_1 in E satisfying $\sum_{i=1}^{n_1} |x_i(j)|^2 \leq 1$ for each coordinate $j = 1,\dots,N$. Fix $\varepsilon > 0$, $\gamma > 0$ and define for each $i = 1,\dots,n_1$ the vector $y_i = \sum_{j\varepsilon A_i} x_i(j)e_j$ where $A_i = \{j \mid |x_i(j)| < \varepsilon\}$. For fixed $1 \leq j \leq n_1$ we may then apply lemma 4 to the sequence $\alpha_i = y_i(j)$ and get with the same notations

$$\mathbb{P}[S \varepsilon \mathcal{S} | \sum_{i\varepsilon S} |y_i(j)| > \gamma] < n^{-\frac{c}{\varepsilon}\gamma}$$

It is now clear that for $\varepsilon \sim \gamma\delta$, there will be $S \subset \{1,\dots,N\}$, $|S| \sim \gamma\sqrt{n}$ such that $\sum_{i\in S} |y_i(j)| \leq \gamma$ for each $1 \leq j \leq N$. Notice also that for any j

$$|S_j| \leq \varepsilon^{-2} \quad \text{where} \quad S_j = \{i \in S \mid j \notin A_i\}$$

Define inductively a sequence j_r as follows

$$s_1 = \sum_{i\in S_{j_1}} |x_i(j_1)| = \max_j \sum_{i\in S_j} |x_i(j)|$$

$$s_{r+1} = \sum_{i\in S_{j_{r+1}}\cap\tilde{S}_r} |x_i(j_{r+1})| = \max_j \sum_{i\in S_j\cap\tilde{S}_r} |x_i(j)| \quad \text{with } \tilde{S}_r = S\setminus(S_1\cup\dots\cup S_r)$$

We let here $r < \varepsilon^2\gamma\sqrt{n}$ so that by the preceding $1 \leq s_r \leq \varepsilon^{-2}$. Let $\xi_r = \sum_{i\in S_{j_r}\cap\tilde{S}_r} \varepsilon_i x_i$ with the signs ε_i chosen so that $\|\xi_r\| \geq |\xi_r(j_r)| = s_r$.

By construction, it follows that for r_1 fixed

$$\sum_{r\geq r_1} |\xi_r(j)| \leq \sum_{i\in\tilde{S}_{r_1-1}} |x_i(j)| \leq s_{r_1} + \sum_{i\in S} |y_i(j)| < s_{r_1} + \gamma \tag{6}$$

Since the s_r form a decreasing sequence, there is a segment R of indices r so that $s_r < (1+\gamma)s_{r'}$, for $r < r'$ in R and $|R| \gtrsim \varepsilon^2\gamma^2(\log\frac{1}{\varepsilon})^{-1}\sqrt{n}$. Now by (6), $\|\sum_R a_r\xi_r\| \leq \max|a_r|\,(1+2\gamma)\min_R \|\xi_r\|$. Similarly, for each $\bar{r} \in R$

$$\|\sum_R a_r\xi_r\| \geq a_{\bar{r}}|\xi_{\bar{r}}(j_{\bar{r}})| - \max|a_r|\left(\sum_R |\xi_r(j_{\bar{r}})| - s_{\bar{r}}\right)$$

$$\geq a_{\bar{r}}|\xi_{\bar{r}}(j_{\bar{r}})| - \max|a_r|(s_{\tilde{r}} + \gamma - s_{\bar{r}}) \qquad (\tilde{r} = \min R)$$

Choosing $\bar{r}$ suitably, we get $\|\sum_R a_r\xi_r\| \geq (1-2\gamma)\min_R \|\xi_r\| \cdot \max a_r$. This proves that $F = [\xi_r;\, r\in R]$ is isomorphic to $\ell^\infty_{|R|}$ up to constant $\frac{1+2\gamma}{1-2\gamma}$. For $\gamma \sim \tau$, it follows that $|R| \gtrsim \tau^4\delta^2(\log \frac{1}{\tau\delta})^{-1}\sqrt{n}$. Hence F satisfies the required conditions.

Remarks.

1. The special case when $n = \dim E \sim N$ was proved by Figiel and Johnson in [10]. They also show that the existence of $\sqrt{N}$-dimensional ℓ^∞-isomorphs in proportional subspaces of ℓ_N^∞, say $\dim E = \frac{N}{2}$, is the best one may hope for.

2. It is a conjecture that if E is a subspace of ℓ_N^∞ and $\frac{\dim E}{\log N} \to \infty$, then E has to contain uniformly ℓ_k^∞-spaces of increasing dimension k. In this context, proposition 8 gives information as long as $\dim E > (\log N)^4$. For "small" dimensions, the technique explained below provides better estimations and will eventually permit us to prove Theorem 2.

3. Proposition 8 admits certain analogous for "large" dimensional subspaces of ℓ_N^p when $p > 2$. They will be presented elsewhere.

The method proposed next is based on an unpublished result of B. Maurey (see [25]). Let us recall the following definitions.

DEFINITION 5: A normed space X is said to have type p $(1 < p \leq 2)$ with constant $T_p(X)$ provided the inequality

$$\{\int \| \sum \varepsilon_i x_i \|^2 d\varepsilon\}^{1/2} \leq T_p(X)(\sum \|x_i\|^p)^{1/p}$$

holds for all finite sequences $\{x_i\}$ in X.

DEFINITION 6: Let X be a (finite dimensional) normed space. Denote K(X) the norm of the "Rademacher projection" $R_1 \otimes Id_X$ acting on $L_X^2(D)$, where $D = \{1,-1\}^{\mathbb{N}}$. Explicitly, if $f \in L_X^2(D)$, then the Rademacher projection of f equals

$\sum \varepsilon_i \hat{f}(i)$ where $\hat{f}(S) = \int f(\varepsilon)w_S(\varepsilon)d\varepsilon$ $(S \subset \mathbb{N})$ are the Walsh-Fourier coefficients

In [17], [24] the number $K(X)$ is called K-convexity constant of X. Following facts are obtained from a simple duality reasoning

PROPOSITION 9: $K(X) = K(X^*)$ and $T_p(X) \leq K(X)C_q(X^*)$ for $\frac{1}{p} + \frac{1}{q} = 1$.

The quantity $K(X)$ can be estimated by methods of harmonic or complex analysis (see [25] for next statement).

PROPOSITION 10: Let X be an n-dimensional normed space and denote d_X its Euclidean distance. Thus d_X: $\inf\{\|T\|\|T^{-1}\|;\ T$ linear one-to-one map from X to $\ell_n^2\}$. Then $K(X) \leq C \log d_X$. Hence $K(X) \leq C \log n$ in general, which is the best possible estimation without additional hypothesis.

PROPOSITION 11: (see [25]): Let X be an n-dimensional normed space and suppose X 2-isomorphic to a subspace of ℓ_N^∞. Then $n \leq C\, T_p(X^*)^q \log N$ $(\frac{1}{p} + \frac{1}{q} = 1)$.

Proof: By hypothesis, there is a map $u: \ell_N^1 \to X^*$ satisfying $\|u\| \leq 1$ and $\frac{1}{2} B_{X^*} \subset \text{conv}(S)$ where $S = \{\pm u(e_i);\ 1 \leq i \leq N\}$ and B_{X^*} stands for the unit ball of X^*. Let now f be a function on a probability space $\Omega, \mathbf{P}$ taking values in S. Let $x = \int_\Omega f$. Denote $f_1, f_2, \ldots$ independent copies of f defined on the product space $\tilde{\Omega} = \Omega_1 \times \Omega_2 \times \ldots$ Then, for each integer J

$$\int_{\tilde{\Omega}} \Big\|\frac{1}{J} \sum_{j=1}^{J} f_j(\omega) - x\Big\| d\omega \leq \frac{2}{J} \int_{\tilde{\Omega}} \int_D \Big\| \sum_{j=1}^{J} \varepsilon_j [f_j(\omega) - x]\Big\| d\omega d\varepsilon$$

$$\leq \frac{2}{p}\, T_p(X^*) \int_{\tilde{\Omega}} \Big(\sum \|f_j(\omega) - x\|^p\Big)^{1/p} d\omega \leq 4T_p(X^*) J^{-1/q} \|u\|$$

It follows that for $J \sim T_p(X^*)^q$, each element of conv(S) can be approximated up to distance $\frac{1}{100}$ by an element of the set

$$\Big\{\frac{1}{J} \sum_{i \in A} \pm u(e_i);\ \ A \subset \{1,\ldots,N\} \text{ and } |A| = J\Big\}$$

Since $B_{X^*} \subset 2\, \text{conv}(S)$, an entropy argument gives now $2^{\dim X} \leq \binom{N}{J} 2^J$. Hence $n \leq C\, T_p(X^*)^q \log N$.

The reader may consult [5] for a different proof. As a consequence of Prop. 9, it follows

PROPOSITION 12: If X is an n-dimensional normed space, 2-isomorphic to a subspace of ℓ_N^∞, then $n \leq C[K(X)C_q(X)]^q \log N$.

This fact may be exploited in an indirect way, using the existence of large Euclidean subspaces. We rely on an improvement due to V. Milman [18] of a theorem proved by Dilworth [7].

PROPOSITION 13: Any n-dimensional normed space X has a subspace Y, $\dim Y = \frac{n}{2}$ satisfying $d_Y \leq CC_2(X) \log C_2(X)$

COROLLARY 9: Let X be an n-dimensional normed space, 2-isomorphic to a subspace of ℓ_N^∞. Then $n \leq C[C_q(X) \log C_2(X)]^q \log N$.

Proof: If Y is as in Proposition 12, then $C_q(Y) \leq C_q(X)$ and $K(Y) \leq C \log C_2(X)$ by Proposition 10. Since Y is 2-isomorphic to a subspace of ℓ_N^∞, Proposition 12 applies to Y and gives the desired estimation.

The latter formula will be exploited in the next section together with Theorem 1 to prove Theorem 2.

4. PROOF OF THE COTYPE DICHOTOMY.

Theorem 2 is again deduced from an a-priori inequality on the Sidon constant of finite subsets of the given set $\Lambda \subset \Gamma$. It follows from the proof of Theorem 1 that

$$\log d(A) \geq \delta|A| \quad \text{if} \quad A \subset \Lambda \quad \Rightarrow \quad S(A) \leq C\delta^{-C} \quad (C = \text{constant})$$

Assume now C_Λ has a finite cotype, say $C_q(C_\Lambda) < \infty$. Let Λ_0 be a fixed finite subset of Λ. For $A \subset \Lambda_0$

$$C_2(C_A) \leq S(A) \leq S(\Lambda_0)$$

and applying Corollary 9 with $X = C_A$

$$|A| \leq C\, C_q(C_\Lambda)^q [\log S(\Lambda_0)]^q \log d(A)$$

This consequently implies

$$S(\Lambda_0) \leq C\, C_q(C_\Lambda)^{C_q} [\log S(\Lambda)]^{C_q}$$

$$S(\Lambda_0) \leq [\log C_q(C_\Lambda)]^{\text{const } q}$$

proving the theorem.

REMARKS

1. Theorem 2 may be considered as the "Banach space" version of the algebra dichotomy mentioned in the introduction. If Λ is not a Sidon set, then C_Λ contains uniform isomorphs of ℓ_k^∞ for arbitrary large dimension k. Our approach however does not give much insight into how those ℓ_k^∞-spaces appear analytically. If Λ is a finite set of characters and $S(\Lambda)$ is large, we define the function

$$f = \frac{1}{|\Lambda|} \sum_{\gamma \in \Lambda} \gamma$$

and consider random translates $f_{x_1}, \ldots, f_{x_k}$ $(x_j \in G)$. In certain cases, for instance if Λ is contained in a finite product $D_1 \times D_2 \times \ldots \times D_t$ with $D_1, D_t, \ldots, D_t$ dissociated (the V-Sidon set setting, see [17]), it can be verified that if k is sufficiently small with respect to $S(\Lambda)$, the system $\{f_{x_j}\}_{1 \leq j \leq k}$ is equivalent to the natural ℓ_k^∞-basis. We do not know if this principle holds in general.

2. It was proved by G. Pisier in [22] that if E is an n-dimensional Banach lattice, then the Euclidean distance may essentially be estimated as $CT_p(E)C_q(E)n^\alpha$ with $\alpha = \max(\frac{1}{p} - \frac{1}{2}, \frac{1}{2} - \frac{1}{q})$. Notice that if $\Lambda \subset \hat{G}$, $|\Lambda| = n$, then

$$d_{C_\Lambda} = n^{1/2}$$

since

$$\|\sum_\Lambda \gamma(x)\gamma\|_{L^2_{C_\Lambda}(G)} = n.$$

A positive solution for the previous problem would imply that if $d_X \sim (\dim X)^{1/2}$ and $C_q(X)$ remains bounded, then for $p = \frac{q}{q-1}$ the constant $T_p(X) \sim n^{1/q}$ and X contains ℓ_m^1-subspaces for m n. This would lead to an alternative proof of Theorem 2. Recently however, it was discovered by the author that the distance problem has a negative solution and the estimation

$$d_X \leq CT_p(X)C_q(X)n^{\frac{1}{p}-\frac{1}{q}}$$

following from results of S. Kwapien and N. Tomczak (see [17] for details) is essentially best possible. The example is natural in the context of harmonic analysis and we sketch it here.

Let $G = \{1,-1\}^n$ be a finite Cantor group and define the norm

$$\| \sum_{S\subset\{1,\dots,n\}} a_S w_S \|_{A_p} = (\sum |a_S|^p)^{1/p} \qquad (1 \leq p \leq 2)$$

Let next $X_p = [L^{p'}(G), A_p]_{1/2,\,2}$ be the real interpolation space $(p' = \frac{p}{p-1})$. It is easily seen that X_p is isometric to its dual space. Also, by the Hausdorff-Young inequality

$$\int \| \sum w_S(\varepsilon)w_S \|_{X_p} d\varepsilon \geq \int \| \sum w_S(\varepsilon)w_S \|_{L^{p'}(G)} d\varepsilon = N^{1/p}, \quad N = 2^n$$

implying $d_{X_p} \geq N^{\frac{1}{p}-\frac{1}{2}}$. By definition, $T_{\bar{p}}(X_p) \leq (p')^{1/4}$ for $\bar{p} = \frac{4p}{2+p}$.

In the limit case, the space X_1 is obtained with extremal Euclidean distance and satisfying $T_{4/3}(X_1) \leq (\log N)^{1/4}$, $T_{4/3}(X_1^*) \leq (\log N)^{1/4}$, which implies that neither X_1 nor X_1^* contain ℓ_k^1-subspaces for $k \sim N^\tau$. In connection with the cotype-cotype problem (cf. [24]) G. Pisier observed that X_1 has no bounded cotype, but the example clearly indicates the limitation of the method [24] (for the case $\frac{1}{p} - \frac{1}{q} < \frac{1}{2}$)

5. THE $B(E) = B_d(E)$ PROBLEM

The material presented in this section was announced in [1]. Again the techniques developed in section 1 will be used. The problem which we discuss appears in the frame of the dichotomy conjecture (discussed in [11], for instance). It is conjectured that if $\Lambda \subset \Gamma = \hat{G}$ either Λ is a Sidon (more generally, Helson-set) or $\sup_{\|f\|_{A(\Lambda)} \leq 1} \|e^{itf}\|_{A(\Lambda)} > e^{ct}$ for all $t > 0$ (c is a constant). It was observed by C. Graham that the latter fact implies only entire functions can operate on $B(\Lambda)$. Since $B_d(\Gamma) = A(\hat{G}_d)$; $G_d = G$ equipped with discrete topology, analytic functions are operating on $B_d(\Gamma)$, hence on $B_d(\Lambda)$ (as a consequence of the Wiener-Levy theorem, see [11]). Assuming the dichotomy true, the equality $B(\Lambda) = B_d(\Lambda)$ implies that Λ is a Sidon set (thus our theorem 3).

The next fact was shown by T. Ramsey using R. Blei's method of sup-norm partitioning (see [11] again)

<u>LEMMA 5</u>: Suppose $B(\Lambda) = B_d(\Lambda)$. Then there is a constant $M < \infty$ such that for each $f \in B(\Lambda)$, $\|f\| \leq 1$ there is $g \in B_d(\Lambda)$ satisfying the conditions

$$\|f - g\|_{B(\Lambda)} < \frac{1}{10}$$

$$g = \hat{\mu}|_{\Lambda} \quad \text{with} \quad \mu \in M_d(G), \quad \|\mu\| \leq M \quad \text{and} \quad |\text{supp}\, \mu| \leq M.$$

<u>Proof of Theorem 3</u>: It follows from Lemma 5 that for a given finite subset A of Λ, each point in the space $B(A)$ is approximable in the corresponding norm up to $\frac{1}{10}$ by an element of the set

$$\delta_x = \text{Dirac measure} \qquad \mathcal{E} = \{ \sum_{i=1}^{M} \lambda_i \hat{\delta}_{x_i}|_A ; \ \sum |\lambda_i| \leq M \ \text{and} \ x_i \in G \ (1 \leq i \leq M)\}$$

From elementary entropy arguments in finite dimensional normed spaces, it follows the existence of a subset $\mathcal{G} = \mathcal{G}_A$ in G satisfying

$$\log |\mathcal{G}| > \varepsilon |A| \quad \text{and} \quad \|\hat{\delta}_x|_A - \hat{\delta}_y|_A\|_{B(A)} > \varepsilon \quad \text{for } x \neq y \text{ in } \mathcal{G}$$

(here $\varepsilon = \varepsilon(M)$). Using the notations appearing in the proof of Theorem 1, one may write

$$\bar{d}_A(x,y) = \|\hat{\delta}_x|_A - \hat{\delta}_y|_A\|_{B(A)}$$

since $B(A)$ and C_A are dual spaces. Hence

$$\bar{N}_A(\varepsilon) > \varepsilon |A|$$

and the proof of Theorem 1 clearly implies that $S(\Lambda) < \varepsilon^{-C}$.

<u>REMARK</u>: If G is a bounded group, that is $G = \prod_{j=1}^{\infty} \mathbb{Z}(n_j)$ with $\sup_j n_j < \infty$, the proof of Theorem 3 simplifies (and does not require several notions of entropy on the group). In this case, the result was obtained independently by T. Ramsey (see [29]).

6. ON SETS SATISFYING THE CONDITION $A(\Lambda) = B_0(\Lambda)$

We consider subsets Λ of $\Gamma = \hat{G}$ with the property that for each $\mu \in M(G)$ with Fourier transform $\hat{\mu} \in c_0(\Gamma)$ (= tending to 0 at infinity), there exists $f \in L^1(G)$ such that $\hat{\mu}|_\Lambda = \hat{f}|_\Lambda$ (i.e. $A(\Lambda) = B_0(\Lambda)$). Let us denote this property by (S). L. Pigno and S. Saeki considered the problem whether or not such sets, which they call Riesz sets of type 0, must be Sidon sets (see [13], p. 165, [11], p. 409 and [19]). A positive solution to this question would solve the tilde problem affirmatively ($A(\Lambda) = \tilde{A}(\Lambda) \Rightarrow \Lambda$ is a Sidon set; see [11] for definitions). Most of the content of this section appears in [4], containing partial solutions. Two aproaches will be described. The first is descriptive and permits us to settle the question in the tensor-product case, for instance if $G = \{1,-1\}^{\mathbb{N}}$ = Cantor group and Λ is contained in a set of Walshes $\{w_S;\ S \subset \mathbb{N}$ with $|S| < r\}$ of bounded length $r^{(*)}$. The second is functional analytic rather than combinatorial and gives information on arithmetical diameters.

Property (S) implies a more quantitative statement, which is the analogue of the compactness property formulated in lemma 5.

<u>LEMMA 6</u>: Assume that $\Lambda \subset \Gamma$ satisfies (S). Then for each $\varepsilon > 0$, there exist $\delta > 0$ and a finite partition of Λ in sets Λ' verifying the condition

(*) This result was obtained independently by N. Varapoulos.

$$\mu \in M_+(G),\quad \|\mu\| = 1 \quad \text{and} \quad \sup_{\gamma \neq 0} |\hat{\mu}(\gamma)| < \delta \Rightarrow \|\hat{\mu}|_{\Lambda'}\|_{B(\Lambda')} < \varepsilon \qquad (*)$$

We then replace Λ by Λ' satisfying (*) for some $\varepsilon > 0$ (chosen small enough)

PROPOSITION 14: If Λ satisfies (*), then $\Lambda \cap D^r$ is a Sidon set, given any dissociated subset D of Γ and defining for $r = 1,2,\ldots$

$$D^r = \{\gamma_1\gamma_2 \cdots \gamma_r;\ \gamma_s \in D \text{ for } 1 \leq s \leq r\}.$$

The simplest case appears when $r = 2$ and $\Lambda \subset D_1 \times D_2$, in which case we want to show that any element of $\ell^\infty(\Lambda)$ is the restriction of an element in the projection tensor algebra $C(D_1) \hat{\otimes} C(D_2)$. These sets are called V-Sidon sets and are characterized by the following descriptive condition (see [32] or [11] again).

PROPOSITION 15: $A \subset D_1 \times D_2$ is a V-Sidon set iff for some constant C

$$|\Lambda \subset (A_1 \times A_2)| \leq C(|A_1| + |A_2|) \qquad (**)$$

holds whenever $A_1 \subset D_1$, $A_2 \subset D_2$ (finite subsets)

Assume (**) violated for large C. One tries to contradict (*) considering probability measures μ_ε with density $\mu_\varepsilon = \prod_{j \in S} (1 + \rho(\gamma_j \otimes \phi_j))$ where S D_1 and the ϕ_j are functions of the form

$$\phi_j = \operatorname{Im} \prod_{k \in A_j} \left(1 \pm i|A_j|^{-1/2} \delta_k\right)$$

where A_j is some (finite) subset of D_2 for each $j \in S$.
In fact $\bigcup_{j \in S} (\{j\} \times A_j)$ is contained in Λ and $\|\mu_\varepsilon\|_{B(\Lambda)} \geq$ const. To estimate $\sup_{\xi \in \Gamma \setminus \{0\}} |\hat{\mu}_\varepsilon(\xi)|$, we are led to evaluating $\|\Phi\|_{PM}$ where Φ is some product of the ϕ_j. This amounts to a combinatorial problem in which the special choice of the sets A plays a role.

Our next purpose is to give some information on the arithmetical diameter of the finite subsets Λ_0 of sets $\Lambda \subset \Gamma$ in general, satisfying $A(\Lambda) = B_0(\Lambda)$.

PROPOSITION 16: If $A(\Lambda) = B_0(\Lambda)$, then $\log d(\Lambda_0) > \exp c(\log |\Lambda_0|)^{1/2}$ whenever Λ_0 is a finite subset of Λ and $|\Lambda_0|$ is large enough (c = numerical).

Let F be a fixed (finite) subset of Γ. For each positive integer m, define $\alpha(n)$ as the largest number satisfying the inequality

$$\int \| \sum_{j=1}^{m} |(\phi_j)_{x_j}| \|_\infty \, dx_1 \ldots dx_m \geq \alpha(m) \sum_{j=1}^{m} \|\phi_j\|_\infty$$

whenever ϕ_j $(1 \leq j \leq m)$ are m functions with $\operatorname{supp} \hat{\phi}_j \subset F$. Again $\phi_x(y) = \phi(y-x)$. It is then a routine exercise to check the submultiplicity property

$$\alpha(m \cdot n) \geq \alpha(m)\alpha(n) \tag{7}$$

This feature will be used later.

The following fact appears as a consequence to lemma 4.

LEMMA 8: Let $f_1, \ldots, f_n \in L_N^\infty$, $n = N^\delta$ satisfy $\left(\sum |f_j|^2 \right)^{1/2} \leq B$. For m $\gamma\sqrt{n}$, ε $\gamma\delta$ and $\overline{f}_j = f_j X_{[|f_j| < \varepsilon]}$, we have

$$\int_{\mathcal{S}_m} \| \sum_{j \in S} |\overline{f}_j| \|_\infty \, dS < \gamma B$$

where $\mathcal{S}_m$ consists of the m-element subsets S of $\{1,\ldots,n\}$ and is equipped with the normalized counting measure.

LEMMA 9: Let A be a finite set of characters and define $f = |A|^{-1} \sum_{\gamma \in A} \gamma$. Then for $n < \frac{1}{10}|A|^{1/2}$, following holds

$$\int_{G^n} \| \sum_{j=1}^{n} |f_{x_j}|^2 \|_\infty^2 \, dx_1 \ldots dx_n \leq 3 \quad \text{where } f_x(y) = f(x+y)$$

Proof: Let $\{c_j\}_{1 \leq j \leq n}$ be a scalar sequence satisfying $\sum |c_j|^2 \leq 1$. Then by definition of f

$$\left|\sum c_j f_{x_j}(t)\right| \leq |A|^{-1} \sum_{j \in A} \left| \sum_{j=1}^{n} c_j \gamma(x_j) \right| \leq |A|^{-1/2} \left(\sum_{j \in A} \left| \sum_{j=1}^{n} c_j \gamma(x_j) \right|^2 \right)^{1/2}$$

Also

$$\sum_{\gamma\varepsilon A}\Big|\sum_{j=1}^{n} c_j\gamma(x_j)\Big|^2 \leq |A| + \sum_{j\neq k}\Big|\sum_{\gamma\varepsilon A}\gamma(x_j - x_k)\Big||c_j||c_k|$$

$$\leq |A| + \{\sum_{j\neq k}\Big|\sum_{\gamma\varepsilon A}\gamma(x_j - x_k)\Big|^2\}^{1/2}$$

Hence

$$\|\sum|f_{x_j}|^2\|_\infty \leq 1 + \frac{1}{|A|}\{\sum_{j\neq k}\Big|\sum_{\gamma\varepsilon A}\gamma(x_j - x_k)\Big|^2\}^{1/2}$$

$$\int_{G^n}\|\sum|f_{x_j}|^2\|_\infty^2\,dx \leq 2 + 2|A|^{-2}\,n|A| < 3 \quad\text{if}\quad n < \frac{1}{10}|A|^{1/2}$$

LEMMA 10: Let A be a finite set of characters. For $\tau > 0$ given, there are functions $\phi_1, \ldots, \phi_r$ with $\operatorname{supp}\hat{\phi}_s \subset A$ such that $\|\phi_s\|_\infty = 1 = \phi_s(0)$ $(1 \leq s \leq r)$ and

$$\int_{G^r}\|\sum_{s=1}^{r}|(\phi_s)_{x_s}|\|_\infty\,dx_1\ldots dx_r \leq 1 + \tau$$

and where

$$\log r \sim \tau\{\log\frac{\log d(A)}{\log|A|}\}^{-1}\log|A|$$

Proof: Let $N = d(A)$, $n \sim |A|^{1/2}$, $\varepsilon \sim \frac{\log|A|}{\log d(A)}$. Take $f = |A|^{-1}\sum_A\gamma$ and $f_j = f_{x_j}$ for $j = 1,\ldots,n$. Apply Lemma 8 to the system $\{f_j\}_{j=1}^n$ and next lemma 9. Letting $m = n^{1/2}$

$$f_\varepsilon = f\chi_{[|f|\geq\varepsilon]} \quad\text{and}\quad f^\varepsilon = f\chi_{[|f|\geq\varepsilon]}$$

it follows for $\gamma = 1/3$

$$\int_{G^m}\|\sum_{1\leq j\leq m}|(f_\varepsilon)_{x_j}|\|_\infty\,dx_1\ldots dx_m = \int_{G^n}\int_{\mathcal{S}_m}\|\sum_{j\varepsilon S}|(f_\varepsilon)_{x_j}|\|_\infty\,dS\,dx$$

$$\leq \gamma\int_{G^m}\|\sum_{1\leq j\leq n}|f_{x_j}|^2\|_\infty^{1/2}\,dx \leq 1$$

Hence, letting t be some positive integer

$$\int_{G^m} \| \sum_{1\le j\le m} |f_{x_j}| \|_\infty \, dx \le \int_{G^m} \| \sum |(f^\varepsilon)_{x_j}| \|_\infty \, dx + \int_{G^m} \| \sum |(f_\varepsilon)_{x_j}| \|_\infty \, dx$$

$$\le \int_{G^m} \sup_{|J|=t} \| \sum_{j\varepsilon J} |f_{x_j}| \|_\infty \, dx + \frac{1}{\varepsilon} \int_{G^m \cap [\|\sum|f_{x_j}|^2\|_\infty > t\varepsilon^2]} \|\sum |(f^\varepsilon)_{x_j}|^2\|_\infty \, dx + 1$$

$$= t + \frac{3}{t\varepsilon^3} + 1 \, .$$

Letting $t \sim \varepsilon^{-3/2}$, we see that $\alpha(m) \le \frac{t}{m}$ where $\log m \sim \log|A|$ and $\log t \sim \log \frac{\log d(A)}{\log |A|}$. The desired result now follows easily from the submultiplicity properties of α (see (7)).

We now return to the proof of Proposition 16. Our purpose is to exhibit $f \varepsilon C_{\Lambda_0}$, $\|f\|_\infty \le 1$ satisfying

$$\text{Re} f > \text{Re} \left(\sum_{\gamma\ne 0} \lambda_\gamma \gamma \right) + \frac{1}{2} \quad \text{whenever} \quad \sum |\lambda_\gamma| \le B \tag{8}$$

A separation argument will then indeed yield a positive measure μ, $\|\mu\| = 1$ such that

$$\|\mu\|_{B(\Lambda_0)} \ge |\langle f,\mu\rangle| \ge B \sup_{\gamma\ne 0} |\hat\mu(\gamma)| + \frac{1}{2}$$

and so $\sup_{\gamma\ne 0} |\hat\mu(\gamma)| \to 0$ as $B \to 0$. This allows us to contradict (*) with $\varepsilon = \frac{1}{2}$ and prove Prop. 16.

In order to deal with the general case, we will rely on the following principle which may be deduced from [6] on invariant approximation

LEMMA 11: Let I be a finite subset of Γ and denote $S(I) = \{ \sum_{\gamma\varepsilon I} \varepsilon_\gamma \gamma ; \varepsilon_\gamma = 0,1,-1\}$ (where $\sum$ refers to the group operation in Γ). There is a positive function p on G satisfying $\hat p \ge 0$, $\int p = 1$ and

$$\log |\text{supp } \hat p| \le |I|^2 \; ; \; \|\phi - (\phi * p)\|_\infty < \frac{9}{10} \text{ if } \text{supp } \hat\phi \subset S(I), \; |\phi| \le 1$$

With the notations of lemma 11, we may then state

LEMMA 12: Define

$$\mathcal{E} = \{ \sum_{\gamma\in V\setminus\{0\}} \lambda_\gamma \gamma; \quad V \subset \operatorname{supp} \hat{p}, \quad |V| < CB^2|I|^2 \quad \text{and} \quad \sum |\lambda_\gamma| \leq B \}$$

If then $\phi \in \{ \sum_{\gamma\neq 0} \lambda_\gamma \gamma; \ \sum |\lambda_\gamma| \leq B\}$, there exists $\psi \in \mathcal{E}$ satisfying

$$\|(\phi * p) - \psi\|_\infty < \frac{1}{100}$$

The proof of lemma 12 is obtained with the same idea as explained in Proposition 11. We leave details to the reader.

If I denote a maximal dissociated subset of Λ_0, then $\Lambda_0 \subset S(I)$ and $\log d(\Lambda_0) \geq |I|$. The negation of (8) leads to the hypothesis that for each $f \in C_{\Lambda_0}$, $\|f\|_\infty \leq 1$, there exists $(\lambda_\gamma)_{\gamma\neq 0}$, $\sum |\lambda_\gamma| \leq B$ such that $\operatorname{Re} f \leq \operatorname{Re}(\sum_{\gamma\neq 0} \lambda_\gamma \gamma) + \frac{1}{2}$.

By Lemma 12, there is a function $\psi \in \mathcal{E}$ so that $\operatorname{Re}(f * p) \leq \operatorname{Re} \psi + \frac{51}{100}$. Let $K \equiv K_\psi = \{x \in G; \operatorname{Re} \psi(x) \leq \frac{1}{10}\}$. Since $0 = \int \operatorname{Re} \psi > \frac{1}{10} - (B+1)|K|$, it follows $|K| > \frac{1}{20B}$. Also, since we may take $\log |\mathcal{E}| \sim B^2(|I|^4)$, this set K will be taken from a family $\mathcal{K}$ of compact subsets of G satisfying again $\log \mathcal{K} \leq CB^2(|I|^4)$. It is clear that we will have the estimation

$$\int \{ \sum_{K\in\mathcal{K}} (1 - \chi_K)(x_1) \ldots (1 - \chi_K)(x_r)\} \, dx_1 \ldots dx_r = \sum_{k\in K} (1 - K)^r < \tau$$

provided $r > CB^3(|I|^4)$. Then a subset $\{x_1,\ldots,x_r\}$ of G will be obtained, intersecting each member of K in at least 1 point. Apply Lemma 10 to the set $A = \Lambda_0$ with some $\tau > 0$ and let r and $\{\phi_s\}_{s=1,\ldots,r}$ be as in the conclusion. Since

$$\operatorname{Re}(\phi_s * p)(0) > \frac{9}{10} \quad \text{and} \quad \int_{G^r} \| \sum_{s=1}^{r} |(\phi_s)_{x_s}| \|_\infty \, dx_1 \ldots dx_r \leq 1+\tau ,$$

the points $\{x_1, \ldots, x_r\}$ may be in addition chosen such that $f = \sum_{s=1}^{r} (\phi_s)_{x_s}$

C_{Λ_0} satisfies $\|f\|_\infty \leq 1+2\tau$, $\mathrm{Re}\,(f*p)(x_s) > \frac{8}{10}(1 - 3\tau)$ for $s = 1,\ldots,r$.

Now, by hypothesis, there is a member $K \in \mathcal{K}$ satisfying $\mathrm{Re}(f*p)|_K \leq \frac{61}{100(1+2\tau)}$.

Hence, for $r > CB^3\ (|I|^4)$, the conclusion

$$\frac{61}{100(1+2\tau)} \geq \sup_K \mathrm{Re}(f*p) \geq \inf_{1\leq s\leq r} \mathrm{Re}(f*p)(x_s) \geq \frac{8}{10}(1 - 3\tau)$$

is contradictory for τ small enough.

This proves that r given Lemma 10 fulfills

$$\log r \leq 4\big(\log B + \log\log d(\Lambda_0)\big)$$

and yields Proposition 16.

REFERENCES

[1] J. Bourgain: Sur les ensembles d'interpolation pour les measures discretes, C. R. Acad. Sc. Paris, t. 296, Sér I, 149-151, 1982.

[2] J. Bourgain: Propriétés de décomposition pour les ensembles de Sidon, Bulletin Soc. Math. de France, T.111, 1983, No. 4.

[3] J. Bourgain: Sidon sets and Riesz products, Annales Fourier 1984, to appear.

[4] J. Bourgain: Some properties of sets satisfying A(E) = B (E), Bulletin Soc. Math. de Belgique 1984, to appear.

[5] J. Bourgain, V. D. Milman: Dychytomie du cotype pour les espaces invariants, C. R. Acad. Sc. Paris, to appear.

[6] Bozejko, A. Pelczynski: A analogue in commutative harmonic analysis of the uniform bounded approximation property of Banach space, Sém. d'Anal. Fonct., Ecole Polytechnique, Exp 9, 1978-79.

[7] S. J. Dilworth: The cotype constant and large Euclidean subspaces of normed spaces, preprint.

[8] S. Drury: Sur les ensembles de Sidon, C. R. Acad. Sci. Paris, 271, 162-163 (1970).

[9] X. Fernique: Régularité des trajectoires des processus gausiens, École d'Été de St.-Flour, Springer LNM 480.

[10] T. Figiel, W. B. Johnson: Large subspaces of ℓ_∞^n and estimates of the Gordon-Lewis constants, Israel J. Math. 37 (1980), 92-112.

[11] C. Graham, C. McGehee: Essays in Commutative Harmonic Analysis, Grundlehren der Math. Wissens. 238, Springer 1979.

[12] C. Graham: Sur un résultat de Katznelsen et McGehee, CRASC Paris 1973.

[13] J. Lopez, K. Ross: Sidon sets, New York, Marcel Dekker, 1975.

[14] S. Kwapien, A. Pelczynski: Absolutely summing operators and translation invariant spaces of functions on compact abelian groups, Math. Nachr. Bd. 94, 303-340 (1980).

[15] L. Lindahl, F. Poulsen: Thin sets in harmonic analysis, New York, Marcel Dekker, 1971.

[16] M. Marcus, G. Pisier: Random Fourier series with applications to harmonic analysis, Annals of Math. Studies n° 101, Princeton Univ. Press, 1981.

[17] B. Maurey, G. Pisier: Séries de variables aléatoires vectorielles indpendantes et propriétis géométriques des espaces de Banach, Studia Math. 58 (1976), 45-90.

[18] V. D. Milman: Volume approach and Iteration Procedures in Local Theory of Normed Spaces, to appear in Proc. Missouri Conf. 1984, Springer LNM.

[19] L. Pigno, S. Saeki: Measures whose transforms vanish at infinity, Bull. AMS 1973, 800-802.

[20] G. Pisier: De nouvelles caracterisations des ensembles de Sidon, Advances in Maths, Supplementary studies, Mathematical Analysis and Applications (Part B), Vol. 7, 1981.

[21] G. Pisier: Conditions d'entropie et caracterisations arithmetiques des ensembles de Sidon, preprint.

[22] G. Pisier: Some applications of the complex method of interpolation to Banach lattices, Journal d'Analyse Math. de Jerusalem 35 (1979), 264-291.

[23] G. Pisier: Holomorphic semi-groups and the geometry of Banach spaces, Annals of Maths. 1982.

[24] G. Pisier: Semi-groupes holomorphes et K-convexité, Séminaire d'Analyse Fonct., Ecole Polytechnique, Exp. 7, 1980-81.

[25] G. Pisier: Remarques sur un résultat non publié de B. Maurey, Seminaire d'Analyse Fonct., Ecole Polytechnique, Exp 5, 1980-81.

[26] G. Pisier: Ensembles de Sidon et espaces de cotype 2, Séminaire géometrie des espaces de Banach, Ecole Polytechnique, Exp. 14, 1977-78.

[27] G. Pisier: Entimations des distances a un espace Euclidien et des constantes de projection des espaces de Banach de dimension finie, Séminaire d'Anal. Fonct., Ecole Polytechnique, Exp. 10, 1978-79.

[28] M. Malliavin-Brameret, P. Malliavin: Caractérisation arithmétique des ensembles de Helson, C.R.A.Sc. Paris, Sér. A, 264 (1967), 192-193.

[29] T. Ramsay: Interpolation sets for almost periodic functions in bounded groups, preprint.

[30] N. Tomczak-Jaegermann: Computing 2 summing norms with few vectors, Arkiv Mat. 17 (1979), 273-279.

[31] N. T. Varopoulos: Sous espaces de C(G) invariant par translation et de type ^{1}, Séminaire Maurey-Schwartz, Exp. 12, 1975-76.

[32] N. T. Varopoulos: Tensor algebras and harmonic analysis, Acta Math., 119, 51-112(1968).

[33] J. Lindenstrauss, L. Tzafriri: Classical Banach Spaces, Springer LNM, 338 (1973).

REPRODUCING KERNEL HILBERT SPACE
FOR SOME NON-GAUSSIAN PROCESSES

by

Murad S. Taqqu and Claudia Czado

School of Operations Research
Cornell University
Ithaca, NY 14853

Keywords: Functional law of the iterated logarithm, multiple Wiener - Itô integrals, self-similar processes, Hermite processes, fractional Brownian motion.

This research was supported by the National Science Foundation grant ECS-84-08524 at Cornell University.

1. Introduction

A sequence $\{f_n(t,\omega),\ n \geq 1\}$ of $C[0,1]$-valued random functions obeys a functional law of the iterated logarithm if there exists a subset K of $C[0,1]$ such that

$$\begin{cases} \lim\limits_{n\to\infty} d(f_n,K) = 0 \quad \text{a.s.} \\ C\{f_n,\ n \geq 1\} = K \quad \text{a.s.} \end{cases}$$

Here d is the sup-norm distance and $C\{X_n,\ n \geq 1\}$ is the set of limit points (cluster set) of f_n.

If f_n is Gaussian, it is often possible to characterize K in two equivalent ways. One can specify K directly in terms of the functions that it contains. One can also characterize it as the unit ball of a suitable reproducing kernel Hilbert space (RKHS), denoted $\mathcal{H}(\Gamma)$, with kernel Γ. The kernel Γ is typically the covariance kernel of a Gaussian process to which f_n, adequately normalized, converges weakly in $C[0,1]$.

For example, let $Y_n(t) = 1/\sqrt{n}\, B(nt)$ where $B(t), t \geq 0$ is standard Brownian motion. The weak limit of $Y_n(t)$ is $B(t)$ whose covariance kernel is $\Gamma(s,t) = \min(s,t)$. If $||\dot{z}||_2^2 = \int_0^1 (dz/dt)^2 dt$, then Strassen's functional law of the iterated logarithm states that the set K of all limit points of

$$f_n(t) = \frac{1}{(2 \log \log n)^{1/2}} \frac{B(nt)}{\sqrt{n}}$$

is a.s.

(1) $K = \{z \in C[0,1]: z(0) = 0$, z absolutely continuous,

$$||\dot{z}||_2^2 \leq 1\}$$

or, equivalently,

(2) K = unit ball of the RKHS $\mathcal{H}(\Gamma)$ with $\Gamma(s,t) = \min(s,t)$.

(See Jain and Marcus (1978), p. 121).

Consider now a sequence $Y_n(t)$ that converges weakly to a <u>non-Gaussian</u> process with a covariance kernel Γ, and suppose that a functional law of the iterated logarithm also holds. Is the set of limit points a.s. identical to the unit ball of the RKHS $\mathcal{H}(\Gamma)$? The answer in general is no. We illustrate this by focusing on a class of self-similar processes given by multiple Wiener - Itô integrals. (The Hermite processes, considered in Section 4, are examples of such processes).

Recall that a process $Y(t)$ is self-similar with index $H > 0$ if $Y(at)$ and $a^H Y(t)$ have identical finite-dimensional distributions. The reader may refer to Taqqu and Czado (1985) for a recent survey on functional laws of the iterated logarithm for self-similar processes.

2. RKHS of multiple Wiener - Itô integrals.

Let $Y^{(m)}(t) \in C[0,1]$ be such that

$$Y^{(m)}(t) = \frac{1}{\sqrt{m!}} \int_{\mathbb{R}^m}' k_t^{(m)}(x_1,\cdots,x_m)\, dB(x_1)\cdots dB(x_m). \qquad (2.1)$$

Here $k_t^{(m)} \in L^2(\mathbb{R}^m)$ for each t, and $\int_{\mathbb{R}^m}'$ means that this is a multiple Wiener - Itô integral of order m. (For a definition see Itô (1951) or the Appendix of Mandelbaum and Taqqu (1984).) Heuristically, the notation $\int_{\mathbb{R}^m}'$ indicates that no integration is performed on the hyperdiagonals of $\mathbb{R}^m$.

The process $Y^{(m)}(t)$ has mean 0 and covariance kernel

$$\begin{aligned}\Gamma^{(m)}(s,t) &= EY^{(m)}(s)Y^{(m)}(t)\\ &= \int_{\mathbb{R}^m} k_s(x_1,\cdots,x_m)k_t(x_1,\cdots,x_m)\, dx_1\cdots dx_m. \qquad (2.2)\end{aligned}$$

No extraneous coefficient appears in (2.2) because of the presence of the coefficient $1/\sqrt{m!}$ in (2.1). Note that $Y^{(m)}(t)$ is Gaussian when $m = 1$ and non-Gaussian when $m > 1$.

Now introduce the Hilbert space

$$\mathcal{H}^{(m)} = \{y(t) = \int_{\mathbb{R}^m} k_t^{(m)}(x_1,\cdots,x_m)g(x_1,\cdots,x_m)dx_1\cdots dx_m : \\ g \in L^2(\mathbb{R}^m)\} \tag{2.3}$$

with inner product

$$< y_1,y_2 >_m = \int_{\mathbb{R}^m} g_1(x_1,\cdots,x_m)g_2(x_1,\cdots,x_m)dx_1\cdots dx_m.$$

We say that $y_1 \in \mathcal{H}^{(m)}$ is represented by $g_1 \in L^2(\mathbb{R}^m)$ and that $y_2 \in \mathcal{H}^{(m)}$ is represented by $g_2 \in L^2(\mathbb{R}^m)$. Note the following:

1) The functions $\Gamma^{(m)}(\cdot,t)$ belong to $\mathcal{H}^{(m)}$ for each $0 \le t \le 1$ because, in view of (2.2), $\Gamma^{(m)}(\cdot,t)$ is represented by $k_t^{(m)} \in L^2(\mathbb{R}^m)$.

2) For $y \in \mathcal{H}^{(m)}$,

$$< y,\Gamma^{(m)}(\cdot,t) >_m = \int_{\mathbb{R}^m} g(x_1,\cdots,x_m)k_t^{(m)}(x_1,\cdots,x_m)dx_1\cdots dx_m \\ = y(t),$$

where $g \in L^2(\mathbb{R}^m)$ is the function that represents y.

In view of 1) and 2), we can apply Theorem 2.2 of Marcus and

Jain (1978) and obtain

Proposition 1. The Hilbert space $\mathcal{H}^{(m)}$ defined by (2.3) with inner product $\langle\ ,\ \rangle_m$, can be identified with the RKHS $\mathcal{H}(\Gamma^{(m)})$ whose kernel $\Gamma^{(m)}$ is given by (2.2).

The following examples illustrate the case $m = 1$. (Interesting cases with $m > 1$ are treated in the next section).

Examples. 1) Suppose $m = 1$ and $k_t^{(1)}(x) = 1_{[0,t]}(x)$. Then $Y^{(1)}(t)$ is the Brownian motion $B(t)$ and

$$\mathcal{H}^{(1)} = \{y(t) = \int_{-\infty}^{\infty} 1_{[0,t]}(x) g(x) dx: \ g \in L^2(\mathbb{R}^1)\}.$$

Changing parameterization by setting $g(x) = \dot{z}(x) \equiv dz/dx$, one gets

$$\mathcal{H}^{(1)} = \{y(t) = \int_{-\infty}^{\infty} 1_{[0,t]}(x) \dot{z}(x) dx: \ \dot{z} \in L^2(\mathbb{R}^1)\}$$

$$= \{z(t): \ z \text{ absolutely continuous}, \ z(0) = 0, \ \dot{z} \in L^2(\mathbb{R}^1)\},$$

which is the RKHS for Brownian motion mentioned in the Introduction.

2) If $m = 1$ and $k_t^{(1)}(x) = e^{-\lambda(t-x)} 1_{\{x < t\}}$ with $\lambda > 0$, then $Y^{(1)}(t)$ is the Ornstein - Uhlenbeck process and its RKHS is

$$\mathcal{H}^{(1)} = \{y(t) = \int_{-\infty}^{t} e^{-\lambda(t-x)} g(x)dx \colon g \in L^2(\mathbb{R}^1)\}.$$

3. <u>Self-similar processes</u>

We now introduce integrands $k_t^{(m)}$ in (2.1) such that the corresponding random processes $Y^{(m)}(t)$ have covariance kernels <u>that do not depend on the order m of the multiple integral</u>. For $t \geq 0$, let

$$k_t^{(m)}(x_1,\cdots,x_m) = \int_0^t q^{(m)}(v-x_1,\cdots,v-x_m)dv \qquad (3.1)$$

where $q\colon \mathbb{R}^m \to \mathbb{R}$ is such that for all $c > 0$,

$$q^{(m)}(cx_1,\cdots,cx_m) = c^{-\lambda} q(x_1,\cdots,x_m) \qquad (3.2)$$

with $\lambda = (m/2) + 1 - H$ and $1/2 < H < 1$. Moreover, we require that

$$\int_{\mathbb{R}^m} [k_1^{(m)}(x_1,\cdots,x_m)]^2 dx_1 \cdots dx_m = 1. \qquad (3.3)$$

These conditions on $k_t^{(m)}, t \geq 0$ ensure that $Y^{(m)}(t)$ defined by (2.1) has stationary increments, is self-similar with index H, has mean zero and variance t^{2H}. We shall take $Y^{(m)}(t)$ to be the version of the process which belongs to $C[0, \infty)$. Such a version exists because $E|Y^{(m)}(s) - Y^{(m)}(t)|^2 = |s - t|^{2H}$ with $2H > 1$. The covariance kernel of $Y^{(m)}(t)$ is

$$\Gamma^{(m)}(s,t) = EY^{(m)}(s)Y^{(m)}(t)$$

$$= \frac{1}{2}\{E[Y^{(m)}(s)]^2 + E[Y^{(m)}(t)]^2 - E[Y^{(m)}(s)-Y^{(m)}(t)]^2\}$$

$$= \frac{1}{2}\{s^{2H} + t^{2H} - |s-t|^{2H}\},$$

and therefore, the covariance kernel does not depend on m. In view of Proposition 1, we get the following result.

<u>Proposition 2</u>. Suppose that $Y^{(m)}(t)$ is defined as (2.1) with $k^{(m)}$ given by (3.1)-(3.3). Then the Hilbert spaces $(\mathcal{H}^m, <\ ,\ >_m)$, $m = 1,2,\cdots$ are all identical to the RKHS $\mathcal{H}(\Gamma)$ with kernel

$$\Gamma(s,t) = \frac{1}{2}\{s^{2H} + t^{2H} - |s - t|^{2H}\}. \tag{3.4}$$

We shall now focus on the process $Y^{(m)}(t)$ defined in (2.1) with kernel $k^{(m)}$ given by (3.1)-(3.3). From now on we let Γ be the kernel (3.4), $\mathcal{H}(\Gamma)$ be the corresponding RKHS, and U be the unit ball of $\mathcal{H}(\Gamma)$.

The sequence $Y^{(m)}(nt)/n^H$ converges weakly, as $n \to \infty$, to $Y^{(m)}(t)$ in $C[0,1]$ because the finite-dimensional distributions of $Y^{(m)}(nt)/n^H$ are identical to those of $Y^{(m)}(t)$ and tightness follows from $E|Y^{(m)}(ns)/n^H - Y^{(m)}(nt)/n^H|^2 =$ $E|Y^{(m)}(s) - Y^{(m)}(t)|^2 = |s - t|^{2H}$ with $2H > 1$. Fox (1981) showed that the limit points of

$$f_n^{(m)}(t) = \frac{Y^{(m)}(nt)}{n^H(2 \log \log n)^{m/2}}$$

are a.s. contained in the unit ball U of $\mathcal{H}(\Gamma)$. Mori and Oodaira (1984), extending a result proved by Taqqu (1977) when $m = 1$, show that $f_n^{(m)}(t)$ satisfies a functional law of the iterated logarithm for all $m \geq 1$, and that the set of its limit points as $n \to \infty$ is a.s.

$$K^{(m)} = \{y(t) = \int_{\mathbb{R}^m} k_t^{(m)}(x_1,\cdots,x_m)g(x_1)\cdots g(x_m)dx_1\cdots dx_m:$$

$$||g||_2^2 \leq 1\}, \tag{3.5}$$

where $||g||_2^2 = \int_{-\infty}^{+\infty} g^2(x)dx$. The set $K^{(m)}$ is clearly a subset of the unit ball of $\mathcal{H}^{(m)}$ since

$$||y||^2_{\mathcal{H}^{(m)}} = \int_{\mathbb{R}^m} g^2(x_1)\cdots g^2(x_m)dx_1\cdots dx_m = (||g||_2^2)^m \leq 1.$$

In view of Proposition 2, we get

<u>Proposition 3</u>. When $m \geq 2$, $K^{(m)}$ is a proper subset of the unit ball U of the RKHS $\mathcal{H}(\Gamma)$. The set $K^{(1)}$ coincides with U.

The next proposition states that the set $K^{(m)}$ is determined by a subset of the unit ball of $L^2(\mathbb{R}^1)$.

Proposition 4. Any $y \in K^{(m)}$ can be represented as

$$y(t) = \int_{-\infty}^{\infty} k_t^{(1)}(x) g_m(x)\, dx \tag{3.6}$$

where g_m is a suitable function in $L^2(\mathbb{R}^1)$ satisfying $||g_m||_2^2 \leq 1$.

Indeed, by Proposition 3, any $y \in K^{(m)}$ can be viewed as a function in $K^{(1)}$. Thus, there must exist a function $g_m \in L^2(\mathbb{R}^1)$ such that

$$\begin{aligned} y(t) &= \int_{\mathbb{R}^m} k_t^{(m)}(x_1,\cdots,x_m) g(x_1),\cdots g(x_m)\, dx_1\cdots dx_m \\ &= \int_{-\infty}^{\infty} k_t^{(1)}(x) g_m(x)\, dx. \end{aligned} \tag{3.7}$$

Since $||y(t)||^2_{\mathcal{H}^{(1)}} = ||y(t)||^2_{\mathcal{H}^{(m)}} \leq 1$, we must have $||g_m||_2^2 = ||g||_2^{2m} \leq 1$.

In principle, the explicit form of g_m can be obtained by solving the functional equation (3.7) in terms of g.

4. The case of the Hermite processes

The functional equation (3.7) simplifies when

$$k_t^{(m)}(x_1,\cdots,x_m) = \frac{1}{A_m}\int_0^t \prod_{i=1}^{m} ((s - x_i)^+)^{-\alpha_m} ds, \tag{4.1}$$

where

$$\frac{1}{2} < \alpha_m < \frac{1}{2} + \frac{1}{2m}$$

and where

$$A_m^2 = \int_{\mathbb{R}^m} \{\int_0^1 \prod_{i=1}^m ((s - x_i)^+)^{-\alpha_m} ds\}^2 dx_1 \cdots dx_m$$

$$= \int_0^1\int_0^1 \{\int_{-\infty}^{s_1 \wedge s_2} (s_1 - x)^{-\alpha_m} (s_2 - x)^{-\alpha_m} dx\}^m ds_1 ds_2$$

$$= \{\int_0^\infty (x + x^2)^{-\alpha_m} dx\}^m \int_0^1 \int_0^1 |s_1 - s_2|^{-2m\alpha_m + m} ds_1 ds_2$$

$$= \{\frac{\Gamma(1 - \alpha_m)\Gamma(2\alpha_m - 1)}{\Gamma(\alpha_m)}\}^m \frac{2}{(1 + m - 2m\alpha_m)(2 + m - 2m\alpha_m)} .$$

This choice of A_m ensures that $||k_1^{(m)}||^2 = 1$. Hence $k_t^{(m)}$ satisfies the Assumptions (3.1)-(3.3). The corresponding process $Y^{(m)}(t)$, defined by (2.1), is

$$Y^{(m)}(t) = \frac{1}{\sqrt{m!}A_m} \int_{\mathbb{R}^m}' \{ \int_0^t \prod_{i=1}^m ((s - x_i)^+)^{-\alpha_m} ds\} dB(x_1) \cdots dB(x_m)$$

$$= \frac{\sqrt{m!}}{A_m} \int_{-\infty}^t dB(x_1) \int_{-\infty}^{x_1} dB(x_2) \cdots \int_{-\infty}^{x_{m-1}} dB(x_m)$$

$$\int_0^t \prod_{i=1}^m ((s - x_i)^+)^{-\alpha_m} ds.$$

It is known as the normalized <u>Hermite process</u> of order m. The process $Y^{(1)}(t)$ is the Gaussian <u>fractional Brownian motion</u> and the process $Y^{(2)}(t)$ is called the <u>Rosenblatt process</u> (see Taqqu (1985)).

When $k_t^{(m)}$ is given by (4.1), the functional equation (3.7) becomes

$$\frac{1}{A_m}\int_{\mathbb{R}^m}\Big[\int_0^t \prod_{i=1}^m ((s-x_i)^+)^{-\alpha_m} ds\Big] g(x_1)\cdots g(x_m)\,dx_1\cdots dx_m$$

$$= \frac{1}{A_1}\int_{\mathbb{R}}\Big[\int_0^t ((s-x)^+)^{-\alpha_1} ds\Big] g_m(x)\,dx. \tag{4.2}$$

Note that the integral $\int_{-\infty}^s (s-x)^{-\alpha_m} g(x)\,dx$ converges for $g \in L^2(\mathbb{R}^1)$ for almost every s. Indeed, fix $T < s < 2T$, and express the integral as $I_1(s) + I_2(s)$ where

$$I_1(s) = \int_{-\infty}^T (s-x)^{-\alpha_m} g(x)\,dx \le \Big(\int_{-\infty}^T (s-x)^{-2\alpha_m} dx\Big)^{1/2} ||g||_2^2 < \infty$$

and where $I_2(s) = \int_T^s (s-x)^{-\alpha_m} g(x)\,dx$. Then

$$\int_T^{2T} |I_2(s)|\,ds \le \int_T^{2T} |g(x)|\Big\{\int_x^{2T} (s-x)^{-\alpha_m} ds\Big\} dx < \infty.$$

Therefore (4.2) becomes

$$\frac{1}{A_m}\int_0^t\Big[\int_{-\infty}^s (s-x)^{-\alpha_m} g(x)\,dx\Big]^m ds = \frac{1}{A_1}\int_0^t\Big[\int_{-\infty}^s (s-x)^{-\alpha_1} g_m(x)\,dx\Big] ds$$

or simply,

$$\left[\int_{-\infty}^{s}(s-x)^{-\alpha_m}g(x)\,dx\right]^m = \frac{A_m}{A_1}\int_{-\infty}^{s}(s-x)^{-\alpha_1}g_m(x)\,dx.$$

This is a functional equation involving fractional integration. Formally, if

$$(I^{1-\alpha}g)(s) = \frac{1}{\Gamma(1-\alpha)}\int_{-\infty}^{s}(s-x)^{-\alpha}g(x)\,dx,$$

and if, $I^{\alpha}(I^{-\alpha}g) = g$, then

$$g_m = \frac{A_1}{A_m}\,\frac{(\Gamma(1-\alpha_m))^m}{\Gamma(1-\alpha_1)}\, I^{\alpha_1-1}\,[I^{1-\alpha_m}g]^m.$$

Observe that by setting this expression for $g_m(x)$ in (3.6) one obviously recovers the set of limit points

$$K_m = \{y(t) = \frac{1}{A_m}\int_0^t ds\left[\int_{-\infty}^{s}(s-x)^{-\alpha_m}g(x)\,dx\right]^m : ||g||_2^2 \le 1\}$$

given in (3.5).

References

[1] R. Fox (1981). "Upper functional laws of the iterated logarithm for non-Gaussian self-similar processes." Technical Report No. 509, School of Operations Research and Industrial Engineering, Cornell University.

[2] K. Itô (1951). "Multiple Wiener integral." J. Math. Soc.

Japan 3, 157-169.

[3] N.C. Jain and M.B. Marcus (1978). "Continuity of subgaussian processes." In: Advances in Probability and Related Topics 4, 84-197. New York: Dekker.

[4] A. Mandelbaum and M.S. Taqqu (1984). "Invariance principles for symmetric statistics." Ann. Stat. 12, 483-496.

[5] T. Mori and H. Oodaira (1984). "The law of the iterated logarithm for self-similar processes represented by multiple Wiener integrals." Preprint.

[6] M.S. Taqqu (1977). "Law of the iterated logarithm for sums of non-linear functions of Gaussian variables that exhibit a long range dependence." Z. Wahrscheinlichkeitstheorie und Verw. Geb. 40, 203-238.

[7] M.S. Taqqu (1985). "Self-similar processes." To appear in: Encyclopedia of Statistical Sciences, New York: Wiley.

[8] M.S. Taqqu and C.Czado (1985). "A survey of functional laws of the iterated logarithm for self-similar process." To appear in Stochastic Models.

An extended Wichura theorem, definitions of Donsker class, and weighted empirical distributions

by R. M. Dudley[1]

Abstract. Let $(X, \mathcal{A}, P)$ be a probability space, $x(1), x(2), \cdots$ coordinates for the countable product of copies of X, and $\mathcal{F} \subset \mathcal{L}^2(X, \mathcal{A}, P)$. Let ν_n be the normalized empirical measure $n^{-1/2}(\delta_{x(1)} + \cdots + \delta_{x(n)} - nP)$. Let $\ell^\infty(\mathcal{F})$ be the set of all bounded real functions G on $\mathcal{F}$ with norm $\|G\|_{\mathcal{F}} := \sup_{f \in \mathcal{F}} |G(f)|$. Let $\mu(f) := \int f d\mu$ for any measure μ. We are interested in central limit theorems where ν_n converges in law for $\|\cdot\|_{\mathcal{F}}$. The limit is a Gaussian process G_P with mean 0 and covariance

$$EG_P(f)G_P(g) = P(fg) - P(f)P(g) := (f,g)_{P,0},$$

$f, g \in \mathcal{F}$. Say $\mathcal{F}$ is G_PBUC if G_P can be chosen to have each sample function bounded and uniformly continuous for $(\cdot,\cdot)_{P,0}$. Say the central limit theorem holds for $\mathcal{F}$ if $\mathcal{F}$ is G_PBUC and we have convergence of upper integrals $\lim_{n\to\infty} \int^* H(\nu_n) dP^n = EH(G_P)$ for every bounded continuous real function H on $\ell^\infty(\mathcal{F})$. (Thus, as noted by J. Hoffmann-Jørgensen, convergence "in law" does not require definition of laws.) In Sec. 4 an extended Wichura's theorem is proved: given such convergence in law, there exist almost surely convergent realizations. Sec. 5 shows that the new definition of central limit theorem holding is equivalent to the previous definition of "functional Donsker class." Secs. 6-7 treat the case where $\mathcal{F} = \{M1_{h \geq M}: 0 < M < \infty\}$ for a random variable h. This case reduces to the study of weighted empirical distribution functions. Conditions for the central limit theorem are collected and made precise.

1. **Partially supported by National Science Foundation Grant MCS-8202122**

1. Introduction. Let (S,d) be a metric space. If (S,d) is separable and complete, Skorohod (1956) shows that if Y_n are S-valued random variables, $n = 0,1,\cdots$, which converge in law to Y_0 as $n \to \infty$, then on some probability space there exist S-valued random variables X_n such that for each n, X_n and Y_n have the same distribution, and $X_n \to X_0$ almost surely. The author (1968) removed the completeness assumption. M. Wichura (1970) and P. J. Fernandez (1974) proved an extension of the theorem where S may be non-separable, only Y_0 is required to have separable range, and the Y_n are measurable with respect to the σ-algebra $\mathcal{B}_b$ generated by all balls $\{y: d(x,y) < r\}$, $x \in S, r > 0$. Here convergence in law is defined as convergence of upper integrals

$$\int^* F(Y_n)dP \longrightarrow \int F(Y_0)dP$$

for every bounded continuous real F on S (Dudley, 1966). Also, Wichura replaced almost sure convergence by almost uniform convergence (see below), which is stronger and more useful in the non-separable case. Hoffmann-Jørgensen (1984) defined convergence of random elements $Y_n \longrightarrow Y_0$ in law as above without requiring laws of Y_n to be defined on any non-trivial σ-algebra. One main result of the paper, stated and proved in Sec. 4, is that, essentially, Wichura's theorem can be extended to hold under Hoffmann's definition. The extension uses the notion of "perfect" function (Sec. 2 below), also due to Hoffmann-Jørgensen.

Let $(X, \mathcal{A}, P)$ be a probability space and f any real-valued function on X (not measurable, in general). Let

$$\begin{aligned}\int^* f dP &:= \inf\{\textstyle\int h dP: h \geq f, \ h \text{ measurable}\}\\ &= \textstyle\int f^* dP, \text{ where}\\ f^* &:= \text{ess. inf}\{g: g \geq f, \ g \text{ measurable}\} \geq f\end{aligned}$$

(Dudley and Philipp, 1983, Lemma 2.1). Also let $P^*(B) := \int^* 1_B dP$ for any set $B \subset X$ and

$$\int_* f dP := -\int^* -f dP = \int f_* dP$$
$$= \sup\{\int g dP\colon g \le f,\ g \text{ measurable}\},$$

where $f_* := -((-f)^*)$.

Let $(S,\mathcal{T})$ be any topological space and β_n probability measures defined on σ-algebras $\mathcal{B}_n$ of subsets of S (not necessarily related to $\mathcal{T}$ for $n > 0$), $n = 0,1,\cdots$, $\beta := \beta_0$.

In 1966 I defined

"$\beta_n \to \beta$ (weak*) if for every bounded continuous real function F on S, $\lim_{n\to\infty} \int^* F d\beta_n = \lim_{n\to\infty} \int_* F d\beta_n = \int F d\beta$."

Results were obtained for this convergence mainly in case $\mathcal{T}$ is metrizable by a metric d and each $\mathcal{B}_n$ includes $\mathcal{B}_b$ (Dudley 1966, 1967a, 1978; Wichura 1970; Fernandez 1974). The following will help to elucidate the notion of almost uniform convergence. It is not claimed as new, but a proof will be given for completeness.

1.1 Proposition. Let $(\Omega,\mathcal{S},P)$ be a probability space, (S,d) a metric space, and X_n any functions from Ω into S, $n = 0,1,\cdots$. Then the following are equivalent:

A) $d(X_n,X_0)^* \to 0$ almost surely;

B) for any $\varepsilon > 0$,
$P^*\{\sup_{n\ge m} d(X_n,X_0) > \varepsilon\} \downarrow 0$ as $m \to \infty$;

C) For each $\delta > 0$, there is some $B \in \mathcal{S}$ with $P(B) > 1-\delta$ such that $X_n \to X_0$ uniformly on B.

D) There exist measurable $h_n \ge d(X_n,X_0)$ with $h_n \to 0$ a.s.

Proof. A) implies that

$$(\sup_{n\ge m} d(X_n,X_0))^* \le \sup_{n\ge m}(d(X_n,X_0)^*) \downarrow 0$$

a.s. as $m \to \infty$, which implies B).

Assuming B), for $k = 1,2,\cdots$ take $m(k)$ such that $P^*(\sup_{n\ge m(k)} d(X_n,X_0) > 1/k) < 2^{-k}$. Take measurable covers B_k,

$$B_k \supset \{\sup_{n\ge m(k)} d(X_n,X_0) > 1/k\}$$

with $B_k \in \mathcal{S}$, $P(B_k) < 2^{-k}$. For $r = 1,2,\cdots$, let $A_r := \Omega \setminus \bigcup_{k>r} B_k$. Then $P(A_r) > 1 - 2^{-r}$ and $X_n \to X_0$ uniformly on A_r, so C) holds.

Now assume C). Take $C_k \in \mathcal{S}$, $k = 1,2,\cdots$, such that $P(C_k) \uparrow 1$ and $X_n \to X_0$ uniformly on C_k. We can take $C_1 \subset C_2 \subset \cdots$. Take m_k such that $d(X_n,X_0) < 1/k$ on C_k for all $n \geq m_k$. Then $d(X_n,X_0)^* \leq 1/k$ on C_k, so $d(X_n,X_0)^* \to C$ a.s., giving A). Clearly, A) and D) are equivalent, Q.E.D.

A sequence X_n satisfying, say, A) in Prop. 1.1 will be said to converge <u>almost uniformly</u> to X_0. This agrees with the definition of Halmos (1950, p.89) for functions such that $d(X_n,X_0)$ is measurable. Then, almost uniform convergence is equivalent to almost sure convergence.

On the other hand, in $[0,1]$ with P = Lebesgue measure let $A(1) \supset A(2) \supset \cdots$ be sets with $P^*(A(n)) = 1$ for all n and $\bigcap_{n=1}^{\infty} A(n) = \emptyset$ (e.g. Cohn, 1980, p.35). Then $1_{A(n)} \to 0$ everywhere and, in that sense, almost surely, but not almost uniformly. To avoid such pathology it will be useful to obtain almost uniform convergence.

Wichura (1970) proved (results stronger than) the following:

<u>Theorem A</u>. Let (S,d) be any metric space and μ_n probability measures on $\mathcal{B}_b$, $n = 0,1,\cdots$. Suppose $\mu_0(S_0) = 1$ for some separable $S_0 \subset S$. If $\mu_n \to \mu_0$ (weak*) then there exist a probability space $(\Omega, \mathcal{S}, P)$ and functions X_n from Ω to S, measurable from $\mathcal{S}$ to $\mathcal{B}_b$, such that $P \circ X_n^{-1} = \mu_n$ for each n and $X_n \to X_0$ almost uniformly.

The measurability for $\mathcal{B}_b$ does not always hold in cases of interest. To obtain it may require replacing S by a suitable subset and/or giving a non-trivial proof. To avoid these difficulties we have the following, due to J. Hoffmann-Jørgensen (1984).

<u>Definition</u>. Let $(S,\mathcal{T})$ be any topological space, $(X_\alpha, \mathcal{A}_\alpha, P_\alpha)_{\alpha \in J}$

probability spaces, where J is a directed index set, $0 \in J$, and f_α are any functions from X_α into S. Say that $f_\alpha \xrightarrow[\mathcal{L}]{} f_0$, or $f_\alpha \to f_0$ in law, iff for every bounded continuous real-valued function G on S, $\int G(f_0)dP_0$ is defined and

$$\int^* G(f_\alpha)dP_\alpha \longrightarrow \int G(f_0)dP_0$$

as $\alpha \to \infty$.

Taking $-G$, we will also have

$$\int_* G(f_\alpha)dP_\alpha \longrightarrow \int G(f_0)dP_0 .$$

If we let $\mathcal{B}_\alpha := \{C \subset S: f_\alpha^{-1}(C) \in \mathcal{Q}_\alpha\}$, a σ-algebra, and set $\beta_\alpha := P \circ f_\alpha^{-1}$ on $\mathcal{B}_\alpha$, Hoffmann's definition is not equivalent to $\beta_\alpha \to \beta_0$ (weak*), as the following shows.

1.2 Example. Let $(X_n, \mathcal{Q}_n, Q_n) = ([0,1], \mathcal{B}, \lambda)$ for all n (λ = Lebesgue measure, $\mathcal{B}$ = Borel σ-algebra). Take sets $C(n) \subset [0,1]$ with $0 = \lambda_*(C(n)) < \lambda^*(C(n)) = 1/n^2$ (Halmos, 1950, p.70). Let S be the two-point space $\{0,1\}$ with usual metric. Then $f_n := 1_{C(n)} \to 0$ in law and almost uniformly, but each $\beta_n := Q_n \circ f_n^{-1}$, $n > 0$, is only defined on the trivial σ-algebra $\{\emptyset, S\}$ and $\beta_n \not\longrightarrow \delta_0$ (weak*). The σ-algebra $\mathcal{B}_b$ generated by balls in this case is 2^S, the only σ-algebra larger than $\{\emptyset, S\}$, but no β_n for $n > 0$ is defined on 2^S.

1.3 Remark. To show that $f_\alpha \xrightarrow[\mathcal{L}]{} f_0$ it is enough to show that

$$\limsup_{\alpha\to\infty} \int^* G(f_\alpha)dP_\alpha \le \int G(f_0)dP_0$$

for every bounded continuous G. For then

$$\begin{aligned} &\limsup \int^* -G(f_\alpha)dP_\alpha \le \int -G(f_0)dP_0, \text{ so} \\ &\int G(f_0)dP_0 \ge \limsup \int^* G(f_\alpha)dP_\alpha \\ &\ge \liminf \int^* G(f_\alpha)dP_\alpha \ge \liminf \int_* G(f_\alpha)dP_\alpha \\ &\ge \int G(f_0)dP_0, \end{aligned}$$

so these terms are all equal.

2. Perfect functions. It will be useful that under some conditions on a measurable function g and general real-valued f, $(f \circ g)^* = f^* \circ g$. Here are some conditions, first found by Hoffmann-Jørgensen and Andersen (1984), given here for completeness.

2.1 Theorem. Let $(X, \mathcal{A}, P)$ be a probability space, $(Y, \mathcal{B})$ any measurable space, and g a measurable function from X to Y. Let Q be the restriction of $P \circ g^{-1}$ to $\mathcal{B}$. For any real-valued function f on Y, define f^* for Q. Then the following are equivalent:

1) for any $A \in \mathcal{A}$ there is a $B \in \mathcal{B}$ with $B \subset g(A)$ and $Q(B) \geq P(A)$;

2) for any $A \in \mathcal{A}$ with $P(A) > 0$ there is a $B \in \mathcal{B}$ with $B \subset g(A)$ and $Q(B) > 0$;

3) for any real function f on Y, $(f \circ g)^* = f^* \circ g$ a.s.;

4) for any $D \subset Y$, $(1_D \circ g)^* = 1_D^* \circ g$ a.s.

Proof. 1) implies 2), clearly.

2) implies 3): note that always $(f \circ g)^* \leq f^* \circ g$. Suppose $(f \circ g)^* < f^* \circ g$ on a set of positive probability. Then for some rational r, $(f \circ g)^* < r < f^* \circ g$ on a set $A \in \mathcal{A}$ with $P(A) > 0$. Let $g(A) \supset B \in \mathcal{B}$ with $Q(B) > 0$. Then $f \circ g < r$ on A implies $f < r$ on B, so $f^* \leq r$ on B a.s., contradicting $f^* \circ g > r$ on A.

3) implies 4), clearly.

4) implies 1): given $A \in \mathcal{A}$, let $D := Y \setminus g(A)$. Then we can take $1_D^* = 1_C$ for some $C \in \mathcal{B}$ ($C = \{1_D^* \geq 1\}$), $D \subset C$, and $1_D \circ g = (1_D \circ g)^* = 0$ a.s. on A. Let $B := Y \setminus C$. Then $B \subset g(A)$, and

$$Q(B) = 1 - Q(C) = 1 - \int 1_D^* d(P \circ g^{-1})$$

$$= 1 - \int 1_D^* \circ g dP = 1 - \int (1_D \circ g)^* dP \geq P(A), \quad \text{Q.E.D.}$$

Following Hoffmann-Jørgensen and Andersen (1984), a function g satisfying any of the four conditions in 2.1 is called <u>perfect</u> or P-<u>perfect</u>.

Call g <u>quasiperfect</u> or P-<u>quasiperfect</u> iff for every $C \subset Y$ with $g^{-1}(C) \in \mathcal{A}$, C is Q-completion measurable. Then $(X, \mathcal{A}, P)$ is called <u>perfect</u> iff every real-valued function g on X, measurable (for the usual Borel σ-algebra on $\mathbb{R}$) is quasiperfect.

2.2 <u>Example</u>. A measurable, quasiperfect function g on a finite set X need not be perfect: let $X = \{a,b,c,d,e,j\}$, $U := \{a,b\}$, $V := \{c,d\}$, $W := \{e,j\}$, $\mathcal{A} := \{\varphi, U, V, W, U \cup V, U \cup W, V \cup W, X\}$, $P(U) = P(V) = P(W) = 1/3$, $Y := \{0,1,2\}$, $g(a) = g(c) = 0$, $g(b) = g(e) = 1$, $g(d) = g(j) = 2$. Let $\mathcal{B} = \{\varphi, Y\}$. For $C \subset Y$, $g^{-1}(C) \in \mathcal{A}$ iff $C \in \mathcal{B}$, so g is quasiperfect. But, $P(U) > 0$ and $g(U)$ does not include any non-empty set in $\mathcal{B}$, so g is not perfect.

2.3 <u>Proposition</u>. Any perfect function g (as in 2.1) is quasiperfect.

<u>Proof</u>. Let $C \subset Y$, $A := g^{-1}(C) \in \mathcal{A}$. By Theorem 2.1 take $B \subset g(A)$ with $B \in \mathcal{B}$, $Q(B) \geq P(A)$. Then $B \subset C$, so $Q(B) = P(g^{-1}B) \leq P(g^{-1}C) = P(A)$, and $Q(B) = P(A)$. Thus the inner measure $Q_*(C) = (P \circ g^{-1})(C)$. Likewise, $Q_*(Y \setminus C) = (P \circ g^{-1})(Y \setminus C)$, so $Q^*(C) = (P \circ g^{-1})(C)$ and C is Q-completion measurable, Q.E.D.

2.4 <u>Example</u>. Let $C \subset [0,1]$ satisfy $0 = \lambda_*(C) < \lambda^*(C) = 1$ (cf. Example 1.2). Let $X = C$, $Y = [0,1]$ with Borel σ-algebra, and $P = \lambda^*$ giving a probability measure on the Borel sets of X. Let g be the identity from X into Y. Then as the range C of g is not Q-completion measurable, g is not quasiperfect, although it is a 1-1, Borel measurable function between metric spaces. If, instead, $X = [0,1]$ with the same P on C and $P(X \setminus C) = 0$, with g the identity, then g is onto $[0,1]$ but still not quasiperfect. (Such examples, if not the terminology, are well known.)

2.5 Proposition. Suppose $A = X \times Y$, P is a product law $v \times m$, and g is the natural projection of A onto Y. Then g is P-perfect.

Proof. Here $P \circ g^{-1} = m$. For any $B \subset A$ let $B_y := \{x: \langle x,y \rangle \in B\}$, $y \in Y$. If B is measurable, then by the Tonelli-Fubini theorem, for $C := \{y: v(B_y) > 0\}$, C is measurable, $C \subset g(B)$, and $P(B) \le m(C)$, Q.E.D.

3. Convergence in outer probability. Let $(X, \mathcal{A}, P)$ be any probability space, (S,d) a metric space, and f_n functions from X into S. Say that $f_n \to f_0$ in outer probability if $d(f_n, f_0)^* \to 0$ in probability as $n \to \infty$, or equivalently, for every $\varepsilon > 0$,

$$\lim_{n\to\infty} P^*(d(f_n, f_0) > \varepsilon) = 0.$$

We have, clearly:

3.1 Proposition. For any probability space $(X, \mathcal{A}, P)$, metric space (S,d), and functions f_n from X into S, $n = 0,1,\cdots$, almost uniform convergence $f_n \to f_0$ implies convergence in outer probability.

The example after Prop. 1.1 shows that $f_n \to f_0$ everywhere pointwise does not imply convergence in outer probability.

3.2 Proposition. Let (S,d) and (Y,e) be two metric spaces and $(X, \mathcal{A}, P)$ a probability space. Let f_n be functions from X into S, $n = 0,1,2,\cdots$, such that $f_n \to f_0$ in outer probability and f_0 has separable range and is Borel measurable. Let g be any continuous function from S into Y. Then $g(f_n) \to g(f_0)$ in outer probability.

Note. If $\int G(f)dP$ is defined for all bounded continuous real G (as it must be if $f_n \xrightarrow{\mathcal{L}} f$, by definition) then $P \circ f^{-1}$ is defined on all Borel subsets of S. Such a law does have a separable support except perhaps in some set-theoretically pathological cases (Marczewski and Sikorski, 1948). Indeed, it is consistent with the usual axioms of set theory (including the axiom of choice) that such pathology never arises (e.g. Drake, 1974, pp.67-68, 177-178). It is apparently unknown whether it can ever, consistently, arise (Drake, 1974, pp.185-186).

<u>Proof of Prop. 3.2</u>. Given $\varepsilon > 0$, $k = 1,2,\cdots$, let $B_k := \{x \in S\colon d(x,y) < 1/k$ implies $e(g(x),g(y)) \le \varepsilon$, $y \in S\}$. Then each B_k is closed and $B_k \uparrow S$ as $k \to \infty$. Fix k large enough so that $P(f_0^{-1}(B_k)) > 1 - \varepsilon$. Then

$$\{e(g(f_n),g(f_0)) > \varepsilon\} \cap f_0^{-1}(B_k) \subset \{d(f_n,f_0) \ge \tfrac{1}{k}\} .$$

Thus

$$P^*\{e(g(f_n),g(f_0)) > \varepsilon\} < \varepsilon + P^*\{d(f_n,f_0) \ge 1/k\} < 2\varepsilon$$

for n large enough, Q.E.D.

On any metric space, the σ-algebra will be the Borel σ-algebra unless stated otherwise.

3.3 <u>Lemma</u>. Let $(X,\mathcal{A},P)$ be a probability space and $\{g_n\}_{n=0}^{\infty}$ a uniformly bounded sequence of real-valued functions on X such that g_0 is measurable. If $g_n \to g_0$ in outer probability then

$$\limsup_{n\to\infty} \int^* g_n dP \le \int g_0 dP.$$

<u>Proof</u>. Let $|g_n(x)| \le M < \infty$ for all n and all $x \in X$. We may assume $M = 1$. Given $\varepsilon > 0$, for n large enough $P^*(|g_n - g_0| > \varepsilon) < \varepsilon$. Let $A(n)$ be a measurable set on which $|g_n - g_0| \le \varepsilon$ with $P(X \setminus A(n)) < \varepsilon$. Then

$$\int g_n^* dP \le \varepsilon + \int^*_{A(n)} g_n dP \le 2\varepsilon + \int_{A(n)} g_0 dP \le 3\varepsilon + \int g_0 dP .$$

Letting $\varepsilon \downarrow 0$ completes the proof. □

3.4 <u>Corollary</u>. If $f_n \to f_0$ in outer probability and f_0 is measurable with separable range then $f_n \xrightarrow{\mathcal{L}} f_0$.

<u>Proof</u>. Apply Prop. 3.2, Lemma 3.3 and Remark 1.3.

3.5 Theorem. Let $(X, \mathcal{A}, P)$ be a probability space, (S,d) a metric space. Suppose that for $n = 0,1,\cdots$, $(Y_n, \mathcal{B}_n)$ is a measurable space, g_n a perfect measurable function from X into Y_n, and f_n a function from Y_n into S, where f_0 has separable range and is measurable. Let $Q_n := P \circ g_n^{-1}$ on $\mathcal{B}_n$ and $f_n \circ g_n \to f_0 \circ g_0$ in outer probability. Then $f_n \xrightarrow[\mathcal{L}]{} f_0$ as $n \to \infty$.

Before proving this, here is:

3.6 Example. Theorem 3.5 can fail without the hypothesis that g_n be perfect. In Example 2.4 let $X = C$, $Y_n := [0,1]$, and let g_n be the identity for all n. Let $S = \{0,1\}$, $f_0 = 0$, and $f_n := 1_{[0,1]\setminus C}$. Then $f_n \circ g_n \equiv 0$ for all n, so $f_n \circ g_n \to f_0 \circ g_0$ in outer probability (and any other sense). Let $\mathcal{B}_n$ be the Borel σ-algebra for each n. Then for G the identity,

$$\int^* G(f_n)\,dQ_n = \int^* f_n\,d\lambda = 1 \quad \text{for} \quad n \geq 1,$$

while $\int G(f_0)\,dQ_0 = 0$, so $f_n \not\xrightarrow[\mathcal{L}]{} f_0$.

Proof of Theorem 3.5. By Cor. 3.4, $f_n \circ g_n \xrightarrow[\mathcal{L}]{} f_0 \circ g_0$. Let G be any bounded, continuous, real-valued function on S. Then

$$\int^* G(f_n(g_n))\,dP \to \int G(f_0(g_0))\,dP = \int G(f_0)\,dQ_0 .$$

Also,

$$\begin{aligned} \int^* G(f_n(g_n))\,dP &= \int G(f_n(g_n))^*\,dP \\ &= \int (G \circ f_n)^*(g_n)\,dP \qquad \text{by Th. 2.1} \\ &= \int (G \circ f_n)^*\,dQ_n \\ &= \int^* G(f_n)\,dQ_n \end{aligned}$$

and the result follows. □

4. An extended Wichura theorem. Here is one of the main results of the paper:

4.1 Theorem. Let (S,d) be any metric space, $(X_n, \mathcal{A}_n, Q_n)$ any probability spaces, and f_n a function from X_n into S for each $n = 0,1,\cdots$. Suppose f_0 has separable range S' and is measurable. Then $f_n \xrightarrow[\mathcal{L}]{} f_0$ if and only if there exists a probability space $(\Omega, \mathcal{S}, Q)$ and perfect measurable functions g_n from $(\Omega, \mathcal{S})$ to $(X_n, \mathcal{A}_n)$ for each $n = 0,1,\cdots$, such that $Q \circ g_n^{-1} = Q_n$ on $\mathcal{A}_n$ for each n and $f_n(g_n) \to f_0(g_0)$ almost uniformly, $n \to \infty$.

Proof. "If" follows from Prop. 3.1 and Theorem 3.5. "Only if" will be proved largely as in Dudley (1968, Theorem 3) and Fernandez (1974). Let $f_n \xrightarrow[\mathcal{L}]{} f_0$.

The space Ω will be taken as the Cartesian product $\prod_{n=0}^{\infty} X_n \times I_n$ where each I_n is a copy of $[0,1]$. Here g_n will be the natural projection of Ω onto X_n for each n.

Let $P := Q_0 \circ f_0^{-1}$ on the Borel σ-algebra of S, concentrated in the separable subset S'. A set $B \subset S'$ will be called a continuity set in S' if $P(\partial B) = 0$ where ∂B is the boundary of B in S'. We then have:

4.2 Lemma. For any $\varepsilon > 0$ there exist $J < \infty$ and disjoint open continuity sets U_j, $j = 1,\cdots,J$, each with diameter

$$\operatorname{diam} U_j := \sup\{d(x,y) \colon x,y \in U_j\} < \varepsilon,$$

and with $\Sigma_{j=1}^{J} P(U_j) > 1 - \varepsilon$.

Proof. This is proved as in Skorohod (1956). Let $\{x_j\}_{j=1}^{\infty}$ be dense in S'. Let $B(x,r) := \{y \in s' \colon d(x,y) < r\}$, $0 < r < \infty$, $x \in S'$. Then $B(x_j,r)$ is a continuity set of P for all but at most countably many values of r. Choose r_j with $\varepsilon/3 < r_j < \varepsilon/2$ such that $B(x_j,r_j)$ is a continuity set of P for each j. The continuity sets form an algebra. Let

$$U_j := B(x_j,r_j) \setminus \cup_{i<j}\{y \colon d(x_i,y) \le r_i\} .$$

Then U_j are disjoint open continuity sets of diameters $< \varepsilon$

with $\Sigma_{j=1}^{\infty} P(U_j) = 1$, so there is a $J < \infty$ with $\Sigma_{j=1}^{J} P(U_j) > 1 - \varepsilon$, Q.E.D.

Now for each $k = 1,2,\cdots$, by Lemma 4.2 take disjoint open continuity sets $U_{kj} := U(k,j)$ of P, $j = 1,2,\cdots, J(k) < \infty$, with $\text{diam}(U_{kj}) < 1/k$, $P(U_{kj}) > 0$, and

$$(4.3) \qquad \Sigma_{j=1}^{J(k)} P(U_{kj}) > 1 - 2^{-k} .$$

For any open set U in S with complement F, let $d(x,F) := \inf\{d(x,y): y \in F\}$. For $r = 1,2,\cdots$, let $F_r := \{x: d(x,F) \geq 1/r\}$. Then F_r is closed and $F_r \uparrow U$ as $r \to \infty$. There is a continuous h_r on S with $0 \leq h_r \leq 1$, $h_r = 1$ on F_r and $h_r = 0$ outside F_{2r}: take $h_r(x) := \min(1, \max(0, 2rd(x,F) - 1))$.

For each j and k, setting $F(k,j) := S \setminus U_{kj}$, take $r = r(k,j)$ large enough so that

$$P(F(k,j)_r) > (1-2^{-k}) P(U_{kj}) .$$

Let h_{kj} be the h_r above for such an r and H_{kj} the h_{2r}.

For n large enough, say for $n \geq n_k$, we have

$$\int_* h_{kj}(f_n) dQ_n > (1-2^{-k}) P(U_{kj}) \quad \text{and}$$

$$\int^* H_{kj}(f_n) dQ_n < (1+2^{-k}) P(U_{kj})$$

for all $j = 1,\cdots,J(k)$. We may assume $n_1 \leq n_2 \leq \cdots$.

For every $n = 0,1,\cdots$, let $f_{kjn} := (h_{kj} \circ f_n)_*$ for Q_n , so that

$$\int_* h_{kj}(f_n) dQ_n = \int f_{kjn} dQ_n, \quad 0 \leq f_{kjn} \leq h_{kj}(f_n),$$

and f_{kjn} is $\mathcal{A}_n$-measurable. For $n \geq 1$ let $B_{kjn} := \{f_{kjn} > 0\} \in \mathcal{A}_n$. Let $B_{kj0} := f_0^{-1}(U_{kj}) \in \mathcal{A}_0$. For each k and n, the $B_{kjn} \subset U_{kj}$ are disjoint, $j = 1,\cdots,J(k)$, and $H_{kj}(f_n) = 1$ on B_{kjn}, so

$$(4.4) \qquad (1-2^{-k}) P(U_{kj}) < Q_n(B_{kjn}) < (1+2^{-k}) P(U_{kj}) .$$

Let $T_n := X_n \times I_n$. Let μ_n be the product law $Q_n \times \lambda$ on T_n where λ is Lebesgue measure on I_n. For each $k \geq 1$, $n \geq n_k$, and $j = 1, \cdots, J_k$, let

$$C_{kjn} := B_{kjn} \times [0, f(k,j,n)] \subset T_n ,$$

$$D_{kjn} := B_{kj0} \times [0, g(k,j,n)] \subset T_0 ,$$

defining f and g here so that

$$\mu_n(C_{kjn}) = \mu_0(D_{kjn}) = \min(Q_n(B_{kjn}), Q_0(B_{kj0})) .$$

Then for each k, j, and $n \geq n_k$, we have

$$\max(f,g)(k,j,n) = 1 \quad \text{and}$$

(4.5) $$\min(f,g)(k,j,n) > 1-2^{-k}, \text{ by (4.4)}.$$

Let $C_{k0n} := T_n \setminus \bigcup_{j=1}^{J(k)} C_{kjn}$, $D_{k0n} := T_0 \setminus \bigcup_{j=1}^{J(k)} D_{kjn}$.

For $k = 0$ let $J_0 := 0$, $C_{00n} := T_n$, $D_{00n} := T_0$, $n_0 := 0$.

For each $n = 1, 2, \cdots$, let $k(n)$ be the unique k such that $n_k \leq n < n_{k+1}$. Then for $n \geq 1$, T_n is the disjoint union of sets $W_{nj} := C_{k(n)jn}$, $j = 0, 1, \cdots, J_{k(n)}$. Also, T_0 is the disjoint union of sets $E_{nj} := D_{k(n)jn}$. If $j \geq 1$ or $k(n) = 0$, then $\mu_n(W_{nj}) = \mu_0(E_{nj}) > 0$.

For x in T_0, and each n, let $j(n,x)$ be the j such that $x \in E_{nj}$. Let

$$L := \{x \in T_0 \colon \mu_0(E_{nj(n,x)}) > 0 \text{ for all } n\} .$$

Then $T_0 \setminus L \subset \bigcup_i E_{n(i)0}$ for some (possibly empty or finite) sequence $n(i)$ such that $\mu_0(E_{n(i)0}) = 0$ for all i. Thus $\mu_0(L) = 1$.

For $x \in L$ and any measurable set $B \subset T_n$ $(B \in \mathcal{A}_n \otimes \mathcal{B})$, let

$$P_{nj}(B) := \mu_n(B \cap W_{nj})/\mu_0(E_{nj}), \quad P_{nx} := P_{nj} ,$$

where $j := j(n,x)$. Then P_{nx} is a probability measure on $\mathcal{A}_n \otimes \mathcal{B}$. Let ρ_x be the product measure $\prod_{n=1}^{\infty} P_{nx}$ on $T := \prod_{n=1}^{\infty} T_n$ (Halmos, 1950, p. 157).

4.6 Lemma. For every measurable set $H \subset T$ $(H \in \times_{n=1}^{\infty}(\mathcal{A}_n \otimes \mathcal{B}))$, $x \to \rho_x(H)$ is measurable on $(T_0, \mathcal{A}_0 \otimes \mathcal{B})$.

Proof. Let $\mathcal{H}$ be the collection of all H for which the assertion holds. Given n, P_{nx} is one of finitely many laws, each obtained for x in a measurable set E_{nj}. Thus if Y_n is the natural projection of T onto T_n and $H = Y_n^{-1}(B)$ for some $B \in \mathcal{A}_n \otimes \mathcal{B}$ then $H \in \mathcal{H}$.

If $H = \bigcap_{i \in F} Y_{m(i)}^{-1}(B_i)$, where $B_i \in \mathcal{A}_{m(i)} \otimes \mathcal{B}$ and F is finite, we may assume the $m(i)$ are distinct. Then

$$\rho_x(H) = \prod_{i \in F} \rho_x(Y_{m(i)}^{-1}(B_i)) ,$$

so $H \in \mathcal{H}$. Then, any finite, disjoint union of such intersections is in $\mathcal{H}$. Such unions form an algebra. If $H_n \in \mathcal{H}$ and $H_n \uparrow H$ or $H_n \downarrow H$, then $H \in \mathcal{H}$. As the smallest monotone class including an algebra is a σ-algebra (Halmos, 1950, p. 27) the Lemma follows.

Now $\Omega = T_0 \times T$. For any product measurable set $C \subset \Omega$ and $x \in T_0$, let $C_x := \{y \in T: \langle x,y\rangle \in C\}$, and $Q(C) := \int \mu_x(C_x) d\mu_0(x)$. Here $x \to \mu_x(C_x)$ is measurable if H is a finite union of products $A_i \times F_i$ where $A_i \in \mathcal{A}_0 \otimes \mathcal{B}$ and F_i is product measurable in T. Thus by monotone convergence, $x \to \mu_x(H_x)$ is measurable for any product measurable set $H \subset \Omega$. Thus Q is defined. It is then clearly a countably additive probability measure.

Let p be the natural projection of T_n onto X_n. Recall that $P_{nx} = P_{nj}$ for all $x \in E_{nj}$. The marginal of Q on X_n, or $Q \circ g_n^{-1}$, is

$$\left(\int P_{nx} d\mu_0(x)\right) \circ p^{-1} = \Sigma_{j=0}^{J(k(n))} \mu_0(E_{nj}) P_{nj} \circ p^{-1} = \mu_n \circ p^{-1} = Q_n ,$$

where $J(k(n)) := J_{k(n)}$. Thus Q has marginal Q_n on X_n for each n as desired.

By (4.3), $\Sigma_{k=1}^{\infty} Q_0(X_0 \setminus \bigcup_{j=1}^{J(k)} f_0^{-1}(U_{kj})) < \Sigma\, 2^{-k} < \infty$. Thus Q_0-almost every $y \in X_0$ belongs to $\bigcup_{j=1}^{J(k)} f_0^{-1}(U_{kj})$ for all large enough

k. Also if $t \in I_0$ and $t < 1$, then by (4.5), $t < g(k,n,j)$ for all $j \geq 1$ as soon as $1 - 2^{-k} > t$. Thus for μ_0-almost all $\langle y,t \rangle$, there is an m such that

$$\langle y,t \rangle \in \cup_{j=1}^{J(k(n))} E_{nj}$$

for all $n \geq m$. Since $\operatorname{diam}(U_{kj}) < 1/k$ for each j,

$$\begin{aligned} &Q^*(d(f_n(g_n), f_0(g_0)) > 1/k(n) \quad \text{for some} \quad n \geq m) \\ &\leq \mu_0(\{\langle g_0, t \rangle \in E_{n0} \quad \text{for some} \quad n \geq m\}) \to 0 \end{aligned}$$

as $m \to \infty$, so $f_n(g_n) \to f_0(g_0)$ almost uniformly.

Lastly, let us see that the g_n are perfect. Suppose $Q(A) > 0$ for some A. Now

$$Q(A) = \int \mu_x(A_x) d\mu_0(x) .$$

Then for some x, $\mu_x(A_x) > 0$. Given $n \geq 1$, if $\mu_0(E_{n0}) = 0$, we take $x \notin E_{n0}$. Now $T = T_n \times T^{(n)}$ where $T^{(n)} := \prod_{1 \leq i \neq n} T_i$. Then on T, $\mu_x = P_{nx} \times Q_{nx}$ for some law $Q_{nx} = \prod_{m \neq n} P_{mx}$ on $T^{(n)}$. Let $A(x) := A_x$. By the Tonelli-Fubini theorem,

$$\mu_x(A_x) = \iint 1_{A(x)}(u,v) dP_{nx}(u) dQ_{nx}(v) .$$

Thus for some v, $\int 1_{A(x)}(u,v) dP_{nx}(u) > 0$. Choose and fix such a v as well as x. Now $P_{nx} = P_{nj}$ for $j = j(n,x)$ with $\mu_0(E_{nj}) > 0$. Let $u = (s,t)$, $s \in X_n$, $t \in I_n$. Then since $P_{nj} = Q_n \times \lambda$ restricted to a set of positive measure and normalized,

$$0 < \iint 1_{A(x)}(s,t,v) dQ_n(s) dt .$$

Choose and fix a t with

$$0 < \int 1_{A(x)}(s,t,v) dQ_n(s) .$$

Let $C := \{s \in X_n : (s,t,v) \in A_x\}$. So $Q_n(C) > 0$. Clearly $C \subset g_n(A)$, so g_n is perfect, $n \geq 1$, by Theorem 2.1, 1), finishing the proof of Theorem 4.1.

5. Definitions and stability of Donsker classes. Recall the notations of the Abstract. Let $\rho_P(f,g) := (f-g,f-g)_{P,0}^{1/2}$. First, some properties of G_PBUC classes will be developed. A process $(\omega,f) \longrightarrow Y(f)(\omega)$, $f \in \mathcal{F}$, $\omega \in \Omega$, for some probability space Ω, which has the finite-dimensional distributions of G_P, will be called a suitable G_P iff for all ω, $f \longrightarrow Y(f)(\omega)$ is bounded and uniformly continuous for ρ_P. Let $Pf := \int f dP$, $f \in \mathcal{L}^1(P)$.

5.1 Theorem. Let Y be a suitable G_P on $\mathcal{F}$. Then a) for almost all ω, the function defined as Y on $\mathcal{F}$ and 0 on constant functions is well-defined and extends uniquely to a linear functional on the linear span of $\mathcal{F}$ and constant functions, ρ_P-uniformly continuous on the symmetric convex hull of $\mathcal{F}$; b) there is a complete separable linear subspace M of $\ell^\infty(\mathcal{F})$ and a Borel probability measure γ on M such that every suitable G_P has its sample functions in M and has distribution γ there.

Proof. First, $\mathcal{F}$ is totally bounded for ρ_P (Dudley, 1967b, Prop. 3.4, p. 295). Let M be the set of all real functions on $\mathcal{F}$ uniformly continuous for ρ_P. Then $Y(\cdot)(\omega) \in M$ for each ω. The functions in M are bounded and it is separable and complete in $\ell^\infty(\mathcal{F})$. Let $\mathcal{H} := \{h_k\}_{k\geq 1}$ be a countable dense set in $\mathcal{F}$ for ρ_P. On M, $\|\cdot\|_{\mathcal{H}} = \|\cdot\|_{\mathcal{F}}$.
For each $g \in M$, $\omega \longrightarrow \|Y-g\|_{\mathcal{H}}$ is measurable. Thus for any Borel set $B \subset M$, $\gamma(B) := \Pr(Y \in B)$ is defined, and does not depend on which suitable G_P was chosen, proving b).

Next, for any $f \in \mathcal{L}^2(P)$ let $\pi f := f-Pf$, $H_0 := \{f \in \mathcal{L}^2(P): Pf = 0\}$, so that π is the usual projection from $\mathcal{L}^2(P)$ onto H_0. For the usual $\mathcal{L}^2$ metric $e_P(\pi u,\pi v) = \rho_P(u,v)$ for any $u,v \in \mathcal{L}^2(P)$. Let $g_k := \pi h_k$, $k = 1,2,\cdots$. Then $\{g_k\}_{k\geq 1}$ is dense in $\pi\mathcal{F}$ for e_P. By Gram-Schmidt orthonormalization, let $\{\varphi_m\}_{m\geq 1}$ be an orthonormal basis of the linear span of $\pi\mathcal{F} \subset H_0$, where for a subsequence $\{j_m\}$ of $\{g_k\}$ and some real a_{mi},

$$\varphi_m = \Sigma_{i=1}^m a_{mi} j_i$$

for each m. Let $W_m := \Sigma_{i=1}^m a_{mi} Y(j_i)$. Then the W_m are i.i.d. $N(0,1)$ ("orthoGaussian") variables. For each $f \in \mathcal{L}^2(P)$,

$$f \sim Pf + \Sigma_{m\geq 1} (f,\varphi_m)\varphi_m,$$

convergent in $\mathcal{L}^2$, and

$$Y(f) = \Sigma_{m\geq 1} (f,\varphi_m)W_m \quad \text{a.s.}$$

by the three-series theorem.

If $\pi f = \pi h$, $f,h \in \mathcal{F}$, then $\rho_P(f,h) = 0$ so $Y(f) \equiv Y(h)$. Thus we can set $W(\pi f) := Y(f)$ for all $f \in \mathcal{F}$ and W is well-defined on $\pi\mathcal{F}$. Since $f \longrightarrow \pi f$ is an isometry for ρ_P, W has each sample function bounded and uniformly continuous on $\pi\mathcal{F}$ for ρ_P. Thus each $W(\cdot)(\omega)$ extends uniquely to the ρ_P-closure B of $\pi\mathcal{F}$, remaining ρ_P-uniformly continuous and bounded. Now B is compact for ρ_P. On B, W is isonormal, i.e. a Gaussian process with mean 0 and covariance equal to the inner product. Now B is a GC-set in H_0 (Dudley 1967b, 1973). It follows that almost surely the series $\Sigma_m (h,\varphi_m)W_m$ converges uniformly in $h \in \pi\mathcal{F}$ (Feldman, 1971; Dudley, 1973, Theorem 0.3), hence to the ρ_P-uniformly continuous function $h \longrightarrow W(h)(\omega)$. So almost surely

$$Y(f) = \Sigma_m (\pi f,\varphi_m)W_m$$

for $f = h_k$ for all k and thus, by ρ_P-uniform continuity of both sides, for all $f \in \mathcal{F}$. The right side is linear in f and convergent for all f in the linear span of $\mathcal{F}$, except on an event of probability 0 not depending on f. The right side is 0 on constants. It follows that it is uniformly continuous for ρ_P on the closed (for ρ_P), symmetric, convex hull of $\mathcal{F}$ (Feldman, 1971, Theorem 3; Dudley, 1967, Theorem 4.6). This proves a), Q.E.D.

We have the countable product $(X^\infty, \mathcal{A}^\infty, P^\infty)$ and let

$$(\Omega, \mathcal{S}, Pr) := (X^\infty, \mathcal{A}^\infty, P^\infty) \times ([0,1], \mathcal{B}, \lambda)$$

where $\mathcal{B}$ is the Borel σ-algebra. $\mathcal{F}$ is called a <u>functional Donsker class</u> (for P) iff it is $G_P BUC$ and there exist processes $Y_j(f,\omega)$, $f \in \mathcal{F}$, $\omega \in \Omega$, where Y_j are independent, suitable G_P processes, such that for every $\varepsilon > 0$

$$\lim_{n\to\infty} Pr^*\{n^{-1/2} \max_{m\leq n} \|\Sigma_{j=1}^m (\delta_{x(j)} - P - Y_j\|_{\mathcal{F}} > \varepsilon\} = 0$$

(Dudley and Philipp, 1983; Dudley, 1984).

5.2 <u>Theorem</u>. Let $\mathcal{F} \subset \mathcal{L}^2(X, \mathcal{A}, P)$ be $G_P BUC$. Then the following are equivalent:

1) $\mathcal{F}$ satisfies the central limit theorem as in Sec. 1, i.e. for every bounded real function H on $\ell^\infty(\mathcal{F})$ continuous for $\|\cdot\|_{\mathcal{F}}$, $\int^* H(\nu_n) dP^n \to EH(G_P)$ as $n \to \infty$;

2) for every $\varepsilon > 0$ there is a $\delta > 0$ and an N such that for $n \geq N$

$$Pr^*\{\sup[|\nu_n(f-g)|: f,g \in \mathcal{F}, \rho_P(f,g) < \delta] > \varepsilon\} < \varepsilon.$$

3) $\mathcal{F}$ is a functional Donsker class.

<u>Proof</u>. 1) implies 2): given $\varepsilon > 0$, take $0 < \delta \leq \varepsilon/3$ such that for any suitable G_P,

$$Pr\{\sup\{|G_P(f) - G_P(g)|: \rho_P(f,g) < \delta\} > \varepsilon/3\} < \varepsilon/2 .$$

Note that such events are measurable by Theorem 5.1, corresponding to open sets in M. By the extended Wichura theorem (4.1), for $n \geq N$ large enough we may assume $Pr^*\{\|\nu_n - G_P\|_{\mathcal{F}} \geq \varepsilon/3\} < \varepsilon/2$. If $\|\nu_n - G_P\|_{\mathcal{F}} < \varepsilon/3$ and $|G_P(f) - G_P(g)| \leq \varepsilon/3$ then $|\nu_n(f) - \nu_n(g)| < \varepsilon$. Thus 2) holds with the δ and N chosen.

2) implies 3): as noted, the $G_P BUC$ assumption implies $\mathcal{F}$ is totally bounded for ρ_P. Then 2) is equivalent to 3)

(Dudley, 1984, Theorem 4.1.1).

3) implies 1): given $\varepsilon > 0$, by Theorem 5.1 and Ulam's theorem take a compact $K \subset M$ with $\gamma(K) > 1-\varepsilon$. Let H be bounded and continuous on $\ell^\infty(\mathfrak{F})$. For some $\delta > 0$, if $\|u-v\|_{\mathfrak{F}} < \delta$, $u \in K$, and $v \in \ell^\infty(\mathfrak{F})$, then $|H(u)-H(v)| < \varepsilon$ (e.g. Dudley, 1966, Lemma 1). Take $n_0 = n_0(\varepsilon)$ such that for $n \geq n_0$,

$$\Pr^*(\|\nu_n - T_n\|_{\mathfrak{F}} \geq \delta) < \varepsilon,$$

where $\nu_n := n^{-1/2} \Sigma_{j=1}^n (\delta_{x(j)} - P)$, $T_n := n^{-1/2} \Sigma_{j=1}^n Y_j$. Then,

$$\Pr^*(|H(\nu_n) - H(T_n)| > \varepsilon) < \varepsilon,$$

$$\int^* H(\nu_n)\, dP^n \leq \varepsilon(1+\sup|H|) + EH(T_n)\ .$$

Now T_n is a suitable G_P, so

$$\limsup_{n\to\infty} \int^* H(\nu_n)\, dP^n \leq EH(G_P) + \varepsilon(1+\sup|H|).$$

Letting $\varepsilon \downarrow 0$ gives 1), Q.E.D.

Next is a stability result.

5.3 Theorem. Let $\mathfrak{F}$ be a functional Donsker class. Let $\mathfrak{G}$ be the symmetric convex hull of $\mathfrak{F}$, and $\mathcal{H}$ the set of all $g \in \mathcal{L}^2(P)$ such that for some $g_m \in \mathfrak{G}$, $g_m(x) \to g(x)$ for all x and $\int (g_m - g)^2 dP \to 0$ as $m \to \infty$. Then $\mathcal{H}$ is a functional Donsker class.

Proof. Take $Y_1, Y_2, \cdots$ as in the definition of functional Donsker class. For any $x(1), \cdots, x(n)$,

$$\|\Sigma_{j=1}^n \delta_{x(j)} - P - Y_j\|_{\mathfrak{G}} = \|\Sigma_{j=1}^n \delta_{x(j)} - P - Y_j\|_{\mathfrak{F}}\ ,$$

using the fact that the $\delta_{x(j)}$, P, and Y_j are all linear on the linear span of $\mathfrak{F}$, by Theorem 5.1(a) for the Y_j. Now if $g_m \to g$ as assumed, then $\delta_x(g_m - g) \to 0$ for all x, $P(g_m) \to P(g)$, and

$\rho_P(g_m, g) \to 0$, so $Y_j(g_m)(\omega) \to Y_j(g)(\omega)$ for all j and ω. Thus

$$\|\Sigma_{j=1}^n \delta_{x(j)} - P - Y_j\|_{\mathcal{H}} = \|\Sigma_{j=1}^n \delta_{x(j)} - P - Y_j\|_{\mathcal{G}},$$

and the result follows. □

6. Weighted empirical distribution functions and related families of functions. Let $(A, \mathcal{a}, P)$ be any probability space and X a nonnegative random variable on A. Let X have distribution $\mathcal{L}(X) = Q$ on $[0,\infty[$. Let

$$\mathcal{G}_X := \{M1_{X \geq M}: \ 0 < M < \infty\}.$$

Let $G(t) := P(X \leq t)$ and

$$G^{-1}(y) := \inf\{t: G(t) \geq y\}, \quad 0 < y < 1.$$

It is well known that for the uniform distribution λ on $]0,1[$, G^{-1} has distribution Q. Let

$$h_X(t) := \lim_{u \uparrow 1-t} G^{-1}(u) := G^{-1}((1-t)^-), \quad 0 < t < 1.$$

Then h_X is a non-increasing, nonnegative function on $]0,1[$, continuous from the left, whose distribution for λ is also Q.

To deal with classes $\mathcal{G}_X$, classes with an exponent on M will also be helpful.

6.1 Lemma. For any $X \geq 0$ in $\mathcal{L}^p(P)$, $2 \leq p < \infty$, let $\mathcal{F}_{X,p} := \{XM^{(p-2)/2}1_{X \geq M}\}_{M>0}$. Then $\mathcal{F}_{X,p}$ is a functional Donsker class.

Proof. For $p = 2$ this follows directly from a central limit theorem of Pollard (1982), see also Dudley (1984, Theorems 11.3.1, 11.1.2). Thus $\{X^{p/2}1_{X \geq K}\}_{K>0}$ is a functional Donsker class. Then by Theorem 5.3, so is

$$\{X^{p/2} \Sigma_{i \geq 1} \lambda_i 1_{X \geq N(i)}: \lambda_i \geq 0, \Sigma \lambda_i = 1\},$$

in particular if $0 \le N(1) < N(2) < \cdots$. Then $\Sigma_{i\ge 0} \lambda_i 1_{X\ge N(i)} = \Sigma_{i\ge 0} c_i 1_{N(i)\le X<N(i+1)}$ where $N(0) := 0$ and $0 \le c_0 < c_1 < \cdots \le 1$. Bounded pointwise limits of such sums give all functions $G(X)$ where G is non-decreasing with $0 \le G \le 1$. Then by Theorem 5.3,

$$\{X^{p/2}G(X)\colon G(0) = 0,\ \text{tot. var}(G) \le 1\},$$

where "tot. var" denotes total variation, is a functional Donsker class.

If $\mathcal{F}$ is a functional Donsker class, so is $\{cf\colon f \in \mathcal{F}\}$ for any constant c, clearly. Now

$$XM^{(p-2)/2}1_{X\ge M} = X^{p/2}G(X)$$

where $G(x) := (M/x)^{(p-2)/2}1_{x\ge M}$, $0 < x < +\infty$. Then G has total variation 2, and the result follows.

6.2 <u>Theorem</u>. For any probability space $(A,\mathcal{Q},P)$ and nonnegative random variable X on it, the following are equivalent:

a) $\mathcal{G}_X$ is a functional P-Donsker class on A;

b) $\mathcal{G}_{id}$ is a functional Q-Donsker class on $[0,\infty[$, where id is the identity function, $Q = \mathcal{L}(X)$;

c) $\mathcal{G}_h$ is a functional λ-Donsker class on $]0,1[$, $h := h_X$.

<u>Proof</u>. As above, let $x(1), x(2), \cdots$, be coordinates on A^∞. Then $X(x(j))$, $j = 1,2,\cdots$, are i.i.d. (Q) in $\mathbb{R}$. For any $M \ge 0$ and rational $q(j) \uparrow M$, $q(j)1_{X\ge q(j)}$ converge boundedly and pointwise, everywhere, to $M1_{X\ge M}$. Thus the supremum over $\mathcal{G}_X$ of any finite signed measure on $(A,\mathcal{Q})$ equals the sup over the countable set $\{q1_{X\ge q}\colon q$ rational, $q \ge 0\}$. The same holds for id on $\mathbb{R}$ or h on $]0,1[$. Thus there is no difficulty about measurability of such suprema for empirical measures.

For any $\omega \in A^\infty$, $n = 1,2,\cdots$, and $0 \le M < \infty$, $P_n(X \ge M) = Q_n(\text{id} \ge M)$ where $Q_n := n^{-1}\Sigma_{j=1}^n \delta_{X(x(j))}$ serves as an empirical measure for Q. Also, $P(X \ge M) = Q(\text{id} \ge M)$ for all M. Distances

ρ_P between functions $Ml_{X \geq M}$ equal ρ_Q distances between the corresponding functions $Ml_{id \geq M}$. Thus the asymptotic equicontinuity criterion (Theorem 5.2,(2)) holds for $\mathcal{G}_X$ if and only if for $\mathcal{G}_{id}$, so a) and b) are equivalent. Applying this to $A =]0,1[$, $P = \lambda$ shows that c) is also equivalent, Q.E.D.

Condition c) is (in effect) a condition on weighted empirical distribution functions (for λ, near 0). An integral condition, f) below, was shown equivalent to a central limit theorem (and thus to (c) in the present formulation) by Chibisov (1964) if h is regularly varying, then by O'Reilly (1974) if h is continuous. (Always, h is non-decreasing at least on some interval $(0,\delta)$, $\delta > 0$.) The following is an effort to put the Chibisov-O'Reilly result in a somewhat more final form by

i) removing the continuity assumption,

ii) adding several more equivalent conditions, all also considered explicitly or implicitly by other previous authors, but perhaps not in the present combinations ((g1) and (g2)),

iii) collecting a more complete and self-contained proof rather than, e.g., citing the "arguments of O'Reilly" who cites "arguments of Chibisov." There has been some confusion in the literature as noted by M. Csörgő (1984).

The past results have allowed a singularity $h \to \infty$ as $t \uparrow 1$ as well as $t \downarrow 0$. This is natural for empirical distribution functions. The two endpoints are symmetrical via $t \leftrightarrow 1-t$. The present formulation, with h non-decreasing on $]0,1[$, is natural in the context of Theorem 6.2. Results for $h \uparrow +\infty$ as $t \uparrow 1$ and/or $t \downarrow 0$ can be written down easily if desired.

Let $W(t)$ be a standard Wiener process.

6.3 Theorem (O'Reilly, Chibisov, et al.). For any non-increasing function $h \geq 0$ on $]0,1[$ the following are equivalent:

c) $\mathcal{G}_h$ is a functional λ-Donsker class;

d) $\{h(t)1_{[0,t]}(\cdot)\colon\ 0 \le t \le 1\}$ is a functional λ-Donsker class;

e) $h(t)W(t) \to 0$ a.s. as $t \downarrow 0$;

e') $h(2^{-k})W(2^{-k}) \to 0$ a.s. as $k \to \infty$;

f') for every $\delta > 0$, $\Sigma_{k=1}^{\infty} \exp(-2^k\delta/h^2(2^{-k})) < \infty$;

f) for every $\varepsilon > 0$, $\int_0^1 t^{-1}\exp(-\varepsilon/(th^2(t)))dt < \infty$;

g) both

g1) $t^{1/2}h(t) \to 0$ as $t \downarrow 0$, and

g2) for every $\varepsilon > 0$,

$$\int_0^1 t^{-3/2}h(t)^{-1}\cdot\exp(-\varepsilon/(th^2(t)))dt < \infty .$$

Proof. First, c) implies d): if $u_1, u_2, \cdots$ are i.i.d. (λ), then almost surely no u_i falls at any jump of h (there are at most countably many jumps). Thus we may assume h is left continuous in this step. The only difficulty is with possible atoms of Q. There are at most countably many such atoms M_k, with $h(x) = M_k$ for $a_k < x \le b_k$ (possibly also at a_k) where we can take a_k minimal, and b_k maximal by left continuity. A function $h(t)1_{[0,t]}(\cdot)$, if not of the form $M1_{h\ge M}$, has $a_k \le t < b_k$ for some k.

Taking finitely many of the intervals $]a_k, b_k]$, say for $k = 1,\cdots,m$, we may number them so that

$$a_1 < b_1 \le a_2 < b_2 \le \cdots \le b_m ,$$

$$M_1 > M_2 > \cdots > M_m .$$

Now

$$M\lambda(h\ge M)^{1/2} \to 0 \quad \text{as} \quad M \to \infty , \tag{6.4}$$

for otherwise the set $\mathcal{G}_h$ is not totally bounded for ρ_λ, contrary to c) (Dudley, 1984, Theorem 4.1.1). Thus, given $\varepsilon > 0$, there is an $M_0 < \infty$ such that $M\lambda(h \ge M)^{1/2} < \varepsilon/16$ for $M \ge M_0$. We may and do choose $M_0 > h(1/2)$. Also, by c) and

Theorem 5.2, let $M_0 := M(0)$ be large enough so that

(6.5) $\Pr\{\sup_{M\geq M(0)} |\nu_n(M1_{h\geq M})| \geq \varepsilon/8\} < \varepsilon/4$.

For any constant c, $\{c1_{[0,t]}: 0 \leq t \leq 1\}$ is a functional Donsker class (essentially by the classical theorem of Donsker (1952), cf. Theorem 5.2 above and Theorem 7.4 of Dudley and Philipp (1983)). Thus by Theorem 5.3, $\{h(t)1_{[0,t]}: \gamma \leq t \leq 1\} := \mathcal{J}_\gamma$ is a functional Donsker class for any $\gamma > 0$, or for $\gamma = 0$ if h is bounded. So we may assume h unbounded and choose $\gamma > 0$ with $h(\gamma) > h(t) \geq M_0$ as $t \downarrow \gamma$. By Theorem 5.2, take $\delta > 0$ and n_1 such that for $n \geq n_1$,

(6.6) $\Pr\{\sup\{|\nu_n(f-g)|: f,g \in \mathcal{J}_\gamma, \operatorname{Var}(f-g) < \delta^2\} > \frac{\varepsilon}{2}\} < \frac{\varepsilon}{2}$.

Now let us restrict to intervals with $b_k \leq \gamma$, so that $b_m \leq \gamma$ and $M_m \geq M_0$. For any given n and ω, let τ be the least t such that for some $k \leq m$, $t \in [a_k, b_k]$ and $M_k|\nu_n([0,t])| \geq \varepsilon/2$, if such a t exists. Note that $(t,\omega) \longrightarrow \nu_n([0,t])(\omega)$ is a strong Markov process. Given τ and $\nu_n([0,\tau])$, the conditional distribution of $\nu_n([0,b_k])$ (an affine function of a binomial variable) has mean $\nu_n([0,\tau])(1-b_k)/(1-\tau)$, whose absolute value is at least $|\nu_n(0,\tau)|/2$. Note that $|\nu_n(M_k1_{[0,\tau]})|/2 \geq \varepsilon/4$. The conditional variance of $M_k\nu_n([0,b_k])$ is

$$M_k^2 n^{-1}(b_k-\tau)(1-b_k)(1-\tau)^{-2}\, n(1-\lambda_n([0,\tau])) \leq M_k^2 b_k .$$

Thus by Chebyshev's inequality there is a conditional probability at least 3/4 that

$$M_k|\nu_n(\{h\geq M_k\})| = |M_k\nu_n([0,b_k])| \geq \frac{\varepsilon}{4} - 2M_k b_k^{1/2} \geq \varepsilon/8$$

by choice of $M(0)$ and γ. Then by (6.5),

$$\Pr\{\text{for some } k = 1,\cdots,m \text{ and } t \in [a_k,b_k],\ M_k|\nu_n([0,t])| \geq \varepsilon/2\} = \Pr\{\tau \text{ exists}\}$$

$\leq \frac{4}{3} \Pr\{M_k |\nu_n([0,b_k])| \geq \varepsilon/8 \quad \text{for some} \quad k\}$

$\leq \frac{4}{3} \cdot \frac{\varepsilon}{4} = \varepsilon/3$.

Now let the number m of intervals increase to obtain all intervals $]a_j,b_j] \subset [0,\gamma]$. (Note that if for some j, $a_j < \gamma$, then $b_j \leq \gamma$ by choice of γ.) Then another use of (6.5) yields

$$\Pr\{|\nu_n(h(t)1_{[0,t]})| \geq \varepsilon/2 \quad \text{for some} \quad t \leq \gamma\} \leq \varepsilon/3 .$$

Then using (6.6),

$$\begin{aligned} (6.7) \quad & \Pr\{|\nu_n(h(t)1_{[0,t]} - h(s)1_{[0,s]})| > \varepsilon \text{ for some} \\ & 0 \leq s < t \leq 1 \quad \text{with} \quad \mathrm{Var}(h(t)1_{[0,t]} - h(s)1_{[0,s]}) < \delta^2\} < \varepsilon, \end{aligned}$$

noting that if $s < \gamma < t$ then

$$\mathrm{Var}(h(\gamma)1_{[0,\gamma]} - h(t)1_{[0,t]}) < \delta^2 .$$

The function $t \longmapsto h(t)1_{[0,t]}(\cdot)$ in $\mathcal{L}^2([0,1],\lambda)$ is left continuous and has limits from the right, with a right limit 0 at 0 by (6.4). Thus $\mathcal{G}_h$ is totally bounded in $\mathcal{L}^2$. This and (6.7), with Theorem 4.1.1 of Dudley (1984), imply d).

Next, d) implies e): clearly d) implies c), giving (6.4) in the last step, which implies g1) in the statement of Theorem 6.3. Since the class in d) is G_PBUC, we have $h(t)B_t \to 0$ a.s. as $t \downarrow 0$, where $B_t := G_P(1_{[0,t]})$ is a standard Brownian bridge process. We can write $W_t = B_t + tG$ where G $(= W_1)$ is a standard normal variable independent of B_t, so e) follows. Clearly e) implies e').

Next, e') implies f'): since $h(2^{-k}) \leq h(2^{-k-1})$, we have $h(2^{-k})W(2^{-k-1}) \to 0$ a.s., so $h(2^{-k})(W(2^{-k})-W(2^{-k-1})) \to 0$ a.s. Since $W(2^{-k}) - W(2^{-k-1})$ are independent and equal to $2^{-(k+1)/2}G_k$ for i.i.d. standard normal variables G_k, this means that for every $\varepsilon > 0$, by the Borel-Cantelli lemma

$$\Sigma_{k=1}^{\infty} \Pr(|G_k| > 2^{(k+1)/2}\varepsilon/h(2^{-k})) < \infty .$$

As $x \to \infty$,

$$\Pr(|G_1|>x) \sim (2/\pi)^{1/2} x^{-1} \cdot \exp(-x^2/2) \geq \exp(-x^2) .$$

Thus

$$\Sigma_{k=1}^{\infty} \exp(-2^{k+1} \varepsilon^2 / h^2(2^{-k})) < \infty .$$

Letting $\varepsilon = (\delta/2)^{1/2}$ proves f').

Now f') implies f): given $\varepsilon > 0$, if $\delta := \varepsilon/2$,

$$\begin{aligned}
\infty &> \Sigma_{k=1}^{\infty} \exp(-\delta 2^k / h^2(2^{-k})) \\
&= \Sigma_{j=0}^{\infty} (2^{-j} - 2^{-j-1}) 2^{j+1} \cdot \exp(-\varepsilon 2^j / h^2(2^{-j-1})) \\
&\geq \Sigma_{j=0}^{\infty} \int_{2^{-j-1}}^{2^{-j}} \frac{1}{t} \exp(-\varepsilon/(t h^2(t)))\, dt \\
&= \int_0^1 t^{-1} \exp(-\varepsilon/(t h^2(t)))\, dt .
\end{aligned}$$

Conversely, f) implies f'): given $\delta > 0$, if $\varepsilon := \delta/2$,

$$\begin{aligned}
\infty &> \int_0^1 \frac{1}{t} \exp(-\varepsilon/(t h^2(t)))\, dt \\
&= \Sigma_{k=0}^{\infty} \int_{2^{-k-1}}^{2^{-k}} \frac{1}{t} \exp(-\varepsilon/(t h^2(t)))\, dt \\
&\geq \Sigma_{k=0}^{\infty} (2^{-k} - 2^{-k-1}) 2^k \exp(-\varepsilon 2^{k+1} / h^2(2^{-k})) \\
&= \frac{1}{2} \Sigma_{k=0}^{\infty} \exp(-\delta 2^k / h^2(2^{-k})) .
\end{aligned}$$

Also, f) implies g): let $q(t) := 1/h(t)$. For any $t > 0$,

$$\begin{aligned}
&\int_{t/2}^{t} s^{-1} \exp(-q^2(s)/(2s))\, ds \\
&\qquad \geq \int_{t/2}^{t} s^{-1} \exp(-q^2(t)/t)\, ds \\
&\qquad = (\log 2) \exp(-q^2(t)/t) .
\end{aligned}$$

Letting $t \downarrow 0$ gives g1). So for any $\varepsilon > 0$ and t small enough,

$$q(t)/t^{1/2} < \exp(\varepsilon q^2(t)/(2t)) .$$

Then the integrand in g2) is smaller than that of f) for $\mathcal{E}/2$, so g2) holds.

Conversely, g) implies f): if g1) holds then as $t \downarrow 0$, $t^{-1} < t^{-3/2}/h(t)$ and f) follows. So f), f') and g) are all equivalent.

It remains to show that these conditions imply c). Assume f), f') and g). Let

$$g(t) := M_k := \max(h(2^{-k-1}),1) \quad \text{for} \quad 2^{-k-1} < t \leq 2^{-k},$$
$$k = 0,1,\cdots .$$

Then $h(t) \leq g(t)$, $0 < t \leq 1$. If c) holds for g, then so does d), then d) for h by Theorem 5.3, then c) for h (recall that d) implies c) directly). So it is enough to prove that

$$\{M_k 1_{[0,2^{-k}]}\}_{k \geq 0}$$

is a functional Donsker class, given that $M_k \leq M_{k+1}$ for all k and

(6.8) $$\Sigma_{j=0}^{\infty} \exp(-2^j\alpha/M_j^2) < \infty$$

for every α $(= \delta/2) > 0$. Then clearly $2^j/M_j^2 \to \infty$ as $j \to \infty$. So for some j_1,

(6.9) $$2^j \geq M_j^2, \ j \geq j_1.$$

The following is essentially Lemma 4 of Chibisov (1964). Let $\nu_n(u) := \nu_n([0,u])$, $0 \leq u \leq 1$.

6.10 <u>Lemma</u>. If f) holds then for any $a < \infty$ and $\mathcal{E} > 0$,

$$\lim_{n\to\infty} \Pr\{\sup_{0<u\leq a/n} |\nu_n(h(u)1_{[0,u]})| \geq \mathcal{E}\} = 0 .$$

<u>Proof</u>. Let $P_{a,n} := \Pr\{|\nu_n(u)| < \mathcal{E}/h(u), \ 0 < u \leq a/n\}$

$$= \Pr\{|\nu_n(v/n)| < \mathcal{E}/h(v/n), \ 0 < v \leq a\}$$

$$= \Pr\{|n\lambda_n(v/n)-v| < \varepsilon n^{1/2}/h(v/n)), \quad 0 < v \le a\}.$$

Let $c_n := \inf_{0<u\le a/n} u^{-1/2}/h(u) \to \infty$ as $n \to \infty$ by g1). Note that $\varepsilon n^{1/2}/h(v/n) = \varepsilon v^{1/2}/(h(v/n)(v/n)^{1/2}) \ge \varepsilon v^{1/2}c_n$. Thus

$$P_{a,n} \ge \Pr(|n\lambda_n(v/n)-v| < \varepsilon v^{1/2}c_n, \ 0 < v \le a).$$

Let $m_n := n\lambda_n(a/n)$. Then as $n \to \infty$, $m := m_n$ converges in law to a Poisson variable with parameter a. In particular, m is bounded in probability.

Let μ be the uniform distribution on $[0,a]$. Then we can write

$$n\lambda_n(v/n) = m\mu_m(v), \quad 0 \le v \le a.$$

Given m,

$$\Pr\{|m\mu_m(v)-v| < \varepsilon c_n v^{1/2}, \quad 0 < v \le a\} \to 1$$

as $n \to \infty$ because a.s. $\mu_m(\{0\}) = 0$ and then $|m\mu_m(v) - v|$ is bounded and equal to v in a neighborhood of 0, so less than $c_n \varepsilon v^{1/2}$ for n large. Thus $P_{a,n} \to 1$, proving Lemma 6.10.

To continue proving c), for $0 \le p \le 1$ let

$$B(x,n,p) := \Sigma_{0\le k\le x}\binom{n}{k}p^k(1-p)^{n-k},$$

$$E(x,n,p) := \Sigma_{x\le k\le n}\binom{n}{k}p^k(1-p)^{n-k},$$

where k runs through integers but x need not be an integer. Then for any $\gamma > 0$,

$$\Pr\{|\nu_n([0,t])| > \gamma) = E(nt+n^{1/2}\gamma,n,t)$$
$$+ B(nt-n^{1/2}\gamma,n,t) .$$

For E and B there are Chernoff-Okamoto bounds (e.g. Dudley, 1984, 2.2.7, 2.2.8): for $0 < t \le 1/2$,

$$(6.11)\qquad B(nt-n^{1/2}\gamma,n,t) \le \exp(-\gamma^2/(2t(1-t)) \le \exp(-\gamma^2/(2t)),$$

$$(6.12)\qquad E(nt+n^{1/2}\gamma,n,t) \le (nt/(nt+n^{1/2}\gamma))^{nt+n^{1/2}\gamma}e^{n^{1/2}\gamma}$$

$$= ((1+y)^{1+y}e^{-y})^{-nt}$$

where $y := \gamma/(n^{1/2}t)$. (Note: Chibisov [1964, Lemma 5] gave a bound of the form (6.12) for Poisson probabilities with $\lambda = nt$, $z = n^{1/2}\gamma$.)

Now for $0 < \beta < 1$ there is a largest $y_\beta > 0$ such that $\log(1+y) \ge \beta y$ for $0 \le y \le y_\beta$. As $\beta \downarrow 0$,

$$(6.13)\qquad y_\beta \uparrow +\infty \text{ and } \log(1+y_\beta) = \beta y_\beta \to +\infty .$$

Following Chibisov [1964, after Lemma 5],

$$(1+y)\cdot\log(1+y) - y = \int_0^y \log(1+x)dx \ge \beta y^2/2 ,\ 0 \le y \le y_\beta .$$

Then

$$(6.14)\qquad E(nt+n^{1/2}\gamma,n,t) \le \exp(-nt\beta y^2/2) = \exp(-\beta\gamma^2/(2t)) .$$

Next,

$$(6.15)\qquad (1+y)\cdot\log(1+y) - y > (1+y)\cdot\log(1+y_\beta) - y$$
$$> y(\beta y_\beta - 1),\ y > y_\beta .$$

Given $\varepsilon > 0$, choose and fix a $\beta > 0$ small enough so that by (6.13)

$$(6.16)\qquad (\beta y_\beta - 1)\varepsilon > 1.$$

Setting $t = 2^{-k}$ and $\gamma = \varepsilon/M_k$, $k = 1,2,\cdots$, we have by (6.11) and (6.8) for some $j(0)$ large enough

$$\Sigma_{k\ge j(0)} \Pr\{M_k\nu_n([0,2^{-k}]) < -\varepsilon\}$$

$$(6.17)\qquad = \Sigma_{k\ge j(0)} B(nt-n^{1/2}\varepsilon/M_k,n,2^{-k})$$

$$\le \Sigma_{k\ge j(0)} \exp(-\varepsilon^2 2^{k-1}/M_k^2) < \varepsilon/4 .$$

For the upper tail terms in (6.14) with $y = \gamma/(n^{1/2}t) = 2^k \varepsilon/(n^{1/2}M_k) \leq y_\beta$, we also have by (6.8)

$$(6.18) \qquad \Sigma_{k\geq j', y\leq y_\beta} \exp(-\beta\varepsilon^2 2^{k-1}/M_k^2) < \varepsilon/4$$

for some $j' \geq \max(j(0), j_1)$.

It remains to treat the upper tail terms E with $y > y_\beta$. Choose $a < \infty$ large enough so that

$$(6.19) \qquad \exp(-a^{1/2})/(1-\exp(-a^{1/2}(\log 2)/2)) < \varepsilon/4 .$$

By Lemma 6.10, we may restrict to those k such that $2^{-k} \geq a/n$, or $k \leq r := r(n) := [\log_2(n/a)]$. Let $M(n)$ be the set of all k with $j' \leq k \leq r(n)$ for which $y \equiv 2^k\varepsilon/(n^{1/2}M_k) > y_\beta$. Then by (6.12) and (6.15) it suffices to bound

$$T_n := \Sigma_{k\in M(n)} \exp(-n2^{-k}(2^k\varepsilon/(n^{1/2}M_k))(\beta y_\beta - 1)).$$

Then

$$\begin{aligned} T_n &\leq \Sigma_{k\in M(n)} \exp(-n^{1/2}/M_k) && \text{by (6.16)} \\ &\leq \Sigma_{k\in M(n)} \exp(-2^{-k/2}n^{1/2}) && \text{by (6.9)} \\ &\leq \Sigma_{s=0}^{\infty} \exp(-2^{(s-r)/2}n^{1/2}) \\ &\leq \Sigma_{s=0}^{\infty} \exp(-2^{s/2}a^{1/2}) \end{aligned}$$

since $2^{-r}n \geq a$ by choice of $r := r(n)$. Now $2^{s/2} = \exp(s(\log 2)/2) \geq 1 + s(\log 2)/2$ for all s, so

$$T_n \leq \exp(-a^{1/2})/(1-\exp(-a^{1/2}(\log 2)/2)) < \varepsilon/4$$

by a geometric series and (6.19). Take n_0 such that the probability in Lemma 6.10 is less than $\varepsilon/4$ for $n \geq n_0$. Combining with (6.17) and (6.18) then gives

$$(6.20) \qquad \Sigma_{k\geq j'} \Pr(M_k|\nu_n([0,2^{-k}])| > \varepsilon) < \varepsilon, \quad n \geq n_0 .$$

Let $\delta := 2^{-(j'+3)/2}$, $f_k := M_k 1_{[0,2^{-k}]}$.
If $i < k$, then

$$\rho_P(f_i,f_k)^2 = \int_0^1 (f_i - M_i 2^{-i} - f_k + M_k 2^{-k})^2\, dx$$
$$\geq \int_{2^{-k}}^{2^{-i}} (M_i - 2^{-i} M_i + M_k 2^{-k})^2\, dx$$
$$\geq 2^{-i-3} M_i^2 \geq 2^{-j'-3} = \delta^2$$

if $i \leq j'$, as $M_i \geq 1$ for all i. Thus if $\rho_P(f_i,f_k) < \delta$, then $i > j'$. So for $n \geq n_0$, by (6.20),

$$\Pr\{\sup\{|\nu_n(f_i - f_k)|: \rho_P(f_i,f_k) < \delta\} > 2\varepsilon\} < \varepsilon.$$

As $i \to \infty$, $\rho_P(f_i,0) \to 0$ by (6.8) and the sentence after it. Thus $\{f_i\}$ is totally bounded for ρ_P, and is a functional Donsker class (Dudley, 1984, Theorem 4.1.1). So c) holds and Theorem 6.3 is proved.

7. Corollaries and remarks on Sec. 6.

7.1 Corollary. If $h \geq 0$ is any non-increasing function on $]0,1[$ with $h(t) = o((t \log\log \frac{1}{t})^{-1/2})$ then the conditions of Theorem 6.3 all hold.

Proof. It is easy to verify f).

7.2 Example. It will be shown that for a positive decreasing h, the sufficient condition in Cor. 7.1 is not necessary for its conclusion. This will be a counter-example to Shorack [1979, (3), (4)] and thus to Shorack and Wellner (1982, Theorem 1.1). A review of the latter by M. Csörgő (1984) also noted the error, stating that S. Csörgő and D. M. Mason also found such examples.

Let $t_k := \exp(-e^k)$, $k = 1,2,\cdots$, and $M_k := (t_k \cdot \log|\log t_k|)^{-1/2} = (kt_k)^{-1/2}$. Let $h(t) := M_k$ for $t_{k+1} < t \leq t_k$, $k = 1,2,\cdots$, $h(t) := M_0 := 1$ for $t_1 < t < t_0 := 1$. Then h is non-increasing and $h(t) \neq o((t \log \log \frac{1}{t})^{-1/2})$ as $t \downarrow 0$.

For any $\mathcal{E} > 0$,

$$\int_0^1 \frac{1}{t} \exp(-\mathcal{E}/(th^2(t)))\, dt =$$

$$\Sigma_{k=0}^{\infty} \int_{t_{k+1}}^{t_k} \frac{1}{t} \exp(-\mathcal{E}/(tM_k^2))\, dt$$

$$= \Sigma_{k=0}^{\infty} \int_{t_{k+1}}^{t_k} \frac{1}{t} \exp(-\mathcal{E}\, k \cdot t_k/t)\, dt .$$

Each term in the sum is clearly finite. Note that $\frac{d}{dt}(\frac{1}{t} \exp(-C/t)) \geq 0$ if $C \geq t$. Thus the sum, for $k \geq 1/\mathcal{E}$, is bounded above by

$$\Sigma_{k \geq 1/\mathcal{E}} (t_k - t_{k+1}) t_k^{-1} e^{-k\mathcal{E}}$$

$$\leq \Sigma_{k \geq 1/\mathcal{E}} e^{-k\mathcal{E}} < \infty .$$

So f) does hold, as do the other conditions of Theorem 6.3.

Nevertheless, Cor. 7.1 is sharp in at least two senses, shown by the next two results.

7.3 __Corollary__. If for some $\delta > 0$, $h(t) \geq \delta(t \log|\log t|)^{-1/2}$ for $0 < t \leq \delta$, then $\mathcal{G}_h$ is not a functional Donsker class.

__Proof__. If $\mathcal{G}_h$ is a functional Donsker class, then so is $\mathcal{G}_j$ where $j(t) := \delta(t \log|\log t|)^{-1/2}$, $0 < t \leq \delta$; $j(t) := 0$, $\delta < t \leq 1$, by Theorem 5.3 (one can take $\delta < e^{-2}$ so that j is non-increasing). But this contradicts Theorem 6.3f) for $0 < \mathcal{E} \leq \delta^2$, Q.E.D.

7.4 __Proposition__. If $h \geq 0$, $h \downarrow$ and $t \longrightarrow t^{1/2}h(t) \uparrow$ (is non-decreasing) for $0 < t \leq 1$, then the conditions of Theorem 6.3 are equivalent to $h(t) = o((t \log \log \frac{1}{t})^{-1/2})$ as $t \downarrow 0$.

__Proof__. Cor. 7.1 gives one direction. For the other, suppose that for some $\delta > 0$, and $t_k \downarrow 0$, $h(t_k) \geq \delta(t_k \log|\log t_k|)^{-1/2}$ for $k = 1,2,\cdots$. Taking a subsequence, it can be assumed that

$t_{k+1} < t_k^2$ for all k. For $t_k \le t \le t_{k-1}$, since $t^{1/2}h \uparrow$, $th^2(t) \ge t_k h^2(t_k) \ge \delta^2/(\log|\log t_k|)$. Then for $\mathcal{E} := \delta^2$,

$$\int_{t_k}^{t_{k-1}} \frac{1}{t} \exp(-\mathcal{E}/(th^2(t)))\, dt$$

$$\ge \int_{t_k}^{t_{k-1}} \frac{1}{t} \exp(-\log|\log t_k|)$$

$$= (\log t_{k-1} - \log t_k)/|\log t_k|$$

$$= 1 - |\log t_{k-1}|/|\log t_k| \ge \frac{1}{2}.$$

Thus the sum of these numbers over k diverges and f) fails, Q.E.D.

7.5 Remarks. Note that in Theorem 6.3 g2) does not imply g1): let $h(t) := 1/t$. Nor does g1) imply g2): let $h(t) := 1/(tLLL(1/t))^{1/2}$ as $t \downarrow 0$ where $Lx := \max(1, \log x)$.

If $t^{1/2}h(t)$ is non-decreasing, then g2) implies e) by the classical Kolmogorov-Petrovskii test, see Petrovskii (1935), Erdős (1942), and Itô and McKean (1974, pp. 33-35). O'Reilly (1974, p.644) states that $t^{1/2}h$ non-decreasing is "not needed for the relevant half of the test". This is correct in the context there, where g1) has been proved, but not in general: if $h(t) = 1/t$, g2) holds but not e).

Itô and McKean (1974, pp.33-35) prove that g) implies e), in effect, although they do not explicitly make the conjunction of g1) with g2). At any rate, integral condition f) seems preferable to g2), as f) is simpler and does not require an extra condition such as g1).

O'Reilly's (1974) assumption that h is continuous would also follow from the two assumptions $h\downarrow$ and $t^{1/2}h\uparrow$. At any rate, continuity is now seen to be unnecessary.

7.6 Corollary. On any probability space $(A, \mathcal{A}, P)$, for any function

$h \in \mathcal{L}^2(P)$, $\mathcal{G}_h$ is a functional Donsker class.

Proof. By Theorem 6.2, one can take $(A, \mathcal{A}, P) = (]0,1[, \mathcal{B}, \lambda)$, and h non-increasing. Then

$$\Sigma_{k=1}^{\infty} 2^{-k-1} h(2^{-k})^2 \leq \int_0^1 h^2(t)\, dt < \infty .$$

Let $a_k := 2^k/h(2^{-k})^2$. Then $\Sigma_{k=1}^{\infty} 1/a_k < \infty$. Thus for every $\varepsilon > 0$, $\Sigma_k \exp(-\varepsilon a_k) < \infty$, which gives f') of Theorem 6.3, hence c), the conclusion.

Note. Cor. 7.6 is not surprising since by a theorem of D. Pollard (1982), $\{h 1_{h \geq M} : 0 \leq M < \infty\}$ is a functional Donsker class and $M 1_{h \geq M} \leq h 1_{h \geq M}$. Still, Cor. 7.6 is sharp in the following senses:

7.7 Proposition. For any sequence $t_k \to 0$ (however slowly) as $k \to \infty$, there exist $M_k \uparrow +\infty$ as $k \to \infty$ such that $\Sigma_{k=1}^{\infty} t_k M_k^2/2^k < \infty$ but $M_k^2/2^k \not\to 0$ as $k \to \infty$, so $\mathcal{F} := \{M_k 1_{[0,2^{-k}]}\}_{k \geq 1}$ is not a functional Donsker class.

For any function $\varphi(x) \downarrow 0$ (however slowly) as $x \to +\infty$, there is a function $h\downarrow$ such that

$$(7.8) \qquad \int_0^1 \varphi(1/t) h^2(t) dt < \infty$$

and $\mathcal{G}_h$ is not a functional Donsker class. Also, there is an $h\downarrow$ such that

$$(7.9) \qquad \int_0^1 \varphi(h(t)) h^2(t) dt < \infty$$

and $\mathcal{G}_h$ is not a functional Donsker class.

Proof. Let $s_k := \sup_{j \geq k} t_j \downarrow 0$ as $k \to \infty$, with $s_k \geq t_k$, so we may assume $t_k \downarrow 0$. Take a subsequence $k(i) \uparrow +\infty$ such that $\Sigma_i t_{k(i)} < \infty$, $k(0) := 1$. Let $M_j := 2^{k(i)/2}$ for $k(i) \leq j < k(i+1)$. Then $M_{k(i)}^2/2^{k(i)} = 1 \not\to 0$. So by g1), $\mathcal{F}$ is not a functional Donsker class. Also

$$\Sigma_k \, t_k M_k^2/2^k \le \Sigma_{i=0}^{\infty} 2^{k(i)} t_{k(i)} \, \Sigma_{k\ge k(i)} 2^{-k}$$

$$\le 2\Sigma_{i=0}^{\infty} t_{k(i)} < \infty,$$

as stated.

For (7.8) let $t_k := \varphi(2^k)$, define the M_j as above, and let

$$h(t) := M_j, \; 2^{-j-1} < t \le 2^{-j}, \; j = 0,1,\cdots .$$

Then $\mathcal{G}_h$ is not functional Donsker class and

$$\int_0^1 \varphi(1/t)h^2(t)dt = \Sigma_{k=0}^{\infty} \int_{2^{-k-1}}^{2^{-k}} \varphi(1/t)M_k^2 \, dt$$

$$\le \Sigma_{k=0}^{\infty} M_k^2 2^{-k-1} t_k < \infty,$$

so (7.8) holds.

For (7.9), let $t_k := \varphi(2^{k/2})$ and apply the previous arguments.

Thus each of the sufficient conditions in Corollaries 7.1 and 7.6 is sharp in its own ways. Neither implies the other: if $h(t) = (t \log(3/t))^{-1/2}$ then h satisfies the conditions of 7.1 but not 7.6. Conversely, let $s_k := \exp(-\exp(e^k))$, $k = 1,2,\cdots$, $M_k := (s_k \log|\log s_k|)^{-1/2} = (e^k s_k)^{-1/2}$. Let $h(t) := M_k$ for $s_{k+1} < t \le s_k$, $k = 1,2,\cdots, h(t) := M_0 := 1$ for $s_1 < t < s_0 := 1$. Then h is non-increasing and $h(t) \ne o((t \log|\log t|)^{-1/2})$ as $t \downarrow 0$. Also

$$\int_0^1 h^2(t)dt = \Sigma_{k=0}^{\infty} M_k^2(s_k - s_{k+1}) \le \Sigma_{k=0}^{\infty} e^{-k} < \infty .$$

For other recent work related to the equivalent conditions in Sec. 6 see Stute (1982), but in light of M. Csörgő (1984).

This paper does not treat laws of the iterated logarithm. See Dudley and Philipp (1983, Theorem 7.5) and references there.

Acknowledgment. I am grateful to M. Talagrand, J. Hoffmann-Jørgensen and N. T. Andersen for communicating to me some of their unpublished work. Talagrand (1984) stated a result close to Theorem 5.2 above.

REFERENCES

Chibisov, D. M. (1964). Some theorems on the limiting behavior of the empirical distribution function. Trudy Mat. Inst. Steklov (Moscow) 71 104-112; Selected Transls. Math. Statist. Prob. 6 147-156.

Cohn, Donald L. (1980). Measure Theory. Birkhäuser, Boston.

Csörgő, M. (1984). Review of Shorack and Wellner (1982). Math. Revs. 84f:60041.

Donsker, M. D. (1952). Justification and extension of Doob's heuristic approach to the Kolmogorov-Smirnov theorems, Ann. Math. Statist. 23 pp. 277-281.

Drake, F. R. (1974). Set Theory: An Introduction to Large Cardinals. North-Holland, Amsterdam.

Dudley, R. M. (1966). Weak convergence of probabilities on nonseparable metric spaces and empirical measures on Euclidean spaces. Illinois J. Math. 10 109-126.

__________ (1967a). Measures on non-separable metric spaces. Ibid. 11 449-453.

__________ (1967b). The sizes of compact subsets of Hilbert spaces and continuity of Gaussian processes. J. Functional Analysis 1 290-330.

__________ (1968). Distances of probability measures and random variables. Ann. Math. Statist. 39 1563-1572.

__________ (1973). Sample functions of the Gaussian process. Ann. Probab. 1 66-103.

__________ (1978). Central limit theorems for empirical measures. Ibid. 6 899-929; Correction, ibid. 7 (1979) 909-911.

__________ (1984). A course on empirical processes. Ecole d'été de probabilités de St.-Flour, 1982. Lecture Notes in Math. 1097 2-142.

__________, and Walter Philipp (1983). Invariance principles for sums of Banach space valued random elements and empirical processes. Z. Wahrsch. verw. Geb. 62 509-552.

Erdös, P. (1942). On the law of the iterated logarithm. Ann. Math. 43 419-436.

Feldman, Jacob (1971). Sets of boundedness and continuity for the canonical normal process. Proc. Sixth Berkeley Symp. Math. Statist. Prob. 2, 357-368. Univ. Calif. Press.

Fernandez, Pedro J. (1974). Almost surely convergent versions of sequences which converge weakly. Bol. Soc. Brasil. Math. 5 51-61.

Halmos, P. (1950). Measure Theory. Princeton, Van Nostrand. 2d. printing, Springer, N. Y. 1974.

Hoffmann-Jørgensen, J., and Niels Trolle Andersen (1984). Personal communication.

________________ (1984). Envelopes and perfect random variables (preprint, section of forthcoming book).

Itô, K., and H. P. McKean Jr. (1974). Diffusion processes and their sample paths. Springer, N. Y. (2d. printing, corrected).

Marczewski, E., and R. Sikorski (1948). Measures in nonseparable metric spaces. Colloq. Math. 1 133-139.

O'Reilly, N. E. (1974). On the weak convergence of empirical processes in sup-norm metrics. Ann. Probab. 2 642-651.

Petrovskii, I. G. (1935). Zur ersten Randwertaufgabe der Wärmeleitungsgleichung. Compositio Math. 1 383-419.

Pollard, D. B. (1982). A central limit theorem for empirical processes. J. Austral. Math. Soc. Ser. A 33 235-248.

Pyke, R. (1969). Applications of almost surely convergent constructions of weakly convergent processes. Lecture Notes in Math. 89 187-200.

Shorack, G. R. (1979). Weak convergence of empirical and quantile processes in sup-norm metrics via KMT-constructions. Stochastic Processes Applics. 9 95-98.

____________, and J. Wellner (1982). Limit theorems and inequalities for the uniform empirical process indexed by intervals. Ann. Probab. 10 639-652.

Skorohod, A. V. (1956). Limit theorems for stochastic processes. Theor. Prob. Appls. 1 261-290 (English), 289-319 (Russian).

Stute, W. (1982). The oscillation behavior of empirical processes, Ann. Probab. 10 86-107.

Talagrand, M. (1984). The Glivenko-Cantelli problem (preprint).

Wichura, M. J. (1970). On the construction of almost uniformly convergent random variables with given weakly convergent image laws. Ann. Math. Statist. 41 284-291.

Room 2-245, MIT
Cambridge, MA 02139
USA

COMPARAISON DE MESURES GAUSSIENNES ET DE MESURES PRODUIT DANS LES ESPACES DE FRECHET SEPARABLES

par

X. FERNIQUE

Département de Mathématiques

7 rue René Descartes

67084 STRASBOURG CEDEX France

Sommaire : Soit P une probabilité gaussienne sur un espace produit $(E,\mathcal{B}) = (\prod_{t\in T} E_t, \bigotimes_{t\in T} \mathcal{B}_t)$ d'espaces de Fréchet séparables indexés par un ensemble arbitraire T ; on montre que si P n'est pas orthogonale à une probabilité produit $Q = \bigotimes_{t\in T} Q_t$, alors P est équivalente au produit $\overline{P} = \bigotimes_{t\in T} P_t$ de ses propres marges. La preuve est basée sur une majoration de la distance de Hellinger $d_H(P,\overline{P})$ à partir de $d_H(P,Q)$ et sur l'utilisation de la propriété de la limite centrale dans les espaces de Fréchet séparables.

0. Introduction, Notations, Rappels.

0.1. Dans un travail précédent [3], nous avons montré que si une v.a. gaussienne à valeurs dans $\mathbb{R}^{\mathbb{N}}$ a une loi non orthogonale à celle d'une v.a. à composantes indépendantes dans le même espace, alors cette loi gaussienne est équivalente au produit de ses propres marges. On se propose d'établir le même résultat dans le cadre vectoriel plus large des espaces de Fréchet séparables. La méthode générale d'étude est semblable à celle de [3], son adaptation à la situation présentée ici a posé de nombreux problèmes techniques liés en particulier à l'absence de mesures de Lebesgue sur les espaces facteurs.

0.2. Notations générales : La notation $(E,\mathcal{B})$ désignera un espace mesurable. Quand E sera un espace de Fréchet (c'est-à-dire un espace vectoriel topologique localement convexe séparé, métrisable et complet) séparable, $\mathcal{B}$

désignera sa tribu topologique qui est aussi engendrée (parce que E est un espace polonais) par son dual topologique que nous noterons E'. Quand $(E_t, \mathcal{B}_t)$ sera une famille d'espaces mesurables indexée par un ensemble T, on notera $(E,\mathcal{B})$ l'espace mesurable produit $(\prod_{t \in T} E_t, \bigotimes_{t \in T} \mathcal{B}_t)$; pour toute partie S de T, on notera $(E_S, \mathcal{B}_S)$ le produit partiel associé et p_S sera l'application canonique de E dans E_S ; l'application p_t sera donc la projection de E sur E_t. Les seules mesures utilisées seront positives et σ-finies ; pour toute mesure μ sur un espace produit $(E,\mathcal{B}) = \prod_{t \in T} (E_t, \mathcal{B}_t)$ et tout élément t de T, on notera μ_t la marge d'indice t de μ qui est donc l'image de μ par p_t ; pour toute probabilité P sur le même espace, on notera $\overline{P}$ le produit $\bigotimes_{t \in T} P_t$ de ses marges. On dira que P est une probabilité produit sur $\prod_{t \in T} (E_t, \mathcal{B}_t)$ si $P = \overline{P}$. Si $(E,\mathcal{B})$ est un espace de Fréchet séparable, on dira que P est une probabilité gaussienne sur $(E,\mathcal{B})$ si pour tout élément y de E', l'image $y.P$ de P par y est une mesure gaussienne centrée sur $(\mathbb{R}, \mathcal{B}(\mathbb{R}))$. Si $(E,\mathcal{B}) = \prod_{t \in T} (E_t, \mathcal{B}_t)$ est un produit d'espaces de Fréchet séparables, on dira que P est une probabilité gaussienne sur $(E,\mathcal{B})$ si pour toute partie au plus dénombrable S de T, $p_S P$ est une probabilité gaussienne sur l'espace de Fréchet séparable $(E_S, \mathcal{B}_S)$.

0.3. <u>Comparaison de mesures</u> : Soient $(E,\mathcal{B})$ un espace mesurable et (P,Q) un couple de mesures sur cet espace ; on dit que P et Q sont orthogonales (ou étrangères) s'il existe un élément A de $\mathcal{B}$ tel que $P(A) = 0$ et $Q(\complement A) = 0$, on note $P \perp Q$; on dit que Q est dominée par P (ou absolument continue pour P) si pour tout élément A de $\mathcal{B}$, $P(A) = 0$ implique $Q(A) = 0$, on note $Q \ll P$; on dit que P et Q sont équivalentes si $P \ll Q$ et $Q \ll P$, on note $P \equiv Q$. Ces trois notions se traduisent simplement en termes de décomposition de Radon-Nikodym ; Q peut en effet s'écrire sous la forme :

$$Q = \rho.P + I_A.Q ,$$

où ρ est une fonction mesurable positive et A un élément de $\mathcal{B}$,

P-négligeable ; ρ est la dérivée de Radon-Nikodym de Q par rapport à P et on a :

$$P \perp Q \rightleftarrows P(\rho > 0) = 0 \quad ; \quad Q \ll P \rightleftarrows Q(A) = 0 \quad ; \quad P \ll Q \rightleftarrows P(\rho = 0) = 0 \; ;$$

on note $\rho = \frac{dQ}{dP}$ (cf. par exemple [1]) .

0.4. <u>L'intégrale de Hellinger</u> : Le maniement des notions ci-dessus utilise l'outil de l'intégrale de Hellinger que nous rappelons maintenant : soient P et Q deux probabilités sur $(E,\mathcal{B})$, l'intégrale $\int\sqrt{(\frac{dP}{d\mu}\frac{dQ}{d\mu})}\,d\mu$ ne dépend pas de la mesure μ choisie pour la définir pourvu qu'elle domine <u>l'une au moins</u> des probabilités P ou Q (il n'est pas nécessaire qu'elle les domine toutes les deux) ; on note alors cette intégrale $\int\sqrt{dP\,dQ}$; pour que P et Q soient orthogonales, il faut et il suffit que $\int\sqrt{dP\,dQ}$ soit nulle. Si la mesure μ ne domine aucune des probabilités P et Q , alors l'intégrale $\int\sqrt{(\frac{dP}{d\mu}\frac{dQ}{d\mu})}\,d\mu$ sera inférieure ou égale à $\int\sqrt{dP\,dQ}$; si P et Q sont orthogonales, elle sera donc encore nulle. Si $\mathcal{C}$ est une sous-tribu de $\mathcal{B}$ et si $P_{\mathcal{C}}$ et $Q_{\mathcal{C}}$ sont les restrictions de P et Q à cette sous-tribu, on a $\int\sqrt{dP\,dQ} \leq \int\sqrt{dP_{\mathcal{C}}\,dQ_{\mathcal{C}}}$. Enfin si $(\mathcal{C}_n\,,\, n \in \mathbb{N})$ est une suite croissante de sous-tribus engendrant $\mathcal{B}$, on a : $\int\sqrt{dP\,dQ} = \lim \int\sqrt{dP_{\mathcal{C}_n}\,dQ_{\mathcal{C}_n}}$ (cf. par exemple [6]) .

0.5. <u>Les théorèmes de Kakutani et de Hajek-Feldman</u> : Dans certaines situations, on sait que les notions d'orthogonalité et d'équivalence sont complémentaires : si P et Q sont des probabilités produit sur un espace produit dont les marges sur chaque espace facteur sont équivalentes, alors <u>un théorème de S. Kakutani</u> indique que P et Q sont soit orthogonales, soit équivalentes.

Nous détaillons davantage le cas gaussien : soit P une probabilité gaussienne sur un espace de Fréchet séparable $(E,\mathcal{B})$. On sait que pour toute semi-norme N continue sur E et tout $a > 0$ assez petit, $\exp(a\,N^2)$ est P-intégrable ; en particulier E' est contenu dans $L^2(E,\mathcal{B},P)$; on note

K l'adhérence de E' dans $L^2(E,\mathcal{B},P)$; K est muni naturellement de la structure hilbertienne K_P induite. Pour tout élément y de K, la v.a. $x \to y(x).x$ à valeurs dans $(E,\mathcal{B})$ est P-intégrable ; son espérance $\int y(x).x \, dP(x)$ est un élément de E noté $\varphi(y)$. L'image de E' par φ est un sous-espace de E qui est appelé le sous-espace autoreproduisant de P ; H est muni naturellement de la structure hilbertienne définie par $<\varphi(y_1),\varphi(y_2)> = \int y_1.y_2 \, dP$. Nous énonçons dans ces conditions et sous la forme où nous l'utiliserons le théorème de Hajek-Feldman :

THEOREME 0.5. ([2],[4] et par exemple [6]) : Soient P et Q deux probabilités gaussiennes sur l'espace de Fréchet séparable $(E,\mathcal{B})$; alors les propriétés suivantes sont équivalentes :

(i) P et Q ne sont pas orthogonales,

(ii) P et Q sont équivalentes,

(iii) Les espaces autoreproduisants associés à P et Q sont identiques (et donc munis de structures hilbertiennes H_P et H_Q équivalentes); de plus il existe une base orthonormale $(b_n, n \in \mathbb{N})$ de K_P telle que :

0.5.1. $$\sum_{n\in\mathbb{N}} \sum_{m\in\mathbb{N}} \left(\int b_n \otimes b_m dP - \int b_n \otimes b_m dQ\right)^2 < \infty$$

ou si l'on préfère

$$\int x \otimes x \, dP(x) - \int x \otimes x \, dQ(x) \in H_P \otimes H_P .$$

0.6. Le plan d'étude sera le suivant : dans un premier paragraphe, nous étudions les variations de $\int\sqrt{dP\,dQ}$, P étant une probabilité donnée sur un produit fini d'espaces mesurables lorsque Q y parcourt l'ensemble des probabilités produit. Dans le deuxième paragraphe, nous étudions de plus près les mêmes variations lorsque P est gaussienne sur un produit fini d'espaces de Fréchet séparables. Le troisième paragraphe énoncera et démontrera les résultats principaux ; le dernier paragraphe traite d'une application.

1. <u>Approximation de mesures par les mesures produit, le cas d'un nombre fini de facteurs.</u>

1.1. Dans tout ce paragraphe, T est un ensemble fini indexant une famille $(E_t, \mathcal{B}_t)$ d'espaces mesurables et $(E,\mathcal{B})$ est leur produit ; on note P une probabilité et $\mu = \bigotimes_{t\in T} \mu_t$ une mesure produit sur $(E,\mathcal{B})$. On suppose que P n'est par orthogonale à μ et on étudie les variations de $\int\sqrt{\frac{dP}{d\mu}\frac{dQ}{d\mu}}\, d\mu$ lorsque Q parcourt l'ensemble des probabilités produit sur $(E,\mathcal{B})$; associant à Q sa dérivée $\frac{dQ}{d\mu} = \bigotimes_{t\in T} \frac{dQ_t}{d\mu_t}$, on pose

$F_\mu = \{f = \bigotimes_{t\in T} f_t : \forall\, t \in T\,,\ f_t \in L^1(E_t,\mathcal{B}_t,R^+)\,,\ \int f_t\, d\mu_t \leq 1\}$

et on étudie les variations de l'application Φ_μ de F_μ dans R^+ définie par :

$$\Phi_\mu(f) = \int(\sqrt{\frac{dP}{d\mu}} \prod_{t\in T} \sqrt{f_t})\, d\mu\ .$$

La topologie naturelle sur F_μ est la topologie associée à la convergence faible de $\sqrt{f}$ dans $L^2(E,\mu)$; elle coïncide d'ailleurs ici avec la topologie produit des topologies faibles des $\sqrt{f_t}$ dans les $L^2(E_t,\mu_t)$. Comme $\frac{dP}{d\mu}$ est μ-intégrable, $\sqrt{\frac{dP}{d\mu}}$ appartient à $L^2(E,\mu)$ de sorte que pour la topologie W, Φ_μ est continue sur F_μ ; F_μ étant W-compacte, l'ensemble $\mathcal{M}_\mu(P) = \{f \in F_\mu : \Phi_\mu(f) = \sup_{F_\mu} \Phi_\mu = M_\mu(P)\}$ est donc un ensemble non vide compact. Dans cette situation, la proposition 1.2. analyse certaines propriétés de $\mathcal{M}_\mu(P)$ (nous y omettons les indices μ).

PROPOSITION 1.2. : (i) $M(P)$ <u>appartient à</u> $]0,1]$,

(ii) <u>Si</u> $\bar{f}$ <u>appartient à</u> $\mathcal{M}(P)$, <u>alors</u> $\bar{f}\, d\mu$ <u>est une probabilité et on a</u> :

1.2.1. $\forall\, t\in T\,,\ \sqrt{\bar{f}_t}(x_t) = \frac{1}{M(P)}\int\sqrt{\frac{dP}{d\mu}}(x) \prod_{s\neq t} \sqrt{\bar{f}_s}(x_s) \bigotimes_{s\neq t} d\mu_s(x_s)$ p.p. dans E_t ,

1.2.2. $\forall\, t\in T\,,\ \bar{f}_t(x_t) \leq \frac{1}{M(P)^2} \frac{dP_t}{d\mu_t}(x_t)$ <u>p.p. dans</u> E_t ,

1.2.3. $\Phi(\bigotimes_{t\in T} \frac{dP_t}{d\mu_t}) \leq M(P) \leq \Phi(\bigotimes_{t\in T} \frac{dP_t}{d\mu_t})(1 + |T|)^{-1}$.

(iii) $\mathcal{M}(P)$ est compact pour la topologie forte de $L^1(E)$ qui y coïncide avec la topologie W .

Remarque : la propriété 1.2.3. a un intérêt particulier, elle indique que le produit $\overline{P}$ des marges de P réalise "presque" la meilleure approximation de P parmi les probabilités produit. On notera pourtant que l'ordre de cette approximation tel qu'il est évalué ici dépend du nombre $|T|$ des facteurs du produit ; on verra dans la suite (proposition 2.4.) que si P est gaussienne, on peut obtenir une meilleure évaluation.

Démonstration : (i) L'inégalité de Cauchy-Schwarz fournit :

$$M(P) = \sup_{f \in F} \Phi(f) \leq \sup_{f \in F} \sqrt{\int \frac{dP}{d\mu}\, d\mu} \sqrt{\prod_{t \in T} \int f_t \, d\mu_t} \leq 1 \ ;$$

prouvons par l'absurde que M(P) est strictement positif : dans le cas contraire $\sqrt{\frac{dP}{d\mu}}$ serait dans $L^2(E)$ orthogonal au sous-espace dense engendré par F ; $\frac{dP}{d\mu}$ serait nul p.p., ce qui est contradictoire avec la non-orthogonalité de P et de μ .

(ii) Le point principal de la démonstration est la preuve de la relation 1.2.1. Soient $\overline{f}$ un élément de $\mathcal{M}(P)$ et f un élément de F ; nous fixons un élément t de T et nous leur associons l'application φ de $[0,1]$ dans $\mathbb{R}$ définie par :

$$\varphi(u) = \Phi(g) \quad \text{où} \quad g_t = u\overline{f}_t + (1-u)f_t \quad \text{et} \quad \forall\, s \neq t \ , \ g_s = \overline{f}_s \ ;$$

puisque $\overline{f}$ appartient à $\mathcal{M}(P)$, φ est maximale en $u = 1$ et $\varphi(1) = M(P)$; nous aimerions écrire que φ est dérivable en $u = 1$ et que sa dérivée, calculable par dérivation sous le signe somme, est nulle en $u = 1$; ceci fournirait effectivement 1.2.1. ; mais ne pouvant justifier ces opérations, nous opérons avec plus de précautions. Pour tout $\epsilon > 0$ et tout $u \in [0,1[$, la propriété $\varphi(u) \leq \varphi(1)$ et l'égalité :

$$\sqrt{\overline{f}_t} - \sqrt{g_t} = \frac{(1-u)(\overline{f}_t - f_t)}{\sqrt{\overline{f}_t} + \sqrt{g_t}}$$

fournissent :

$$\int \frac{2\,\bar f_t \prod_{s\neq r} \sqrt{\bar f_s}\,\sqrt{\frac{dP}{d\mu}}}{\sqrt{\bar f_t}+\sqrt{u\,\bar f_t+(1-u)f_t}}\,d\mu \geq \Big(\int_{\{\bar f_t=0\}}+\int_{\{\bar f_t>\varepsilon\, f_t\}}\Big) \frac{2\,f_s \prod_{s\neq t} \sqrt{\bar f_s}\,\sqrt{\frac{dP}{d\mu}}}{\sqrt{\bar f_t}+\sqrt{u\,\bar f_t+(1-u)f_t}}\,d\mu\,.$$

Quand u tend vers 1, le théorème de convergence dominée par $2 \prod_{s\in T} \sqrt{\bar f_s}\,\sqrt{\frac{dP}{d\mu}}$ qui est μ-intégrable montre que le premier membre a la limite finie $M(P)$; le premier terme du second membre a donc une limite supérieure finie, ce qui tenant compte de $\{\bar f_t = 0\}$ s'écrit :

$$\forall\, f_t \in L^1(E_t,\mu_t;R^+)\ ,\int \sqrt{f_t}\Big\{I_{\{\bar f_t=0\}}\int \prod_{s\neq t} \sqrt{\bar f_s}\,\sqrt{\frac{dP}{d\mu}} \bigotimes_{s\neq t} d\mu_s\Big\}d\mu_t = 0$$

En reprenant le deuxième argument de (i), on en déduit :

$$I_{\{\bar f_t=0\}}(x_t)\Big\{\int \prod_{s\neq t} \sqrt{\bar f_s}\,\sqrt{\frac{dP}{d\mu}} \bigotimes_{s\neq t} d\mu_s\Big\} = 0 \quad ,\ \text{p.p. dans } E_t\ ,$$

et ceci fournit la relation 1.2.1. quand son premier membre est nul.

On peut aussi appliquer au second terme du second membre le théorème de convergence dominée par $\varepsilon^{-\frac{1}{2}}(1+\sqrt{u})^{-1}\sqrt{f_t}\prod_{s\neq t} \sqrt{\bar f_s}\,\sqrt{\frac{dP}{d\mu}}$ qui est μ-intégrable ; il donne :

$$M(P) \geq \sup_{\varepsilon>0} \int_{\{\bar f_t>\varepsilon\, f_t\}} \frac{f_t \prod_{s\neq t} \sqrt{\bar f_s}\,\sqrt{\frac{dP}{d\mu}}}{\sqrt{\bar f_t}}\,d\mu\ ,$$

et donc puisque f_t est fini μ_t-p.p. :

$$M(P) \geq \int_{\{\bar f_t>0\}} \Big\{\prod_{s\neq t} \sqrt{\bar f_s}(x_s)\,\sqrt{\frac{dP}{d\mu}}(x) \bigotimes_{s\neq t} d\mu(x_s)\Big\}\bar f_t(x_t)^{-\frac{1}{2}}\Big\{f_t(x_t)d\mu_t(x_t)\Big\}\,.$$

Cette dernière inégalité est vraie pour tout élément f_t de la boule unité de $L^1(E_t,\mu_t\,;R^+)$; en l'appliquant à des multiples d'indicatrices d'ensembles bien choisis, on en déduit :

$$M(P)\,\sqrt{\bar f_t}(x_t) \geq I_{\{\bar f_t>0\}}(x_t)\int \prod_{s\neq t} \sqrt{\bar f_s}(x_s)\,\sqrt{\frac{dP}{d\mu}}(x) \bigotimes_{s\neq t} d\mu_s(x_s) \quad ,\mu_t\text{-p.p.}$$

On multiplie par $\sqrt{\bar{f}_t(x_t)}$ et on intègre pour $d\mu_t(x_t)$; puisque $\bar{f}$ appartient à $\mathcal{M}(P)$, l'intégrale du second membre est $M(P)$ et celle du premier est $(\int \bar{f}_t \, d\mu_t)\, M(P)$; ceci n'est compatible avec l'ordre de ces deux membres que si $\int \bar{f}_t \, d\mu_t = 1$, c'est-à-dire si $(\bar{f}\, d\mu)$ est une probabilité et si aussi l'inégalité qui a été intégrée est en fait p.p une égalité : ceci démontre la relation 1.2.1. quand son premier membre n'est pas nul et donc finalement dans tous les cas.

La relation 1.2.2. se déduit immédiatement de 1.2.1. par l'inégalité de Cauchy-Schwarz. Enfin la relation 1.2.3. s'obtient en reportant les majorations des différents $\sqrt{\bar{f}_t}$ par 1.2.2. dans l'intégrale définissant $\Phi(\bar{f}) = M(P)$.

(iii) Soit $(f^k , k \in \mathbb{N})$ une suite d'éléments de $\mathcal{M}(P)$; la W-compacité de $\mathcal{M}(P)$ permet d'en extraire une suite partielle indexée par $\mathbb{N}'$ inclus dans $\mathbb{N}$ telle que pour tout $s \in T$, la suite $(f_s^k , k \in \mathbb{N}')$ converge faiblement dans $L^2(E_s,\mu_s)$; dans ces conditions, pour tout $t \in T$, la suite $(\otimes_{s \neq t} f_s^k , k \in \mathbb{N}')$ converge aussi faiblement dans $L^2(\prod_{s\neq t} E_s)$ de sorte que la relation 1.2.1. montre que $(f_t^k , k \in \mathbb{N}')$ converge p.p. dans (E_t,μ_t) ; nous notons f_t la limite μ_t-p.p. de $(f_t^k , k \in \mathbb{N}')$. La relation 1.2.2. majore $(f_t^k , k \in \mathbb{N}')$ par $M(P)^{-2} \frac{dP_t}{d\mu_t}$ qui est μ_t-intégrable ; le théorème de convergence dominée assure donc la convergence forte dans $L^1(E_t,\mu_t)$ de $(f_t^k , k \in \mathbb{N}')$ vers f_t . A fortiori la suite $(\sqrt{f^k} , k \in \mathbb{N}')$ converge faiblement dans $L^2(E)$ vers $\sqrt{f} = \otimes_{t\in T} \sqrt{f_t}$ si bien que $\Phi(f)$ est égale à $M(P)$: de toute suite d'éléments de $\mathcal{M}(P)$, on a finalement extrait une suite partielle convergeant fortement dans $L^1(E)$ vers un élément de $\mathcal{M}(P)$: $\mathcal{M}(P)$ est compact pour la topologie forte de $L^1(E)$ plus fine que la topologie séparée W qui y est donc identique. La proposition est démontrée.

COROLLAIRE 1.3. <u>Soient</u> $(E,\mathcal{B})$ <u>un produit fini d'espaces mesurables et</u> P <u>une probabilité sur</u> $(E,\mathcal{B})$; <u>dans ces conditions</u> :

(i) <u>s'il existe une mesure produit équivalente à</u> P , P <u>est équivalente au produit de ses propres marges,</u>

(ii) <u>s'il existe une mesure produit non orthogonale à</u> P , P <u>n'est pas non plus orthogonale au produit de ses propres marges.</u>

<u>Démonstration</u> : La première affirmation est triviale, la seconde le semble moins, elle est pourtant immédiate à partir de la proposition ; en effet sous l'hypothèse indiquée, la relation 1.2.3. montre que $\int \sqrt{\frac{dP}{d\mu}\cdot\frac{d\overline{P}}{d\mu}}\, d\mu$ n'est pas nulle ; a fortiori $\int \sqrt{dP\, d\overline{P}}$ n'est pas nul de sorte que P et $\overline{P}$ ne sont pas orthogonales (cf. 0.4) .

1.4. Dans cet alinéa, nous utilisons plusieurs fois les structures ci-dessus. L'index T est fixe et fini ; les espaces $(E^1,\mathcal{B}^1) = \prod_{t\in T} (E_t^1,\mathcal{B}_t^1)$ et $(E^2,\mathcal{B}^2) = \prod_{t\in T} (E_t^2,\mathcal{B}_t^2)$ sont deux produits finis d'espaces mesurables. On forme suivant le même index le produit fini $(E,\mathcal{B}) = \prod_{t\in T} (E_t,\mathcal{B}_t) = \prod_{t\in T} (E_t^1 \times E_t^2 , \mathcal{B}_t^1 \otimes \mathcal{B}_t^2)$. On note P^1 et P^2 deux probabilités, μ^1 et μ^2 deux mesures produit sur les espaces produit d'indices correspondants ; $P = P^1 \otimes P^2$ est une probabilité et $\mu = \mu^1 \otimes \mu^2$ est une mesure produit sur $(E,\mathcal{B})$. On suppose que P^1 et μ^1 , P^2 et μ^2 ne sont pas orthogonales.

PROPOSITION 1.4. <u>Dans les conditions ci-dessus,</u> P <u>n'est pas orthogonale à</u> μ ; <u>on a l'inclusion</u> $\mathcal{M}(P^1) \otimes \mathcal{M}(P^2) \subset \mathcal{M}(P)$ <u>et l'égalité</u> $M(P) = M(P^1)M(P^2)$. <u>De plus pour tout</u> $f \in \mathcal{M}(P)$, <u>les intégrales</u> $\int f_t(x_t^1,x_t^2)\,d\mu_t^2(x_t^2)$ <u>définissent un élément de</u> $\mathcal{M}(P^1)$.

<u>Démonstration</u> : Soient f^1 et f^2 des éléments de $\mathcal{M}(P^1)$ et $\mathcal{M}(P^2)$; on a :

$$0 < M(P^1)M(P^2) = \int \sqrt{f^1 \otimes f^2}\sqrt{\frac{dP}{d\mu}}\, d\mu \quad \text{et donc} \quad \mu\{\frac{dP}{d\mu} > 0\} \neq 0 \quad ;$$

P n'est donc pas orthogonale à μ et on a $M(P^1) \times M(P^2) \leq M(P)$. Soit maintenant f un élément de $\mathcal{M}(P)$, on peut écrire la suite d'inégalités :

$$M(P^1)M(P^2) \leq M(P) = \int \prod_{t\in T} \sqrt{f_t(x_t^1,x_t^2)}\, \sqrt{\frac{dP}{d\mu}(x)}\, d\mu^1(x^1)\,d\mu^2(x^2) \;;$$

en notant qu'on peut aussi bien intégrer seulement dans l'ensemble où les f_t sont strictement positifs, le dernier terme peut s'écrire :

$$\int\Big\{\prod_{t\in T}\sqrt{\int f_t(x_t^1,x_t^2)\,d\mu_t^2(x_t^2)}\Big\}\Big\{\prod_{t\in T}\sqrt{\frac{f_t(x_t^1,x_t^2)}{\int f_t(x_t^1,x_t^2)\,d\mu_t^2(x_t^2)}}\sqrt{\frac{dP^2}{d\mu^2}(x^2)\,d\mu^2(x^2)}\Big\}\sqrt{\frac{dP^1}{d\mu^1}(x^1)\,d\mu^1(x$$

En utilisant dans l'intégrale interne la définition de $M(P^2)$, ceci se majore par :

$$M(P^2)\int\prod_{t\in T}\sqrt{\int f_t(x_t^1,x_t^2)\,d\mu_t^2(x_t^2)}\,\sqrt{\frac{dP^1}{d\mu^1}(x^1)}\;d\mu^1(x^1)\quad ;$$

les éléments du produit de facteurs définissent maintenant, puisque f est un élément de $F_\mu(E)$, un élément f^1 de $F_{\mu^1}(E^1)$; en utilisant la définition de $M(P^1)$, on majore donc finalement par le minorant initial $M(P^1)\,M(P^2)$ si bien que les inégalités successives de ce calcul sont toutes des égalités et f^1 est un élément de $\mathcal{M}(P^1)$. Tous les résultats énoncés s'ensuivent.

2. Approximation des mesures gaussiennes par des mesures produit, le cas d'un nombre fini de facteurs.

2.1. Dans tout ce paragraphe, $(E_t,\mathcal{B}_t)$, $t \in T$, est une famille finie d'espaces de Fréchet séparables et P une probabilité gaussienne sur leur produit. Dans ce cas, le théorème de Hajek et Feldman permet de renforcer les résultats 1.2. et 1.3. :

PROPOSITION 2.1. Dans les conditions ci-dessus, les trois propriétés suivantes sont équivalentes :

(i) il existe une mesure produit non orthogonale à P ,

(ii) P est équivalente au produit de ses marges,

(iii) L'espace hilbertien $H_{\overline{P}}$ est inclus dans H_P et on a :

$$\forall\, s \in T\ ,\ \forall\, t \in T\ ,\ \int p_s \circ p_t \; dP \in H_{P_s} \otimes H_{P_t}\ ;$$

(p_s est la projection de E sur E_s , cf. 0.2.)

Démonstration : Sous l'hypothèse (i) , le corollaire 1.3. montre que les probabilités gaussiennes P et $\bar{P}$ ne sont pas orthogonales ; le théorème de Hajek et Feldman conclut donc à leur équivalence. Le même théorème, par un calcul simple explicitant 0.5.1., montre l'équivalence de (ii) et (iii) .

2.2. La proposition 2.3. utilisera la notion de limite centrale dans $(E,\mathcal{B})$. Comme $(E,\mathcal{B})$ est un espace polonais, la convergence en loi des v.a. ou la convergence étroite des probabilités y ont les mêmes caractéristiques que dans un espace de Banach séparable ; nous y serons ici dans le cas le plus simple :

PROPOSITION 2.2. Soit $(A^k , k \in \mathbb{N})$ une suite de v.a. indépendantes et de même loi Q sur $(E,\mathcal{B})$, on suppose que la suite $(Q^N , N \in \mathbb{N})$ des lois de $(N^{-\frac{1}{2}} \sum_{k=1}^{N} A^k , N \in \mathbb{N})$ est relativement compacte pour la topologie de la convergence étroite. Alors cette suite converge vers une loi gaussienne sur $(E,\mathcal{B})$ ayant même covariance que Q .

Démonstration : Soit μ un point d'adhérence de la suite (Q^N) et y un élément de E' ; puisque y est continue, la suite $(N^{-\frac{1}{2}} \sum_{k=1}^{N} \langle A^k,y\rangle, N \in \mathbb{N})$ des images converge en loi vers $y.\mu$; la propriété de la limite centrale implique donc que $\int y\, dQ = 0$, $\int y^2\, dQ < \infty$ et la loi $y.\mu$ est $\mathcal{N}(0,\sqrt{\int y^2\, dQ})$. Ceci implique en particulier que $\int \exp(iy) d\mu$ est $\exp(-\frac{1}{2} \int y^2\, dQ)$ de sorte que μ est une probabilité gaussienne (cf. 0.2.) sur $(E,\mathcal{B})$ de même covariance que Q ; soient μ^1 et μ^2 deux points d'adhérence de la suite, elles coincideront donc sur l'algèbre engendrée par E' (c'est le théorème de Bochner en dimension finie) et sur la tribu engendrée $\mathcal{B}$; la suite $(Q^N , N \in \mathbb{N})$ a un point adhérent et un seul, elle converge vers cette limite, c'est le résultat.

PROPOSITION 2.3. On suppose réalisées les conditions 2.1. ; on suppose P gaussienne équivalente au produit $\bar{P}$ de ses marges ; soient f un élément de $\mathcal{M}_{\bar{P}}(P)$ et A une v.a. à valeurs dans $(E,\mathcal{B})$ de loi $(f\, d\bar{P})$. Alors A vérifie la propriété de la limite centrale dans $(E,\mathcal{B})$ et la loi gaussienne ayant même covariance est une loi produit $(g\, d\bar{P})$ où g est aussi un élément de $\mathcal{M}_{\overline{P}}(P)$.

<u>Démonstration</u> : Nous appliquons le schéma et la proposition 1.4. dans le produit $(E \times E, \mathcal{B} \otimes \mathcal{B})$ à la probabilité $P \otimes P$ relativement à la probabilité produit $\mu \otimes \mu = \overline{P} \otimes \overline{P}$. La proposition 1.4. indique (en omettant les indices) que $M(P \otimes P)$ est égal à $M(P)^2$ et aussi que pour tout couple (f^1, f^2) d'éléments de $\mathcal{M}(P)$, $f^1 \otimes f^2$ appartient à $\mathcal{M}(P \otimes P)$; on a donc :

$$M(P)^2 = \int \prod_{t \in T} \sqrt{f_t^1(x_t^1) f^2(x_t^2)} \sqrt{\frac{dP}{d\overline{P}}(x^1) \frac{dP}{d\overline{P}}(x^2)} \; d\overline{P}(x^1) d\overline{P}(x^2) \quad ;$$

soit $\theta \in [0, 2\pi[$, onl lui associe le changement de variables défini dans $(E \times E)$ par :

$$\begin{pmatrix} x^1 \\ x^2 \end{pmatrix} = \begin{pmatrix} \cos\theta , & \sin\theta \\ -\sin\theta , & \cos\theta \end{pmatrix} \begin{pmatrix} y^1 \\ y^2 \end{pmatrix} , \begin{pmatrix} y^1 \\ y^2 \end{pmatrix} = \begin{pmatrix} \cos\theta , & -\sin\theta \\ \sin\theta , & \cos\theta \end{pmatrix} \begin{pmatrix} x^1 \\ x^2 \end{pmatrix} ;$$

il est choisi pour conserver les mesures $d\overline{P}(x^1)\, d\overline{P}(x^2)$ aussi bien que $dP(x^1) dP(x^2)$ associées à des couples de v.a. gaussiennes indépendantes de mêmes lois : il conserve aussi de ce fait la dérivée $\frac{dP}{d\overline{P}}(x^1) \frac{dP}{d\overline{P}}(x^2)$. On obtient donc :

$$M(P)^2 = \int \prod_{t \in T} \sqrt{f_t^1(y_t^1)\; f_t^2(y_t^2)} \sqrt{\frac{dP \otimes P}{d\overline{P} \otimes \overline{P}}(x^1, x^2)} \; d\overline{P}(x^1)\; d\overline{P}(x^2) \quad ;$$

la proposition 1.4. montre alors que g définie par :

$$g_t(x_t^1, x_t^2) = f_t^1(y_t^1)\; f_t^2(y_t^2)$$

est un élément de $\mathcal{M}(P \otimes P)$ et la même proposition implique que la suite $\overline{f} = \int g(\;, x^2) d\overline{P}(x^2)$ est un élément de $\mathcal{M}(P)$. Ceci se traduit plus simplement en introduisant deux v.a. A^1 de loi $(f^1\, d\overline{P})$ et A^2 de loi $(f^2\, d\overline{P})$ indépendantes à valeurs dans $(E, \mathcal{B})$; on constate en effet que $(f\, d\overline{P})$ est la loi de $A^1 \cos\theta - A^2 \sin\theta$; cette loi définit donc un élément de $\mathcal{M}(P)$.

Soit maintenant f un élément arbitraire de $\mathcal{M}(P)$; nous notons A une v.a. de loi $(f\, d\overline{P})$ et $(A^k , k \in \mathbb{N})$ une suite de copies indépendantes de A . Le résultat ci-dessus montre par induction que la loi

Q^N de $(N^{-\frac{1}{2}} \sum_{k=1}^{N} A^k)$ a une dérivée $\frac{dQ^N}{d\overline{P}}$ qui appartient à $\mathcal{M}(P)$. Les propriétés de compacité de $\mathcal{M}(P)$ (proposition 1.2. (iii)) montrent que la suite $(Q^N , N \in \mathbb{N})$ est relativement compacte pour la topologie de la convergence étroite. La propriété de la limite centrale s'applique donc (proposition 2.2.) à la loi de A : $(Q^N , N \in \mathbb{N})$ converge étroitement vers une loi gaussienne Q ; comme A a une loi produit, les Q^N sont des lois produit et Q une loi gaussienne produit. La compacité de $\mathcal{M}(P)$ montre que la dérivée $\frac{dQ}{d\overline{P}}$ appartient à $\mathcal{M}(P)$. La proposition est démontrée.

PROPOSITION 2.4. <u>Dans les conditions 2.1., soit</u> P <u>une probabilité gaussienne équivalente au produit de ses marges ; alors la borne supérieure de</u> $\int \sqrt{dP\, dQ}$ <u>où</u> Q <u>parcourt l'ensemble des probabilités produit se majore indépendamment du nombre des facteurs</u> :

2.4.1. $\sup \int \sqrt{dP\, dQ} \leq (1 + 2 \ln 1 / \int \sqrt{dP\, d\overline{P}})^{-\frac{1}{4}}$.

<u>Remarque</u> : dans un cas plus général, l'inégalité 1.2.3. fournit une majoration du même type, mais dépendant du nombre d'éléments de T ; on notera que le second membre de l'inégalité 2.4.1. croit de 0 à 1 en même temps que $\int \sqrt{dP\, d\overline{P}}$; on obtient une inégalité significative dans tous les cas.

<u>Démonstration</u> : puisque P est équivalente à $\overline{P}$, le premier membre de 2.4.1. est $M_{\overline{P}}(P)$; puisque P est gaussienne, la proposition 2.3. montre que ce premier membre est plus simplement la borne supérieure de $\int \sqrt{dP\, dQ}$ où Q parcourt l'ensemble des seules probabilités gaussiennes produit ; c'est ce que nous allons évaluer : pour tout $t \in T$, on note $(b_t^k , k \in \mathbb{N})$ une base orthonormale de l'espace hilbertien K_{P_t} engendré par le dual E_t' de E_t dans $L^2(E_t,P_t)$(cf. 0.5.) . Alors les tribus $\mathcal{B}^k$ engendrées par $(b_t^j \circ p_t , j \in [1,k], t \in T)$ engendrent elles-mêmes la tribu $\mathcal{B}$ sur E (aux ensembles P-négligeables près). Dans ces conditions, le dernier argument 0.4. et la proposition 2.3. montrent qu'on aura établi la proposition si on prouve que pour toute probabilité gaussienne produit Q , on a :

2.4.2. $\forall k \in \mathbb{N} \quad \int \sqrt{dP_{\mathcal{B}^k}\, dQ_{\mathcal{B}^k}} \leq (1 + \ln 1/\int \sqrt{dP_{\mathcal{B}^k}\, d\overline{P}_{\mathcal{B}^k}})^{-\frac{1}{2}}$;

les deux membres de 2.4.2. s'écrivent simplement (si le premier membre n'est pas nul). On note Γ la matrice de covariance de $(b_t^j \circ p_t,\ 1 \leq j \leq k,\ t \in T)$ dans $L^2(E,P)$ et X sa matrice de covariance dans $L^2(E,Q)$. Puisque $(b_t^j \circ p_t,\ 1 \leq j \leq k,\ t \in T)$ est orthonormale dans $K_{\overline{P}}$, elle est libre dans l'espace K commun aux trois probabilités gaussiennes équivalentes $\overline{P}$, P et Q ; les matrices Γ et X ont donc des inverses G et Y ; on note I la matrice unité. Pour chacune de ces matrices, on note par l'indice $t \in T$, la sous-matrice des lignes et colonnes possédant cet indice ; le choix particulier des familles (b_t^k) implique donc que Γ_t est égale à I_t. On notera que Y_t est l'inverse de X_t parce que Q est une probabilité produit. Par contre G_t n'est pas nécessairement l'inverse de Γ_t ; on a pourtant :

$$\forall x \in R^{k|T|},\ < x,x >^2 \leq < \Gamma x,x > < Gx,x > ,$$

et donc en particulier :

$$\forall x \in R^k,\ \forall t \in T,\ < x,x >^2 \leq < \Gamma_t x,x > < G_t x,x > ,$$

qui fournit en fait :

$$\forall x \in R^k,\ \forall t \in T,\ <x,x > \leq < G_t x,x > .$$

Ceci montre que G_t qui est une matrice symétrique positive a toutes ses valeurs propres supérieures ou égales à 1 ; on note $\sqrt{G_t}$ sa racine carrée symétrique positive elle a ces mêmes propriétés.

Par ailleurs, un calcul simple montre :

$$(\int \sqrt{dP_{\mathcal{B}^k}\, dQ_{\mathcal{B}^k}})^{-4} = \text{dét}(\frac{\Gamma Y + xG + 2I}{4}) \geq 1 + \text{Tr}(\frac{\Gamma Y + XG - 2I}{4}) ,$$

$$(\int \sqrt{dP_{\mathcal{B}^k}\, dQ_{\mathcal{B}^k}})^{-4} \geq 1 + \sum_{t \in T} \text{Tr}(\frac{Y_t + X_t G_t - 2I_t}{4}) ;$$

pour tout $t \in T$, on a :

$$\text{Tr}(Y_t + X_t G_t - 2I_t) = \text{Tr}(Y_t - 2\sqrt{G_t} + \sqrt{G_t} X_t \sqrt{G_t}) + 2\text{tr}(\sqrt{G_t} - I_t) ,$$

le premier terme du second membre est positif ou nul, on en déduit :

$$\left(\int \sqrt{dP_{\mathcal{B}^k}\, dQ_{\mathcal{B}^k}}\right)^{-4} \geq 1 + \frac{1}{2} \sum_{t \in T} \operatorname{Tr}(\sqrt{G_t} - I_t) \quad .$$

Un calcul très semblable fournit :

$$\left(\int \sqrt{dP_{\mathcal{B}^k}\, d\overline{P}_{\mathcal{B}^k}}\right)^{-4} \leq \exp\left(\sum_{t \in T} \operatorname{Tr}\left(\frac{G_t - I_t}{2}\right)\right) \quad .$$

En comparant ces deux résultats et en utilisant pour cela les propriétés des valeurs propres de $\sqrt{G_t}$, on obtient la relation 2.4.2. ; la proposition est démontrée.

3. <u>Comparaison de mesures gaussiennes et de mesures produit, le cas d'un produit arbitraire d'espaces de Fréchet séparables</u>. On énonce maintenant le résultat principal de ce travail :

THEOREME 3.1. <u>Soit</u> T <u>un ensemble arbitraire indexant une famille</u> $(E_t, \mathcal{B}_t)$ <u>d'espaces de Fréchet séparables ; soient de plus</u> P <u>et</u> Q <u>deux probabilités sur l'espace produit</u> $(E, \mathcal{B})$, <u>on suppose que</u> P <u>est gaussienne et que</u> Q <u>est une probabilité produit. Dans ces conditions, les propriétés suivantes sont équivalentes</u> :

(i) <u>les probabilités</u> P <u>et</u> Q <u>ne sont pas orthogonales</u>

(ii) P <u>est équivalente au produit</u> $\overline{P}$ <u>de ses marges</u> ; Q <u>n'est pas orthogonale à</u> $\overline{P}$.

<u>De plus, si pour tout</u> $t \in T$, <u>les marges</u> P_t <u>et</u> Q_t <u>sont équivalentes, alors les propriétés</u> (i) <u>et</u> (ii) <u>sont équivalentes à</u> :

(iii) <u>les probabilités</u> P <u>et</u> Q <u>sont équivalentes</u>.

<u>Démonstration</u> : (a) cas d'un index au plus dénombrable. Les implications (ii) $\Rightarrow$ (i) et (iii) $\Rightarrow$ (i) sont triviales ; nous prouvons (i) $\Rightarrow$ (ii) : sous l'hypothèse (i) , posons $m = \int \sqrt{dP\, dQ} > 0$; soit de plus S une partie finie de T , on aura (cf. 0.4) :

$$\int \sqrt{dP_{\mathcal{B}_S}\, dQ_{\mathcal{B}_S}} \geq m \quad .$$

Le premier membre de cette inégalité se calcule dans l'espace produit fini E_S : l'application de la proposition 2.4. fournit :

$$\int \sqrt{dP_{\mathcal{B}_S}\ d\bar{P}_{\mathcal{B}_S}} \geq \exp(-\frac{1-m^4}{2m^4}) .$$

On en déduit (cf. 0.2) la même minoration pour $\int \sqrt{dP\ d\bar{P}}$; P et $\bar{P}$ ne sont pas orthogonales ; comme elles sont gaussiennes, elles sont équivalentes et Q non orthogonale à P n'est pas non plus orthogonale à $\bar{P}$.

Nous démontrons maintenant (i) $\Rightarrow$ (iii) sous l'hypothèse additionnelle indiquée ; P et $\bar{P}$ sont alors deux probabilités produit vérifiant les hypothèses du théorème de Kakutani, elles sont donc équivalentes de sorte que P et Q toutes deux équivalentes à $\bar{P}$ sont équivalentes entre elles. Le théorème est démontré dans ce premier cas.

(b) <u>cas d'un index arbitraire</u>. On sait alors que la tribu $\mathcal{B}$ est la réunion $(p_S^{-1}\ \mathcal{B}_S)$ où S parcourt l'ensemble $\mathcal{N}(T)$ des parties dénombrables de T. On vérifie pour tout couple (P,Q) de probabilités sur $(E,\mathcal{B})$ P est équivalente (resp. non orthogonale) à Q si et seulement si pour tout $S \in \mathcal{N}(T)$, $p_S P$ et $p_S Q$ le sont. De la même manière, Q est une probabilité produit sur $(E,\mathcal{B})$ si et seulement si pour tout $S \in \mathcal{N}(T)$, $p_S P$ est une probabilité produit sur $(E_S,\mathcal{B}_S)$. La nature des probabilités gaussiennes (cf. 0.2) montre donc que le théorème énoncé dans le cas général est strictement équivalent à sa restriction à un index au plus dénombrable. Le théorème est démontré dans tous les cas.

4. <u>Exemple d'application</u>.

4.1. Dans ce paragraphe, nous étudierons le problème suivant qui nous a été proposé par S.D. Chatterji : soit $(X_n\ , n \in \mathbb{N})$ une suite gaussienne de v.a. réelles ; on note P la loi de la suite et pour tout $n \in \mathbb{N}$, $\mathcal{T}_n$ la tribu engendrée par $(X_m\ , m \geq n)$; $\mathcal{T} = \bigcap_{n \in \mathbb{N}} \mathcal{T}_n$ est la tribu terminale de la suite $(X_n\ , n \in \mathbb{N})$. Dans ces conditions, on cherche les relations entre les deux propriétés suivantes :

(i) la tribu terminale $\mathcal{T}$ est dégénérée,

(ii) la loi P est équivalente à une loi produit $Q = \bigotimes_{n \in \mathbb{N}} Q_n$.

Notre étude sera basée sur le théorème 3.1. et sur la proposition :

PROPOSITION 4.2. Les deux propriétés suivantes sont équivalentes :

(i) la tribu terminale est P-dégénérée

(ii) la suite X est supérieurement triangulable : il existe une suite gaussienne centrée réduite $(Y_k, k \in \mathbb{N})$ et une $\mathbb{N} \times \mathbb{N}$-matrice A telles que :

$$\forall n \in \mathbb{N}, X_n = \sum_{k=1}^{\infty} a_{n,k} Y_k, \text{ p.s. et } \forall j < n, a_{n,j} = 0 .$$

Démonstration : Sous l'hypothèse (ii), la tribu $\mathcal{T}$ est contenue dans la tribu terminale $\mathcal{T}'$, de la suite à composantes indépendantes Y, (i) est donc vérifiée. Réciproquement, si (i) est vérifiée, alors pour tout $n \in \mathbb{N}$, $E(X_n/\mathcal{T})$ est nulle de sorte que $E(X_n/\mathcal{T}_m)$ tend vers zéro dans $L^2(P)$ et presque sûrement quand m tend vers l'infini. Pour tout $m \in \mathbb{N}$, nous notons K_m le sous-espace de $L^2(P)$ engendré par $(X_m, n \geq m)$ et $(m_k, k \in \mathbb{N})$ la suite des entiers définie par :

$$m_o = 1 \quad , \quad m_{k+1} = \inf(m > m_k : K_m \neq K_{m_k}) \quad ;$$

enfin pour tout $k \geq 1$, notons Y_k une base réduite de $K_{m_{k-1}} \ominus K_{m_k}$. On vérifie que X est supérieurement triangulable suivant la suite Y.

La proposition suivante résoud le problème énoncé :

PROPOSITION 4.3. Si la loi P de la suite gaussienne $(X_n, n \in \mathbb{N})$ est équivalente à une loi produit, Q, alors on a :

(i) la tribu terminale $\mathcal{T}$ est P-dégénérée ; de plus, il existe une loi produit Q telle que pour tout entier n, la loi de $(X_k, k \leq n)$ soit équivalente à $\bigotimes_{k=1}^{n} Q_k$.

Par contre, il existe des suites gaussiennes $(X_n, n \in \mathbb{N})$ de loi P vérifiant la propriété ci-dessus, P étant pourtant orthogonale à toute loi produit.

<u>Démonstration</u> : la première affirmation est triviale, nous justifions la seconde par l'exemple suivant : $Y = (Y_k , k \in \mathbb{N})$ est une suite gaussienne réduite indépendante et X est définie par :

$$\forall n \in \mathbb{N} , X_n = \sum_{k=n}^{\infty} \frac{1}{k+1} Y_k .$$

Alors la proposition 4.2. montre que $\mathcal{T}$ est P-dégénérée ; pour tout $n \in \mathbb{N}$, la loi de $(X_k , k \leq n)$ n'est pas dégénérée et n'est donc pas orthogonale à la loi de $(Y_k , k \leq n)$; la proposition 2.1. montre qu'elle est équivalente au produit de ses propres marges de sorte que la propriété (i) est vérifiée avec $Q = \overline{P}$. Par contre si la loi P était équivalente à une loi produit Q , elle serait aussi (théorème 3.1.) équivalente à $\overline{P}$ et la double série des coefficients de corrélation des composantes d'indices différents serait convergente ; or on vérifie que le terme général de cette série ne tend pas vers zéro, le résultat s'ensuit.

4.3. L'auteur remercie S. Ramaswany dont le preprint [7] a éveillé son intérêt pour les problèmes étudiés ici ; il remercie S.D. Chatterji pour ses conseils.

REFERENCES

[1] S.D. CHATTERJI et V. MANDREKAR : Equivalence and singularity of gaussian measures and applications. Probabilistic Analysis and related topics, Vol. 1 , Academic Press, N.Y., 1978, 169-197.

[2] J. FELDMAN : Equivalence and perpendicularity of gaussian processes. Pacific J. Math., 9, 1958, 699-708 .

[3] X. FERNIQUE : Comparaison de mesures gaussiennes et de mesures produit. Ann. Inst. Henri Poincaré, 20, 1984, 165-175.

[4] J. HAJEK : On a property of normal distribution of any stochastic processes. Math., Statist. Prob., 1, 1958-1961, 245-252.

[5] S. KAKUTANI : On equivalence of infinite product measures. Ann. of Math., 49, 1948, 214-224 .

[6] J. NEVEU : Processus aléatoires gaussiens. Univ. of Montréal Press, Canada, 1968 .

[7] S. RAMASWANY : Gaussian measures and product measures. Preprint, 1983

ON CONVERGENCE AND DEMICONVERGENCE OF BLOCK MARTINGALES AND SUBMARTINGALES

Nikos E. Frangos and Louis Sucheston*

Department of Mathematics
The Ohio State University
Columbus, Ohio 43210

Let (Ω,F,P) be a probability space; I a directed set filtering to the right, $(F_t, t \in I)$ an increasing filtration of sub-sigma-fields of F. Our main results concern processes (X_t,F_t), $t \in I = J^m = J_1 \times J_2 \times \ldots \times J_m$, where J_k are directed sets filtering to the right, and the order on J^m is determined by $s = (s_1,\ldots,s_m) \le t = (t_1,\ldots,t_m)$ if $s_i \le t_i$ for all $i \le m$. In the first phases of the multiparameter martingale theory begun by R. Cairoli [4], each J_k was N, and P was the product of probability measures defined on the coordinate filtrations (F_{t_i}). Later this assumption was relaxed to conditional independence (F4 in Cairoli-Walsh [5]) or, equivalently, commutation (cf. Meyer [21], p. 3). Here we typically assume that the first filtration $(F_{t_1})_{t_1 \in J_1}$ has enough order for L_1-bounded martingales to converge, or equivalently has a weak maximal inequality involving $\limsup X_t$, and the other filtrations $(F_{t_i})_{t_i \in J_i}$, $2 \le i \le m$ have a (stronger) weak maximal inequality involving $\sup X_t$. For this it suffices that the first filtration has the covering condition (C), introduced in [27], and the other filtrations satisfy a new regularity condition (MR_α); we then call (F_t) regular. Conditions (C) and (MR_α) are stopping conditions involving multivalued stopping times. We do not assume commutation, considering instead of martingales <u>block martingales</u>. This notion seems new if $m \ge 3$, but in two parameters a block martingale is exactly a 1-martingale in the sense of [28]. Under commutation, every martingale is necessarily a block martingale. The interest of block martingales and submartingales

*The research of this author is in part supported by the National Science Foundation Grant MCS-8301619.

is that they arise in the natural context of laws of large numbers where (F4) may fail; see the end of Section 3.

A typical result is that a block martingale bounded in $L \log^{m-1} L$ and taking values in a Banach space with the Radon-Nikodym property converges essentially. Essential convergence is not very essential in this paper, because in all applications there is a countable cofinal subset, so that one obtains almost sure convergence. On the other hand, replacement of $J_k = N$ by more general index sets is important in applications to differentiation.

If there is convergence of a process indexed by N, then typically the analogous process indexed by J^m converges in probability without additional integrability assumptions, but not essentially. This gives rise to the notion of demiconvergence, first introduced in the two-parameter case for $J = -N$ in [9], and for $J = N$ in [29]. The case $J = N$ is more difficult, and the proofs given here are simpler than those in [29].

The stochastic limit is denoted by $s \lim X_t = X$; the letter e stands for essential. We say that a real valued process X_t demiconverges if $e \limsup X_t = X$ (upper demiconvergence) or $e \liminf X_t = X$ (lower demiconvergence). We show that $L \log^{m-1} L$ bounded block submartingales upper demiconverge, and positive block martingales lower demiconverge. The convergence results are applied to obtain a Banach-valued form of a theorem of Zygmund [39] about differentiation of integrals in m dimensions, along a net of rectangles with sides of no more than $s \leq m$ different lengths. Instead of rectangles we consider "substantial sets," which is slightly more general. Zygmund's theorem is a generalization of the Jessen-Marcinkiewicz-Zygmund theorem [15] in which $s = m$. A demiconvergence version of Zygmund's theorem is also given.

Section 1 gives basic definitions. In section 2 we prove maximal inequalities for positive submartingales under (C) and (MR_α). Under (C) these results are known for martingales [27], and the same article shows that (C) is sufficient for convergence of L_1-bounded martingales. Recently Talagrand [38] proved that (C) is also necessary, if there is cofinal countable subset. We believe that (MR_α) may be necessary for our maximal inequalities, but this is here not discussed. Section 2 also proves maximal inequalities for positive block submartingales. Convergence and demiconvergence of block processes is studied in section 3. This section contains the most general result of the paper (Theorem 3.3), a convergence theorem about real functions of Banach-valued processes, which implies simultaneously the convergence and demiconvergence results (Theorems 3.4, 3.5, and 4.2). An extension to Banach lattices is given in section 4. Under the commutation assumption, we reduce in section 5 demiconvergence of block submartingales to convergence to block martingales. The last, the sixth, section contains applications to differentiation of integrals.

The main results of this paper were presented by the second-named author at the Fifth International Conference on Probability in Banach spaces, held at Tufts University, in July 1984.

I. Definitions and basic notions. Let J be a directed set filtering to the right. For a fixed $m \in N$, define $J^m = J_1 \times J_2 \times \ldots \times J_m$, $J_k = J$; with the order $s = (s_1,\ldots,s_m) \leq t = (t_1,\ldots,t_m)$ if $s_k \leq t_k$, $k = 1,2,\ldots,m$. The set J^m is then filtering to the right. Let (Ω, F, P) be a complete probability space, and let $(F_t, t \in J^m)$ be an increasing net of sub-σ-fields of F. For any integers k,ℓ, $1 \leq k \leq \ell \leq m$, $F_t^{k-\ell}$ is defined as the σ-field obtained by lumping together the σ-fields on all the axes except for the k-th, $k+1$-th,$\ldots$,ℓ-th ones. That is,

$$F_t^{k-\ell} = \bigvee_{\substack{t_1,\ldots,t_{k-1},\\ t_{\ell+1},\ldots,t_m}} F_{(t_1,\ldots,t_{k-1},t_k,\ldots,t_\ell,\ldots,t_m)} .$$

If $k = \ell$ then $F_t^{k-\ell}$ is denoted by F_t^k. Obviously, if $k = 1$ and $\ell = m$, then $F_t^{k-\ell} = F_t$. We often denote $E(\ |F_t)$ by $E^t(\cdot)$ and $E(\ |F_t^{k-\ell})$ by $E_t^{k-\ell}(\cdot)$.

An integrable process $(X_t, F_t, t \in J^m)$ is a <u>(sub)martingale</u> whenever $E^s(X_t) = X_s$ $(\geq)$ for $s \leq t$. An integrable process $(X_t, F_t, t \in J^m)$ is a <u>block k-(sub)martingale</u> for a fixed $k \leq m$ if

$$E_s^{1-k}(X_t) = X_{(s_1,\ldots,s_k,t_{k+1},\ldots,t_m)} \; (\geq) \quad \text{for} \quad s \leq t \; .$$

An integrable process is a block (sub)martingale if it is a block k-(sub)martingale for all $k \leq m$. Thus a block (sub)martingale is necessarily a (sub)martingale.

Let $\mathcal{J}$ denote the set of finite subsets of J. An (incomplete) multivalued stopping time is a map τ from Ω (from a subset of Ω denoted $D(\tau)$) to $\mathcal{J}$ such that $R(\tau) = \bigcup_{\omega} \tau(\omega)$ is finite, and for every $t \in J$,

$$\{\tau = t\} \overset{def}{=} \{\omega \in \Omega : t \in \tau(\omega)\} \in F_t \; .$$

Denote by $M(IM)$ the set of (incomplete) multivalued stopping times. A simple stopping time is an element $\tau \in M$ such that $\tau(\omega)$ is a singleton for every ω; the set of simple stopping times is denoted by T. For $\sigma,\tau \in IM$, we say that $\sigma \leq \tau$ if for every s and t such that $\{\sigma = s\} \cap \{\tau = t\} \neq \emptyset$, one has $s \leq t$. With this order, IM is a directed set filtering to the right.

The excess function of $\tau \in IM$ is

$$e_\tau = \sum_t 1_{\{\tau=t\}} - 1_{D(\tau)} \; .$$

Let $\tau \in IM$; for a positive stochastic process $(X_t, t \in J)$, we set

$$X(\tau) = \sup_t (1_{\{\tau=t\}} X_t) \; .$$

If (A_t) is an adapted family of sets, i.e. $A_t \in F_t$, we set

$$A(\tau) = \bigcup_t (\{\tau=t\} \cap A_t) \; .$$

Hence if $X_t = 1_{A_t}$ then $X(\tau) = 1_{A(\tau)}$.

The letter e means "essential." Thus $e \lim X_t$ is the essential limit of X_t. In most applications, there is a countable cofinal subset and in this case the word "essential" can be replaced by "almost sure".

A filtration $(F_t, t \in J)$ satisfies the covering condition C, introduced in [27], if for every $\varepsilon > 0$ there exists a constant $M_\varepsilon > 0$ such that for every adapted family of sets $(A_t, t \in J)$, there exists $\tau \in IM$ with $e_\tau \leq M_\varepsilon$ and

(1) $$P[A(\tau)] \geq P[e \limsup A_t] - \varepsilon .$$

Let α be a fixed number, $0 < \alpha \leq 1$. A filtration $(F_t, t \in J)$ satisfies the <u>regularity condition</u> MR_α if for every $\varepsilon > 0$ there exists a constant $M = M(\varepsilon,\alpha) \geq 0$ such that for every adapted family of sets $(A_t, t \in J)$ there exists $\tau \in IM$ with $e_\tau \leq M$ and

(2) $$P[A(\tau)] \geq \alpha(1-\varepsilon) P[e \sup A_t] .$$

The set of α, $0 < \alpha \leq 1$ such that (F_t) satisfies MR_α has a maximum, unless it is empty. Indeed, let α be the supremum of all α_i such that MR_{α_i} holds. Given $\varepsilon > 0$, apply (2) with an ε_i and α_i such that $\alpha_i(1-\varepsilon_i) > \alpha(1-\varepsilon)$.

A stricter regularity condition R_α was introduced in [25]: the stopping times τ are required to be single-valued (that is, $e_\tau = 0$). An example given in [23] (for a different purpose) shows that MR_α is strictly weaker than R_α. It was observed in [29] that a filtration (F_t) totally ordered by set-inclusion satisfies R_1.

The condition MR_α implies condition C. Indeed, let $\varepsilon > 0$ and (A_t) be an adapted family of sets. Choose $s \in J$ such that $P[e \sup_{t \geq s} A_t] - P[e \limsup A_t] \leq \varepsilon$. Let $\tau \in IM$, $\tau \geq s$, then $P[e \limsup A_t \cap A(\tau)] \geq P[e \sup_{t \geq s} A_t \cap A(\tau)] - \varepsilon = P[A(\tau)] - \varepsilon$. Applying now MR_α to (B_t), $B_t = A_t$ for $t \geq s$ and $B_t = \emptyset$ otherwise, one obtains an $M = M(\varepsilon,\alpha)$, and a $\tau \in IM$ with $e_\tau \leq M$, $\tau \geq s$ and $P[A(\tau)] \geq \alpha(1-\varepsilon) P[e \sup_{t \geq s} A_t]$. Since $P[e \sup_{t \geq s} A_t] \geq P(e \limsup A_t]$, one has

$$P[A(\tau)] \geq \alpha(1-\varepsilon) P[e \limsup A_t] \geq \alpha P[e \limsup A_t] - \varepsilon .$$

Hence

$$P[e \limsup A_t \cap A(\tau)] \geq \alpha P[e \limsup A_t] - 2\varepsilon ,$$

which is equivalent with condition C ([27], theorem 1.1.)

A filtration $(F_t, t \in J^m)$ is called <u>regular</u> if $(F_t^1, t_1 \in J_1)$ satisfies condition C , and for each k , $2 \leq k \leq m$, there exists α_k, $0 < \alpha_k \leq 1$ such that for each fixed $t_1 \in J_1, \dots, t_{k-1} \in J_{k-1}$, $(F_t^{1-k}, t_k \in J_k)$ satisfies condition MR_{α_k} .

If the probability space is of product type, i.e.,

$(\Omega, F, P) = \prod_{k=1}^{m} (\Omega_k, F_k, P_k)$ and $F_t = F_{t_1} \otimes \dots \otimes F_{t_m}$, then the filtration $(F_t, t \in J^m)$ is regular if $(F_{t_1}, t_1 \in J_1)$ satisfies C , and for each $k \geq 2$, $(F_{t_k}, t_k \in J_k)$ satisfies MR_{α_k} .

If $J_k = N$, $k = 1,2,\dots,m$, then necessarily the filtration (F_t) is regular.

II. Maximal inequalities. We now prove weak and strong maximal inequalities for positive submartingales.

Theorem 2.1. (Maximal inequality under MR_α) . Let $(F_t, t \in J)$ be a filtration satisfying MR_α . Fix $v \in J$. There is a constant c depending only on (F_t) such that for every positive submartingale (X_t, F_t) and every $\lambda > 0$ one has, letting $X_v^* = e \sup_{t \leq v} X_t$,

(1) $$P[X_v^* \geq \lambda] \leq \frac{c}{\alpha\lambda} \int_{\{X_v^* \geq \lambda\}} X_v \, dP .$$

If there is an upper bound β for the excess $M(\varepsilon,\alpha)$ as $\varepsilon \to 0$, then

(2) $$P[X_v^* \geq \lambda] \leq \frac{\beta+1}{\alpha\lambda} \int_{\{X_v^* \geq \lambda\}} X_v \, dP .$$

Proof. As shown in the previous section, one can assume that α is the largest number such that MR_α holds. Fix $\varepsilon > 0$, $\lambda > 0$ and $0 < \delta < \lambda$. Let $A_t = \{X_t > \lambda-\delta\}$ for $t \leq v$; $A_t = \emptyset$ otherwise. There exist $M = M(\varepsilon,\alpha)$ and $\tau \in IM$ such that $e_\tau \leq M$ and $P[A(\tau)] \geq \alpha(1-\varepsilon)P[e \sup A_t]$. Hence

$$P[X_v^* > \lambda-\delta] \le P[e \sup A_t]$$

$$\le \frac{1}{\alpha(1-\varepsilon)} P[A(\tau)]$$

$$= \frac{1}{\alpha(1-\varepsilon)} P[\bigcup_t \{\tau = t\} \cap A_t]$$

$$\le \frac{1}{\alpha(1-\varepsilon)} \sum_t P[\{\tau = t\} \cap A_t]$$

$$\le \frac{1}{\alpha(1-\varepsilon)} \frac{1}{\lambda-\delta} \sum_t \int_{\{\tau = t\} \cap A_t} X_t dP$$

$$\le \frac{1}{\alpha(1-\varepsilon)} \frac{1}{\lambda-\delta} \sum_t \int_{\{\tau = t\} \cap A_t} X_v dP$$

$$\le \frac{1}{\alpha(1-\varepsilon)} \frac{1}{\lambda-\delta} \int_{\{X_v^* > \lambda-\delta\}} X_v \sum_t 1_{\{\tau = t\}} dP$$

$$\le \frac{1}{\alpha(1-\varepsilon)} \frac{M+1}{\lambda-\delta} \int_{\{X_v^* > \lambda-\delta\}} X_v dP .$$

Let $\delta \to 0$. The relation (1) follows on letting $\varepsilon = \frac{1}{2}$ and $c = 2[M(\frac{1}{2}, \alpha) + 1]$. If $M(\varepsilon,\alpha) \le \beta$ for all ε, (2) follows on letting $\varepsilon \to 0$. ///

We recall some facts about Orlicz spaces (see also [16] and [31]). Let $\phi : \underline{R}^+ \to \underline{R}^+$ be an increasing, left continuous function which is zero at the origin. Assume also that $\lim_{t \to \infty} \phi(t) = \infty$. Let $\Phi(t) = \int_0^t \phi(s)ds$. For a random variable X, let

$$\|X\|_\Phi = \inf\{a > 0 : E[\Phi(\frac{|X|}{a})] \le 1\} .$$

We denote by L_Φ the Orlicz space of all those X for which $\|X\|_\Phi < \infty$. A process (X_t) is bounded in L_Φ if $\sup_t \|X_t\|_\Phi < \infty$. The function $\Phi(t)$ satisfies condition Δ_2 (at infinity) if $\limsup_{t \to \infty} \frac{\Phi(2t)}{\Phi(t)} < \infty$. If Δ_2 is satisfied, then $E[\Phi(|X|)] < \infty$ if and only if $\|X\|_\Phi < \infty$. In general $\|X\|_\Phi \leq \max(1, E[\Phi(|X|)])$. Let $\Phi_m(t) = t(\log^+ t)^m$. The space L_{Φ_m} is denoted $L \log^m L$. Since $\Phi_m(t)$ satisfies the Δ_2 condition, a process (X_t) is bounded in $L \log^m L$ if and only if $\sup_t E[|X_t|(\log^+|X_t|)^m] < \infty$. We set $\Phi_0(t) = t$ for all $t \geq 0$ and consider $L_{\Phi_0} = L_1$ as an Orlicz space.

Lemma 2.2. Let $(F_t, t \in J)$ be a filtration satisfying MR_α. Fix $v \in J$. Let $(X_t, F_t, t \in J)$ be a positive submartingale. Then there is a constant c depending on (F_t) such that for every η, and every $k \geq 0$, one has

(3) $$E[e \sup_{t \leq v} \Phi_k(X_t)]$$
$$\leq \frac{e}{e-1} [\eta + |\log \eta| E[\Phi_k(\frac{c}{\alpha} X_v)] + (k+1) E[\Phi_{k+1}(\frac{c}{\alpha} X_v)]]$$

Proof. Since the function $\Phi_k(t)$ is convex and increasing, by Jensen's inequality $Y_t = \Phi_k(X_t)$ is a positive submartingale with respect to (F_t). Let $Y_v^* = e \sup_{t \leq v} Y_t$. From theorem 2.1., one has

$$P[Y_v^* \geq \lambda] \leq \frac{c}{\alpha\lambda} \int_{\{Y_v^* \geq \lambda\}} Y_v \, dP .$$

Hence applying Fubini and using the elementary inequality $a \log^+ b \leq a \log^+ a + \frac{b}{e}$, one obtains (see also [28], page 23)

$$E[Y_v^*] = \int_0^\infty P(Y_v^* \geq \lambda) d\lambda$$

$$\leq \eta + \int_\eta^\infty \frac{c}{\alpha\lambda} \int_{\{Y_v^* \geq \lambda\}} \Phi_k(X_v) dP d\lambda$$

$$\leq \eta + \int \Phi_k(\frac{c}{\alpha} X_v) \int_\eta^{Y_v^*} \frac{1}{\lambda} d\lambda dP \qquad \text{(because } \frac{c}{\alpha} \geq 1)$$

$$= n + \int \Phi_k(\tfrac{c}{\alpha} X_v)[\log Y_v^* - \log n]dP$$

$$\leq n + \int [\Phi_k(\tfrac{c}{\alpha} X_v)\log^+\Phi_k(\tfrac{c}{\alpha} x_v) + \frac{Y_v^*}{e} + |\log n|\ \Phi_k(\tfrac{c}{\alpha} X_v)]dP .$$

Therefore

(4) $$E[Y_v^*]$$

$$\leq \frac{e}{e-1}\ [n + |\log n| E[\Phi_k(\tfrac{c}{\alpha} X_v)] + E[\Phi_k(\tfrac{c}{\alpha} X_v)\log^+\Phi_k(\tfrac{c}{\alpha} X_v)]] .$$

Now applying

$$t(\log^+ t)^k \log^+(t(\log^+ t)^k) \leq (k+1)\ t(\log^+ t)^{k+1}$$

one obtains

(5) $$\Phi_k(\tfrac{c}{\alpha} X_v) \log^+\Phi_k(\tfrac{c}{\alpha} X_v) \leq (k+1)\ \Phi_{k+1}(\tfrac{c}{\alpha} X_v) .$$

Inequality (3) now follows from (4) and (5). ///

<u>Proposition 2.3</u>. Let $(F_t, t \in J^m)$ be a filtration such that for each k, $1 \leq k \leq m$, there is α_k, $0 < \alpha_k \leq 1$, such that $(F_t^{1-k}, t_k \in J_k)$ satisfies MR_{α_k} for each $t_1, t_2, \ldots, t_{k-1}$ fixed. Fix $v \in J^m$. Let $(X_t, F_t, t \in J^m)$ be a positive block submartingale. Then there exists a constant C_m (depending on the filtration), such that for every $\delta > 0$ there exists a constant $A(m,\delta) > 0$ with

(6) $$E[e \sup_{t \leq v} X_t] \leq \delta + A(m,\delta) \sup_{0 \leq k \leq m} E[\Phi_k(C_m x_v)] .$$

C'_m may be chosen equal to $\dfrac{c_1 \ldots c_m}{\alpha_1 \ldots \alpha_m}$ where $c_i = 2[M(\frac{1}{2}, \alpha_i)+1]$, $1 = 1,2,\ldots,m$.

<u>Proof</u>. We use induction on m to prove (6). For $m = 1$ (6) holds with

$A(1,\delta) = \frac{e}{e-1}(|\log \frac{\delta}{2}| + 1)$: apply lemma 2.2. with $k = 0$ and $\eta = \frac{\delta}{2}$. Suppose that (6) holds for $m = n$. Let $(X_t, F_t, t \in J^{n+1})$ be a positive block submartingale. For $t = (t_1,\ldots,t_n,t_{n+1}) \in J^{n+1}$, denote $(t_1,\ldots,t_n)$ by t'. Let $\mathcal{G}_{t'} = V_{t_{n+1}} F_t = F_t^{1-n}$. Since $\mathcal{G}_{t'}^{1-k} = F_t^{1-k}$ for all $t \in J^{n+1}$, it follows that for each fixed t_{n+1} J_{n+1}, the process $(X_{t'}, \mathcal{G}_{t'}, t' \in J^n)$ is an n-parameter block martingale. Let $Y_{t'} = \sup_{t_{n+1} \leq v_{n+1}} X_t$. Then $(Y_{t'}, \mathcal{G}_{t'}, t' \in J^n)$ is an n-parameter block submartingale. The filtration $(\mathcal{G}_{t'}, t' \in J^n)$ satisfies the assumptions of the theorem with $m = n$ since $(F_t, t \in J^{n+1})$ does with $m = n+1$. We have

$$X_v^* = e \sup_{t \leq v} X_t = e \sup_{t' \leq v'} (e \sup_{t_{n+1} \leq v_{n+1}} X_t) = e \sup_{t' \leq v'} Y_{t'} = Y_{v'}^* .$$

Applying the induction hypothesis to the block submartingale $(Y_{t'}, \mathcal{G}_{t'})$, one obtains

$$(7\ \ ) \qquad E[X_v^*] = E[Y_{v'}^*] \leq \frac{\delta}{2} + A(n, \frac{\delta}{2}) \sup_{0 \leq k \leq n} E[\Phi_k(C_n\ Y_{v'})] .$$

The stochastic basis $(F_{(v_1,\ldots,v_n,\ t_{n+1})}, t_{n+1} \in J_{n+1})$ satisfies condition $MR_{\alpha_{n+1}}$. Thus theorem 2.1 can be applied to the one-parameter positive submartingale $(C_n\ X_{(v_1,\ldots,v_n,t_{n+1})}, t_{n+1} \in J_{n+1})$. From lemma 2.2 one has

$$(8)\quad E[\Phi_k(C_n\ Y_{v'})] = E[e \sup_{t_{n+1} \leq v_{n+1}} \Phi_k(C_n\ X_{(v_1,\ldots,v_n,t_{n+1})})]$$

$$\leq \frac{e}{e-1}[\eta + |\log \eta|\ E[\Phi_k(C_{n+1}\ X_v)] + (k+1)\ E[\Phi_{k+1}(C_{n+1}\ X_v)]] .$$

It follows that

$$(9)\quad \sup_{0 \leq k \leq n} E[\Phi_n(C_n\ Y_{v'})] \leq \frac{e}{e-1}[\eta + (|\log \eta| + n+1) \sup_{0 \leq k \leq n+1} E[\Phi_k(C_{n+1} X_v)]] .$$

Choose now η_o so that $A(n, \frac{\delta}{2}))\frac{e}{e-1}\eta_o \leq \frac{\delta}{2}$ and apply (9) with $\eta = \eta_o$ to (7), to obtain

$$E[X_v^*] \leq \delta + A(n+1,\delta) \sup_{0 \leq k \leq n} E[\Phi_k(C_{n+1}X_v)]$$

where $A(n+1,\delta) = A(n, \frac{\delta}{2})(|\log \eta_o| + n+1)\frac{e}{e-1}$. Thus (6) is true for all m. ///

<u>Lemma 2.4.</u> Let $(F_t, t \in J)$ be a filtration satisfying C. For every positive submartingale (X_t, F_t) and every $\lambda > 0$ one has

(11) $$P[e \lim \sup X_t \geq \lambda] \leq \frac{1}{\lambda} \lim E[X_t] .$$

The lemma is proved in Section 5, following theorem 5.2.

<u>Proposition 2.5.</u> Let $(F_t, t \in J^m)$, $m > 1$, be a regular filtration. Let $(X_t, F_t, t \in J^m)$ be a positive block submartingale. There there exists a constant C'_m (depending on the filtration), such that for every $\delta > 0$ there is a constant $A(m-1,\delta)$ with

(12) $$P[e \lim \sup X_t \geq \lambda] \leq \frac{1}{\lambda} [\delta + A(m-1,\delta) \sup_{0 \leq k \leq m-1} \lim_{J^m} E[\Phi_k(C'_m X_t)]] ,$$

C'_m may be chosen equal to $\dfrac{c_2 \cdots c_m}{\alpha_2 \cdots \alpha_m}$ where $c_i = 2[M(\frac{1}{2}) + 1]$, $i = 2,\ldots,m$.

<u>Proof.</u> For fixed $t_2 \in J_2, \ldots, t_m \in J_m$, the process $(X_{(t_1,t_2,\ldots,t_m)}, F_t, t_1 \in J_1)$ is a one parameter submartingale. Hence $(e \sup_{t_2,\ldots,t_m} X_t, F^1_t, t_1 \in J_1)$ is also a one parameter positive submartingale. The filtration $(F^1_t, t_1 \in J_1)$ satisfies C since (F_t) is regular (Section I). Therefore, by lemma 2.4., one has

$$P[e \lim \sup_{t_1} (e \sup_{t_2,\dots,t_m} X_t) \geq \lambda] \leq \frac{1}{\lambda} \lim_{t_1} E[e \sup_{t_2,\dots,t_m} (X_t)] .$$

But $e \lim \sup_{J^m} X_t \leq e \lim \sup_{t_1} (e \sup_{t_2,\dots,t_m} X_t)$, hence

$$P[e \lim \sup X_t \geq \lambda] \leq \frac{1}{\lambda} \lim_{t_1} E[e \sup_{t_2,\dots,t_m} (X_t)] . \tag{13}$$

For $t_1 \in J_1$ fixed, the process $(X_t, F_t, (t_2,\dots,t_m) \in J^{m-1})$ is a positive (m-1)-parameter block submartingale. Indeed, $E[X_t | F^{1-k}_{(t_1,s_2,\dots,s_m)}] \geq X_{(t_1,s_2,\dots,s_k,t_{k+1},\dots,t_m)}$. Since (F_t) is regular, the filtration $(F_t,(t_2,\dots,t_m) \in J^{m-1})$ satisfies the assumptions of proposition 2.3. for $m - 1$. Hence

$$E[e \sup_{t_2,\dots,t_m} (X_t)] \leq \delta + A(m-1,\delta) \sup_{0 \leq k \leq m-1} \lim_{t_2,\dots,t_m} E[\Phi_k(C'_m X_t)] . \tag{14}$$

The proposition now follows from inequalities (13) and (14). ///

III. <u>Convergence and Demiconvergence</u>. In this section we will apply maximal inequalities to obtain some convergence and demiconvergence results. We first recall the following basic lemma ([31], page 96).

<u>Lemma 3.1</u>. In order that the net $(x_t, t \in J)$ converge in a complete metric space on which it is defined, it suffices that $(x_{t_n}, n \in N)$ be convergent for all increasing sequences $(t_n, n \in N)$ in J .

<u>Lemma 3.2</u>. Let $(X_t, F_t, t \in J^m)$, $m > 1$, be a martingale or positive submartingale bounded in $L \log^{m-1} L$. Let $X = s \lim X_t$. Then $X_t \to X$ in $L \log^k L$ for all $k \leq m-1$.

Proof. Since convergence in probability is determined by a complete metric, it follows that the (sub)martingale (X_t) converges in probability to a random variable X. Bounded in $L \log^{m-1} L$, $m > 1$, (X_t) is necessarily uniformly integrable and therefore it converges to X in L_1. Fatou's lemma holds also for directed index sets, hence $X \in L \log^{m-1} L$. Since $\int_A X_s dP = \int_A X_t dP$ $(\leq)$ for all $s \leq t$, $A \in F_s$, it follows that $\int_A X_s dP = \int_A X\, dP$ $(\leq)$ for all $A \in F_s$. Consequently $X_t = E^t(X)$ $(\leq)$. Observe that $|X_t| \leq E^t(|X|)$ if (X_t) is a martingale or a positive submartingale. By Jensen's inequality, one has

$$\begin{aligned}\Phi_k(|X_t - X|) &\leq \frac{1}{2}[\Phi_k(2|X|) + \Phi_k(2|X_t|)] \\ &\leq \frac{1}{2}[\Phi_k(2|X|) + \Phi_k(2E^t(|X|))] \\ &\leq \frac{1}{2}[\Phi_k(2|X|) + E^t(\Phi_k(2|X|))],\ k \geq 0 .\end{aligned}$$

Thus $(\Phi_k(|X_t - X|), t \in J^m)$ is uniformly integrable for all $k \leq m-1$. But $\Phi_k(|X_t - X|)$ converges to 0 in probability, since $|X_t - X|$ does, and therefore $\Phi_k(|X_t - X|)$ converges to 0 in L_1 for all $k \leq m-1$, that is $X_t \to X$ in $L \log^k L$ for all $k \leq m-1$. ///

The following is in a sense the main result of the paper.

Theorem 3.3. Let $(F_t, t \in J^m)$, $m > 1$, be a regular filtration. Let E be a Banach space and $\pi : E \to \underline{R}^+$ a continuous map with $\pi(0) = 0$, $\pi(x+y) \leq \pi(x) + \pi(y)$ for all $x,y \in E$. Assume that there is a random variable X such that $\pi(X_t - X)$ converges to 0 in $L \log^k L$ for all $k \leq m-1$. Suppose that for each $t_0 \in J^m$, $(\pi(X_t - X_{t_0}), F_t, t \geq t_0)$ is a positive block submartingale. Then $e \lim_t \pi(X_t - X) = 0$.

Proof. Fix $\varepsilon > 0$, $\lambda > 0$. Choose $\delta > 0$ such that $\delta < \frac{\varepsilon\lambda}{6}$. Then choose

η, $0 < \eta < \inf(\frac{\varepsilon}{3}, \frac{\varepsilon\lambda}{6})$, such that $A(m-1,\delta)\eta < \delta$, where $A(m-1,\delta)$ is the constant from proposition 2.5. Next choose X_{t_0} so that

$$\sup_{0 \le k \le m-1} \lim_t E[\Phi_k(C'_m \pi(X_t - X_{t_0}))] = \sup_{0 \le k \le m-1} E[\Phi_k(C'_m \pi(X - X_{t_0}))] \le \eta .$$

Note that this implies $E[\Phi_0(C'_m \pi(X - X_{t_0}))] = C'_m \| \pi(X - X_{t_0})\|_1 \le \eta$, hence $\| \pi(X - X_{t_0})\|_1 \le \eta$ since $C'_m \ge 1$. Then

$$P[e \lim\sup_t \pi(X_t - X) \ge \lambda]$$

$$\le P[e \lim\sup_t (\pi(X_t - X_{t_0}) + \pi(X_{t_0} - X)) \ge \lambda]$$

$$\le P[e \lim\sup_t \pi(X_t - X_{t_0}) \ge \frac{\lambda}{2}] + P[\pi(X_{t_0} - X) \ge \frac{\lambda}{2}]$$

$$\le \frac{2}{\lambda}[\delta + A(m-1,\delta) \sup_{0 \le k \le m-1} \lim_t E[\Phi_k(C'_{m-1} \pi(X_t - X_{t_0}))]$$

$$+ \frac{2}{\lambda} \| \pi(X_{t_0} - X)\|_1 \qquad \text{(proposition 2.5.)}$$

$$\le \frac{\varepsilon}{3} + \frac{\varepsilon}{3} + \frac{\varepsilon}{3} = \varepsilon .$$

Hence $e \lim_t \pi(X_t - X) = 0$. ///

<u>Theorem 3.4</u>. Let $(F_t, t \in J^m)$, $m > 1$, be a regular filtration.

(i) Let $(X_t, F_t, t \in J^m)$ be a block submartingale bounded in $L \log^{m-1} L$ and let $X = s \lim X_t$. Then $e \lim\sup_t X_t = X$ (upper demiconvergence).

(ii) Let $(X_t, F_t, t \in J^m)$ be a positive (integrable) block martingale and let $X = s \lim X_t$. Then $e \lim\inf_t X_t = X$ (lower demiconvergence).

(iii) Let $(X_t, F_t, t \in J^m)$ be a block martingale bounded in $L \log^{m-1} L$. Let $X = s \lim X_t$. Then $e \lim X_t = X$.

Proof. (i) Assume first that (X_t) is positive block submartingale. Then by lemma 3.2., $X_t \to X$ in $L \log^k L$ for all $k \leq m-1$. Now apply theorem 3.3. to the process (X_t, F_t) and $\pi : \underline{R} \to \underline{R}^+$, $\pi(x) = x^+$ to obtain $e \lim(X_t - X)^+ = 0$. This shows that $e \lim\sup X_t \leq X$, and the equality follows from the general inequality $X = s \lim X_t \leq e \lim\sup X_t$. The theorem is thus proved for positive block submartingales, and consequently for block submartingales which are bounded below by a constant. Let (X_t) be a (not necessarily positive) block submartingale. Fix a real number a, then the process $(X_t \vee a, F_t)$ is also a block submartingale and thus $e \lim\sup X_t \vee a = X \vee a$ by the first part of the proof. Since $e \lim\sup X_t \vee (-n) = X \vee (-n)$ for all $n > 0$ and since $e \lim\sup X_t > -\infty$ by Fatou's lemma, it follows that $e \lim\sup X_t = X$.

(ii) Let (X_t, F_t) be a positive integrable block martingale. Then (X_t) converges in probability to a random variable X. Let $U_t = e^{-X_t}$; then (U_t, F_t) is a block submartingale bounded in L_∞, which converges in probability to $U = e^{-X}$. By part (i), $e \lim\sup U_t = U$. Fatou's lemma implies $e \lim\inf X_t < \infty$, thus $e \lim\sup(-X_t) = (-X)$ or $e \lim\inf X_t = X$.

(iii) Let (X_t) be a block martingale bounded in $L \log^{m-1} L$. By lemma 3.2. $X_t \to X$ in $L \log^k L$ for all $k \leq m-1$. Now apply theorem 3.3. to the process (X_t, F_t) and $\pi : \underline{R} \to \underline{R}^+$, $\pi(x) = |x|$ to obtain $e \lim |X_t - X| = 0$. ///

Recall that if $J_k = N$, $k = 1,2,\ldots,m$, then the filtration (F_t) is regular (section I). The case $m = 2$, $J_1 = J_2 = N$ was obtained in [29], theorem 2.1. and [28] theorem 1.1.

Theorem 3.5. Let $(F_t, t \in J^m)$, $m > 1$, be a regular filtration. Let $(E, \|\cdot\|)$ be a Banach space with the Radon-Nikodym property. Let (X_t, F_t) be an E-valued block martingale bounded in $L \log^{m-1} L(E)$, i.e., such that $\sup_t E[\|X_t\|(\log^+ \|X_t\|)^{m-1}] < \infty$. Then (X_t) converges essentially and in $L \log^{m-1} L(E)$ to a random variable X.

Proof. Since (X_t) is uniformly integrable and E has the Radon-Nikodym property, (X_t) admits a representation $X_t = E^t(X)$ for an E-valued random variable X and $E[\|X\|(\log^+\|X\|)^{m-1}] < \infty$. By lemma 3.2., $X_t \to X$ in $L \log^k L(E)$ for all $k \leq m-1$. Then theorem 3.3. applied to the process (X_t, F_t) and $\pi : E \to \underline{R}^+$, $\pi(x) = \|x\|$, gives $e \lim_t \|X_t - X\| = 0$. ///

There is also a version of our results corresponding to the case where the directed set is filtering to the left. For the sake of simplicity we assume that $J_k = -N = \{\ldots -3,-2,-1\}$ for all $k \leq m$. For any integers k, ℓ, $1 \leq k \leq \ell \leq m$,

$$F_t^{k-\ell} = F_{(-1,-1,\ldots,-1,t_k,t_{k+1},\ldots,t_\ell,-1,-1,\ldots,-1)}$$

$$F_t^k = F_{(-1,-1,\ldots,-1,t_k,-1,\ldots,-1)} .$$

An integrable process $(X_t, F_t, t \in J^m)$ is a reversed block k (sub)martingale for a fixed $k \leq m$, if $E_s^{1-k}(X_t) = X_{(s_1,\ldots,s_k,t_{k+1},\ldots,t_m)}$ $(\geq)$ for $s \leq t$. An integrable process is a reversed block (sub)martingale if it is a reversed block k (sub)martingale for all $k \leq m$.

Theorem 3.6. (i) Let $(X_t, F_t, t \in J^m)$ be a reversed block submartingale such that $X_{(-1,-1,\ldots,-1)}$ is $L \log^{m-1} L$ integrable (hence (X_t) is $L \log^{m-1} L$ bounded). Let $X = s \lim X_t$. Then $\limsup X_t = X$ a.s..

(ii) Let $(X_t, F_t, t \in J^m)$ be a reversed positive (integrable) block martingale. Let $X = s \lim X_t$. Then $\liminf X_t = X$ a.s..

(iii) Let $(X_t, F_t, t \in J^m)$ be a reversed block martingale such that $X_{(-1,-1,\ldots,-1)}$ is $L \log^{m-1} L$ integrable. Let $X = s \lim X_t$. Then $e \lim X_t = X$ a.s.. ///

The proof is similar to the case filtering to the right, but simpler, because the process $(X_t - X, t \in J)$, where $X = s\lim X_t$, is now adapted and converges to 0 in $L\log^k L$ for all $k \leq m-1$.

An application to the multiparameter Marcinkiewicz theorem for $p < 1$, similar to the one given for $m = 2$ in [9], is possible. In this application the σ-fields are not of product type, and they do not satisfy the conditional independence assumption (F4) because the conditional expectations are with respect to fewer and fewer sums: one applies a multiparameter version of the classical Doob reversed martingale argument, extended in [9] to submartingales.

IV. Banach lattices. In this section we extend the demiconvergence results to random variables taking values in a separable Banach lattic $(E,\|\cdot\|)$. We at first consider the case $E = L_1(\Omega_1,F_1,P_1)$ where (Ω_1,F_1,P_1) is a fixed probability space.

The following lemma is part of a more general theory developed in [3].

Lemma 4.1. Let $(X_t,F_t,t \in J)$ be an E-valued positive Bochner integrable martingale. Let $Y_t : \Omega \times \Omega_1 \to \underline{R}$, $Y_t(\omega,\omega_1) = X_t(\omega)(\omega_1)$ P - a.s. . Then $(Y_t,F_t \otimes F_1, t \in J)$ is a positive martingale. Moreover, if $X = e\lim\inf X_t$, $Y = e\lim\inf Y_t$, then P - a.s. $X(\omega) \in E$ and $X(\omega)(\cdot) = Y(\omega,\cdot)$.

Proof. Since X_t is strongly measurable, Y_t is measurable with respect to $F_t \otimes F_1$. If $s \leq t$ and $A \in F_s$ then $\int_A X_s dP = \int_A X_t dP$. Thus for $A \in F_s$, $B \in F_1$

$$\begin{aligned}\mu_s(A \times B) &\overset{def}{=} \int_{A\times B} Y_s\, d(P \otimes P_1) = \int_B\int_A Y_s(\omega,\omega_1)dP(\omega)dP_1(\omega)\\ &= \int_B\int_A X_s(\omega)(\omega_1)dP(\omega)dP_1(\omega) = \int_B\int_A X_t(\omega)(\omega_1)dP(\omega)dP_1(\omega)\\ &= \int_{A\times B} Y_t d(P \otimes P_1) = \mu_t(A \times B)\ .\end{aligned}$$

Therefore μ_s, μ_t are defined on the semialgebra $\{A \times B : A \in F_s, B \in F_1\}$ of measurable rectangles, and by their definition are bounded and σ-additive. Thus μ_s, μ_t can be extented to measures on $F_s \otimes F_1$. Therefore $(Y_t, F_t \otimes F_1)$ is a martingale. ///

In Banach lattices with the Radon-Nikodym property, L_1-bounded positive submartingales indexed by N converge a.s., as proved by Heinich [14], hence in probability. In the multiparameter case we have the following.

Theorem 4.2. Let $(F_t, t \in J^m), m > 1$, be a regular filtration. Let $(E, \|\cdot\|)$ be a Banach lattice with the Radon-Nikodym property.
(i) Let $(X_t, F_t, t \in J^m)$ be an E-valued, $L \log^{m-1} L$ bounded, positive block submartingale. Then the stochastic limit $s \lim X_t = X$ exists and $e \lim \|(X_t - X)^+\| = 0$.
(ii) Let $(X_t, F_t, t \in J^m)$ be an E-valued, L_1-bounded, positive block martingale. Then the stochastic limit $s \lim X_t = X$ exists and $e \lim\inf X_t = X$.

Proof. (i) Let (t_n) be an increasing sequence in J^m. Then (X_{t_n}) is a positive submartingale, hence converges in probability. By lemma 3.1., the net (X_t) converges in probability, say to X. By lemma 3.2. $X_t \to X$ in $L \log^k L(E)$ for all $k \leq m-1$. Then theorem 3.3. applied to the process (X_t, F_t), $\pi : E \to R^+$, $\pi(x) = \|x^+\|$ gives $e \lim \|(X_t - X)^+\| = 0$.
(ii) Let (X_t) be a positive L_1-bounded block martingale. As in (i) one shows that $s \lim X_t = X$ exists. Let E have the Radon-Nikodym property then c_0 is not contained in E ([7], pp. 60 and 81) and therefore E is weakly sequentially complete ([20], p. 34) hence order continuous. Thus E is order isometric to an ideal of an $L_1(\Omega_1, F_1, P_1)$ ([20], p. 25). Then the real-valued, positive block martingale $(Y_t, F_t \otimes F_1, t \in J^m)$ necessarily converges to Y in probability, where $Y(\omega, \omega_1) = X(\omega)(\omega_1)$, and $e \lim\inf Y_t = Y$. (Theorem 3.4.) By lemma 4.1, $e \lim\inf X_t = X$. ///

The case $m = 2$, $J_k = N$, of theorem 4.2 (i) was obtained in [29].

V. Relations between Convergence and Demiconvergence. We show here that in general convergence of martingales is equivalent with demiconvergence of submartingales. Under commutation, this extends to block martingales. It is also remarked that under commutation the proof of theorem 3.4. simplifies.

The following is joint version of Krickeberg and Riesz decomposition theorems in the directed index set case.

Proposition 5.1. Let L_Φ be an arbitrary Orlicz space. Let $(X_t, F_t, t \in J)$ be a submartingale bounded in L_Φ. Then $X_t = Y_t^1 - Y_t^2 - S_t$ where (Y_t^1, F_t), (Y_t^2, F_t) are positive martingales bounded in L_Φ and (S_t, F_t) is a positive supermartingale with $E(S_t) \to 0$. If (X_t) is a martingale then $S_t = 0$ for all t.

Proof. Let $s \in J$ be fixed but arbitrary. Set $U_t = E^s(X_t)$, $t \geq s$. Then $t \leq t'$ implies $E^t(X_{t'}) \geq X_t$, consequently $U_{t'} = E^s E^t(X_{t'}) \geq E^s(X_t) = U_t$. Thus $(U_t, t \geq s)$ is increasing and therefore converges essentially to Y_s. By Jensen's inequality

$$E[\Phi(|U_t|)] = E[\Phi(|E^s(X_t)|)] \leq E[\Phi(|X_t|)] .$$

Since $Y_s = \lim_{t \geq s} \uparrow U_t$ by Fatou's lemma $E[\Phi(|Y_s|) \leq \lim_t E[\Phi(|X_t|)]$. Since s is arbitrary, $\sup_{s \in J^m} E[\Phi(|Y_s|) < \infty$. If $s \leq s'$ then

$E^s(Y_{s'}) = E^s(\lim_{t \geq s'} \uparrow E^{s'}(X_t)) = \lim_{t \geq s'} E^s(X_t) = Y_s$, hence (Y_t, F_t) is a martingale $Y_t \geq X_t$. We write $S_t = Y_t - X_t$; S_t is a positive supermartingale. For $t \geq s$ $E^s(S_t) = E^s(Y_t) - E^s(X_t) = Y_t - E^s(X_t)$, hence $\lim_{t \geq s} \downarrow E^s(S_t) = 0$. It follows that $\lim_{t \geq s} \downarrow E[E^s(S_t)] = 0$, thus $E(S_t) \to 0$. Define

$Y_s^1 = \lim_{t \geq s} \uparrow E^s(Y_t^+)$, $Y_s^2 = \lim_{t \geq s} \uparrow E^s(Y_t^-)$. Since both (Y_t^+) and (Y_t^-) are positive submartingales bounded in L_Φ the above argument shows that each (Y_t^i, F_t), $i = 1,2$ is a positive martingale bounded in L_Φ . Clearly $Y_t = Y_t^1 - Y_t^2$. ///

Proposition 5.1. remains valid if L_Φ is replaced by L_∞ .

<u>Theorem 5.2.</u> Let $(F_t, t \in J)$ be a stochastic basis.

(i) Assume that all martingales bounded in L_Φ essentially converge. Then all submartingales bounded in L_Φ essentially upper demiconverge.

(ii) Assume that all positive submartingales bounded in L_Φ essentially upper demiconverge. Then all martingales bounded in L_Φ essentially converge.

<u>Proof.</u> (i) Let (X_t, F_t) be a submartingale bounded in L_Φ . By proposition 5.1., $X_t = Y_t - S_t$ with $Y_t = Y_t^1 - Y_t^2$ a martingale and $E(S_t) \to 0$. Let $X = s \lim X_t$, then $s \lim S_t = 0$ implies $s \lim Y_t = X$. The martingale (Y_t, F_t) is bounded in L_Φ and therefore essentially converges to X . Hence $e \lim\sup X_t \leq X$. Since $s \lim X_t \leq e \lim\sup X_t$ always, $e \lim\sup X_t = X$.
(ii) Let (Y_t, F_t) be a positive martingale bounded in L_Φ . Let $Y = s \lim Y_t$.

Since (Y_t) is also a positive submartingale $e \lim\sup Y_t = Y$. Let $X_t = e^{-Y_t}$;

then (X_t) is an L_∞-bounded positive submartingale. Hence $e \lim\sup e^{-Y_t} = e^{-Y}$ and $e \lim\inf Y_t = Y$. Therefore $e \lim Y_t = Y$. If (Y_t) is not positive apply proposition 5.1. to represent Y_t as a difference of two positive martingales bounded in L_Φ . ///

Since condition C is known to imply convergence of L_1-bounded martingales, ([27], theorem 3.3.), we have the following:

Corollary 5.3. Let $(F_t, t \in J)$ be a stochastic basis satisfying condition C. Let (X_t, F_t) be an L_1-bounded submartingale and let X be its limit in probability. Then $e \limsup X_t = X$.

Let $(F_t, t \in J)$ be any stochastic basis and (X_t, F_t) a positive submartingale. It was shown in [25], theorem 1.4., that

$$P(s \limsup X_t \geq \lambda] \leq \frac{1}{\lambda} \lim E[X_t] .$$

The proof of lemma 2.4 now follows.

Observe that L_1-bounded submartingales need not converge not only under C, but even under the stronger Vitali condition V ([18], [13]). We also note that since the covering conditions V_p, $1 \leq p < \infty$, are necessary and sufficient for the convergence of L_q-bounded, martingales, $\frac{1}{p} + \frac{1}{q} = 1$, (Krickeberg [17] and A. Millet [22]), they are also necessary and sufficient for the essential upper demi-convergence of L_q-bounded submartingales. Analogous results hold for classes of Orlicz spaces ([19], [38]).

We now sketch an alternative proof of theorem 3.4. assuming commutation. Recall that

$$F_t^k = \bigvee_{\substack{t_1,\ldots,t_{k-1},\\ t_{k+1},\ldots,t_m}} F_{(t_1,\ldots,t_{k-1},t_k,t_{k+1},\ldots,t_m)}$$

and E_t^k is the conditional expectation given F_t^k.

The commutation assumption is that the L_1-operators E_t^k commute, [21] page 3. Observe that E_t^k commute if for every martingale $X_t = E^t(X)$ and all $k \leq m$.

$$(1) \qquad E_s^k(X_t) = X_{(t_1,\ldots,t_{k-1},s_k,t_{k+1},\ldots,t_m)} .$$

Indeed for $s \leq t$

$$X_{(s_1,\ldots,s_m)} = E_s^{\pi(1)} \cdots E_s^{\pi(m)}(X_t)$$

for any permutation π of $(1,2,\ldots,m)$. Since (X_t) converges to X in L_1, one obtains

$$X_{(s_1,\ldots,s_m)} = E_s^{\pi(1)} \ldots E_s^{\pi(m)}(X) .$$

This implies that the operators E_t^k, $k = 1,2,\ldots,m$, commute for all t. Conversely, if $\bigcap_{k=1}^{m} F_t^k = F_t$ for each t, and E_t^k commute, then (1) holds. Indeed, then $E_t^1 \ldots E_t^m(X) = E_t^{\pi(1)} \ldots E_t^{\pi(m)}(X)$ for any permutation π of $(1,2,\ldots,m)$, hence $E_t^1 \ldots E_t^m(X)$ is F_t^k-measurable for all k, and therefore F_t-measurable. Consequently $E^t(X) = E_t^1 \ldots E_t^m(X) = E_t^{\pi(1)} \ldots E_t^{\pi(m)}(X)$. Let now (X_t, F_t) be a martingale. Then, replacing X by X_t and t by $(t_1,\ldots,t_{k-1},s_k,t_{k+1},\ldots,t_m)$, one has

$$\begin{aligned} X_{(t_1,\ldots,t_{k-1},s_k,t_{k+1},\ldots,t_m)} &= E[X_t | F_{(t_1,\ldots,t_{k-1},s_k,t_{k+1},\ldots,t_m)}] \\ &= E_t^1 \ldots E_t^{k-1} E_s^k E_t^{k+1} \ldots E_t^m(x) \\ &= E_s^k E_t^1 \ldots E_t^{k-1} E_t^{k+1} \ldots E_t^m(X_t) = E_s^k(X_t) . \end{aligned}$$

Relation (1) for $k = 1,2,\ldots,\ell \le m$ implies

$$E_s^1 E_s^2 \ldots E_s^\ell(X_t) = X_{(s_1,\ldots,s_\ell,t_{\ell+1},\ldots,t_m)} .$$

Applying $E_s^{1-\ell}$ to both sides, one obtains that (X_t) is a block ℓ-martingale. In particular, from the commutation assumption, which implies (1) for all k, it follows that every martingale is a block martingale; similarly every submartingale is a block submartingale. Therefore demiconvergence of block submartingales reduces to convergence of block martingales. We may observe that the simple proof of convergence given in [37] assuming that each $J_k = N$ extends to the case of Section III. It suffices to apply the lemma 2.2. above in the induction step in [37].

VI. Applications. In the present section we give applications to differentiation of integrals in $\underline{R}^m$ $m \geq 1$. We first show that the regularity condition R_α is satisfied in the setting of differentiation in m-dimentional Euclidean space. Let μ denote the Lebesgue measure on $[0,1]^m$. Given a countable partition t of $[0,1]^m$, by the diameter $d(t)$ of t we mean the supremun of the diameters of the elements (atoms) of t.

Proposition 6.1. Let $\mathcal{C}$ be a collection of measurable subsets C of $[0,1]^m$. Assume that $\mathcal{C}$ is a family of substantial sets, i.e., there exists a constant M such that every C in $\mathcal{C}$ is contained in an open ball B with $\mu(B) \leq M\,\mu(C)$. Let J be a non-empty family of countable partitions (modulo sets of measure 0) of $[0,1]^m$ into elements of $\mathcal{C}$. J is order by refinement, i.e., if $s,t \in J$, $s \leq t$, then every element (atom) in s is a union of atoms in t. J is assumed filtering to the right. Then the filtration (F_t) of σ-fields generated by the partitions t satisfies the regularity condition R_α with $\alpha = M^{-1}3^{-m}$.

Remark. A simple example of such a family J is a family, ordered by refinement, of countable partitions of $[0,1]^m$ into parallelepipeds such that the ratio between the largest and shortest edges is bounded, say by a. (If $m = 2$, one can choose $M = \frac{a\pi}{2}$.)

Proof. Let (A_t) be an adapted family of sets and let $A = \sup A_t$. Since $A_t = \bigcup_{i=1}^{\infty} C(t,i)$, $C(t,i) \in t$, A is covered by the family $\{C(t,i)\}$ except for a set of measure zero. Let $\{B(t,i)\}$ be the corresponding family of open balls, i.e., $C(t,i) \subseteq B(t,i)$ and $\mu(B(t,i)) \leq M\,\mu(C(t,i))$. Let $B_t = \bigcup_{i=1}^{\infty} B(t,i)$, $B = \sup B_t$. Then $A \subseteq B$. For every $\varepsilon > 0$, one can choose a finite collection of disjoint open balls $B(t_1,\ell_1),\dots,B(t_1,\ell_{t_1}),\dots,$

$B(t_n,\ell_1),\dots,B(t_n,\ell_{t_n})$, such that

$$3^{-m}(1-\varepsilon)\mu(B) \leq \sum_{1 \leq j \leq n} \sum_{1 \leq i \leq t_j} \mu(B(t_j,\ell_i)) .$$

(For this result, due to J. Serrin, see W. Rudin [33], page 164.) Hence

$$3^{-m}(1-\varepsilon)\mu(A) \leq M \sum_{1 \leq j \leq n} \sum_{1 \leq i \leq t_j} \mu(C(t_j,\ell_i)) .$$

Let $A_{t_j} = \bigcup_{1 \leq i \leq \ell_{t_j}} C(t_j,\ell_i)$, $j = 1,2,\dots,n$. Then the sets A_{t_j} are disjoint and $A_{t_j} \in F_{t_j}$. Define

$$\tau = \begin{cases} t_j & \text{on} \quad A_{t_j}, \; J = 1,2,\dots,m \\ t_{n+1} & \text{on} \quad (\bigcup_{1 \leq j \leq n} A_{t_j})^c \end{cases} .$$

Now τ is the desired single-valued stopping time τ . ///

<u>Remark</u>. Call sets $B \subseteq \underline{R}^m$, <u>V-balls</u> if there is number α_m such that given $\varepsilon > 0$ and any collection $\{B_i\}$ of V-balls, there is a number M_ε and a finite subcollection $B_{k_1},\dots,B_{k_n}$ such that

$$P(\bigcup_{i=1}^{n} B_{k_i}) \geq \alpha_m(1-\varepsilon)P(B_i) \quad \text{and} \quad \sum_{i=1}^{n} 1_{B_{k_i}} - 1 \leq M_\varepsilon .$$

Let $\mathcal{C}'$ be a collection of measurable sets, called V-substantial sets, such that there exists a constant M so that every $C \in \mathcal{C}'$ is contained in a V-ball B with $\mu(B) \leq M\mu(C)$. Then the filtration (F_t) generated by the collection $\mathcal{C}'$ satisfies the condition MR_α with $\alpha = \alpha_m M^{-1}$. The following theorem is still true if each $\mathcal{C}_i$ is a collection of V-substantial sets.

Theorem 6.2. Let $1 \leq s \leq m$ and consider positive numbers $k_1, k_2, \ldots, k_s$ such that $k_1 + \ldots + k_s = m$. For each $i \leq s$, let $\mathcal{C}_i$ be a collection of substantial subsets of $[0,1]^{k_i}$. Let J_i be a family of countable partitions of $[0,1]^{k_i}$ into elements of $\mathcal{C}_i$ such that for each $\varepsilon > 0$ there exists $t_i \in J_i$ with $d(t_i) < \varepsilon$. On $[0,1]^m$ define

$$I_R(x) = \frac{\int_R f \, d\mu}{\mu(R)}$$

where $R = C_1 \times C_2 \times \ldots \times C_s$, $C_i \in \mathcal{C}_i$, $x \quad R$ and $f \in L_1([0,1]^m)$.

(i) If f is positive, then

$$\liminf I_R(x) = f(x) \quad \text{a.s.}$$

as R shrinks to x, i.e., $d(R) \to 0$.

(ii) If f is Banach-space valued, strongly measurable and $L \log^{s-1} L$ integrable, then

$$\lim I_R(x) = f(x) \quad \text{a.s.}$$

as R shrinks to x.

Proof. For each $t_i \in J_i$ let F_{t_i} be the σ-field generated by the partition t_i. By the definition of conditional expectation, one has

$$E[f | F_{t_1} \otimes F_{t_2} \otimes \ldots \otimes F_{t_s}](x) = \frac{\int_R f \, d\mu}{\mu(R)}, \quad x \in R .$$

For each i, the filtration $(F_{t_i}, t_i \in J_i)$ satisfies condition R_{α_i} with $\alpha_i = M_i^{-1} 3^{-k_i}$. Since $\lim_{J_i} d(t_i) = 0$, $\bigvee_{t_i} F_{t_i}$ coincides with the σ-field of al Lebesgue measurable sets on $[0,1]^{k_i}$.

(i) If f is positive, then by theorem 3.4 (ii)

$$\liminf E[f|F_{t_1} \otimes \ldots \otimes F_{t_s}] = E[f|\vee F_{t_1} \otimes \ldots \otimes \vee F_{t_s}] = f \quad \text{a.s.}\ .$$

(ii) If f is $L \log^{s-1} L$ integrable, then by theorem 3.5. for $s > 1$ and by the martingale convergence theorem under C proved in [27], for $s = 1$

$$\lim E[f|F_{t_1} \otimes \ldots \otimes F_{t_s}] = E[f|\vee F_{t_1} \otimes \ldots \otimes \vee F_{t_s}] = f \quad \text{a.s.}\ . \quad ///$$

A classical case ($s = m = 2$ and C_k intervals) of part (i) is due to Besicovich (see e.g. [12], page 100.)

We now state Zymund's theorem [39] on differentiation of integrals, a generalization of the theorem of Jessen-Marcinkiewicz-Zygmund ([15] or [12], page 51.)

<u>Theorem 6.3.</u> (Zygmund). Let $1 \le s \le m$ and consider only intervals R in $[0,1]^m$ whose sides have no more than s different sizes. If f is $L \log^{s-1} L$ integrable then $\lim I_R(x) = f(x)$ a.s. as R shrinks to x .

<u>Proof.</u> For each rectangle R we have at most s different sizes. Without loss of generality we can assume that the first k_1 coordinates are equal, then the next k_2 are equal, finally the last k_s coordinates are equal, $k_1 + k_2 + \ldots + k_2 = m$. Let J_i denotes the family of all partitions of $[0,1]^{k_i}$ into cubes. Then each J_i is a collection of substantial sets. There are only finitely many possible orderings of coordinates to be considered. Therefore theorem 6.2. implies convergence. ///

References

1. Astbury, K., (1980). The order convergence of martigales indexed by directed sets. Trans. Amer. Math. Sec. 265, 495-510.

2. Bagchi, S., (1983). On almost sure convergence of classes of multi-valued asymptotic martingales. Ph.D. dissertation, Department of Mathematics Ohio State University.

3. Bru, B., Heinich, H., (1983). Conditional martingales (preprint).

4. Cairoli, R., (1970). Une inégalité pour martingales à indices multiples. Seminaire de Probabilité IV. Université de Strasbourg. Lecture notes in Math. 124, 1-27. Springer-Verlag.

5. Cairoli, R., Walsh, J. B., (1975). Stochastic integrals in the plane. Acta Math. 134, 111-183.

6. Chatterji, S., (1976). Vector-valued martingales and their applications. Lecture notes in Math. 526, 33-51. Springer-Verlag.

7. Diestel, J., Uhl, J., Jr., (1977). Vector Measures. AMS Mathematical Surveys 15, Providence, Rhode Island.

8. Doob, J. L., (1953). Stochastic Processes. Wiley, New York.

9. Edgar, G. A., Sucheston, L., (1981). Démonstrations de lois des grands nombres par les sous-martingales descendantes, C. R. Acad. Sci. Paris, Ser. A 292, 967-969.

10. Fölmer, H., (1983). Almost sure convergence of multiparameter martingales for Markov random fields. Ann Prob. 12, 133-140.

11. Ghoussoub, N., Talagrand, M., (1978). A generalized Chacon's inequality and order convergence of processes. Seminaire Choquet, 17e annee.

12. Guzman, M. de, (1975). Differentiation of Integrals in R^n. Lecture Notes in Math 481. Springer-Verlag.

13. Hayes, C. A., Pauc, C. Y., (1970). Derivation and Martingales. Springer-Verlag.

14. Heinich, H., (1978). Convergence de sous-martingales positives dans un Banach réticulé. C. R. Acad. Sci. Paris, Ser. A 286, 279-280.

15. Jessen, B., Marcinkievicz, J., Zygmund, A., (1935). Note on the differentiability of multiple integrals. Fund Math. 25, 217-234.

16. Krasnosel'ski, M. A., Rutickii, Ya.B., (1961). Convex Functions and Orlicz Spaces. Gordon and Breach Science Publishers, New York.

17. Krickeberg, K., (1956). Convergence of martingales with a directed index set. Trans. Amer. Math. Soc. 83, 313-337.

18. Krickeberg, K., (1957). Stochastische Konvergenz von Semimartingalen. Math. Z. 66, 470-486.

19. Krickeberg, K., Pauc, C. Y., (1963). Martingales et dérivation. Bull. Soc. Math. France 91, 455-543.

20. Lindenstraus, J., Tzafriri, L., (1979). Classical Banach Spaces II, Springer-Verlag.

21. Meyer, P. A., (1981). Theorie élementaire des processus à deux indices. Lectures Notes in Math. 863, 1-39. Springer-Verlag.

22. Millet, A., (1978). Sur la caractérisation des conditions de Vitali par la convergence essentielle des martingales, C. R. Acad. Sci. Paris, Ser. A 287, 887-890.

23. Millet, A., Sucheston, L., (1979). La convergence essentielle des martingales bornées dans L^1 n'implique pas la condition de Vitali V, C. R. Acad. Sci. Paris, Ser. A 288, 595-598.

24. Millet, A., Sucheston, L., (1979). On covering conditions and convergence. Proceedings of the 1979 Oberwolfach Conference in Measure theory. Lecture Notes in Math. 794, 431-454. Springer-Verlag.

25. Millet, A., Sucheston, L., (1980). A characterization of Vitali conditions in terms of maximal inequalities. Ann. Prob. 8, 339-349.

26. Millet, A., Sucheston, L., (1980). Convergence of classes of amarts indexed by directed sets. Canad. J. Math. 32, 86-125.

27. Millet, A., Sucheston, L., (1980). On convergence of L_1-bounded martingales indexed by directed sets. J. Prob. Math. Statist. 1, 151-189.

28. Millet, A., Sucheston, L., (1981). On regularity of multiparameter amarts and martingales. Z. Wahrscheinlichkeitstheorie Verw. Gebiete 56, 21-45.

29. Millet, A., Sucheston, L., (1983). Demiconvergence of processes indexed by two indices. Ann. Inst. Henri Poincaré XIX, no 2, 175-187.

30. Neveu, J., (1965). Mathematical Fountations of the Calculus of Probability. Holden-Day, Inc.

31. Neveu, J., (1975). Discrete Parameter Martingales. Amsterdam, North Holland.

32. Royden, H. L., (1968). Real Analysis. Macmillen Company, N. Y.

33. Rudin, W., (1970). Real and Complex Analysis. McGraw-Hill.

34. Shieh, N., (1982). Strong differentiation and martingales in product spaces. Math. Rep. Toyama Univ. 5, 29-36.

35. Smythe, R. T., (1973). Strong laws of large numbers for r-dimensional arrays of random variables. Ann. Prob. 1, 164-170.

36. Smythe, R. T., (1976). Multiparameter subadditive processes. Ann. Prob. 4, 772-782.

37. Sucheston, L., (1983). One one-parameter proofs of almost sure convergence of Multiparameter processes. Z. Wahrscheinlichkeitstheorie Verw. Gebiete 63, 43-49.

38. Talagrand, M., (1984). Derivation, L^ψ-bounded martingales and covering conditions. (preprint).

39. Zygmund, A., (1967). A note on differentiability of integrals. Colloquiun Mathematicum XVI, 199-204.

M-INFINITELY DIVISIBLE RANDOM COMPACT CONVEX SETS

Evarist Giné[1]
Texas A&M University
College Station, Texas 77843 USA

Marjorie G. Hahn[2]
Tufts University
Medford, MA 02155 USA

§0. Introduction

Some limit theorems for Minkowski sums of independent identically distributed random compact convex (c.c.) subsets of a separable Banach space B were obtained in [13]. (See also [6], [16], [21], [22] and [24] for the classical Gaussian case.) The central limit theorems proved there did not presuppose the existence of a limiting probability law on random sets: in fact we only considered convergence in distribution of the real random variables $\{a_n \delta(S_n/n, EX)\}_{n=1}^{\infty}$. (Here δ is Hausdorff distance, $S_n = \sum_{j=1}^{n} X_i$ is the Minkowski sum of the i.i.d. random c.c. sets X_i with law $L(X_i) = L(X)$, EX is the expectation of the random set X and $a_n \in \mathbb{R}$, $a_n \to \infty$. See these and other definitions below.) In other words, those were speed of convergence results for the law of large numbers. The next natural question is whether the laws of the sums $L(S_n/c_n)$ can be approximated by Gaussian, stable and, in general, by M-infinitely divisible c.c. sets (with M for Minkowski addition). This question is obviously related to (in fact dependent upon) the existence of a Lévy-Khinchin representation for M-infinitely divisible (M-i.d.) random c.c. sets.

In this paper we survey recent work on M-i.d. c.c. sets by Lyashenko [17], Mase [18], Vitale [23] and ourselves [11], [12], and present as well several partial results and examples about general M-i.d. c.c. sets in infinite-dimensional Banach spaces.

The situation is as follows: (1) the M-i.d. c.c. sets of $\mathbb{R}^d$ are completely determined; (2) the p-stable c.c. sets of any separable infinite-dimensional Banach space B are also completely characterized (and the results show that, curiously enough, often a CLT for $\{a_n \delta(S_n/n, EX)\}$ holds but no non-degenerate limiting stable c.c. set exists for the sums); and (3), very little is known about general M-i.d. c.c. sets in infinite dimensions.

[1]Research partially supported by NSF grant No. DMS-8318610
[2]Research partially supported by NSF grant No. MCS 8101895 01

By using support processes these problems reduce to special questions about infinitely divisible laws in the Banach space of continuous functions on a compact metric space. The available general theory in this setting turns out to be very useful.

Our interest on this subject stems from the facts that it is a fertile field for application of Banach space probability theory, and that some of the results are quite elegant. Moreover, we know of at least one interesting application of limit theorems for random sets: see Artstein and Hart [3] and Artstein [2], where a central limit theorem and a law of large numbers for Minkowski addition are applied in an optimization problem of interest in economics. (However, it may be argued that this theory has not yet had significant interactions with other areas.)

This paper is organized as follows. Section 1 provides a brief survey of the known results including support processes, M-i.d. c.c. subsets of $\mathbb{R}^d$ and p-stable c.c. subsets of separable Banach spaces. Section 2 contains new results on M-i.d. c.c. sets of infinite-dimensional Banach spaces essentially showing the inadequacy of the methods used in $\mathbb{R}^d$. Some necessary conditions for M-i.d. are provided as well as some examples and counterexamples. It is shown in particular that it is impossible, in infinite dimensions, to choose a point from every compact convex subset in a linear and uniformly continuous way i.e., there are no "Steiner points" defined for all c.c. sets in infinite dimensions. (The Steiner point is an important tool in finite dimensions.)

Now we describe the notation and basic definitions used throughout. Let B denote a separable Banach space with norm $\|\cdot\|$, and let $K(B)$ be the collection of nonempty compact subsets of B. Define on $K(B)$ two basic operations:

$$A+C := \{a+c:\ a \in A,\ \ c \in C\} \quad \text{(Minkowski addition)}$$

$$\alpha A := \{\alpha a:\ \ a \in A\},\ \ \alpha \geq 0 \quad \text{(positive homothetics)}$$

where $A, C \in K(B)$. $K(B)$ is not a vector space since generally $A + (-A) \neq \{0\}$. However, $K(B)$ becomes a complete separable metric space when endowed with the Hausdorff distance δ,

$$\delta(A,C) := \max\{\sup_{a\in A}\inf_{c\in C}\|a-c\|,\ \sup_{c\in C}\inf_{a\in A}\|a-c\|\}$$

$$= \inf\{\varepsilon > 0:\ A \subset C^{\varepsilon},\ C \subset A^{\varepsilon}\}$$

where $D^{\varepsilon} := \{x \in B:\ \delta(x,D) < \varepsilon\}$. Let

$$\|A\| := \delta(\{0\}, A) \ , \quad A \in K(B) \ .$$

Two relevant subsets of $K(B)$ are

$$\text{co } K(B) := \{A \in K(B)\colon A \text{ is convex}\} = \text{convex sets in } K(B)$$

and

$$\text{co } K_0(B) := \{A \in \text{co } K(B)\colon 0 \in A\} = \text{convex sets in } K(B) \text{ which contain } 0.$$

A <u>random compact set</u> K is a Borel measurable function from an abstract probability space into $K(B)$. If $K \in \text{co } K(B)$ a.s., then K is called a <u>random compact convex (c.c.) set</u>.

With a view towards statistical applications of random sets, Artstein and Vitale [4] adapt Aumann's definition of the integral of a set-valued function to provide the following notion of the <u>expectation of a random set</u>

$$EK := \{Eg\colon g \in L_1(\Omega,B) \text{ and } g(\omega) \in K(\omega) \text{ a.s.}\}$$

where Eg denotes the expectation of the B-valued random variable g. Other definitions of expectation exist, for example, via the Bochner integral. However, they coincide with the above if $E\|K\| < \infty$ ([5]) in which case $EK \in K(B)$. It is worthwhile to recall here that the statistical interest of Minkowski addition of random sets arises from the need to estimate EK: the natural method is via the law of large numbers for Minkowski addition.

M-infinitely divisible and p-stable sets are defined as follows.

0.1 <u>Definition</u>. A random compact convex set K is <u>M-infinitely divisible</u> (M for Minkowski) if for each $n \in \mathbb{N}$, there exist $K_{n1},\ldots,K_{nn}$ i.i.d. such that

$$L(K) = L(K_{n1} + \ldots + K_{nn}) \ .$$

0.2 <u>Definition</u>. A random compact convex set K is <u>p-stable</u>, $0 < p \leq 2$, if for K, K_1, K_2 i.i.d. random compact convex sets there exist $C, D \in \text{co } K(B)$ such that for all $\alpha, \beta \geq 0$

$$L(\alpha K_1 + \beta K_2 + C) = L((\alpha^p + \beta^p)^{1/p} K + D) \ .$$

As usual, 2-stable random sets are called Gaussian.

The two sets C and D rather than a single set arise because of the inadequacies of subtraction in co $K(B)$. When $p = 2$, the notion of Gaussian according to this definition is equivalent to the definitions of Gaussian proposed by Lyashenko [17] and Vitale [23]. (See [11] for further discussion about this point.)

Finally, a notion of centered Poisson random c.c. set will be required.

0.3 Definition. Let R be a finite measure on co $K(B)$. As in linear spaces the law of the compound Poisson random compact convex set with measure R, denoted by Pois R, is defined by the equation

$$\text{Pois } R = e^{-|R|} \sum_{k=0}^{\infty} \frac{R^k}{k!}$$

where $R^k = R * \dots^{k)} * R$ is the k-fold convolution of R with itself and $|R| = R(\text{co } K(B))$. By the convolution of finite measures R_1, R_2 on co $K(B)$ we mean $R_1 * R_2 = |R_1||R_2| L(X_1 + X_2)$ where X_1 and X_2 are independent random sets with laws $R_1/|R_1|$ and $R_2/|R_2|$ respectively.

0.4 Definition. A σ-finite measure μ on co $K(B)$ is a Lévy measure, $\mu \in$ Lévy(co $K(B)$), if there exist finite measures μ_n on co $K(B)$ and compact convex sets M_n such that $\mu_n(C) \uparrow \mu(C)$ for all Borel subsets C of co $K(B)$ and the sequence of probability measures

$$\{\delta_{M_n} * \text{Pois } \mu_n\}_{n=1}^{\infty}$$

converges weakly (i.e. weak-star, w^*). c Pois μ, a centered Poisson compact convex set with Lévy measure μ, will denote any such limit. If $M_n = \{0\}$ for all n we call Pois μ this limit.

§1. A brief survey of previous results

The fact that a compact convex subset of a Banach space is the intersection of its supporting hyperplanes, leads to an isomorphism between convex sets and support functions. It is this isomorphism which allows the characterization of M-infinitely divisible or p-stable c.c. sets to be reformulated as an equivalent Banach space characterization problem. Support functions were first used in proving probabilistic limit theorems by Artstein and Vitale [4].

1.1 Definition. The support function of a compact subset A of B is the function $A^{\#}$ defined on $B_1^* := \{x \in B^* : \|x\| \leq 1\}$ by the equation

$$A^{\#}(f) = \sup_{x \in A} f(x), \quad f \in B_1^* .$$

Support functions are useful because the map $s: A \to A^{\#}$ is isometric $(\delta(A,B) = \|A^{\#} - B^{\#}\|_{\infty})$ and preserves both addition and multiplication by positive scalars. Moreover, if A is convex, $A^{\#}$ uniquely determines A.

The following characterization of support functions is essential for our proof of the Lévy-Khinchin formula for M-infinitely divisible and p-stable c.c. sets. It is obtained in [11] and the proof is similar to Hormander's [14] proof of the characterization of support functions of closed sets.

1.2 Theorem. The support functions of compact convex subsets of B form the closed cone $\mathcal{V}$ of $C(B_1^*, w^*)$ consisting of all the functions $H: B^* \to \mathbb{R}$ continuous for the weak-star (w^*) topology of B_1^* which satisfy both

(1) H is subadditive, i.e. $H(f+g) \leq H(f) + H(g)$, $f, g, f+g \in B_1^*$

and

(2) H is positively homogeneous, i.e. $H(\lambda f) = \lambda H(f)$, $\lambda > 0$, and $f, \lambda f \in B_1^*$.

1.3 Corollary. The support functions of compact convex sets of B which contain zero form the closed cone $\mathcal{V}_0$ of $C(B_1^*, w^*)$ consisting of all the w^*-continuous, subadditive, positively homogeneous non-negative functions on B_1^*.

The random analogue of the support function is given by the support process.

1.4 Definition. The support process corresponding to a random compact set K is the process $K^{\#}$ defined by

$$K^{\#}(f,\omega) = (K(\omega))^{\#}(f), \quad f \in B_1^* .$$

The induced 1-1 correspondence between K and $K^{\#}$, for K convex, also preserves addition and multiplication by positive scalars. Consequently, Theorem 1.2 leads to the following equivalence:

(1.5) K is an M-infinitely divisible (p-stable) random compact convex set [containing 0] iff $K^{\#}$ is an infinitely divisible (p-stable) $C(B_1^*,w^*)$-valued random variable with support in V [V_0].

The above correspondence suggests the existence of a Lévy Khinchin representation for M-infinitely divisible c.c. sets which is analogous to the representation for infinitely divisible laws in $C(B_1^*,w^*)$. The objective is to show that K is an M-infinitely divisible c.c. set in B if and only if K assumes the form

$$(1.6) \qquad L(K) = \delta_M * \gamma * c \text{ Pois } \mu$$

for some $M \in \text{co } K(B)$, Gaussian law γ on co $K(B)$ and Lévy measure μ and then identify γ, μ and the centering. This program is carried out in Mase [18] for random c.c. sets of $\mathbb{R}^d$ which contain 0. (See also Cor. 4.5.1 in [12].) The general problem is solved in Giné and Hahn [12].

We now summarize these results closely following [12] and indicating the features which are dependent upon dimension.

Restrict attention to $B = \mathbb{R}^d$ and let $U_d := \{x \in \mathbb{R}^d: \|x\| \leq 1\}$. The basic approach is to use correspondence (1.5), which requires the identification of those infinitely divisible laws on $C(U_d)$ whose support is contained in V. The starting point is the Lévy-Khinchin representation for an i.d. law η on $C(U_d)$:

$$(1.7) \qquad \eta = \delta_y * \xi * c \text{ Pois } \nu$$

where $y \in C(U_d)$, ν is a Lévy measure on $C(U_d)$ and ξ is a Gaussian law on $C(U_d)$. Therefore, it suffices to determine precisely which y, ξ, and ν correspond to supp $\eta \subset V$.

If supp $\eta \subset V_0$, it is substantially easier to identify ξ and ν using two facts about V_0. First, the only affine subspace contained in V_0 is 0, hence Gaussian laws which are supported by affine subspaces must degenerate to 0 in V_0. Second, the fact that V_0 consists solely of nonnegative functions allows the deduction that all 1-dimensional projections of c_1Pois ν are supported by half-lines. Since Lévy measures on $\mathbb{R}$ which are supported by half-lines must integrate $\min(1, |z|)$, an estimate depending on the dimension d easily gives that if ν is Lévy on $C(U_d)$ with supp c_1 Pois $\nu \subset V_0$ then

$$\int \min(1, \|x\|)d\nu(x) < \infty \tag{1.8}$$

But it is well known that (1.8) is sufficient for ν to be a Lévy measure in any Banach space.

Converting back to sets, the above yields our version of Mase's theorem ([12], Proposition 1.12).

1.9 Theorem. A random compact convex set which contains 0 a.s. is M-infinitely divisible if and only if there exist

(i) $M \in \text{co}\, K_0(\mathbb{R}^d)$

and

(ii) a σ-finite measure μ on $\text{co}\, K_0(\mathbb{R}^d)$ with

$$\int \min(1, \|A\|)d\mu(A) < \infty$$

such that

$$L(K) = \delta_M * \text{Pois}\, \mu .$$

If supp $\eta \subset V$, support properties combined with the fact that V consists solely of subadditive, positively homogeneous functions allows the deduction that the Gaussian measure ξ must give mass 1 to linear functions, i.e. supp $\xi \subset \mathbb{R}^d \subset C(U_d)$. Thus, ξ is really the law of a point set.

The lack of simple characterizations of Lévy measures on infinite-dimensional spaces constitutes the main obstacle to identifying the Lévy measures ν for which c Pois $\nu \in V$. The basic trick is to utilize the characterizations of Lévy measures on R^d and V_0. The following lemma from [12] provides a first step in the appropriate direction.

1.11 <u>Lemma</u>. If B is a Banach space and $T: B \to B$ is a continuous linear map, then ν is a Lévy measure on B if and only if $\nu \circ T^{-1}$ and $\nu \circ (I - T)^{-1}$ are Lévy measures.

To use this lemma, we define our choice of T:

1.12 <u>Definition</u>. Let $S^{\#}: C(U_d) \to \mathbb{R}^d$ be the function defined by

$$S^{\#}(x) = \sigma_d^{-1} \int_{\partial U_d} u\, x(u)du, \quad x \in C(U_d)$$

where $\partial U_d := \{x \in \mathbb{R}^d: \|x\| = 1\}$, σ_d = volume of U_d and du denotes the element of surface area on ∂U_d. If $x = A^{\#}$ for $A \in co\, K(\mathbb{R}^d)$, then $S^{\#}(x) \in A$, and is called the <u>Steiner point</u> $S(A)$ of A.

The important features about $S^{\#}$ are that it is linear and uniformly continuous, in fact $\|S^{\#}\| \le d$. Now $\nu \circ (S^{\#})^{-1}$ and $\nu \circ (I-S^{\#})^{-1}$ are σ-finite measures on $\mathbb{R}^d$ and V_0 respectively (since ν lives on support functions and $(I-S^{\#})(A^{\#}) = (A + \{-S(A)\})^{\#} \in V_0$). A σ-finite measure on $\mathbb{R}^d$ is Lévy if and only if it integrates $\min(1,\|x\|^2)$. Theorem 1.9 and its proof show that $\nu\circ(I-S^{\#})^{-1}$ is a Lévy measure with $\operatorname{supp} \operatorname{Pois}\left(\nu\circ(I-S^{\#})^{-1}|_{\|x\|>1/n}\right) \subset V_0$ for all n if and only if the former integrates $\min(1,\|x\|)$. These two integrability properties together with Lemma 1.11 allow the deduction that $\operatorname{supp}(c \operatorname{Pois} \nu) \subset V$ for an appropriate centering, as desired. The converse requires some work with supports but is based on the same principles. Upon converting to sets, the above discussion leads to the Lévy Khinchin representation ([12]), Theorem 1.17).

1.13 <u>Theorem</u>. A random compact convex set K is M-infinitely divisible if and only if there exist

(i) $M \in co\, K(\mathbb{R}^d)$ nonrandom ;

(ii) a centered Gaussian measure γ on $\mathbb{R}^d$; and

(iii) a σ-finite measure μ on $co\, K(\mathbb{R}^d)$ satisfying both

$$\int \min(1,\|S(A)\|^2)d\mu(A) < \infty \tag{1.14}$$

and

(1.15) $$\int \min(1,\|(I-S)(A)\|)d\mu(A) < \infty ,$$

such that

$$L(K) = \delta_M * \gamma * c_S\text{Pois}\,\mu$$

where

$$c_S \text{ Pois } \mu := w^* - \lim_{n\to\infty} \delta_{\{-\int_{n^{-1}<\|A\|\leq 1} S(A)d\mu(A)\}} * \text{Pois}(\mu|_{\|A\|>n^{-1}})$$

It is obvious that if K is M-infinitely divisible then $S(K)$ is infinitely divisible in $\mathbb{R}^d$ and $(I-S)(K)$ is M-infinitely divisible in co $K_0(\mathbb{R}^d)$. However, the converse is <u>not</u> true (although Lemma 1.11 holds).

1.16 <u>Example</u>. The converse fails even in $\mathbb{R}^1$.

Let Y_1, Y_2 be nonnegative infinitely divisible random variables on $\mathbb{R}$ such that $Y_1 + Y_2$ is not infinitely divisible. Set $K = [Y_1-Y_2, Y_1+Y_2]$. Observe that

(i) $S(K)$ = midpoint of $K = Y_1$ is infinitely divisible on $\mathbb{R}$;

(ii) $(I-S)(K) = [-Y_2,Y_2]$ is M-infinitely divisible on co $K(\mathbb{R})$ since $Y_2 \overset{\mathcal{D}}{=} c_{n1} + \ldots + c_{nn}$ with c_{ni} i.i.d. implies that $[-Y_2,Y_2] \overset{\mathcal{D}}{=} [-c_{n1},c_{n1}] + \ldots + [-c_{nn},c_{nn}]$;

(iii) K is not M-infinitely divisible on co $K(\mathbb{R})$ because if $K \overset{\mathcal{D}}{=} [a_{n1},b_{n1}] + \ldots + [a_{nn},b_{nn}]$ for all n with $[a_{ni},b_{ni}]$ i.i.d. then $Y_1 + Y_2 \overset{\mathcal{D}}{=} b_{n1} + \ldots + b_{nn}$ which contradicts the non-infinite divisibility of $Y_1 + Y_2$.

For a specific example, let $r_0 = 0 < r_1 < \ldots < r_n < \ldots < 1$ be the increasing heights of the cumulative distribution function of a Poisson random variable with parameter 1, i.e. $r_n - r_{n-1} = e^{-1}/(n-1)!$ $n \geq 1$. Define on $\Omega = [0,1]$,

$$Y_1(\omega) = n-1 \quad \text{for} \quad r_{n-1} \leq \omega < r_n$$

and

$$Y_2(\omega) = \begin{cases} 0 & \text{for } e^{-1}/2 \leq \omega < 3e^{-1}/2 \\ 1 & \text{for } 0 \leq \omega < e^{-1}/2 \text{ or } 2e^{-1} \leq \omega \leq 5e^{-1}/2 \\ 2 & \text{for } 3e^{-1}/2 \leq \omega < 2e^{-1} \\ Y_1(\omega) & \text{for } r_3 \leq \omega . \end{cases}$$

Now Y_1 and Y_2 are each Poisson random variables with parameter 1. Furthermore, $Y_1 + Y_2$ assumes the values $\{0, 1, 3, 2n, n \geq 3\}$ which is not a semigroup. Since the support of an infinitely divisible law must be a semigroup (μ^t has the same support for all $t > 0$, Tortrat [19] Cor. p. 32), $Y_1 + Y_2$ is not infinitely divisible.

The integrability conditions in Theorem 1.13 put constraints on the random set's distance to the origin and its diameter. Let

$$D(0,A) := \inf \{\|a\| : a \in A\}$$

$$D(A) := \sup \{\|x-y\| : x, y \in A \} .$$

Then

$$D(A)/2 \leq \|(I-S)A\| \leq D(A)$$

and

$$D(0,A) \leq \|S(A)\| \leq D(0,A) + D(A) .$$

Consequently, conditions (1.14) and (1.15) may be reformulated to give

1.17 <u>Proposition</u>: μ is a Lévy measure on co $K(\mathbb{R}^d)$ if and only if

(1.18) $$\int \min(1, D^2(0,A))d\mu(A) < \infty$$

and

(1.19) $$\int \min(1, D(A))d\mu(A) < \infty .$$

Furthermore, (1.19) is necessary and sufficient for μ to be a Lévy measure on co $K_0(\mathbb{R}^d)$.

The following two criteria for Lévy measures on $C(U_d)$ may be of interest. A σ-finite measure ν on $C(U_d)$ is

(1) a Lévy measure with supp $\nu \subset V$ if and only if

$$\int \min(1, \|x\|^2) d\nu(x) < \infty . \tag{1.20}$$

(See Proposition 2.30 of [12]);

(2) a Lévy measure with supp $\nu \subset V$ and supp c Pois $\nu \subset V$ for some centering if and only if

$$\int \min(1, \|S^{\#}(x)\|^2 d\nu(x) < \infty \tag{1.21}$$

and

$$\int \min(1, \|(I-S^{\#})(x)\| d\nu(x) < \infty . \tag{1.22}$$

(This is obtained in proving Theorem 1.13.)

In finite-dimensional Banach spaces the Lévy measures which give rise to p-stable laws can be identified using Proposition 1.17 (see [12]). Let $K_d^1 = \{A \in \text{co } K(R^d): \|A\| = 1\}$. Then μ is the Lévy measure of a p-stable law if and only if μ is defined on co $K_d \backslash \{0\}$ by

$$\mu\{\lambda A: A \in C, \ \lambda_1 < \lambda \leq \lambda_2\} = \sigma(C) \frac{1}{p} (\lambda_1^{-p} - \lambda_2^{-p}), \ C \in B(K_d^1),$$
$$0 < \lambda_1 < \lambda_2 < \infty ,$$

where

(1) σ is any finite measure on K_d^1 if $0 < p < 1$

and

(2) σ is concentrated on singletons if $1 \leq p < 2$.

In particular, K is p-stable for $1 \leq p < 2$ if and only if $K \overset{\mathcal{D}}{=} M + \{\xi\}$ where $M \in \text{co } K(\mathbb{R}^d)$ is nonrandom and ξ is a p-stable real-valued random variable. Similarly, Theorem 1.13 implies that the same is true for $p = 2$, i.e. in the Gaussian case. This last observation is first made in Lyashenko [17]. Vitale [23] provides a simpler proof using the Steiner point together with positivity.

Although it seems impossible to obtain complete results on the Lévy-Khinchin representation in infinite dimensions, such as in Theorem 1.13, it is however possible to describe the p-stable random compact convex subsets of any separable Banach space B. The exact analogues of the finite-dimensional results hold in B and a very specific construction of the p-stable compact convex sets can be given. This is done in [11].

Just as for M-infinitely divisible laws, one passes to support processes via (1.5) and attempts to identify those p-stable laws η on $C(B_1^*,w^*)$ for which $\operatorname{supp}\eta \subset V$. The linear functionals on $C(B_1^*,w^*)$,

$$h_1(F) := h_1(F,\lambda,x^*) := F(\lambda x^*) - \lambda F(x^*)$$

$$\text{for } \lambda \in R^+,\ x^*,\ \lambda x^* \in B_1^*,\ F \in C(B_1^*,w^*)$$

and

$$h_2(F) := h_2(F,x^*,y^*) := F(x^*) + F(y^*) - F(x^* + y^*)$$

$$\text{for } x^*,\ y^*,\ x^*+y^* \in B_1^*,\ F \in C(B_1^*,w^*),$$

satisfy $h_1(F) = 0$ and $h_2(F) \geq 0$ for all $F \in V$, by Theorem 1.2. Using this and the fact that the only positive p-stable random variables on $\mathbb{R}$ are constants if $1 \leq p \leq 2$, the following is obtained:

1.23 Theorem. K is a p-stable compact convex set in B with $1 \leq p \leq 2$ if and only if

$$K = M + \{\xi\} \quad \text{a.s.}$$

for some $M \in \text{co}\,K(B)$ and p-stable B-valued random variable ξ.

However, if $0 < p < 1$, there are plenty of p-stable measures on $\mathbb{R}$ supported by half-lines. Consequently, it is easy to construct nondegenerate p-stable c.c. sets. Just let $A \in \text{co}\,K(B)$ and let θ be a positive real p-stable random variable. Then θA easily satisfies Definition 0.2 and is therefore a p-stable set. These are the building blocks of a stochastic integral construction which yields all of the p-stable c.c. sets for $0 < p < 1$.

Let $K_1 = \{A \in \text{co}\,K(B) : \|A\| = 1\}$ and let σ be a finite Borel measure on K_1. Without loss of generality we may assume $\sigma(K_1) = 1$. Let θ be a real positive p-stable random variable. Define an independently scattered random measure M_p on the Borel sets of K_1 by

(i) $M_p(A) \stackrel{\mathcal{D}}{=} (\sigma(A))^{1/p}\theta$;

(ii) if $\{A_i\}$ are disjoint then

$$M_p\left(\bigcup_{i=1}^{n} A_i\right) = \sum_{i=1}^{n} M_p(A_i) \text{ a.s.}, \quad n < \infty$$

and

$$M_p\left(\bigcup_{i=1}^{\infty} A_i\right) = \text{pr} - \lim_{n\to\infty} \sum_{i=1}^{n} M_p(A_i) ;$$

(iii) for $n < \infty$, $M_p(A_1), \ldots, M_p(A_n)$ are independent if $\{A_i\}$ are disjoint.

M_p is called a positive p-stable independently scattered random measure on K_1 with spectral measure σ.

The stochastic integrals

$$\int_{K_1} A \, dM_p(A)$$

basically denote all sets of the form $\sum_{i=1}^{n} \theta_i A_i$ where θ_i are i.i.d real nonnegative p-stable random variables and $A_i \in \text{co } K(\mathbb{R}^d)$ and limits of such which exist in the Λ_p metric, where $\Lambda_p(\eta) := \left(\sup_{t>0} t^p P(\eta > t)\right)^{1/p}$. (For an exact construction see Section 3 of [11].)

1.24 Theorem. K is a p-stable compact convex set in B with $0 < p < 1$ if and only if

$$K = M + \int_{K_1} A \, dL(A) \quad \text{a.s.}$$

where $M \in \text{co } K(B)$ and L is a positive p-stable independently scattered random measure on K_1 with finite spectral measure σ.

For $1 \leq p \leq 2$, domain of attraction questions make no sense. This can be seen in two ways. First, the degeneracy obtained in Theorem 1.23 shows that no interesting limit laws are possible. Second, centering the partial sums is essential for $1 \leq p \leq 2$ and this is impossible in the case of random sets. In this case it is more appropriate to consider the rate of convergence to zero of the Hausdorff distance between the Minkowski averages and the expectation ([13] and references therein). But for $0 < p < 1$, domains of attraction and domains of normal attraction for the p-stable c.c. sets do make sense. In [11] we completely characterize all domains of normal attraction, provide sufficient conditions for a random set to be in a given domain of attraction, and construct specific examples.

§2. Some remarks on infinitely divisible measures on co $K(B)$.

Even if attention is restricted to infinitely divisible random c.c. subsets K of a separable Banach space B with $0 \in K$ a.s., the characterization of M-infinite divisibility that holds in $\mathbb{R}^d$, completely fails if $\dim B = \infty$. Moreover, if $\dim B = \infty$, there exists no "Steiner point" on co $K(B)$. These statements are proven below. It is also shown that $\int \min(1,\|A\|)d\mu(A) < \infty$ is the weakest possible integrability assumption on a σ-finite measure μ defined on co $K_0(B)$ which will ensure that μ is a Lévy measure in general.

The following is all that can be salvaged from Theorem 1.9 in the case of infinite-dimensional B.

2.1 Proposition. Let B be a separable Banach space, and let $K(\omega)$ be a random compact convex set of B that contains 0. Then:

a) There exists a (nonrandom) compact convex subset $M \subset B$ and a σ-finite measure μ on co $K_0(B)$ satisfying

(i) $\mu\{\|A\| > r\} < \infty$ for all $r \in (0,\infty)$,

(ii) the sequence $\{\int_{n^{-1}<\|A\|\le 1} A \, d\mu(A)\}_{n=1}^{\infty}$ converges in the Hausdorff distance, and

(iii) for all $\phi \in L^+(\text{co } K(B), \mathbb{R}^d) := \{f\colon \text{co } K(B) \to \mathbb{R}^d$ which are positively linear and Lipschitz $\}$, $\int \min(1,\|\phi(A)\|^2)d\mu(A) < \infty$,

such that

$$L(K) = \delta_M * \text{Pois } \mu . \tag{2.2}$$

b) If μ is a σ-finite measure on co $K_0(B)$ (or co $K(B)$) satisfying $\int \min(1,\|A\|)\, d\mu(A) < \infty$, then Pois μ exists and defines an M-infinitely divisible random set.

Proof. Part (b) follows from the proof of the corresponding statement in [1], Theorem 3.6.3, with the obvious modifications. Thus, we turn to the proof of part (a).

Gaussian c.c. sets are degenerate even in infinite dimensions (Theorem 1.23), and (B_1^*, w^*) is a compact metric space. Therefore, the finite-dimensional proof of the characterization of M-infinitely divisible c.c. sets which contain 0 ([12], proof of Proposition 1.12) shows here that $K^\#$ has a Lévy-Khinchin decomposition whose Gaussian part is δ_0 and whose Lévy measure $\mu^\#$ satisfies the following properties:

(1) $\operatorname{supp} \mu^\# \subset V_0$;

(2) for all $\lambda \in C' := C^*(B_1^*, w^*)$,

$$\int_{\|h\| \leq 1} |\lambda(h)|\, d\mu^\#(h) < \infty \; ; \tag{2.3}$$

(3) Conditions (i) and (iii) (with $\mu^\#$ instead of μ).

This last condition and the isometry s give (i) and (iii). (ii) remains to be proven.

Define $\mu_n^\# = \mu^\#\big|_{n^{-1} < \|x\| \leq 1}$ and $T_n : C' \to \mathbb{R}$ by

$$T_n(\lambda) = \int \lambda(h)\, d\mu_n^\#(h), \qquad \lambda \in C'.$$

We claim that

(2.4) the sequence $\{T_n|_{\{\|\lambda\| \leq 1\}}\}$ is w*-equicontinuous at $0 \in C'$.

To prove the claim (2.4), first note that

$$\sup_n |T_n(\lambda)| = \sup_n \left|\int \lambda(h)\, d\mu_n^\#(h)\right|$$

$$\leq \int \lambda^+(h)\, d\mu_n^\#(h) + \int \lambda^-(h)\, d\mu_n^\#(h)$$

where λ^+ and λ^- are the positive and negative variations of the measure λ respectively. The inequality holds because $\mu^\#$ is concentrated in V_0 which consists solely of non-negative functions h. Hence, (2.3) implies

(2.5) $$\sup_n \left|T_n(\lambda)\right| < \infty \quad \text{for all} \quad \lambda \in C' .$$

Also, if $\lambda_k(h) \to 0$ for all $h \in C(B_1^*, w^*)$ and $\|\lambda_k\| \le c < \infty$, then by bounded convergence

$$\lim_{k\to\infty} \int \lambda_k(h) d\mu_n^{\#}(h) = 0 \qquad \text{for all} \quad n \in \mathbb{N}.$$

Therefore,

(2.6) each T_n is continuous for the bounded-weak-star topology of C'.

Finally, observe that C', with the bounded weak-star topology, is a Fréchet space (if $\{h_n\}$ is a countable dense subset of the unit ball of $C(B_1^*, w_1^*)$, then this topology is metrized by $\rho(\lambda,\mu) = \Sigma\, 2^{-n}\left|\int h_n d(\lambda-\mu)\right|$). Hence by (2.5) and (2.6), the equicontinuity principle (see e.g. Dunford and Schwartz [8], p. 52) implies that

$$\lim_{\lambda\to 0} T_n(\lambda) = 0 \quad \text{uniformly in} \quad n$$

(where $\lambda \to 0$ in the sense of the bounded weak-star topology). This proves claim (2.4).

Since $\mu^{\#}$ is a Lévy measure, the sequence

$$\{\int (e^{i\lambda(h)} - 1 - i\lambda(h))d\mu_n^{\#}(h)\Big|_{\{\|\lambda\|\le 1\}}\}_{n=1}^{\infty}$$

is also w^*-equicontinuous at $0 \in C'$, because its terms are the logarithms of the characteristic functionals of a tight sequence (see e.g. [1], exercise 8(c), page 34). This, together with claim (2.4), shows that the sequence

$$\{(\text{Pois}\ \mu_n^{\#})^{\wedge}\Big|_{\{\|\lambda\|\le 1\}}\}_{n=1}^{\infty} = \{\exp(\int (e^{i\lambda(h)} - 1)\ d\mu_n^{\#}(h))\ \Big|_{\{\|\lambda\|\le 1\}}\}_{n=1}^{\infty}$$

is w^*-equicontinuous at zero. Since $\{\text{Pois}\ \mu_n^{\#}\}$ is relatively shift compact it follows ([1], Theorem 1.4.16) that $\{\text{Pois}\ \mu_n^{\#}\}_{n=1}^{\infty}$ is relatively compact. On the other hand, $\{c \text{ Pois } \mu_n^{\#}\}_{n=1}^{\infty}$ is also relatively compact (in fact it converges weakly). Therefore, the shifts of this last sequence, namely $\{\delta_{-\int h d\mu_n^{\#}(h)}\}_{n=1}^{\infty}$, form a w^*-relatively compact sequence. Equivalently, the sequence $\{\int h \, d\mu_n^{\#}(h)\}_{n=1}^{\infty}$ is relatively compact in $C(B_1^*, w^*)$. But any limit x satisfies

$$\lambda(x) = \int_{\|k\| \leq 1} \lambda(h) \, d\mu^{\#}(h)$$

by (2.3) and bounded convergence. So, the sequence converges. By the isometry between V_0 and co $K_0(B)$, condition (ii) follows, and so does (2.2).

Condition (ii) of the previous proposition, in the case $B = \mathbb{R}^d$, implies $\int_{\|A\| \leq 1} \|A\| d\mu(A) < \infty$ (note that for A in co $K_0(\mathbb{R}^d)$, $\|A\| \leq d^{1/2} \max_{i \leq d} |\langle A^{\#}, \pm e_i \rangle|$). But the situation is very different in infinite dimensions: Theorem 1.9 fails to hold by far even in (infinite-dimensional) Hilbert space, as the following proposition shows.

2.7 <u>Proposition</u>. Let H be an infinite-dimensional separable Hilbert space. Then, for every $\delta > 0$, there exists a measure μ on co $K_0(H)$ for which Pois μ exists but such that

$$\int_{\|A\| \leq 1} \|A\|^{2-\delta} \, d\mu(A) = \infty .$$

<u>Proof</u>. Let the measure μ on co $K_0(H)$ be defined as

$$\mu = \sum_{i=1}^{\infty} \delta_{\{\lambda e_i : \, 0 \leq \lambda \leq i^{-\alpha}\}}$$

with $\alpha = (2-\delta)^{-1}$. Let $\{e_i\}_{i=1}^{\infty}$ be a complete orthonormal sequence of H and let $H^+ = \{x \in H: \langle x, e_i \rangle \geq 0$ for all $i\}$ be its associated positive cone. Define

$$T: H^+ \to \text{co } K_0(H)$$

by

$$T(\sum_i a_i e_i) = \sum_i [0,a_i]e_i$$

where for simplicity of notation we set $[0,a_i]e_i :=$ $\{\lambda e_i: 0 \leq \lambda \leq a_i\}$. The map T is positive linear (i.e. T commutes with sums and with multiplication by nonnegative scalars). Moreover,

$$\delta(Tx,Ty) \leq \|x-y\|, \quad \delta(0,Tx) = \|x\|.$$

Finally, let

$$\tilde{\mu} = \sum_{i=1}^{\infty} \delta_{i^{-\alpha} e_i} .$$

Note that $\tilde{\mu}$ is the measure on H^+ such that $\mu = \tilde{\mu} \circ T^{-1}$. Then

$$\int \min(1, \|x\|^{2-\delta}) d\tilde{\mu}(x) = \sum_{i=1}^{\infty} i^{-1} = \infty , \tag{2.8}$$

$$\int \|x\|^2 \, d\tilde{\mu}(x) = \sum_{i=1}^{\infty} i^{-2/(2-\delta)} < \infty \tag{2.9}$$

and

$$\int x \, d\tilde{\mu}(x) = \sum_{i=1}^{\infty} i^{-1/(2-\delta)} e_i \in H \tag{2.10}$$

in the Pettis sense. By (2.9), $\tilde{\mu}$ is a Lévy measure on H, (see e.g. [1], Theorem 3.7.6) and by (2.10), Pois $\tilde{\mu}$ (without centering) exists. This last assertion follows from weak convergence of the sequence $\{\delta_{-\int_{\|x\|>k^{-1}} x d\tilde{\mu}(x)} * \text{Pois}(\tilde{\mu}|_{\|x\|>k^{-1}})\}$ and from convergence of the shifts $\int_{\|x\|>k^{-1}} x d\tilde{\mu}(x)$ (to $\int x \, d\tilde{\mu}$). Now the properties of T imply that Pois μ exists and that $\int \min(1, \|k\|^{2-\delta}) d\mu(k) = \infty$. □

A. Araujo showed that in general Banach spaces no integrability condition weaker than $\int \min(1, \|x\|) \, d\mu(x) < \infty$ implies that μ is a Lévy measure. (See e.g. [1], Theorem 3.6.3). Here is a similar result for Lévy measures on co $K_0(B)$.

2.11 Proposition. Let $B = C[0,1]$. For every $\delta > 0$ there exists a measure μ on co $K_0(B)$ satisfying:

(i) $\mu\{\|A\| > 1\} < \infty$

(ii) $\int_{\|A\|\le 1} \|A\|^{1+\delta}\, d\mu(A) < \infty$

(iii) $\int_{\|A\|\le 1} A\, d\mu(A)$ exists as the limit of $\int_{M^{-1}<\|A\|\le 1} A\, d\mu(A)$ in $\mathrm{co}\ K(B)$,

and such that Pois μ does not exist (in the sense that no subsequence of $\{\mathrm{Pois}\,(\mu|_{\|A\|>n^{-1}})\}_{n=1}^{\infty}$ is tight).

Proof. By the universality of $C[0,1]$, it is enough to consider a different ℓ_p for each δ. Given $\delta > 0$, let $1 < p < 1+\delta \le 2$ (only small values of δ are interesting). We will construct a measure μ on $\mathrm{co}\ K_0(\ell_p)$ satisfying the conditions of the proposition. In fact, by use of the map $T: \ell_p^+ \to \mathrm{co}\ K_0(\ell_p)$ defined as in the previous proposition for H, it suffices to construct a measure $\tilde{\mu}$ on ℓ_p satisfying

(i) $\tilde{\mu}\{\|x\| > 1\} = 0$,

(ii) $\int x\, d\tilde{\mu}(x)$ exists as the limit of $\int_{\|x\|>n^{-1}} x d\tilde{\mu}(x)$,

(iii) $\int \|x\|^{p+\tau}\, d\tilde{\mu}(x) < \infty$ for all $\tau > 0$,

(iv) $\{\mathrm{Pois}(\tilde{\mu}|_{\|x\|>n^{-1}})\}_{n=1}^{\infty}$ has no tight subsequence.

We claim that the measure

$$\tilde{\mu} = \sum_{i=1}^{\infty} (\ell n\ i)^{-1}\, \delta_{i^{-1/p} e_i}$$

satisfies these conditions (as usual, $\{e_i\}$ is the canonical basis of ℓ_p). In fact, (i) is trivial, and (ii) and (iii) follow by observing that $\sum_{i=2}^{\infty} i^{-1/p}(\ell n\ i)^{-1} e_i \in \ell_p$ and $\sum_{i=1}^{\infty} i^{-(p+\tau)/p}\ (\ell n\ i)^{-1} < \infty$ for all $\tau > 0$. To prove (iv) it is necessary to show divergence of the series

$$\sum_{i=2}^{\infty} i^{-1/p} N_i e_i \tag{2.12}$$

where $\{N_i\}$ is a sequence of independent real Poisson random variables with expectations $EN_i = (\ln i)^{-1}$, $i = 1,2,\ldots$. We just apply the three series theorem to the series

(2.13) $$\sum_{i=2}^{\infty} i^{-1} N_i^p .$$

Since $\sum_{i=2}^{\infty} i^{-1} EN_i^p \geq \sum_{i=2}^{\infty} i^{-1} EN_i = \infty$ and

$$\sum_{i=2}^{\infty} i^{-1} EN_i^p I_{\{N_i > i^{1/p}\}} \leq \sum_{i=2}^{\infty} i^{-1} (\ln i)^{-i^{1/p}} ([i^{1/p} - 2]!)^{-1} < \infty ,$$

divergence of (2.13) follows. □

The other main tool in the characterization of M-infinitely divisible c.c. sets in $\mathbb{R}^d$ is the existence of the Steiner point. No Steiner point exists in infinite dimensions, even if it is defined in a very broad sense.

2.14 Definition. A map S: co $K(B) \to B$, B a separable Banach space, is a generalized Steiner functional on co $K(B)$ if

(i) S commutes with Minkowski addition and with multiplication by nonnegative scalars,

(ii) S is uniformly continuous with respect to Hausdorff distance, and

(iii) $S(\{x\}) = x$ for all $x \in B$ (or (iii)', $S(K) \in K$ for all $K \in$ co $K(B)$).

The following result, formally new, is essentially contained in Lindenstrauss [15]. The present proof has been obtained in collaboration with Y. Benyamini.

2.15 Proposition. If B is an infinite dimensional Banach space, there exist no generalized Steiner functionals on co $K(B)$.

Proof. We show first that if S: co $K(\mathbb{R}^d) \to \mathbb{R}^d$ is a Steiner functional, then S is Lipschitzian with Lipschitz norm $\|S\| \geq \alpha d^{1/2}$, for a constant $\alpha > 0$ independent of d. S is obviously Lipschitzian by linearity. Let V be any metric space that contains $\mathbb{R}^d$ (i.e. $\mathbb{R}^d$ is isometrically embedded in V). If V itself does not contain co $K(\mathbb{R}^d)$, it is at least contained in a metric space U that contains co $K(\mathbb{R}^d)$ (take U to be the disjoint union

$U := (\text{co } K(B)\backslash\mathbb{R}^d) \cup \mathbb{R}^d \cup (V\backslash\mathbb{R}^d)$ with the natural distance on each component and with $d(x,y) = \inf_{p\in\mathbb{R}^d} (d(x,p) + d(y,p))$ if $x \in \text{co } (\mathbb{R}^d)$, $y \in V$ and at most one of them is in $\mathbb{R}^d$). By a refinement of Lindenstrauss ([15], Lemma 3) of a Lemma of Isbell, there exists an absolute constant η_0 and a Lipschitz projection (retract) $P: U \to \text{co } K(\mathbb{R}^d)$ such that $\|P\| \leq \eta_0$ ($\|\cdot\|$ refers to the Lipschitz norm of P). Then $\|S \circ P\| \leq \eta_0\|S\|$. This shows that from every metric space V that contains $\mathbb{R}^d$ there is a Lipschitz projection onto $\mathbb{R}^d$ of norm no larger than $\eta_0\|S\|$. But by Theorem 5 in Lindenstrauss (loc. cit.) and a result of B. Grunbaum (see e.g. Garling and Gordon [10], page 349), the greatest lower bound of all the numbers λ with the property that for all $V \supseteq \mathbb{R}^d$ there exists a Lipschitz projection onto $\mathbb{R}^d$ with norm at most λ is $\approx (2d/\pi)^{1/2}$ (as $n \to \infty$). Hence $\|S\| > \left(2d/\eta_0^2\pi\right)^{1/2}$, or $\alpha \gtrsim (2/\pi)^{1/2}/\eta_0$.

If B is infinite dimensional and S is a Steiner functional on co $K(B)$, then S is Lipschitzian. Let $c \doteq \|S\|$. By Dvoretzsky's theorem [9] there exists in B a subspace B_d of dimension $d \geq 32c^2\alpha^{-2}$ at a distance not larger than 1 from $\mathbb{R}^d$. Let $T: R^d \to B_d$ satisfy $\|T\| \leq 2$, $\|T^{-1}\| \leq 2$. The induced map $\tilde{T}: \text{co } K(\mathbb{R}^d) \to \text{co } K(B_d)$ has norm $\|\tilde{T}\| \leq 2$ and therefore the projection $\tilde{T} \circ S \circ T^{-1}: \text{co } K(\mathbb{R}^d) \to \mathbb{R}^d$ is a Steiner functional with norm at most $2^{-1/2}\alpha d^{1/2}$, in contradiction with the first part of the proof.

Acknowledgements. We are indebted to Y. Benyamini for a very useful conversation whose outcome was the proof of Proposition 2.15. We thank also J. Lindenstrauss for pointing out his work on nonlinear projections.

References

1. Araujo, A. and Giné, E. (1980). The central limit theorem for real and Banach valued random variables. Wiley, New York.

2. Artstein, Z. (1984). Limit laws for multifunctions applied to an optimization problem. Preprint.

3. Artstein, Z. and Hart, S. (1981). Law of large numbers for random sets and allocation processes. Math. Operations Res. 6, 485-492.

4. Artstein, Z. and Vitale, R. A. (1975). A strong law of large numbers for random compact sets. Ann. Prob. 3, 879-882.

5. Byrne, C. L. (1978). Remarks on the set-valued integrals of Debreu and Aumann. J. Math. Analysis and Appl. 62, 243-246.

6. Cressie, N. (1979). A central limit theorem for random sets. Z. Wahrscheinlichkeitstheorie 49, 37-47.

7. Debreu, G. (1966). Integration of correspondences. Proc. Fifth Berkeley Symp. Math. Statist. and Probability 2, 351-372. Univ. of California Press.

8. Dunford, N. and Schwartz, J. T. (1958). Linear Operators. Part I. Interscience Publishers, Inc. N.Y.

9. Dvoretzsky, A. (1961). Some results on convex bodies and Banach spaces, Proc. Symp. on Linear Spaces, 123-160. Jerusalem.

10. Garling, D.J.H. and Gordon, Y. (1971). Relations between some constants associated with finite dimensional Banach spaces. Israel J. Math. 9, 346-361.

11. Giné, E. and Hahn, M. G. (1984). Characterization and domains of attraction of p-stable random compact convex sets. Ann. Probability 14.

12. Giné, E. and Hahn, M. G. (1984). The Lévy-Khinchin representation for random compact convex subsets which are infinitely divisible under Minkowski addition. To appear in Z. Wahrscheinlichkeitstheorie.

13. Giné, E.; Hahn, M. G and Zinn, J. (1983). Limit theorems for random sets: an application of probability in Banach space results. Lect. Notes in Math. 990, 112-135.

14. Hormander, L. (1954). Sur la fonction d'appui des convexes dans un espace localement convexe. Arkiv. för Matematik 3, 181-186.

15. Lindenstrauss, J. (1964). On nonlinear projections in Banach spaces. Michigan Mathematical Journal 11, 263-287.

16. Lyashenko, N. N. (1982). Limit theorems for sums of independent compact random subsets of Euclidean space. J. Soviet Math. 20, 2187-2196.

17. Lyashenko, N. N. (1983). Statistics of random compacts in Euclidean space. J. Soviet Math. 21, 76-92.

18. Mase, S. (1979). Random convex sets which are infinitely divisible with respect to Minkowski addition. Adv. Appl. Prob. 11, 834-850.

19. Tortrat, A. (1977). Sur le support des lois indéfiniment divisibles dans les espaces vectoriels localement convexes. Ann. Inst. Henri Poincaré 13, 27-43.

20. Trader, D. A. (1981) Infinitely divisible random sets. Thesis, Carnegie-Mellon University.

21. Trader, D. A. and Eddy, W. F. (1981). A central limit theorem for Minkowski sums of random sets. Technical Report No. 228, Carnegie-Mellon University.

22. Vitale, R. A. (1981). A central limit theorem for random convex sets. Technical report, Claremont Graduate School.

23. Vitale, R. A. (1983). On Gaussian random sets. To appear in Proceedings of the conference on stochastic geometry, geometric statistics, and stereology. Oberwolfach. (R. V. Ambartzumian and W. Weil eds.) Teubner-Verlag 222-224.

24. Weil, W. (1982). An application of the central limit theorem for Banach space-valued random variables to the theory of random sets. Z. Wahrscheinlichkeitstheorie 60, 203-208.

ON BRUNK'S LAW OF LARGE NUMBERS IN SOME TYPE 2 SPACES

Bernard HEINKEL

Département de Mathématique

7, Rue René Descartes

67084 STRASBOURG Cédex (France)

Let $(B, \| \|)$ be a real separable Banach space. A sequence (X_n) of B-valued random variables (r.v.) is said to satisfy the weak law of large numbers ($(X_n) \in$ WLLN) if :

$$(1/n) \| X_1 + \dots + X_n \| \to 0 \text{ in probability ;}$$

it is said to satisfy the strong law of large numbers ($(X_n) \in$ SLLN) if :

$$(1/n) \| X_1 + \dots + X_n \| \to 0 \text{ a.s. } .$$

During the last years there have been a lot of papers on sufficient conditions for the WLLN or the SLLN in the Banach space setting. The leading idea in the results and in their proofs is to mimic the scalar situation by replacing absolute values by norms. As an example of such results let's recall the Banach space version of Brunk's theorem [2] :

THEOREM 1 ([6] for q=1 , [10] for q ≥ 1) : Let $p \in [1,2]$ and $(B, \| \|)$ be a real separable Banach space, of type p. If (X_n) is a sequence of independent centered B-valued r.v. such that there exists $q \geq 1$ with :

$$(1) \qquad \sum_{n \geq 1} E(\|X_n\|^{pq} / n^{pq+1-q}) < +\infty ,$$

then $(X_n) \in$ SLLN.

By making only norm-assumptions on a sequence (X_n) of r.v., one loses a lot of information on it, and so these norm-assumptions have to be very strong to compensate that loss. So it is not surprising that the hypotheses of general results involving only norm-assumptions are sometimes too strong to allow to conclude that a sequence of r.v. satisfies the SLLN, even if they take their values in a space as nice as a Hilbert space. We begin with such an example.

1. AN EXAMPLE OF FAILURE OF " NORM-ASSUMPTION " STRONG LAWS OF LARGE NUMBERS.

Let's consider a triangular array (λ_j^n , $j = 1, 2, \ldots, [n^{1/6}]$) of real valued r.v., whose lines are independent. For every n, the λ_j^n are i.i.d., the common law being the one of a Cauchy r.v. truncated at the level $n^{5/6} / L_3 n$, where the function L_3 is defined by :

$L_3 x = \text{Log Log Log } x \quad \text{if } x \geq e^e ,$

$L_3 x = 1 \quad \text{otherwise.}$

Denoting by (e_n) the canonical basis of the Hilbert space l^2, one defines the following sequence of centered, independent, l^2-valued r.v. :

$$X_n = \sum_{1 \leq j \leq [n^{1/6}]} \lambda_j^n e_j \ .$$

It is easy to see that $(X_n) \in$ WLLN ; does $(X_n) \in$ SLLN also ?

The space l^2 being of type 2, a natural idea for trying to answer to this question is to check if Theorem 1 applies. Unfortunately for every $q \geq 1$ there exists a positive constant $C(q)$ such that :

$$\forall n \quad E \|X_n\|^{2q} \geq C(q) \, (n/ L_3 n)^q$$

and so (1) doesn't hold.

As polynomial assumptions like (1) aren't fulfilled in our example, we will try if weaker hypotheses - exponential ones - work. More precisely, we will check if we are in the domain of application of the following extension of Prohorov's theorem :

THEOREM 2 (J. Kuelbs and J. Zinn [7]) : <u>Let</u> (X_k) <u>be a sequence of independent r.v. with values in a real separable Banach space</u> $(B, \| \ \|)$. <u>Suppose that</u>

the following conditions hold :

(i) *There exists a positive constant M such that* :

$$\forall k \qquad \|X_k\| \le M\,(k/\,L_2k) \quad \text{a.s.}$$

(*where* $L_2x = \text{Log Log sup}\,(x,e)$) .

(ii) *If for every integer n one defines* $I(n) = (\,2^n+1, \ldots, 2^{n+1}\,)$ *and* :

$$\Lambda(n) = 2^{-2n} \sum_{j \in I(n)} E\,\|X_j\|^2 \quad ,$$

then :

$$\forall \varepsilon > 0 \,, \quad \sum_{n \ge 1} \exp\,(\,-\,\varepsilon/\,\Lambda(n)\,) < +\infty \,.$$

Under these hypotheses $(X_n) \in$ SLLN *if and only if* $(X_n) \in$ WLLN.

It is easy to see that this result also doesn't apply to our example because there exists a positive constant C such that :

$$\forall n \,, \quad \Lambda(n) \ge (\,C/\,L_2n\,) \,.$$

The failure of our two attempts for checking if $(X_n) \in$ SLLN is due to the too strong integrability assumptions which are required by the general results that we tried to apply. So if we want our example to belong to the domain of application of a general sufficient condition for the SLLN, this condition will have to require less integrability ; this weaker requirement will have to be compensated by hypotheses on the finite dimensional projections of the r.v. .But if we make finite dimensional hypotheses, we are led to restrict our interest to spaces in which an efficient use of finite dimensional projections is possible. What are such reasonable spaces ? Recent work done by A. de Acosta and J. Kuelbs [1] and M. Ledoux [8] shows that in the study of the law of the iterated logarithm finite dimensional projections can be handled well in 2-uniformly smooth spaces. The law of the iterated logarithm being a kind of limiting i.i.d. SLLN it is reasonable to bet that for the non i.i.d. SLLN also 2-uniformly smooth spaces are nice spaces. The sequel will show that this bet is a good one.

From now we will only consider r.v. taking their values in a real separable Banach space $(B, \|\ \|)$ which is 2-uniformly smooth, that means it has dimension $n \geq 2$ and there exists a positive constant K such that :

$$\forall\ (x,y) \in B^2 \qquad \|x+y\|^2 + \|x-y\|^2 \leq 2\ \|x\|^2 + K\ \|y\|^2 \quad .$$

2. A " BRUNK TYPE SLLN " IN 2-UNIFORMLY SMOOTH BANACH SPACES.

Brunk's SLLN can be extended in the following way to a 2-uniformly smooth Banach space $(B, \|\ \|)$:

THEOREM 3 : Let (X_n) be a sequence of independent, centered B-valued r.v., such that :

1) $(X_n) \in$ WLLN .

2) $\exists\ p \geq 1 : \sum_{n \geq 1} E\left(\|X_n\|^{4p} / n^{3p+1} \right) < +\infty$.

3) $n^{-2} \sum_{1 \leq k \leq n} \|X_k\|^2 \to 0$ in probability .

4) If one defines for every integer n :

$$\lambda(n) = 2^{-2n} \sum_{j \in I(n)} \sup\left(E\ f^2(X_j)\ ,\ \|f\|_{B'} \leq 1 \right) \quad ,$$

then there exists an integer $k \geq 1$ such that :

$$\sum_{n \geq 1} \left(\lambda(n) \right)^k < +\infty \quad .$$

Under these assumptions $(X_n) \in$ SLLN.

Let's make some comments on Theorem 3.

a) It is easy to check that our example fulfils the hypotheses of Theorem 3 ; condition (2) holds for $p = 1$ and condition (4) for $k = 1$. So Theorem 3 can reach situations that Theorem 1 cannot.

b) It is interesting to notice that Theorem 3 is sharp even in the scalar case : it allows for instance to show that the SLLN holds for sequences for which results as strong as Teicher's extension of Kolmogorov's SLLN are unable to conclude. Before to give an example of such a situation, I will recall Teicher's result [9] :

PROPOSITION 1 : Let (X_n) be a sequence of centered, independent, real valued r.v., such that the following conditions hold :

1) $\sum_{k \geq 2} (E X_k^2 / k^4) \sum_{1 \leq j \leq k-1} E X_j^2 < +\infty$,

2) $n^{-2} \sum_{1 \leq k \leq n} E X_k^2 \to 0$,

3) There exists a sequence (c_k) of positive constants such that :

(a) $\sum_{k \geq 1} P(|X_k| > c_k) < +\infty$,

(b) $\sum_{k \geq 1} (c_k^2 / k^4) E X_k^2 < +\infty$.

Then $(X_n) \in$ SLLN.

An example for which Proposition 1 doesn't apply but Theorem 3 does is easy to construct : consider (X_n) a sequence of independent real valued r.v. such that for every n, X_n is a Cauchy r.v. truncated at the level $n / (\mathrm{Log}\ n)^{\frac{1}{2}}$. Then condition (1) above doesn't hold, but all the hypotheses of Theorem 3 are fulfilled (with for instance $p = 1$, $k = 3$) .

Now we give the proof of Theorem 3.

The starting point is the idea that Teicher used for proving Proposition 1 : From the identity :

(3) $(\sum_{1 \leq k \leq n} X_k)^2 = \sum_{1 \leq k \leq n} X_k^2 + 2 \sum_{2 \leq k \leq n} (\sum_{1 \leq j \leq k-1} X_j) X_k = Y_n + Z_n$,

he obtained the SLLN by showing that the sequences (Y_n / n^2) and (Z_n / n^2) both converge a.s. to 0.

Of course, in an infinite dimensional space this cannot be done as straight, but in a 2-uniformly smooth space there is a good substitute for the identity (3). This substitute was obtained by M. Ledoux [8] in his work on the law of the iterated logarithm. The result is as follows :

PROPOSITION 2 : If $(B, \| \|)$ is a 2-uniformly smooth Banach space, then there exists $F : B \to B'$, such that :

(i) $F(0) = 0$,

(ii) $F(x)(x) = \|x\|^2$,

(iii) $\|F(x)\|_{B'} = \|x\|$,

(iv) there exists a positive constant C such that for every finite sequence $(x_1, \ldots, x_n)$ of elements of B one has :

(4) $\| \sum_{1 \leq k \leq n} x_j \|^2 \leq 2 \sum_{2 \leq j \leq n} F(x_1 + \ldots + x_{j-1})(x_j) + C \sum_{1 \leq j \leq n} \|x_j\|^2$,

(v) B is a type 2 space and the type inequality is fulfilled with constant

A classical argument [7] shows that it suffices to prove Theorem 3 for r.v. X_n which are symmetrically distributed, so we will consider only that case.

Let's notice that there exists a sequence of positive numbers $\alpha_n \uparrow +\infty$, such that :

2)' $\sum_{n \geq 1} (\alpha_n^{4p} / n^{3p+1}) E \|X_n\|^{4p} < +\infty$.

So, if one puts for every k :

$Y_k = X_k I(\|X_k\| \leq (k / \alpha_k))$,

it is clear that :

$(X_n) \in SLLN \Leftrightarrow (Y_n) \in SLLN$.

Another classical argument [7] shows that this latter property holds if and only if :

$\forall \varepsilon > 0 \quad \sum_{n \geq 1} P(\| \sum_{j \in I(n)} Y_j \| > \varepsilon 2^n) < +\infty$.

Now we put $\gamma = \sup (k,p)$. By symmetry, the application of Lemma 4.4 of [5] shows that in fact it suffices to check :

(5) $\forall \varepsilon > 0 \quad \sum_{n \geq 1} P^{\gamma}(\| \sum_{j \in I(n)} Y_j \| > \varepsilon 2^n) < +\infty$.

By applying now inequality (4), one sees that (5) holds if the following two conditions are fulfilled :

(a) $\forall\, \varepsilon > 0 \quad \sum_{n \geq 1} P^{Y}\big(\sum_{j \in I(n)} \|Y_j\|^2 > \varepsilon\, 2^{2n} \big) < +\infty$,

(b) $\forall\, \varepsilon > 0 \quad \sum_{n \geq 1} P^{Y}\big(\sum_{j \in J(n)} F(S(n,j))(Y_j) > \varepsilon\, 2^{2n} \big) < +\infty$,

where : $J(n) = (2^n + 2, \ldots, 2^{n+1})$, and :

$S(n,j) = Y_{2^n+1} + \ldots + Y_{j-1}$.

For proving property (a), one notices that by an elementary symmetrization argument there exists $n(\varepsilon) \in \mathbb{N}$ such that :

$$\forall\, n \geq n(\varepsilon) \quad P^{Y}\big(\sum_{j \in I(n)} \|Y_j\|^2 > \varepsilon\, 2^{2n} \big) \leq 2^{-n(3p+1)}(32/\varepsilon^2)^p \sum_{j \in I(n)} E\,\|Y_j\|^{4p} \;;$$

and therefore property (a) follows by hypothesis 2).

On the other hand, one obtains by type 2 and symmetry :

(6) $$P\big(\sum_{j \in J(n)} F(S(n,j))(Y_j) > \varepsilon\, 2^{2n} \big) \leq (2\lambda(n)C/\varepsilon^2 2^{2n}) \sum_{j \in I(n)} E\,\|Y_j\|^2 \;.$$

Another symmetrization argument showing that :

$$\lim_{n \to +\infty} \big(2^{-2n} \sum_{j \in I(n)} E\,\|Y_j\|^2 \big) = 0 \;,$$

property (b) then follows from inequality (6) and hypothesis 4).

The gain in norm integrability hypotheses in Theorem 3 with respect to Theorem 1 is clear : the denominator n^{2p+1} which would have been required by Theorem 1 has been weakened in n^{3p+1} in Theorem 3. Is it possible to obtain a gain of the same importance in a Prohorov type SLLN ? The next section answers to this question.

3. A " PROHOROV TYPE SLLN " IN 2-UNIFORMLY SMOOTH BANACH SPACES.

In [3] and [4] Prohorov's SLLN has been extended to a 2-uniformly smooth Banach space $(B, \|\;\|)$ in the following way :

THEOREM 4 : Let (X_n) be a sequence of independent, centered, B-valued r.v., such that :

(i) There exists a positive constant M such that :

$$\forall k \quad \|X_k\| \leq M\,(K / L_2 k) \quad \text{a.s.} \quad ,$$

(ii) $n^{-2} \sum_{1 \leq k \leq n} \|X_k\|^2 \to 0$ in probability,

(iii) $\forall \varepsilon > 0 \quad \sum_{n \geq 1} \exp(-\varepsilon / \lambda(n)) < +\infty$.

Then $(X_n) \in$ SLLN.

Let's make some comments on this result :

- Our example didn't fulfil the hypotheses of Theorem 2 but it fulfils the ones of Theorem 4 ; this is not surprising because condition (iii) above is much weaker than (ii) in Theorem 2.
- In the special case of a Hilbert space, Theorem 4 takes the following nice form :

COROLLARY [4] : Let (X_n) be a sequence of independent, centered, r.v. with values in a real separable Hilbert space $(H, < , >)$, such that hypotheses (i) and (iii) of Theorem 4 hold. Then $(X_n) \in$ SLLN if and only if $(X_n) \in$ WLLN.

- We don't give the proof of Theorem 4 which is rather long and technical. Its idea is again to use inequality (4), but the treatment of the non-square part is more complicated as in the proof of Theorem 3 ; it involves iterative use of exponential inequalities for real valued martingales.

REFERENCES

[1] DE ACOSTA, A. and KUELBS, J. : Some results on the cluster set $C(S_n / a_n)$ and the LIL.

Ann. Prob. 11 (1983), 102 - 122

[2] BRUNK, H. D. : The strong law of large numbers.
Duke Math. J. 15 (1948), 181 - 195

[3] HEINKEL, B. : The non i.i.d. strong law of large numbers in 2-uniformly smooth Banach spaces.
Probability Theory on Vector Spaces III - Lublin 1983 -
Lecture Notes in Math 1080, 90 - 118

[4] HEINKEL, B. : Une extension de la loi des grands nombres de Prohorov.
Z. Wahrscheinlichkeitstheorie 67 (1984), 349 - 362

[5] HOFFMANN-JØRGENSEN, J. : Probability in Banach space.
Ecole d'Eté de Probabilités de St Flour 6 - 1976 -
Lecture Notes in Math 598, 1 - 186

[6] HOFFMANN-JØRGENSEN, J. and PISIER, G. : The law of large numbers and the central-limit theorem in Banach spaces.
Ann. Prob. 4 (1976), 587 - 599

[7] KUELBS, J. and ZINN, J. : Some stability results for vector valued random variables. Ann. Prob. 7 (1979), 75 - 84

[8] LEDOUX, M. : Sur les théorèmes limites dans certains espaces de Banach lisses. Probability in Banach spaces 4 - Oberwolfach 1982
Lecture Notes in Math 990, 150 - 169

[9] TEICHER, H. : Some new conditions for the strong law.
Proc. Nat. Acad. Sci. USA 59 (1968), 705 - 707

[10] WOYCZYNSKI, W. A. : On Marcinkiewicz-Zygmund laws of large numbers in Banach spaces and related rates of convergence.
Prob. and Math. Stat. 1 (1980), 117 - 131

NECESSARY AND SUFFICIENT CONDITION FOR THE UNIFORM LAW OF LARGE NUMBERS

J. Hoffmann-Jørgensen

1. Introduction. The law of large number has played a central role in probability theory ever since 1695, when it was first discovered by James Bernoulli (published in "Ars Conjectandi" seven years after James Bernoulli's death 1706). In the last decades it has been generalized to random variables taking values in Banach space, see [2], [5] and [8]. And recently it has been generalized to non-separable and non-measurable random variables, see [1], [4] and [7], this latter extension may seem exotic, but this generality is needed in order to study the Glivenko-Cantelli theorem for empirical distributions, see [4], [7] and [9].

Throughout this paper T denotes a set and $(S,\mathcal{S},\mu)$ denote a probability space, and we put

$$(\Omega,F,P) = (S^{\mathbb{N}},\mathcal{S}^{\mathbb{N}},\mu^{\mathbb{N}}), \quad \xi_j(\omega) = \omega_j \quad \forall\, \omega = (\omega_n) \in \Omega$$

Let f be a function from $S \times T$ into $\mathbb{R}$, we shall then give a series of necessary and sufficient condition for the following law of large number to hold:

$$(1) \qquad \sup_{t\in T} \Big|\frac{1}{n}\sum_{j=1}^{n} (f(\xi_j,t)-\varphi(t))\Big| \to 0 \quad \text{a.s. and in probability}$$

for some bounded function $\varphi: T \to \mathbb{R}$.

We shall use the notation of [6] and [7], but for the sake of completeness we shall remind the reader about some definitions in [6] and [7].

P^* denotes the outer P-measure, E^* and $\int^*$ denotes the upper P-expectation respectively the upper μ-integral. If $\varphi:\Omega \to \overline{\mathbb{R}}$, then φ^* denotes the upper P-envelope of φ.

If M is a topological space and $\{X_n\}$ is a sequence of functions from Ω into M (not necessarily measurable) we say that $X_n \to X_o$ a.s. if there exist a P-nullset $N \in \mathcal{F}$ so that $X_n(\omega) \to X_o(\omega)$ for all $\omega \in \Omega \setminus N$, and we say that $X_n \to X_o$ in probability if

(2) $$\lim_{n\to\infty} E^*|f(X_n)-f(X_o)| = 0 \qquad \forall f \in C(M)$$

where $C(M)$ is the set of all bounded continuous real valued functions on M. And we say that $\{X_n\}$ is eventually tight on M if

(3) $$\forall \varepsilon > 0 \; \exists K \in K(M): \limsup_{n\to\infty} E^*f(X_n) < \varepsilon \qquad \forall f \in C(T): 0 \le f \le 1_{T\setminus K}$$

where $K(M)$ is the set of all compact subsets of M.

$\Gamma(T)$ denotes the set of all finite covers of T. If φ is a function from T into $\mathbb{R}$ and ρ is a pseudo metric on T we define the following oscillation functions:

$$w_\rho^o(\varphi,r,t) = \sup_{\rho(u,t)<r} |\varphi(t)-\varphi(u)| \qquad \forall r > 0 \;\; \forall t \in T$$

$$w_\rho(\varphi,r) = \sup_{\rho(u,v)<r} |\varphi(u)-\varphi(v)| \qquad \forall r > 0$$

$$W_A(\varphi) = \sup_{u,v\in A} |\varphi(u)-\varphi(v)| \qquad \forall A \subseteq T$$

$$W_\alpha(\varphi) = \max_{A\in\alpha} W_A(\varphi) \qquad \forall \alpha \in \Gamma(T)$$

$B(T)$ denote the set of all bounded real valued functions on T with its supremum norm:

$$\|\varphi\| = \sup_{t\in T} |\varphi(t)|$$

Then $(B(T), \|\cdot\|)$ is a Banach space, whose dual space equals the set of all bounded real valued finitely additive setfunctions defined on all subsets of T, denoted $ba(T)$. If φ is a function from T into $\mathbb{R}$ and $A \subseteq T$ we put.

$$\|\varphi\|_A = \sup_{t \in A} |\varphi(t)|$$

2. <u>The law of large numbers</u>. Let $f: S \times T \to \mathbb{R}$ be a function, then we put

$$S_n = S_n(f) = \sum_{j=1}^{n} f(\xi_j, \cdot)$$

Then S_n maps Ω into $\mathbb{R}^T$, and we shall study uniform convergence of the averages $\{n^{-1} S_n\}$.

$LLN = LLN(B(T), \mu)$ denotes the set of functions $f: S \times T \to \mathbb{R}$ such that (see [7])

(4) $$\exists\, \varphi \in B(T): \|\varphi - \tfrac{1}{n} S_n\| \to 0 \text{ a.s.}$$

and $LLN_0 = LLN_0(B(T), \mu)$ denotes the set of functions $f: S \times T \to \mathbb{R}$ satisfying (1) for some $\varphi \in B(T)$. Notice that a.s. convergence does <u>not</u> imply convergence in probability, when the functions are non-measurable. Hence a priori we only have that LLN_0 is a subspace of LLN. However M. Talagrand has proved that the two sets actually are equal, see [9].

In analogy with the definition of "totally bounded in mean" in [7], we say that f is <u>eventually totally bounded in mean</u> if

(5) $$f(\cdot, t) \text{ is } \mu\text{-measurable } \forall t \in T$$

(6) $$\int^* \|f(s)\| \mu(ds) < \infty$$

(7) $$\forall \varepsilon > 0\ \exists\, \alpha \in \Gamma(T): \inf_n \tfrac{1}{n} E^* W_A(S_n) < \varepsilon \qquad \forall A \in \alpha$$

Then clearly "totally bounded in mean" implies "eventually totally bounded in mean".

Our next lemma explores the fact that $\{\frac{1}{n}S_n\}$ is (or rather ought to be) a backward martingale.

Lemma 1. Let $f: S \times T \to \mathbb{R}$ be a map and $q: \mathbb{R}^T \to \overline{\mathbb{R}}_+$ a seminorm on $\mathbb{R}^T$. Let

(1.1) $$S_n = \sum_{j=1}^{n} f(\xi_j,\cdot) \quad \text{and} \quad \beta = \inf_n \frac{1}{n} E^*q(S_n)$$

If $q(S_n)^*$ is the upper P-envelope of $q(S_n)$ then we have

(1.2) $$\limsup_{n\to\infty} \frac{1}{n}E^*q(S_n) = \beta$$

(1.3) $$\limsup_{n\to\infty} \frac{1}{n}q(S_n)^* \le \beta \qquad P\text{-a.s.}$$

(1.4) If $\beta < \infty$ then $\sup_n \frac{1}{n} E^*q(S_n) \le E^*q(S_1) < \infty$

Proof. Let $m,n \in \mathbb{N}$ and put $p = p(n,m) = [\frac{n}{m}]$ and $c=c(n,m)=(mp)/n$, where $[x]$ denotes the integer part of x. Then putting

$$S_{mj} = \frac{1}{m} \sum_{i=1}^{m} f(\xi_{i+(j-1)m}) \qquad m \ge 1,\ j \ge 1$$

$$R_{nm} = \frac{1}{n} \sum_{mp<j\le n} f(\xi_j) \qquad m \ge 1,\ n \ge 1$$

we have

(i) $$\frac{1}{n}S_n = \frac{c}{p} \sum_{j=1}^{p} S_{mj} + R_{nm}$$

Now let $w_m(\omega_1,\dots,\omega_m)$ be the outer μ^m-envelope of

$$q(\frac{1}{m} \sum_{j=1}^{m} f(\omega_j))$$

and let g be the outer μ-envelope of $q(f(s))$. Then if

$$w_{mj} = w_m(\xi_{1+(j-1)m},\dots,\xi_{jm}) \quad \text{for} \quad j,m \geq 1$$

$$g_j = g(\xi_j) \qquad \text{for} \quad j \geq 1$$

we have that $w_{m1}, w_{m2}, \dots$ are independent and identically distributed and so are $g_1, g_2, \dots$. Moreover by (i) we have (see Proposition 3.8 in [1])

$$\text{(ii)} \qquad q(\tfrac{1}{n}S_n)^* \leq \frac{c}{p}\sum_{j=1}^{p} w_{mj} + \frac{1}{n}\sum_{mp<j\leq n} g_j$$

$$\text{(iii)} \qquad E\, w_{m1} = \tfrac{1}{m}E^*q(S_m), \quad Eg_1 = E^*q(S_1)$$

$$\text{(iv)} \qquad \tfrac{1}{n}E^*q(S_n) \leq \tfrac{c}{m}E^*q(S_m) + \tfrac{n-mp}{n}E^*q(S_1)$$

Now if $\beta = \infty$ then (1.2) and (1.3) are trivially satisfied. So suppose that $\beta < \infty$, then there exist $m \geq 1$ so that $E^*q(S_m) = Ew_{m1} < \infty$, and so

$$\infty > \int_{S^m}^{*} q(\sum_{j=1}^{m} f(\omega_j))\mu^m(d\omega_1,\dots,d\omega_m)$$
$$\geq \int_{S^{m-1}}^{*} \mu^{m-1}(d\omega_2,\dots,d\omega_m)\int_{S}^{*} q(\sum_{j=1}^{m} f(\omega_j))\mu(d\omega_1)$$

hence there exist $\omega_2^*,\dots,\omega_m^* \in S$, so that

$$\text{(v)} \qquad E^*q(S_1+x^*) = \int_S^* q(f(\omega_1)+x^*)\mu(d\omega_1) < \infty$$

where $x^* = f(\omega_2^*)+\dots+f(\omega_m^*)$. Hence we have

$$E^*q(S_m+mx^*) \leq mE^*q(S_1+x^*) < \infty$$

Thus $q(S_m+mx^*) < \infty$ P-a.s. and $q(S_m) < \infty$ P-a.s. and so there exist $y^* = S_m(\omega^*)$ satisfying $q(y^*) < \infty$ and $q(y^*+mx^*) < \infty$, and so

$$q(x^*) = \tfrac{1}{m}q(mx^*) \leq \tfrac{1}{m}(q(y^*+mx^*)+q(y^*)) < \infty$$

Hence by (v) we find that

$$E^*q(S_1) \leq q(x^*) + E^*q(S_1+x^*) < \infty$$

That is $Eg_1 < \infty$ and $\sup \frac{1}{n} E^*q(S_n) \leq E^*q(S_1) < \infty$.

Now since $p(n,m) \to \infty$ and $c(n,m) \to 1$ when $m \in \mathbb{N}$ is fixed and $n \to \infty$ we have

$$\frac{c}{p} \sum_{j=1}^{p} w_{mj} \to Ew_{m1} \quad \text{a.s.} \quad \text{when} \quad n \to \infty$$

$$\frac{1}{n} \sum_{mp<j\leq n} g_j = \frac{1}{n} \sum_{j=1}^{n} g_j - \frac{c}{mp} \sum_{j=1}^{mp} g_j \to 0 \quad \text{a.s.} \quad \text{when} \quad n \to \infty$$

by the real valued law of large numbers. Hence by (ii) (iii) and (iv) (note that $c \leq 1$ and $0 \leq n-mp \leq m$) we have:

$$\limsup_{n\to\infty} \frac{1}{n} E^*q(S_n) \leq \frac{1}{m} E^*q(S_m) \qquad \forall\, m \geq 1$$

$$\limsup_{n\to\infty} \frac{1}{n} q(S_n)^* \leq \frac{1}{m} E^*q(S_m) \quad \text{a.s.} \quad \forall\, m \geq 1$$

Thus (1.2)-(1.4) are proved. □

Theorem 2. *Let* $f: S \times T \to \mathbb{R}$ *be a function, and let* $S_n = f(\xi_1)+\ldots+f(\xi_n)$. *Then the following eight statement are equivalent*

(2.1) $\qquad f \in LLN_0(B(T),\mu)$

(2.2) $\qquad E^*\|S_1\| < \infty$, *and* $\exists \varphi \in B(T): \|\varphi - \frac{1}{n} S_n\| \to 0$ *in probability*

(2.3) $\qquad E^*\|S_1\| < \infty$, *and* $\exists \varphi \in B(T): \inf_n P^*(\|\varphi - \frac{1}{n} S_n\| > \varepsilon) < \varepsilon \quad \forall\, \varepsilon > 0$

(2.4) $\qquad \exists \varphi \in B(T): \inf_n E^*\|\varphi - \frac{1}{n} S_n\| = 0$

(2.5) $\qquad \exists \varphi \in B(T): \lim_{n\to\infty} E^*\|\varphi - \frac{1}{n} S_n\| = 0$

(2.6) $\qquad E^*\|S_1\| < \infty$, $f(\cdot,t) \in L^1(\mu) \quad \forall\, t \in T$ *and* $\{\frac{1}{n} S_n\}$ *is*

<u>eventually tight on</u> $(B(T), \|\cdot\|)$.

(2.7) f <u>is eventually totally bounded in mean</u>

(2.8) $E^*\|S_1\| < \infty$ <u>and if</u> $\lambda \in ba(T)$, <u>then</u>

(a) $s \sim \int_T f(s,t)\lambda(dt)$ <u>is</u> μ-<u>integrable</u>

(b) $\int_S \mu(ds)\int_T f(s,t)\lambda(dt) = \int_T \lambda(dt)\int_S f(s,t)\mu(ds)$

<u>If</u> $\varphi(t) = \int_S f(s,t)\mu(ds)$, <u>then</u> $\varphi \in B(T)$ <u>and</u>

(c) $\|\varphi - \frac{1}{n}S_n\|^* \to 0$ P-<u>a.s. and in</u> $L^1(P)$

<u>where</u> ba(T) <u>is the set of all bounded, real valued, finitely additive set functions defined on all subsets of</u> T, <u>and</u> $\|\varphi - \frac{1}{n}S_n\|^*$ <u>is the upper</u> P-<u>envelope of</u> $\|\varphi - \frac{1}{n}S_n\|$.

<u>Proof</u>. (2.1) ⇒ (2.2): Let $\varphi \in B(T)$ so that $\|\varphi - \frac{1}{n}S_n\| \to 0$ a.s. and in probability, then $E^*\|S_1\| < \infty$ by Theorem 2.4 in [7], so (2.2) follows

(2.2) ⇒ (2.3): Evident!

(2.3) ⇒ (2.4): Let g be the upper μ-envelope of $\|\varphi - f(s)\|$ and let W_n be the upper P-envelope of $\|\varphi - \frac{1}{n}S_n\|$. Then we have

$$0 \le W_n \le \frac{1}{n}\sum_{j=1}^{n} g(\xi_j)$$

and $g(\xi_1), g(\xi_2), \ldots$ are independent identically distributed real random variable with finite expectation. hence $\{W_n\}$ is uniformly integrable so if $\varepsilon > 0$ is given there exist $0 < \delta < \frac{1}{2}\varepsilon$, so that

(i) $$F \in \mathcal{F} \text{ and } P(F) < \delta \Rightarrow \int_F W_n dP < \tfrac{1}{2}\varepsilon \quad \forall n \ge 1$$

By assumption there exist $k \ge 1$ so that $P(W_k > \delta) < \delta$ and so

$$E^*\|\varphi - \frac{1}{k}S_k\| = EW_k \le \delta + \int_{W_k > \delta} W_k dP < \varepsilon$$

since $\delta < \frac{1}{2}\varepsilon$. Thus (2.4) holds.

(2.4) $\Rightarrow$ (2.5): Immediate consequence of Lemma 1 applied to $[f(s,t)-\varphi(t)]$ and $q = \|\cdot\|$.

(2.5) $\Rightarrow$ (2.6): By Lemma 1 we have that $E^*\|S_1-\varphi\| < \infty$ and so $E^*\|S_1\| < \infty$ since $\|\varphi\| < \infty$. Since

$$\frac{1}{n}\sum_1^n f(\xi_j,t) \to \varphi(t) \quad \text{in probability} \quad \forall t \in T$$

we have that $f(\cdot,t)$ is μ-measurable by Proposition 6.3 in [1], so $f(\cdot,t) \in L^1(\mu)$. By Corollary 7.16 in [6] we have that $\frac{1}{n}S_n \tilde{\to} \varphi$ in $(B(T),\|\cdot\|)$, and so by Proposition 7.17 in [6], we have that $\{\frac{1}{n}S_n\}$ is eventually tight. Thus (2.6) holds.

(2.6) $\Rightarrow$ (2.7): If $\alpha \in \Gamma(T)$ and $n \in \mathbb{N}$ we put

$$W_n^\alpha = \max_{A\in\alpha} W_A(\tfrac{1}{n}S_n)^*$$

Since $W_A(\psi) \leq 2\|\psi\|$, we have

$$0 \leq W_n^\alpha \leq 2\|\tfrac{1}{n}S_n\|^* \leq \frac{2}{n}\sum_{j=1}^n g(\xi_j)$$

where g is the upper μ-envelope of $\|f(s)\|$. Then $\{g(\xi_n)\}$ are independent, identically distributed random variables with finite mean (since $Eg(\xi_1) = E^*\|S_1\| < \infty$). Hence $\{W_n^\alpha \mid n \in \mathbb{N}, \alpha \in \Gamma(T)\}$, is uniformly integrable. So if $\varepsilon > 0$ is given there exist $0 < \delta < \frac{1}{2}\varepsilon$ so that

(ii) $$F \in F, P(F) < \delta \Rightarrow \int_F W_n^\alpha dP < \tfrac{1}{2}\varepsilon \quad \forall n \geq 1 \quad \forall \alpha \in \Gamma(T)$$

By Example 7.28 (in particular (7.28.3)), there exist a finite cover $\alpha \in \Gamma(T)$, and $k \geq 1$ so that

(iii) $$P^*(W_n^\alpha > \delta) < \delta \quad \forall n \geq k$$

But then by (ii) we have

$$E\,W_n^\alpha \le \delta + \int_{W_n^\alpha > \delta} W_n^\alpha dP < \varepsilon \qquad \forall n \ge k$$

Thus we find

$$\inf_n \frac{1}{n} E^* W_A(S_n) \le \limsup_n E\,W_n^\alpha < \varepsilon$$

for all $A \in \alpha$, and so f is eventually totally bounded in mean. Thus (2.7) holds.

(2.7) $\Rightarrow$ (2.8): By assumption we have that $E^*\|S_1\| < \infty$ and $f(\cdot,t) \in L^1(\mu)$ $\forall t \in T$. Hence if

$$\varphi(t) = \int_S f(s,t)\mu(ds) \qquad \forall t \in T$$

Then $\varphi \in B(T)$. For each $k \in \mathbb{N}$ we choose a finite cover $\alpha_k \in \Gamma(T)$ of T so that

(i) $$\inf_n \frac{1}{n} E^* W_A(S_n) < 2^{-k} \qquad \forall A \in \alpha_k \ \forall k \ge 1$$

(ii) $$W_A(\varphi) < 2^{-k} \qquad \forall A \in \alpha_k \quad k \ge 1$$

(recall that φ is bounded). For each $A \in \cup\alpha_k$ we choose a point $t_A \in A$, since α_k is a cover of T we have

$$\|\varphi - \tfrac{1}{n}S_n\| = \max_{A\in\alpha_k} \|\varphi - \tfrac{1}{n}S_n\|_A$$

$$\le \max_{A\in\alpha_k} \{W_A(\varphi) + W_A(\tfrac{1}{n}S_n) + |\varphi(t_A) - \tfrac{1}{n}S_n(t_A)|\}$$

$$\le \max_{A\in\alpha_k} \{2^{-k} + W_A(\tfrac{1}{n}S_n)^* + |\varphi(t_A) - \tfrac{1}{n}S_n(t_A)|\}$$

$$\le 2^{-k} + \max_{A\in\alpha_k} W_A(\tfrac{1}{n}S_n)^* + \max_{A\in\alpha_k} |\varphi(t_A) - \tfrac{1}{n}S_n(t_A)|$$

$$= w_{nk}$$

Then w_{nk} is measurable and by Lemma 3.1 applied to the seminorm $W_A(\cdot)$ for $A \in \alpha_k$ we find

$$\limsup_{n\to\infty} \{\max_{A\in\alpha_k} W_A(\tfrac{1}{n}S_n)^*\} = \max_{A\in\alpha_k} \{\limsup_{n\to\infty} W_A(\tfrac{1}{n}S_n)^*\}$$

$$\leq \max_{A\in\alpha_k} \{\inf_n E^* W_A(\tfrac{1}{n}S_n)\}$$

P-a.s., and by the real valued law of large numbers we have that

$$\max_{A\in\alpha_k} |\varphi(t_A) - \tfrac{1}{n}S_n(t_A)| \underset{n\to\infty}{\to} 0 \quad \text{P-a.s.}$$

Hence by (i) we conclude that

(iii) $$\limsup_{n\to\infty} w_{nk} \leq 2^{-k+1} \quad \text{P-a.s.}$$

Now let $w_n = \inf_k w_{nk}$, then w_n is measurable, and by the computation above we have

$$\|\varphi - \tfrac{1}{n}S_n\|^* \leq w_n \qquad \forall n \geq 1$$

and so by (iii) we have that $\|\varphi - \frac{1}{n}S_n\|^* \to 0$ P-a.s. Moreover if g is the upper μ-envelope of $\|\varphi - f(s)\|$ then

$$0 \leq \|\varphi - \tfrac{1}{n}S_n\|^* \leq \tfrac{1}{n}\sum_{j=1}^{n} g(\xi_j)$$

and $g(\xi_1), g(\xi_2), \ldots$ are independent identically distributed random variables with finite expectation $(Eg = E^*\|S_1\| < \infty)$, so $\{\|\varphi - \frac{1}{n}S_n\|^* \mid n \in N\}$ is uniformly integrable. Thus we have that $\|\varphi - \frac{1}{n}S_n\|^* \to 0$ in $L^1(P)$.

In particular we see that $f \in LLN(B(T),\mu)$, so (2.8.a) and (2.8.b) holds by Theorem 2.4 and Example 3.5 in [7]. Thus (2.8) holds.

(2.8) $\Rightarrow$ (2.1): Evident! □

Combining condition (2.6) and Example 7.28 in [6] we have four more statements equivalent to (2.1)-(2.8):

Theorem 3. Let $f: S \times T \to \mathbb{R}$ be a function satisfying

(3.1) $$f(\cdot,t) \in L^1(\mu) \quad \forall t \in T$$

(3.2) $$\int_S^* \|f(s)\| \mu(ds) < \infty$$

If $S_n = f(\xi_1) + \ldots + f(\xi_n)$, then the following five conditions are equivalent:

(3.3) $$f \in LLN_0(B(T),\mu)$$

(3.4) $$\forall \varepsilon > 0 \ \exists \alpha \in \Gamma(T): \ \limsup_n E^* \frac{1}{n} W_\alpha(S_n) < \varepsilon$$

(3.5) $$\forall \varepsilon > 0 \ \exists \alpha \in \Gamma(T): \ \inf_n P^*(W_A(\tfrac{1}{n}S_n) > \varepsilon) < \varepsilon \qquad \forall A \in \alpha$$

(3.6) $\forall \varepsilon > 0 \ \exists r > 0 \ \exists \rho$ a totally bounded pseudo metric on T, so that $$\inf_n P^*(w_\rho^o(\tfrac{1}{n}S_n, r, t) > \varepsilon) < \varepsilon \qquad \forall t \in T$$

(3.7) $\exists \rho$ a totally bounded pseudo ultrametric on T, such that $$\lim_{r\to 0}[\limsup_{n\to\infty} E^* w_\rho(\tfrac{1}{n}S_n, r)] = 0$$

Remark (1) A pseudo metric is totally bounded if we can cover the whole space with finitely many balls of any given radius r, and a pseudo metric ρ is an ultrametric if ρ satisfies the following strong form of the triangle inequality:

$$\rho(u,v) \leq \max\{\rho(u,w), \rho(w,v)\} \qquad \forall u,v,w$$

Proof. Let $B \subseteq T \times T$, and put

$$V_B(\varphi) = \sup_{(u,v)\in B} |\varphi(u)-\varphi(v)| \qquad \forall \varphi \in \mathbb{R}^T$$

$$V_n(B) = V_B(\tfrac{1}{n}S_n)$$

$$V_n^*(B) = \text{the upper P-envelope of } V_n(B)$$

$$g = \text{the upper } \mu\text{-envelope of } \|f(\cdot)\|$$

Since $V_B(\varphi) \leq 2\|\varphi\|$ we have

$$0 \leq V_n^*(B) \leq \frac{2}{n}\sum_{j=1}^{n} g(\xi_j) \qquad \forall n \geq 1 \quad \forall B \subseteq T \times T$$

Hence by (3.2) we have

(i) $\{V_n^*(B) \mid n \in \mathbb{N}, B \subseteq T \times T\}$ is uniformly integrable and

by Lemma 1 we have

(ii) $\limsup_{n\to\infty} E^*V_n(B) = \inf_n E^*V_n(B)$

Hence we find (see the proof of (2.6) ⇒ (2.7))

(iii) $\forall \varepsilon > 0 \; \exists \delta > 0$ so that $\limsup_{n\to\infty} E^*V_n(B) < \varepsilon$ whenever $\inf_n P^*(V_n(B) > \delta) < \delta$

(3.3) ⇒ (3.7): By Example 7.28 in [6] and Theorem 2 there exist a totally bounded, pseudo ultrametric ρ on T so that

$$\limsup_n P^*(w_\rho(\tfrac{1}{n}S_n, r) > r) < r \qquad \forall\, 0 < r < 1$$

Putting $B(r) = \{(u,v) | \rho(u,v) < r\}$, we see that (iii) gives

$$\limsup_n E^* w_\rho(\frac{1}{n}S, r) < \varepsilon \qquad \forall\, 0 < r < \delta$$

and so (3.7) holds.

(3.7) ⇒ (3.6): Let ρ be the pseudo metric from (3.7). Then

$$\inf_n P^*(w_\rho^o(\frac{1}{n}S_n, r, t) > \varepsilon) \leq \limsup_n P^*(w_\rho(\frac{1}{n}S_n, r) > \varepsilon)$$

$$\leq \varepsilon^{-1} \limsup_n E^* w_\rho(\frac{1}{n}S_n, r)$$

so if we choose $r > 0$ so that

$$\limsup_n E^* w_\rho(\frac{1}{n}S_n, r) < \varepsilon^2$$

we see that (3.6) holds.

(3.6) ⇒ (3.5): Let $\varepsilon > 0$ be given and choose $r > 0$ and ρ according to (3.6). Since ρ is totally bounded we can cover T by finitely many ρ-balls $B_1, \ldots, B_m$ of radius r. So choosing $\alpha = \{B_1, \ldots, B_m\}$ we see that (3.5) holds.

(3.7) ⇒ (3.4): Mutatis mutandis as (3.6) ⇒ (3.5)

(3.4) ⇒ (3.5): Evident!

(3.5) ⇒ (3.3): Let $\varepsilon > 0$ be given and choose $\delta > 0$ according to (iii). Now choose $\alpha \in \Gamma(T)$ so that

$$\inf_n P^*(W_A(\frac{1}{n}S_n) > \delta) < \delta \qquad \forall\, A \in \alpha$$

then by (iii) we have

$$\limsup_{n\to\infty} E^* W_A(\frac{1}{n}S_n) < \varepsilon \qquad \forall\, A \in \alpha$$

Thus f is eventually totally bounded in mean, and so (3.3) follows from Theorem 2.

Thus (3.3)-(3.7) are equivalent and the theorem is proved. □

In [4] E. Giné and J. Zinn has given necessary and sufficient conditions for $f \in LLN_0$ provided that f satisfied a certain measurability condition (condition NSM(P) in [4]). The conditions in [4] are all expressed in terms of the random metrics:

$$d_{np}(u,v) = \{\frac{1}{n} \sum_{j=1}^{n} |f(\xi_j,u) - f(\xi_j,v)|^p\}^{1 \wedge (1/p)}$$

and their random entropies:

$$H_{np}(r) = \log N_{np}(r)$$

where $N_{np}(r)$ is the covering number of d_{np}, i.e. the smallest integer N so that T may be covered by N balls of radius r. However as shown by W. Dobric [3], there exist $f \notin LLN_0$ which do satisfies the entropy conditions in [4], so the measurability condition in [4] seem indispensable for the entropy conditions to imply the law of large numbers. A close look at the proofs in [4], shows that their methods works if the function

$$\| \sum_{j=1}^{n} \pm f(\xi_j) \|$$

is P-measurable for all $n \geq 1$ and all choices of $\pm$.

REFERENCES

[1] N.T. Andersen, A central limit theorem for non-separable valued functions, Mat. Inst. Aarhus University 1984, Preprint Series No.

[2] A. Beck, On the strong law of large numbers, Ergodic Theory Proc. Int. sympt. New Orleans 1961 (ed. F.B. Wright), Academic Press

[3] V. Dobric, Counter examples for the law of large numbers in some Banach spaces, Mat. Inst. Aarhus University, Preprint Series No.

[4] E. Giné and J. Zinn, Some limits theorems for empirical processes, (to appear in Ann.Prob.)

[5] J. Hoffmann-Jørgensen, Probability in B-spaces, Functional Analysis, Proc. Conf. Dobrovnik Yugoslavia 1981, Springer Verlag 1982 (LNS 948) (ed. Butkovic et al.

[6] J. Hoffmann-Jørgensen, Stochastic Processes on Polish spaces, (to appear)

[7] J. Hoffmann-Jørgensen, The law of large numbers for non-measurable and non-separable random elements, (to appear in Astērique)

[8] J. Hoffmann-Jørgensen and G. Pisèer, The law of large number and the central limit theorem in Banach spaces, Ann. Prob. 4 (1976), p. 587-599

[9] M. Talagrand, The Glivenko-Cantelli Problem, Preprint Ohio State University, Urbana, 1984.

AN INTRODUCTION TO LARGE DEVIATIONS

Naresh C. Jain(1)
University of Minnesota
Minneapolis, MN 55455/USA

Introduction. There have been phenomenal developments in the theory of large deviations over the last two decades. Most significant is the work of M.D. Donsker and S.R.S. Varadhan who have developed a powerful machinery in a series of papers to deal with many old and new problems in probability where precise estimates of large deviation (away from the central part of the distribution) probabilities play an important role. To give an introduction to this formidable subject in four lectures in any intelligible way is quite difficult, to say the least. Fortunately, there are several very interesting and readable accounts of these developments now available in the literature. The expository article by Varadhan [A.5] and the lecture notes by Stroock [A.4] should be very helpful to someone trying to understand these developments. There is also a book on the subject by Ellis [A.2]. One application of the large deviation theory is the beautiful work of Freidlin and Ventzell on small random perturbations of dynamical systems. This is all contained in their book [A.3]; see also [A.1], [A.4] and [A.5].

Instead of trying to give a complete survey of the various developments on the subject (my own knowledge of which is quite limited) I will attempt to make this introduction rather concrete with as few technicalities as possible. The basic aim here is to ease the reader into some of the more authoritative accounts of the subject mentioned earlier. To this end, the lectures are organized as follows: Cramér's problem and Varadhan's formulation of the large deviation principle will be discussed in the first lecture. A large deviation principle for a Gaussian measure in a separable Banach space will be derived in the second lecture as a consequence of some well known facts about Gaussian measures in separable Banach spaces. This contains a large deviation principle for the Wiener measure in $C[0,T]$ (a theorem of Schilder [23]) and one can then deduce a large deviation principle for a class of diffusions ([A.1],[A.4], [A.5]) and apply it to prove a celebrated theorem of Freidlin and Ventzell; this will be sketched in lecture 3. The Sanov problem and the Donsker-Varadhan theory will be introduced in the last lecture. Further references will be made as we go along.

1. Cramér's problem and Varadhan's formulation of the large deviation principle

Let $X_1, X_2, \ldots$ be a sequence of real-valued independent identically distributed (i.i.d.) random variables with a common nondegenerate distribution F and mean m. By Kolmogorov's strong law of large numbers

---

(1) Partially supported by NSF.

$$\lim_{n\to\infty} \frac{S_n}{n} = m \text{ , a.s.} \tag{1.1}$$

where

$$S_n = X_1 + \dots + X_n \text{ .}$$

If $a<b$ and $m \notin [a,b]$, then

$$\lim_{n\to\infty} P\{\frac{S_n}{n} \in (a,b)\} = 0 \text{ .} \tag{1.2}$$

Cramér (1937) studied the problem of finding the rate at which the probability in (1.2 tends to zero. Of course one cannot say very much if nothing more than the existence of the first moment is assumed. If one assumes the existence of the moment generating function, then one has the following theorem of Cramér, whose proof (both real-valued and $\mathbb{R}^d$-valued cases) can be found in [A.4], [A.5].

Theorem 1.1. (Cramér). Suppose

$$M(t) = E\, e^{tX_1} < \infty \text{ , } t \in \mathbb{R} \text{ ,} \tag{1.3}$$

and for $x \in \mathbb{R}$ let

$$I(x) = \sup_{t\in\mathbb{R}} \{tx - \log M(t)\} \text{ ,} \tag{1.4}$$

then for every closed set $F \subset \mathbb{R}$

$$\limsup_n \frac{1}{n} \log P\{\frac{S_n}{n} \in F\} \le - \inf_{x\in F} I(x) \text{ ,} \tag{1.5}$$

and for every open set $G \subset \mathbb{R}$

$$\liminf_n \frac{1}{n} \log P\{\frac{S_n}{n} \in G\} \ge - \inf_{x\in G} I(x) \text{ .} \tag{1.6}$$

Remark 1.1. The function $I\colon \mathbb{R} \to [0,\infty]$ is convex and $I(x) = 0$ iff $x = m$. This theorem tells us that if

$$\lambda_0 = \inf_{x\in(a,b)} I(x) \quad , \quad \lambda_1 = \inf_{x\in[a,b]} I(x)$$

then given $\epsilon > 0$, there exists n_0 such that for $n \ge n_0$

$$e^{-n(\lambda_0+\varepsilon)} \le P\{\frac{S_n}{n} \in (a,b)\} \le P\{\frac{S_n}{n} \in [a,b]\} \le e^{-n(\lambda_1+\varepsilon)} , \tag{1.7}$$

i.e., if $\lambda_0 = \lambda_1$ and $m \notin [a,b]$ then the probability decays exponentially fast and the function I gives the rate of decay in the exponent.

Example 1.1. If X_1 is $N(0,1)$, then $I(x) = x^2/2$.

Example 1.2. If $P\{X_1 = 1\} = P\{X_1 = 0\} = 1/2$, then

$$\begin{aligned} I(x) &= \log 2 + x \log x + (1-x)\log(1-x) , \quad 0 \le x \le 1 , \\ &= \infty , \quad \text{otherwise.} \end{aligned}$$

Remark 1.2. Cramér's theorem deals with deviations of S_n of order n from the mean nm. There are interesting questions involving more moderate deviations but we are not concerned with those here.

Remark 1.3. Independence plays an important role. If $X_1, X_2, \ldots$ is a stationary ergodic sequence of real-valued random variables with $EX_1 = m$, then (1.1) holds by the ergodic theorem. If (1.3) is assumed, the rate of convergence in (1.2) may be very different from the one indicated in (1.7) if independence is not assumed. The following example should make this clear.

Example 1.3. Let $X_1, X_2, \ldots$ be a stationary Gaussian sequence with spectral density for the covariance function given by

$$\begin{aligned} f(t) &= c\left(\sin^2 \frac{t}{2}\right) t^{-1-\alpha} , \quad |t| \le \pi , \\ &= 0 , \quad \text{otherwise,} \end{aligned}$$

where $0 < \alpha < 2$ and c is chosen so $\int_{-\infty}^{\infty} f(t)dt = 1$. Then

$$\rho(n) = \int_{-\infty}^{\infty} \cos nt \, f(t)dt \to 0 \quad \text{as} \quad n \to \infty$$

by the Riemann-Legesgue lemma, so the sequence is ergodic. Also, there exists $C > 0$ such that (assuming $EX_1 = 0$)

$$E\, S_n^2 = \int_{-\infty}^{\infty} \frac{\sin^2 \frac{nt}{2}}{\sin^2 \frac{t}{2}} f(t)dt \sim C\, n^{\alpha}$$

as $n \to \infty$. Thus if $a > 0$, $\sigma_n = (ES_n^2)^{1/2}$, then

$$P\{\frac{S_n}{n} > a\} = P\{\frac{S_n}{\sigma_n} > \frac{an}{\sigma_n}\} = P\{X_1 > \frac{an}{\sigma_n}\} .$$

Therefore by the usual tail probability estimate for a $N(0,1)$ random variable we get

$$\frac{1}{n} \log P\{\frac{S_n}{a_n} > a\} \sim -a^2 n^{1-\alpha}/2C .$$

This shows that very different rates are possible when independence is not assumed.

In the context of Cramér's Theorem, let μ_n denote the measure induced on $\mathbb{R}$ by S_n/n. Then (1.5) and (1.6) can be restated as

$$\limsup_n \frac{1}{n} \log \mu_n(F) \leq - \inf_{x \in F} I(x) \quad , F \text{ closed,} \tag{1.5'}$$

$$\liminf_n \frac{1}{n} \log \mu_n(G) \geq - \inf_{x \in G} I(x) \quad , G \text{ open .} \tag{1.6'}$$

This should motivate Varadhan's formulation (1966) of the large deviation principle:

Large deviation principle: Let S be a Polish space and $\mathcal{S}$ its Borel sets. A family of probability measures $\{\mu_\varepsilon , \varepsilon > 0\}$ on $(S,\mathcal{S})$ is said to satisfy the large deviation principle with rate function I if

(i) $I: S \to [0,\infty]$ is lower semi-continuous, $I \not\equiv \infty$.

(ii) Given $\ell < \infty$, the set $\{x \in S : I(x) \leq \ell\}$ is compact.

(iii) For a closed set $F \subset S$

$$\limsup_{\varepsilon \to 0} \varepsilon \log \mu_\varepsilon(F) \leq - \inf_{x \in F} I(x) .$$

(iv) For an open set $G \subset S$

$$\liminf_{\varepsilon \to 0} \varepsilon \log \mu_\varepsilon(G) \geq - \inf_{x \in G} I(x) .$$

Remark 1.4. The function I defined in (1.4) is a rate function which satisfies (i)-(iv). Typically $\mu_\varepsilon \xrightarrow{w} \delta_{x_0}$ as $\varepsilon \to 0$, where " $\xrightarrow{w}$ "indicates the weak convergence of probability measures and δ_{x_0} is the probability measure which gives all the mass to the singleton x_0 . Then for a set A disjoint from a neighborhood of x_0 we have $\mu_\varepsilon(A) \to 0$ as $\varepsilon \to 0$. If the large deviation principle holds for $\{\mu_\varepsilon\}$, then (iii) and (iv) tell us that $\mu_\varepsilon(A)$ decays exponentially fast and I determines the constant in the exponent. If for $A \in \mathcal{S}$,

$$\lambda = \inf_{x \in A^o} I(x) = \inf_{x \in \bar{A}} I(x) ,$$

then (iii) and (iv) can be combined to read $\lim_{\varepsilon \to 0} \varepsilon \log \mu_\varepsilon(A) = -\lambda$. Also, when $\varepsilon = \frac{1}{n}$

it is more convenient to write μ_n rather than $\mu_{1/n}$ in place of μ_ϵ .

Some consequences of the large deviation principle. The following consequences are very useful; they are noted in [A.5].

Proposition 1.1. (Contraction principle). Let S_1 and S_2 be Polish spaces with $\mathcal{S}_1$ and $\mathcal{S}_2$ the corresponding Borel σ-algebras. Let $\pi : S_1 \to S_2$ be a continuous map. Let $\{\mu_\epsilon\}$ on $(S_1,\mathcal{S}_1)$ satisfy the large deviation principle with rate function I. Then the family $\{\mu_\epsilon \pi^{-1}\}$ on $(S_2,\mathcal{S}_2)$ satisfies the large deviation principle with rate function

$$\begin{aligned} J(y) &= \inf\{I(x) : \pi(x) = y\} \\ &= \infty \text{, if the set is empty.} \end{aligned} \tag{1.8}$$

The proof of this proposition is a simple exercise. As an application consider Example 1.2 and let $\pi(x) = 2x-1$. Then we get the rate function J for the symmetric case $P\{X_1 = 1\} = P\{X_1 = -1\} = 1/2$ given by

$$\begin{aligned} J(y) &= \log 2 + \frac{1+y}{2} \log\left(\frac{1+y}{2}\right) + \frac{1-y}{2} \log \frac{1-y}{2} \quad , -1 \leq y \leq 1 , \\ &= \infty \text{ , otherwise.} \end{aligned}$$

In some applications the following variant of Proposition 1.1 is useful.

Proposition 1.1'. Suppose $\{\mu_\epsilon\}$ on $(S_1,\mathcal{S}_1)$ satisfies the large deviation principle with rate function I and $(S_2,\mathcal{S}_2)$ is another Polish space. Let $\pi_\eta : S_1 \to S_2$, $\eta > 0$, be continuous maps such that $\lim_{\eta\to 0} \pi_\eta = \pi$ uniformly on compacts. Then we have

$$\limsup_{\substack{\epsilon\to 0 \\ \eta\to 0}} \epsilon \log \mu_\epsilon \circ \pi_\eta^{-1}(F) \leq - \inf_{y\in F} J(y) \quad , F \text{ closed} \subset S_2 ,$$

$$\liminf_{\substack{\epsilon\to 0 \\ \eta\to 0}} \epsilon \log \mu_\epsilon \circ \pi_\eta^{-1}(G) \geq - \inf_{y\in G} J(y) \quad , G \text{ open} \subset S_2 ,$$

where $J(y)$ is given by (1.8).

For the proofs of this proposition and the following theorem we refer to Varadhan [A.5].

Theorem 1.2. (Varadhan). Suppose $\{\mu_\epsilon\}$ on $(S,\mathcal{S})$ satisfies the large deviation principle with rate function I . Then for any bounded continuous $f : S \to \mathbb{R}$ we have

$$\lim_{\varepsilon \to 0} \varepsilon \log \int_S e^{f(x)/\varepsilon} \, d\mu_\varepsilon(x) = \sup_{x \in S} \{f(x) - I(x)\} .$$

This theorem gives us asymptotic evaluation of expectations of certain functionals when the large deviation principle holds.

2. A large deviation principle for a Gaussian probability measure in a separable Banach space

Cramér's theorem in higher dimensions is no easy generalization. Donsker and Varadhan [6,III] proved the analogue of Cramér's theorem when S is a separable Banach space. In this context

$$I(x) = \sup_{f \in S^*} \{f(x) - \log M(f)\}$$

where S^* = dual of S, and

$$M(f) = \int_S e^{f(x)} d\mu(x) .$$

Theorem 2.1 (Donsker and Varadhan). If $X_1, X_2, \ldots$ are i.i.d. random variables taking values in a separable Banach space S such that for all $t \in \mathbb{R}$, $\int \exp(t\|x\|) d\mu(x) < \infty$, where μ = distribution of X_1, then $\{\mu_n\}$ satisfies the large deviation principle with $\varepsilon = 1/n$, where μ_n is the distribution of $(X_1 + \ldots + X_n)/n$.

Donsker and Varadhan derive this theorem as a corollary of their large deviation results for Markov processes. A direct proof is given by Stroock [A.4] which is based on the approach of Bahadur and Zabell [1]. See also [18]. The important thing about this approach, when it works, is that it allows one to prove the existence of a rate function first and then the problem reduces to identifying the rate function.

Instead of going into the proof of this theorem, we will prove a large deviation principle for a Gaussian measure in a Banach space. This of course is a special case of Theorem 1. The proof given below, is not new (see [A.1], e.g.) and is based on well known facts about Gaussian measures in separable Banach spaces such as Fernique's estimate [11] and the Marcus-Shepp argument [19] leading to the asymptotic estimate of the upper tail of the norm distribution. Schilder [23] first proved this result for Wiener measure in $C[0,1]$ and Pincus [22] extended it to a class of Gaussian processes.

Let μ be a centered Gaussian probability measure on $(S, \mathcal{S})$, where S is a separable Banach space with norm $\|\cdot\|$, i.e. given $f_1, \ldots, f_k$ in S^*, the distribution of the vector $(f_1, \ldots, f_k) : S \to \mathbb{R}^k$ is a k-dimensional Gaussian distribution with zero mean and covariance Γ given by

$$\Gamma(f_i,f_j) = \int_S f_i(x)f_j(x)d\mu(x) .$$

The following facts about such μ are well-known.

(a) There exists a separable Hilbert space $H \subset S$ such that the inclusion map $i : H \to S$ is continuous and

$$\operatorname{supp} \mu = \bar{H} : = \text{the closure of } H \text{ in } S .$$

<u>Without loss of generality we assume</u> $\bar{H} = S$.

(b) For any $\ell < \infty$, the ball $\{x : \|x\|_H \leq \ell\}$ is a compact subset of S .

(c) Let S^* denote the dual of S and let $\hat{H}$ be the closure of S^* in $L^2(S,\mathcal{S},\mu) : = L^2(\mu)$. There exists an isometric isomorphism $\psi : H \to \hat{H}$ such that if $\psi(x) = f \in S^*$, then $\langle \psi^{-1}f,y\rangle_H = f(y)$ for all $y \in H$.

(d) There exist $\{f_1,f_2,\ldots\} \subset S^*$ such that $\{e_1,e_2,\ldots\}$, $e_j : = \psi^{-1}f_j$, $j \geq 1$, is an orthonormal basis in H and

$$H = \{x \in S : \sum_{n=1}^{\infty} f_n(x)^2 < \infty\} .$$

If $x \in H$, $\|x\|_H^2 = \sum_1^{\infty} f_n(x)^2$.

(e) (Ito-Nisio). Let $\{f_n\}$, $\{e_n\}$ be as in (d), then

$$x = \sum_{n=1}^{\infty} f_n(x)e_n \text{ , a.e. } (\mu)x ,$$

where the series is to converge in S when equality holds.

(f) For $x \in S$, $A \in \mathcal{S}$, let $\mu^x(A) : = \mu(A+x)$. Then $\mu \sim \mu^x$ iff $x \in H$; if $x \in H$, then

$$\frac{d\mu^x}{d\mu}(y) = \exp\{\psi(x)(y) - \|x\|_H^2/2\} \text{ , a.e. } (\mu)y ,$$

where ψ is the map in (c).

(g) (Fernique [11]). Let $X : S \to S$ be a measurable map. For $s \geq 0$ let

$$\varphi(s) : = \mu\{x : \|X(x)\| > s\} .$$

Then for $\lambda \geq s$

$$\varphi(\lambda) \leq \exp\{-\frac{\lambda^2}{24s^2} \log \frac{1-\varphi(s)}{\varphi(s)}\} .$$

We will use the following simple corollary of this theorem:

Corollary: Let X_N , $N \geq 1$, be a sequence of measurable maps from S into S such that $\|X_N\| \to 0$ as $N \to \infty$ a.e. (μ) . Then there exist $\gamma_N \to \infty$ and for $\lambda > 0$ an N_0 such that for $N \geq N_0$

$$\mu\{x : \|X_N(x)\| > \lambda\} \leq e^{-\gamma_N \lambda^2} .$$

Proof. We can pick $s_N \to 0$ such that $\varphi_N := \mu\{x : \|X_N(x)\| > s_N\} \leq 1/4$. Given $\lambda > 0$, pick N_0 so for $N \geq N_0$ we have $s_N \leq \lambda$. Set $\gamma_N = (24\, s_N^2)^{-1} \log(\frac{1-\varphi_N}{\varphi_N})$ We have

$$M(f) = \int_S e^{f(x)} d\mu(x) = \exp\{\frac{1}{2} \int_S f^2(x) d\mu(x)\} = e^{\frac{1}{2} \|\psi^{-1} f\|_H^2}$$

where ψ is the map given in (c). Therefore I defined in (2.1) is given by

$$I(x) = \sup_{f \in S^*} \{f(x) - \frac{1}{2} \|\psi^{-1} f\|_H^2\} . \tag{2.2}$$

Theorem 2.2. Let μ be a centered Gaussian measure on $(S, \mathcal{S})$ and for $A \in \mathcal{S}$ let $\mu_\varepsilon(A) = \mu\{x : \sqrt{\varepsilon}\, x \in A\}$. Then $\{\mu_\varepsilon\}$ satisfies the large deviation principle with rate function I given by (2.2). Furthermore,

$$\begin{aligned} I(x) &= \frac{1}{2} \|x\|_H^2 , \quad \text{if } x \in H \\ &= \infty , \quad \text{otherwise.} \end{aligned} \tag{2.3}$$

Remark 2.1. If $X_1, X_2, \ldots$ are i.i.d. S-valued random variables with distribution μ , then $(X_1 + \ldots + X_n)/n$ has the same distribution as $X_1/\sqrt{n}$ since μ is centered Gaussian and this distribution is μ_ε with $\varepsilon = 1/n$.

Proof. The proof is carried out in steps.

Step 1. First we prove (2.3). Let $x \in H$. By (d) we can find $g_n \in S^*$ such that $\|\psi^{-1} g_n - x\|_H \to 0$. Then by (2.2) and (c)

$$\begin{aligned} I(x) &\geq g_n(x) - \frac{1}{2} \|\psi^{-1} g_n\|_H^2 = \langle \psi^{-1} g_n, x\rangle_H - \frac{1}{2} \|\psi^{-1} g_n\|_H^2 \\ &\to \frac{1}{2} \|x\|_H^2 , \text{ as } n \to \infty . \end{aligned}$$

Thus $I(x) \geq \|x\|_H^2/2$. To get the opposite inequality, if $f \in S^*$, by (c)

$$f(x) - \frac{1}{2}\|\psi^{-1}f\|_H^2 = \langle \psi^{-1}f, x\rangle_H - \frac{1}{2}\|\psi^{-1}f\|_H^2$$
$$\leq \|\psi^{-1}f\|_H \|x\|_H - \frac{1}{2}\|\psi^{-1}f\|_H^2 \leq \frac{1}{2}\|x\|_H^2$$

since for nonnegative a,b we have $ab - a^2/2 \leq b^2/2$. Thus $I(x) \leq \|x\|_H^2/2$. If $x \notin H$, let $g_N = \sum_{n=1}^{N} f_n(x) f_n$, where $\{f_n\}$ is as in (d). Then

$$g_N(x) - \frac{1}{2}\|\psi^{-1}g_N\|_H^2 = \frac{1}{2}\sum_{n=1}^{N} f_n(x)^2 \to \infty$$

as $N \to \infty$, so $I(x) = \infty$. Thus (2.3) is established.

Step 2. We now check that I is l.s.c. and for $\ell < \infty$ the set $\{x \in S : I(x) \leq \ell\}$ is compact. The second assertion is an immediate consequence of (2.3) and (b). Clearly $I \not\equiv \infty$. Now let $x_n \to x$ in S, then by (2.2)

$$I(x_n) \geq f(x_n) - \frac{1}{2}\|\psi^{-1}f\|_H^2 \quad , \text{ for any } f \in S^* .$$

Thus $\liminf_n I(x_n) \geq f(x) - \frac{1}{2}\|\psi^{-1}f\|_H^2$ for all $f \in S^*$, so $\liminf_n I(x_n) \geq I(x)$, which proves lower semicontinuity.

It remains to check that I given by (2.3) satisfies properties (iii) and (iv) of a rate function. We need a technical lemma for that.

Lemma 2.1. Suppose F is a closed subset of S and for $\delta > 0$, $F^\delta = \{x \in S : \|x-y\| \leq \delta$ for some $y \in F\}$, then

$$\inf_{x \in F} I(x) = \lim_{\delta \to 0} \inf_{x \in F^\delta} I(x) ,$$

where I is given by (2.3).

Proof. Clearly the left side is larger than the right side, so it suffices to prove

$$\inf_{x \in F} I(x) \leq \lim_{\delta \to 0} \inf_{x \in F^\delta} I(x) .$$

Let the right side equal $\ell < \infty$. Then there exist $\delta_n \downarrow 0$ and $x_n \in F^{\delta_n}$ such that $\lim_n I(x_n) = \ell$. Since $I(x_n) < \infty$ we have $x_n \in H$, so $\|x_n\|_H^2 \leq 2\ell + \varepsilon$ for $n \geq$ some n_0. By (b) the set $\{x_n\}$ is precompact, so along a subsequence $x_n \to x_0$ in S, and by the lower semicontinuity of I we have $\ell \geq I(x_0)$. Clearly $x_0 \in F$, so $\ell \geq \inf_{x \in F} I(x)$, and the lemma is proved.

Step 3. (lower bound). Let G be an open subset of S. Clearly $\inf_{x\in G} I(x) = \inf_{x\in G\cap H} I($
Let $x_o \in G\cap H$ and let $\delta>0$ be such that $B_\delta(x_o)=\{x\in S: \|x-x_o\|<\delta\}\subset G$. Then by (f), for any $\alpha>0$

$$\mu_\epsilon(B_\delta(x_o)) = \mu\{x: \|x-x_o/\sqrt{\epsilon}\|<\delta/\sqrt{\epsilon}\} = \mu^{x_o/\sqrt{\epsilon}}(B_{\delta/\sqrt{\epsilon}}(0))$$

$$= \int_{B_{\delta/\sqrt{\epsilon}}(0)} \exp\{\psi(x_o/\sqrt{\epsilon})(y) - \|x_o\|_H^2/2\epsilon\}d\mu(y)$$

$$\geq \int_{B_{\delta/\sqrt{\epsilon}}(0)\cap\Delta} \exp\{\psi(x_o/\sqrt{\epsilon})(y) - \|x_o\|_H^2/2\epsilon\}d\mu(y)$$

where $\Delta=\{y: \psi(x_o/\sqrt{\epsilon})>-\alpha/\sqrt{\epsilon}\}$. Therefore

$$\mu_\epsilon(B_\delta(x_o)) \geq e^{-\alpha/\sqrt{\epsilon}}\, e^{-\|x_o\|_H^2/2\epsilon}\, \mu\{B_{\delta/\sqrt{\epsilon}}(0)\cap\Delta\}.$$

Since $\psi(x_o)$ is a Gaussian random variable, we can pick α large so for all sufficiently small ϵ

$$\mu\{B_{\delta/\sqrt{\epsilon}}(0)\cap\Delta\} = \mu\{B_{\delta/\sqrt{\epsilon}}(0)\cap[y: \psi(x_o)(y)>-\alpha]\}\geq \eta>0.$$

Then

$$\epsilon \log \mu_\epsilon(G) \geq -\sqrt{\epsilon}\,\alpha - \tfrac{1}{2}\|x_o\|_H^2 + \epsilon\log\eta$$

so for $x_o\in G\cap H$

$$\liminf_{\epsilon\to 0} \epsilon\log\mu_\epsilon(G) \geq -\tfrac{1}{2}\|x_o\|_H^2$$

and this proves the lower bound.

Step 4. (Upper bound). Let $F\subset S$ be closed and for $\delta>0$ let F^δ be as in Lemma 2.1. Then for any N

$$\mu_\epsilon(F) \leq \mu\{x: \|\sqrt{\epsilon}\,x - \sum_{n=1}^N f_n(\sqrt{\epsilon}\,x)e_n\|\geq\delta\} + \mu\{x: \sum_{n=1}^N f_n(\sqrt{\epsilon}\,x)e_n\in F^\delta\},$$

where $\{f_n\}$ and $\{e_n\}$ are as in (d). Let

$$\ell_\delta = \inf_{x\in F^\delta} I(x), \quad \ell = \inf_{x\in F} I(x).$$

Using the linearity of f_n

$$\mu_\epsilon(F) \leq \mu\{x : \|x - \sum_{n=1}^{N} f_n(x)e_n\| > \delta/\sqrt{\epsilon}\} + \mu\{x : \|\sum_{n=1}^{N} f_n(\sqrt{\epsilon}x)e_n\| \geq \ell_\delta\}$$

$$= \mu\{x : \|x - \sum_{n=1}^{N} f_n(x)e_n\| > \delta/\sqrt{\epsilon}\} + \mu\{x : \sum_{n=1}^{N} f_n(x)^2 \geq \frac{2\ell_\delta}{\epsilon}\} .$$

Let $X_N(x) = x - \sum_{n=1}^{N} f_n(x)e_n$. By (e), $\|X_N\| \to 0$ a.e. (μ) as $N \to \infty$, so using (g) and the chi-square tail estimate we get

$$\mu_\epsilon(F) \leq \exp\{-\gamma_N^2 \delta^2/\epsilon\} + C(\ell_\delta/\epsilon)^{\frac{N}{2}-1} e^{-\ell_\delta/\epsilon} ,$$

where $\gamma_N \to \infty$ as $N \to \infty$. Therefore

$$\epsilon \log \mu_\epsilon(F) \leq \epsilon \log 2 + \max\{-\gamma_N^2 \delta^2, \epsilon \log C + (\tfrac{N}{2}-1)\epsilon \log(\ell_\delta/\epsilon) - \ell_\delta\} .$$

Note that $\ell_\delta < \infty$ since F^δ contains nonempty interior. Thus

$$\limsup_{\epsilon \to 0} \epsilon \log \mu_\epsilon(F) \leq \max\{-\gamma_N^2\delta^2, -\ell_\delta\} .$$

Now letting $N \to \infty$, and then $\delta \to 0$, the result follows from Lemma 2.1.

Remark 2.2. The Marcus-Shepp theorem [19], which says

$$\lim_{\lambda \to \infty} \frac{1}{\lambda^2} \log \mu\{x : \|x\| > \lambda\} = -\frac{1}{2\sigma^2} ,$$

where μ is centered Gaussian, and $\sigma^2 = \sup_{\|f\|_{S^*} \leq 1} \|\psi^{-1}f\|_H^2$, can be derived from this large deviation principle by checking that $\inf_{\|x\| \geq 1} I(x) = 1/2\sigma^2$.

As a corollary we state Schilder's theorem [23] which will be used below. Let $S = C([0,T]; \mathbb{R}^d)$, the space of continuous $\mathbb{R}^d$-valued functions on $[0,T]$. As usual, $\|x\| = \sup_{0 \leq t \leq T} |x(t)| = \sup_{0 \leq t \leq T} (x_1(t)^2 + \ldots + x_d(t)^2)^{1/2}$. Let μ be the standard Wiener measure on S . Then

$$H = \{x \in S : x \text{ absolutely continuous}, \int_0^T |\dot{x}(t)|^2 dt < \infty\}$$

$$\langle x,y \rangle = \int_0^T \langle \dot{x}(t) , \dot{y}(t) \rangle_{\mathbb{R}^d} dt .$$

Let

$$I^T(x) = \begin{cases} \frac{1}{2}\|x\|_H^2 = \frac{1}{2}\int_0^T |\dot{x}(t)|^2 dt \ , \text{ if } x \in H \ , \\ \infty \ , \text{ otherwise.} \end{cases} \tag{2.4}$$

Theorem 2.3 (Schilder). If μ is standard Wiener measure in $C([0,T] ; \mathbb{R}^d)$ and $\mu_\epsilon(A) = \mu(A\,\epsilon^{-1/2})$ for $A \in \mathcal{S}$, then $\{\mu_\epsilon\}$ satisfies the large deviation principle with rate function I^T .

3. Random perturbations of dynamical systems - a theorem of Freidlin and Ventzell

Theorem 2.3 can be used to derive rather easily a large deviation principle for a class of diffusion processes which forms the major part of the proof of a theorem of Freidlin and Ventzell on the asymptotics of the exit disbribution of a randomly perturbed dynamical system. The book by Freidlin and Ventzell [A.3] of course contains this theorem and a lot more; the exposition here follows that of Varadhan [A.5]. See also [A.1], [A.4].

Let $X_\epsilon^x(t)$ be a diffusion process which satisfies the Ito stochastic differential equation

$$\begin{aligned} dX_\epsilon^x(t) &= \sqrt{\epsilon}\,\sigma(X_\epsilon^x(t))\,d\beta(t) + b(X_\epsilon^x(t))\,dt \ , \quad 0 \le t \le T \\ X_\epsilon^x(0) &= x \ , \end{aligned} \tag{3.1}$$

where $X_\epsilon^x(t) \in \mathbb{R}^d$, $\sigma : \mathbb{R}^d \to d \times d$ matrix-valued, and $b : \mathbb{R}^d \to \mathbb{R}^d$; β is d-dimensional standard Brownian motion. If one could view X_ϵ^x as a continuous transformation of the Brownian path $\sqrt{\epsilon}\,\beta$, one would expect to use the contraction principle to derive the large deviation principle for X_ϵ^x from that of $\sqrt{\epsilon}\,\beta$. Unfortunately, one cannot define X_ϵ^x as a continuous transformation of $\sqrt{\epsilon}\,\beta$; however, the idea can be made to work via an approximation procedure. We will now describe it.

Let $a(x) = \sigma(x)\sigma^*(x)$, where σ^* is the transpose of σ . Assume that a is uniformly elliptic, i.e. there exists $\lambda > 0$ such that for all x,y

$$y^* a(x) y \ge \lambda |y|^2 \ . \tag{3.2}$$

Also assume that σ and b are bounded and both satisfy a global Lipschitz condition

$$|\sigma(x) - \sigma(y)| + |b(x) - b(y)| \le M|x-y|$$

for some $M < \infty$, all x,y in $\mathbb{R}^d$. Then it is well-known that (3.1) has a (pathwise) unique solution $X_\epsilon^x(t)$ and the corresponding differential operator is given by

$$L_\epsilon = \frac{\epsilon}{2} \sum_{i,j=1}^{d} a_{ij}(x) \frac{\partial^2}{\partial x_i \partial x_j} + \sum_{j=1}^{d} b_j(x) \frac{\partial}{\partial x_j} . \tag{3.3}$$

Let $P_{\epsilon,x}$ be the measure induced on $C([0,T]; \mathbb{R}^d)$ by X_ϵ^x. We have

$$P_{\epsilon,x}\{X_\epsilon^x(0) = x\} = 1 .$$

For $x \in \mathbb{R}^d$, $f : [0,T] \to \mathbb{R}^d$, let

$$J_x^T(f) = \frac{1}{2} \int_0^T \langle \dot{f}(t) - b(f(t)), a^{-1}(f(t))(\dot{f}(t) - b(f(t))) \rangle_{\mathbb{R}^d} dt \tag{3.4}$$

if f is absolutely continuous and $f(0) = x$, let $J_x^T(f) = \infty$, otherwise.

Remark 3.1. One can check that $J_x^T(f) \geq 0$ and $J_x^T(f) = 0$ iff f is the solution of the differential equation

$$\begin{aligned} \dot{x}(t) &= b(x(t)) , \quad 0 < t \leq T \\ x(0) &= x . \end{aligned} \tag{3.5}$$

Thus $J_x^T(f)$ provides a measure of how far a trajectory f is from the solution of (3.5).

The following theorem is a slightly stronger version of the usual large deviation principle for the class of diffusions described above. This form is suitable for the study of the Freidlin-Ventzell problem.

Theorem 3.1. For any closed F and open G in $C([0,T]; \mathbb{R}^d)$

$$\limsup_{\substack{\epsilon \to 0 \\ y \to x}} \epsilon \log P_{\epsilon,y}(F) \leq - \inf_{f \in F} J_x^T(f) ,$$

$$\liminf_{\substack{\epsilon \to 0 \\ y \to x}} \epsilon \log P_{\epsilon,y}(G) \geq - \inf_{f \in G} J_x^T(f) .$$

To prove this theorem via the contraction principle (Propositions 1.1, 1.1') one must go through an approximation procedure to get a continuous map. For $\delta > 0$ define

$$X_{\epsilon,\delta}^x(t) = x + \sigma(x)\sqrt{\epsilon}\, \beta(t) + b(x)t , \quad 0 \leq t \leq \delta , \tag{3.6}$$

and assuming that $X_{\epsilon,\delta}^x(t)$ has been defined for $t \leq (j-1)\delta$, $1 \leq j \leq [T/\delta]$, writing t_j for $j\delta$, let

$$X^x_{\epsilon,\delta}(t) = X^x_{\epsilon,\delta}(t_{j-1}) + \sqrt{\epsilon}\,\sigma(X^x_{\epsilon,\delta}(t_{j-1}))(\beta(t)-\beta(t_{j-1}))$$
$$+ b(X^x_{\epsilon,\delta}(t_{j-1}))(t-t_{j-1}) \;,\; t_{j-1} \le t < t_j \;. \tag{3.7}$$

If we write $\psi_\delta(t)$ for the step function $[t/\delta]\,\delta$, then (3.6) and (3.7) are the same as

$$dX^x_{\epsilon,\delta}(t) = \sqrt{\epsilon}\,\sigma(X^x_{\epsilon,\delta}(\psi_\delta(t)))d\beta(t) + b(X^x_{\epsilon,\delta}(\psi_\delta(t)))dt$$
$$X^x_{\epsilon,\delta}(0) = x \;. \tag{3.8}$$

For $x \in \mathbb{R}^d$, $\delta > 0$ fixed, (3.6) and (3.7) define a 1-1 continuous transformation $F_{\delta,x} : C([0,T];\mathbb{R}^d) \to C([0,T];\mathbb{R}^d)$ as follows: suppose $g \in C([0,T];\mathbb{R}^d)$, and let $F_{\delta,x}(g) = f$ be given by

$$f(t) = f(t_{j-1}) + \sigma(f(t_{j-1}))(g(t) - g(t_{j-1})) + b(f(t_{j-1}))(t-t_{j-1}) \;,\; t_{j-1} \le t \le t_j \;. \tag{3.9}$$

$F_{\delta,x}$ is clearly invertible since σ is invertible and f is absolutely continuous iff g is, indeed

$$\dot f(t) = \sigma(f(t_{j-1}))\dot g(t) + b(f(t_{j-1})) \;,\; t_{j-1} \le t \le t_j$$

which is the same as

$$\dot f(t) = \sigma(f(\psi_\delta(t)))\dot g(t) + b(f(\psi_\delta(t))) \;.$$

Therefore

$$|\dot g(t)|^2 = \langle \dot f(t) - b(f(\psi_\delta(t))), a^{-1}(f(\psi_\delta(t)))(\dot f(t) - b(f(\psi_\delta(t))))\rangle_{\mathbb{R}^d} \;.$$

Thus to apply the contraction principle, with I^T defined in (2.4) and $f \in C([0,T];\mathbb{R}^d)$, f absolutely continuous, $f(0) = x$, let

$$J^T_{\delta,x}(f) = \inf_{g:F_{\delta,x}(g) = f} I^T(g) = I^T(g) = \frac{1}{2}\int_0^T |\dot g(t)|^2 dt$$
$$= \frac{1}{2}\int_0^T \langle \dot f(t) - b(f(\psi_\delta(t))), a^{-1}(f(\psi_\delta(t)))(\dot f(t) - b(f(\psi_\delta(t))))\rangle_{\mathbb{R}^d} dt \;. \tag{3.10}$$

$J^T_{\delta,x}(f) = \infty$, if f does not satisfy the above conditions. Note that

$$\lim_{\delta \to 0} J^T_{\delta,x}(f) = J^T_x(f) \;,$$

where J^T_x is defined in (3.4). For $\delta > 0$ fixed, and μ_ϵ the distribution of $\sqrt{\epsilon}\,\beta$

in $C([0,T];\mathbb{R}^d)$ we would like to apply the contraction principle of Proposition 1.1' with $\pi_\eta = F_{\delta,y}$, $y \to x$. We need to check that $F_{\delta,y} \to F_{\delta,x}$ uniformly on compacts. For this it suffices to observe that $F_{\delta,x}$ viewed as a map from $C([0,T];\mathbb{R}^d) \times \mathbb{R}^d$ to $C([0,T];\mathbb{R}^d)$ is continuous. We thus get

Theorem 3.2. Let $\delta > 0$ be fixed. If F is closed and G is open in $C([0,T];\mathbb{R}^d)$, and $P_{\epsilon,\delta,x} = \mu_\epsilon F^{-1}_{\delta,x}$, then

$$\limsup_{\substack{\epsilon \to 0 \\ y \to x}} \epsilon \log P_{\epsilon,\delta,y}(F) \le - \inf_{f \in F} J_{\delta,x}(f) ,$$

$$\liminf_{\substack{\epsilon \to 0 \\ y \to x}} \epsilon \log P_{\epsilon,\delta,y}(G) \ge - \inf_{f \in G} J_{\delta,x}(f) .$$

To get Theorem 3.1 from Theorem 3.2 we have to let $\delta \to 0$. To do that one needs an estimate on the difference between $X^x_{\epsilon,\delta}$ and X^x_ϵ. This is given in the following proposition:

Proposition 3.1. For any $\eta > 0$, $T < \infty$,

$$\lim_{\delta \to 0} \limsup_{\epsilon \to 0} \epsilon \sup_x \log \mu_1\{ \sup_{0 \le t \le T} |X^x_{\epsilon,\delta}(t) - X^x_\epsilon(t)| \ge \eta \} = -\infty .$$

For a proof of this we refer to Varadhan ([A.5], Lemma 6.2). Once Theorem 3.2 and Prop. 3.1 are available, Theorem 3.1 follows readily.

We now discuss the exit problem of Freidlin and Ventzell. Let G be a bounded connected subset of $\mathbb{R}^d$ $(d \ge 3)$ with a smooth boundary. Let L_ϵ be as in (3.3) and let σ, a and b satisfy the conditions specified earlier. The corresponding diffusion X^x_ϵ satisfies the Ito equation (3.1). Consider the differential equation (3.5) and suppose that there is a unique equilibrium point $\bar{x} \in G$ such that for all $x \in \bar{G}$ the solution $f_o(t)$ of (3.5) lies in G for $t > 0$ and $f_o(t) \to \bar{x}$ as $t \to \infty$. Consider the Dirichlet problem: h is a continuous boundary function,

$$\begin{aligned} L_\epsilon u_\epsilon &= 0 \quad \text{in } G \\ u_\epsilon &= h \quad \text{on } \partial G . \end{aligned} \tag{3.12}$$

We know that under our conditions if $x \in G$, then $X^x_\epsilon(t)$, the solution of (3.1), must exit from G. Let τ^x_ϵ be the (first) exit time. Then

$$\mathcal{U}_\epsilon(x) = Eh(X^x_\epsilon(\tau^x_\epsilon))$$

solves (3.12). By Remark 3.1 and Theorem 3.1 it is easy to conclude that $P_{\epsilon,x} \overset{w}{\to} \delta_{f_o}$

as $\epsilon \to 0$ (in the sense of weak convergence in $C([0,T];\mathbb{R}^d)$ for any $T>0$); where $P_{\epsilon,x}$ is the measure induced by X_ϵ^x. This means that for small ϵ the trajectories of X_ϵ^x are close to f_0 on $[0,T]$ with high probability, but they must still exit from G. The problem of Freidlin and Ventzell is to determine the asymptotic behavior of the exit distribution as $\epsilon \to 0$. For x,y in $\mathbb{R}^d$, define

$$\varphi_t(x,y) = \inf_{\substack{f:f(t)=y \\ f(0)=x}} J_x^t(f)$$

$$\varphi(x,y) = \inf_{t>0} \varphi_t(x,y) .$$

φ is jointly continuous in x,y and $\varphi(x,y) \leq C|x-y|$. Suppose there exists a point $y_0 \in \partial G$ such that

$$\varphi(\bar{x},y_0) < \varphi(\bar{x},y) \quad \text{for all } y \in \partial G .$$

Theorem 3.3. (Freidlin and Ventzell). If f is a continuous function on ∂G and (3.13) holds, then

$$\lim_{\epsilon \to 0} E\{f(X_\epsilon^x(\tau_\epsilon^x))\} = f(y_0) . \tag{3.14}$$

Since (3.14) holds true for all continuous f on ∂G, it is clear that for small ϵ exit occurs close to y_0 with a high probability. After the necessary probability estimates are available in Theorem 3.1 the proof of this theorem still takes some work. The details can be found in [A.5]. Of course, [A.3] contains a great deal more information on many variants of the problem.

4. The Sanov problem and the large deviation theory of Donsker and Varadhan for occupation times of Markov processes

As before, let S be a Polish space and $\mathcal{S}$ its Borel sets. Let $\mathcal{M}(S)$ denote the set of all probability measures on $(S,\mathcal{S})$. Under the topology of weak convergence the space $\mathcal{M}(S)$ becomes a Polish space.

Let $X_1, X_2, \ldots$ be a sequence of i.i.d. S-valued random variables on a probability space $(\Omega,\mathcal{F},P)$, and let $\mu \in \mathcal{M}(S)$ be the common distribution. Then $\delta_{X_1}, \delta_{X_2}, \ldots$ are i.i.d. $\mathcal{M}(S)$-valued random variables. For $\omega \in \Omega$ let

$$L_n(\omega,\cdot) = \frac{1}{n}\{\delta_{X_1(\omega)} + \ldots + \delta_{X_n(\omega)}\}(\cdot) , \tag{4.1}$$

then for any bounded measurable $f: S \to \mathbb{R}$

$$\int_S f(x) L_n(\omega,dx) = \frac{1}{n}\{f(X_1(\omega)) + \ldots + f(X_n(\omega))\} , \tag{4.2}$$

and by the strong law of large members almost surely

$$\lim_n \int_S f(x) L_n(\omega,dx) = \int f(x)\,d\mu(x) \, , \tag{4.3}$$

where the null set depends on f. In particular, (4.3) holds for a coutable number of bounded continuous functions f on S. This suffices for weak convergence, so

$$L_n \overset{w}{\to} \mu \, , \text{ a.s. in } \mathcal{M}(S) \text{ as } n \to \infty \tag{4.4}$$

If $f = \chi_A$, $A \in \mathcal{S}$, then $L_n(\omega,A) = \frac{1}{n} \sum_{j=1}^{n} \chi_A(X_j(\omega))$, so $L_n(\omega,\cdot)$ is the normalized occupation time measure for the $\{X_n\}$ for each $\omega \in \Omega$. If Γ is a measurable subset of $\mathcal{M}(S)$, disjoint from a neighborhood of μ, then by (4.4)

$$P\{\omega : L_n(\omega,\cdot) \in \Gamma\} \to 0 \quad \text{as} \quad n \to \infty$$

and Sanov's problem is to determine the rate at which this probability tends to zero.

Remark 4.1. Sanov's problem may be viewed as the Cramér problem in an appropriate locally convex Hausdorff topological vector space.

A generalized form of Sanov's theorem is

Theorem 4.1. Let $X_1, X_2, \ldots$ be i.i.d. S-valued random variables on $(\Omega,\mathcal{F},P)$ with a common distribution $\mu \in \mathcal{M}(S)$. If L_n is defined by (4.1) and for $\nu \in \mathcal{M}(S)$

$$I(\nu) = \sup_{f \in B_c(S)} \left\{ \int_S f\,d\nu - \log \int_S e^{f(x)}\,d\mu(x) \right\} ,$$

where $B_c(S)$ = space of bounded continuous functions on S, then I is a rate function and for F closed and G open in $\mathcal{M}(S)$

$$\limsup_n \frac{1}{n} \log P\{\omega : L_n(\omega,\cdot) \in F\} \le - \inf_{\nu \in F} I(\nu)$$

$$\liminf_n \frac{1}{n} \log P\{\omega : L_n(\omega,\cdot) \in G\} \ge - \inf_{\nu \in G} I(\nu) \, .$$

Furthermore,

$$\begin{aligned} I(\nu) &= \int \varphi \log \varphi \, d\mu \, , \text{ if } \nu << \mu \text{ and } \varphi = \frac{d\nu}{d\mu} \, , \\ &= \infty \, , \text{ otherwise.} \end{aligned} \tag{4.6}$$

Remark 4.2. If μ_n is the distribution $(\mu_n \in \mathcal{M}(\mathcal{M}(S)))$ of the $\mathcal{M}(S)$-valued random variable L_n, then the theorem says that $\{\mu_n\}$ satisfies the large deviation principle with rate function I.

A nice proof of Theorem 4.1 can be found in Stroock [A.4]. See also Bahadur and

Zabell [1] and Groeneboom et al [12] for many more results in this direction. The theorem is also a corollary of general results of Donsker and Varadhan.

The Donsker-Varadhan Theory : The problem becomes considerably more complicated even when $S = \mathbb{R}$ if instead of looking at the occupation time measures of the i.i.d. sequence $\{X_n\}$ one considers the occupation time measures of the sequence $\{S_n\}$, where $S_n = X_1 + \dots + X_n$. Since $\{S_n\}$ is a time homogeneous process it is natural to consider a time homogeneous Markov process $\{X_n\}$ and consider the large deviation principle for $L_n = \frac{1}{n}(\delta_{X_1} + \dots + \delta_{X_n})$. The Donsker-Varadhan theory deals with this problem. The most interesting part of this theory is that it is not motivated by technical generalizations but by very interesting and concrete problems in physics and probability. It would obviously be impossible to go into any details of the theory or its applications in less than an hour's time that is available but I will try to give a flavor of both.

Let β_t, $t \geq 0$, be standard real Brownian motion. For $A (\subset \mathbb{R})$ a Borel set, $t > 0$, let

$$L_t(\omega, A) = \frac{1}{t} \int_0^t \chi_A(\beta_s(\omega))\, ds , \tag{4.7}$$

so formally

$$L_t(\omega, \cdot) = \frac{1}{t} \int_0^t \delta_{\beta_s}(\omega)\, ds$$

is the natural analogue of L_n in continuous time. Clearly $L_t(\omega,\cdot) \in \mathcal{M}(\mathbb{R})$ for each $\omega \in \Omega$, $t > 0$. One can check that $L_t(\omega,\cdot) \xrightarrow{v} \theta$, a.s., as $t \to \infty$, in the sense of vague convergence of subprobability measures; here θ denotes the null measure. Let L denote the infinitesimal generator of Brownian motion and for $\nu \in M(\mathbb{R})$ = set of all subprobability measures on $\mathbb{R}$, let

$$I(\nu) = \inf_{\substack{u > 0 \\ u \in D(L)}} \int \left(\frac{Lu}{u}\right)(x)\, d\nu(x) ,$$

where $D(L)$ = domain of L. Donsker and Varadhan (1975) proved the following large deviation principle:

Theorem 4.2. Let F be a closed and G be an open subset of $M(R)$ (topology of vague convergence), then

$$\limsup_{t\to\infty} \frac{1}{t} \log P\{\omega : L_t(\omega,\cdot) \in F\} \leq - \inf_{\nu \in F} I(\nu)$$
$$\liminf_{t\to\infty} \frac{1}{t} \log P\{\omega : L_t(\omega,\cdot) \in G\} \geq - \inf_{\nu \in G} I(\nu) .$$

Furthermore, if $\nu \in \Sigma$, i.e. ν is a probability measure and has a density $f > 0$ on (a,b), where $-\infty \leq a < b \leq \infty$, $f = 0$ outside of $[a,b]$, and f is continuously differentiable on $\mathbb{R}$, then

$$I(\nu) = \frac{1}{8} \int_a^b \frac{(f'(y))^2}{f(y)} dy .$$

In fact, for applications Donsker and Varadhan prove this theorem in stronger forms. In a series of papers [6, I-IV] they prove analogues of such theorems for a large class of Markov processes (both discrete and continuous time). The applications of these results are extensive and the end results are simply beautiful. Here is a brief listing of some of these applications.

(i) In [5] it is shown (cf. Theorems 4.2 and 1.2 above) that if Φ is a nonnegative functional on $\mathcal{M}(\mathbb{R})$ satisfying certain conditions then

$$\lim_{t \to \infty} \frac{1}{t} \log E^x\{e^{-t\,\Phi(L_t(\omega,\cdot))}\} = \inf_{\nu \in \Sigma} \{I(\nu) + \Phi(\nu)\} , \tag{4.9}$$

where Σ is defined in Theorem 4.2 and E^x corresponds to Brownian motion started at x. This can be used to show via Kac's formula that if $V \geq 0$ and continuous on $\mathbb{R}$ and λ_1, is the least eigenvalue of

$$\frac{1}{2} \psi''(y) - V(y)\psi(y) = -\lambda\psi(y)$$

then λ_1 is given by the well known formula

$$\lambda_1 = \inf\{\int V(y)\psi^2(y)dy + \frac{1}{2} \int_{-\infty}^{\infty} (\psi'(y))^2 dy\} , \tag{4.10}$$

where the inf is taken over $\psi \in L^2(\mathbb{R})$ such that $\int \psi^2(y)dy = 1$. In fact, this was the motivation for establishing (4.9), a formula which turns out to have useful applications in many diverse situations.

(ii) Wiener sausage problem ([6,II], [8]). Let β_t be d-dimensional Brownian motion starting at the origin. Let $\varepsilon > 0$ and let

$$C_t = \{x : x = \beta_s \text{ for some } 0 \leq s \leq t\} ,$$

$$C_t^\varepsilon = \bigcup_{x \in C_t} S(x,\varepsilon) = \{x : |x - \beta_s| < \varepsilon \text{ for some } 0 \leq s \leq t\} ,$$

C_t^ε is the sausage of radius ε around C_t. Then there exists $k(d,v) \neq 0$ such that

$$\lim_{t \to \infty} \frac{1}{t^{d/(d+2)}} \log E\, e^{-v\, \mathrm{vol}(C_t^\varepsilon)} = k(d,v) .$$

A discrete analogue of this is given in [9]. The problem arose in a certain context in physics.

(iii) The Plaron problem ([6,IV], [10]). Let β_s be three dimensional tied down Brownian motion, $0 \leq s \leq t$, then

$$\lim_{t\to\infty} \frac{1}{t} \log E \exp\{\alpha \int_0^t \int_0^t \frac{e^{-|u-v|}}{|\beta_u - \beta_v|} du dv\} = g(\alpha)$$

and

$$\lim_{\alpha\to\infty} \frac{g(\alpha)}{\alpha^2} = \sup_{\substack{\varphi \in L_2(\mathbb{R}^d) \\ \|\varphi\|_2 = 1}} \{2 \iint \frac{\varphi^2(x)\varphi^2(y)}{|x-y|} dx dy - \frac{1}{2} \int |\nabla\varphi|^2 dx\} .$$

This was a conjecture of Pekar and the problem arose in statistical mechanics.

(iv) Local times of stable processes ([6,III], [7]). Let X_t, $t \geq 0$, be a symmetric stable process of index α, $0 < \alpha \leq 2$. Let L be the infinitesimal generator of the process and for $\nu \in M(\mathbb{R})$ let

$$I(\nu) = \inf_{\substack{u>0 \\ u\in D(L)}} \int (\frac{Lu}{u})(x) d\nu(x) \; ; \; C = \{\nu \in M(\mathbb{R}) : I(\nu) \leq 1\} . \tag{4.11}$$

For A a Borel subset of $\mathbb{R}$, let

$$\hat{L}_t(\omega, A) = \frac{1}{t} \int_0^t \chi_A\{X_s(\omega) (\frac{\log \log t}{t})^{1/\alpha}\} ds .$$

Then a.s. (closure in terms of the topology of vague convergence)

$$\bigcap_T \overline{\bigcup_{t\geq T} \{\hat{L}_t(\omega, \cdot + x) : x \in \mathbb{R}\}} \subset C$$

and

$$\bigcap_T \overline{\bigcup_{t\geq T} \{\hat{L}_t(\omega, \cdot)\}} \supset C .$$

This result is used to derive many interesting facts about the "small" values of the process X_s, e.g. the lim inf behavior of $\sup_{s\leq t} |X_s|$. The remarkable thing about this approach is that one always gets formulas for the limit constants which were not available before.

If $\alpha > 1$ then let ℓ_t denote the local time of the process, i.e.

$$\int_0^t \chi_A(X_s(\omega)) ds = \int_A \ell_t(\omega, x) dx .$$

Let

$$\hat{\ell}_t(\omega,y) = \rho_t \ell_t(\omega,\rho_t y) \ , \ \rho_t = (t/\log\log t)^{1/\alpha} \ .$$

Let

$$\mathcal{G} = \{f : f \geq 0 \ , \int_{-\infty}^{\infty} f(y)dy \leq 1 \ , \ f \text{ unif. cont. on } \mathbb{R}\} \ . \tag{4.12}$$

Then a.s. (closure in terms of the topology of uniform convergence on compacts)

$$\bigcap_T \overline{\bigcup_{t \leq T} \{\hat{\ell}_t(\omega,\cdot+x) : x \in \mathbb{R}\}} \subset \{f \in \mathcal{G} : I(fdx) \leq 1\}$$

and

$$\bigcap_T \overline{\bigcup_{t \leq T} \{\hat{\ell}_t(\omega,\cdot)\}} \supset \{f \in \mathcal{G} : I(fdx) \leq 1\} \ .$$

This result leads to a number of very interesting asymptotic facts about the large values of the local time and one always gets formulas for the limit constants. For instance, if $E\, e^{iuX_t} = e^{-|t|^{\alpha}}$, then a.s.

$$\limsup_{t\to\infty} \left(\frac{t}{\log\log t}\right)^{1/\alpha} \frac{1}{t} \ell_t(\omega,x) = \limsup_{t\to\infty} \left(\frac{t}{\log\log t}\right)^{1/\alpha} \frac{1}{t} \sup_z \ell_t(\omega,z) = d_\alpha \ ,$$

where

$$d_\alpha = \{\Gamma(1/\alpha)\Gamma(1-1/\alpha)\} / \pi(\alpha-1)^{1-1/\alpha} \ .$$

For $\alpha = 2$ this contains an earlier result of Kesten [17]. Of course one also gets a number of new results.

These results and their applications have inspired a great deal of work by other authors. In the context of real-valued i.i.d. random variables $\{X_n\}$ Strassen's results [24] concerning the large values of $S_n = X_1 + \ldots + X_n$ when $EX_1^2 < \infty$ are well known. It was also known that such results could not be expected to hold if $EX_1^2 = \infty$. However, some results concerning the small values of $\{S_n\}$ were known [14], [15], but there really was no counterpart of Strassen's invariance principle [24]. Using the machinery of Donsker and Varadhan an invariance principle is established in [13] which gives precise counterparts of Strassen's results for small values of $\{S_n\}$ when X_1 is in the domain of attraction of a stable law. To describe this approach briefly, let H be the common distribution which is in the domain of attraction of stable law $\hat{H}$ of index α. Assume $EX_1 = 0$ when it exists and that

$$\frac{S_n}{a(n)} \overset{w}{\to} \hat{H} \ .$$

Then $a(n) = n^{1/\alpha} \ell(n)$, where ℓ is a slowly varying function. Suppose $\hat{H}$ has an everywhere positive density. Let

$$L_n(\omega,A) = \frac{1}{n} \sum_{j=0}^{n-1} \chi_A(S_j(\omega) / a(n /\log\log n)) .$$

Then a large deviation principle for $L_n(\omega,\cdot)$ is proved in [13] for which the rate function is the same as that for the limit stable process. Using this large deviation principle it is shown that the set of vague limit points of $\{L_n(\omega,\cdot) : n \geq 1\}$ is a.s. C given in (4.11). Some sample applications are the following: (a) Let $A_n = \max_{j \leq n} |S_j|$; $c(n) = a(n/\log\log n)$. Then

$$\liminf_n \frac{A_n}{c_n} = c_1 \text{ , a.s.,} \tag{4.13}$$

where $0 < c_1 < \infty$ and depends only on $\hat{H}$. It is determined by I. This contains the result of [15]. (b) For $c > 0$, there exists k_c (depending on $\hat{H}$) such that

$$\limsup_n \frac{1}{n} \sum_{j=0}^{n-1} \chi_{[0,c]}(|S_j| / c(n)) = k_c \text{ , a.s.,} \tag{4.14}$$

and for $c \geq c_1$, $k_c = 1$, where c_1 is given in (4.12). (c) For $a > 0$ there exists A_a (depending on $\hat{H}$) such that

$$\liminf_n \frac{1}{n} \sum_{j=0}^{n-1} (|S_j| / c(n))^a = A_a \text{ , a.s.,}$$

where $0 < A_a \leq c_1^a$.

There is also an invariance principle [16] for the local time of the random walk (S_n) . Let $\alpha > 1$ and $EX_1 = 0$. Assume X_1 to take values in the integer lattice. Let

$$L_n(\omega,k) = \sum_{j=0}^{n-1} \chi_{\{k\}}(S_j) \text{ , } k \in \mathbb{Z} ,$$

and

$$g_n(\omega,u) = \frac{c(n)}{n} L_n(\omega,k) \text{ , } u = \frac{k}{c(n)}$$

$$= \text{linear in between.}$$

Then the invariance principle says that a.s. the set $\{g_n(\omega,\cdot) : n \geq 1\}$ is precompact (topology of uniform convergence on compacts) in $\mathcal{G}$, where $\mathcal{G}$ is defined in (4.12), and its set of limit points equals $\mathcal{G} \cap \{f : I(f dx) \leq 1\}$. An application of this is the following: let $\hat{H}$ have characteristic function $e^{-|u|^\alpha}$, $1 < \alpha \leq 2$. Then

$$\limsup_n \frac{c(n)}{n} L_n(\omega,k) = \limsup_n \frac{c(n)}{n} \sup_k L_n(\omega,k) = d_\alpha \text{ , a.s.,} \tag{4.16}$$

where d_α is defined earlier. As another application, let $\{k_n\} \subset \mathbb{Z}$, $k_n/c(n) \to a \in \mathbb{R}$, $\varphi: \mathbb{R} \to \mathbb{R}$ continuous. Then

$$\limsup_n \varphi(\frac{c(n)}{n} L_n(\omega,k_n)) = \sup_{0 \le t \le d_\alpha} \varphi(t) \text{ .} \tag{4.17}$$

For other applications one can refer to the papers mentioned above.

Before closing I would like to mention some very interesting results obtained by T.-S. Chiang [3],[4], C. Mueller [20], and S. Orey [21] listed in the references. This by no means is an exhaustive list; some of the references will provide further references to the literature on large deviations.

References

A. Books, survey articles, and lecture notes

[A.1] R. Azencott: Grandes deviations et applications, Lecture Notes in Math. 774, Springer-Verlag, New York, 1-249 (1980).

[A.2] R. Ellis: Entropy, Large Deviations, and Statistical Mechanics, Springer-Verlag, New York (1984).

[A.3] M.I. Freidlin and A.D. Wentzell: Random Perturbations of Dynamical Systems, Springer-Verlag, New York,(1984).

[A.4] D. Stroock: An Introduction to the Theory of Large Deviations, Springer-Verlag, New York (1984).

[A.5] S.R.S. Varadhan: Large deviations and applications, C.B.M.S. Lectures (1982).

B. Papers

[1] R.R. Bahadur and S.L. Zabell: Large deviations of the sample mean in general vector spaces, Ann. Prob. 7, 537-621 (1979).

[2] H. Cramér: On a new limit theorem in the theory of probability:colloquium on the Theorey of Probability, Hermann, Paris (1937).

[3] T.-S. Chiang: Large deviations of some Markov processes on compact metric spaces, Z. Wahr. Verw. Geb. 61, 271-281 (1982).

[4] T.-S. Chiang: A lower bound of the asymptotic behavior of some Markov processes, Ann. Prob. 10, 955-967 (1982).

[5] M.D. Donsker and S.R.S. Varadhan: Asymptotic evaluation of certain Wiener integrals for large time, Functional Integration and its Applications (Ed. A.M. Arthur), Oxford Press, 15-33 (1975).

[6] M.D. Donsker and S.R.S. Varadhan: Asymptotic evaluation of certain Markov expectations for large time, I,II,III,IV, Communications in Pure and Applied Math. 28 (1975), 1-47; 28 (1975), 279-301; 29(1976), 389-461; 36 (1983), 525-565.

[7] M.D. Donsker and S.R.S. Varadhan: On laws of the iterated logarithm for local times, Comm. Pure and Applied Math. 30, 707-753 (1977).

[8] M.D. Donsker and S.R.S. Varadhan: Asymptotics for the Wiener sausage, Comm. Pure and Applied Math. 28, 525-565 (1975).

[9] M.D. Donsker and S.R.S. Varadhan: On the number of distinct sites visited by a random walk, Comm. Pure and Applied Math. 32, 721-747 (1979).

[10] M.D. Donsker and S.R.S. Varadhan: Asymptotics for the Polaron, Comm. Pure and Applied Math. 36, 505-528 (1983).

[11] X. Fernique: Integrabilite des processus gaussiens, C.R. Acad. Sci. Paris Ser. A 270, 1698-1699 (1970).

[12] P. Groeneboom, J. Oosterhoff and F.H. Ruymgaart: Large deviation theorems for empirical probability measures, Ann. Prob. 7, 553-586 (1979).

[13] N.C. Jain: A Donsker-Varadhan type of invariance principle, Z. Wahr. Verw. Geb. 59, 117-138 (1982).

[14] N.C. Jain and W.E. Pruitt: Maxima of partial sums of independent random variables, Z. Wahr. Verw. Geb. 27,141-151 (1973).

[15] N.C. Jain and W.E. Pruitt: The other law of the iterated logarithm, Ann. Prob. 3, 1046-1049 (1975).

[16] N.C. Jain and W.E. Pruitt: An invariance principle for the local time of a recurrent random walk, Z. Wahr. Verw. Geb. 66, 141-156(1984).

[17] H. Kesten: An iterated logarithm law for local time, Duke Math. J. 32 (3) 447-456 (1965).

[18] O.E. Lanford: Entropy and equilibrium states in classical statistical mechanics, Statistical Mechanics and Mathematical Problems, Lectures Notes in Physics, Vol. 20, Springer-Verlag, New York, 1-113 (1971).

[19] M.B. Marcus and L.A. Shepp: Sample behavior of Gaussian processes, Proc. Sixth Berkeley Symp. Math. Statist. Prob. 2, 423-442 (1971).

[20] C. Mueller: Strassen's law for local time, Z. Wahr. Verw. Geb. 63, 29-41 (1983).

[21] S. Orey: Large deviations in ergodic theory, Symposium on Stochastic Processes, Evanston (1985). (Ed. E. Cinlar). To appear.

[22] M. Pincus: Gaussian processes and Hammerstein integral equations, Trans. Amer. Math. Soc. 134, 193-216 (1968).

[23] M. Schilder: Some asymptotic formulae for Wiener integrals, Trans. Amer. Math. Soc. 125, 63-85 (1966).

[24] V. Strassen: An invariance principle for the law of the iterated logarithm, Z. Wahr. Verw. Geb. 3, 211-226 (1964).

[25] S.R.S. Varadhan: Asymptotic probabilities and differential equations, Comm. Pure and Applied Math. 19, 261-286 (1966).

RANDOM INTEGRAL REPRESENTATION FOR ANOTHER CLASS OF LIMIT LAWS

Zbigniew J. Jurek*
Department of Mathematics, Tufts University
Medford, Mass. 02155, U.S.A.

In the theory of limit distributions for sequences of independent random variables there are two basic classes of limit laws: stable laws and selfdecomposable laws (so called class L_0). The first class and its generalizations were intensively investigated during the last fifty years: characterizations in terms of characteristic functions, domains of attraction for stable measures, stable processes, stable random measures, integrals with respect to stable random measures, Banach spaces of stable type, stability with respect to groups; cf. for references Weron (1984).

For the second class, class L_0, the Lévy characterization was determined around 1937; cf. Lévy (1937). Thirty years later, Urbanik (1968) described the class L_0 in the terms of characteristic functions using Choquet's Theorem on extreme points. Another proof is given in Jurek (1982b). Kubik (1962) found elements that are dense in L_0 upon taking convolutions and weak limits. Finally, Yamazato (1978) solved the unimodality problem. This problem seemed to be the only class L_0 problem that attracted much attention during the last forty years. Recently the class L_0 and its generalizations have received more interest. Wolfe (1982) and Jurek-Vervaat (1983) proved that the class L_0 is an image of the class, $ID_{\log}$, of all infinitely divisible measures with finite logarithmic moments, by some random integral mapping I. In fact, I is an isomorphism between L_0 and $ID_{\log}$. The only fixed points of I are the stable measures. The topological properties of I and I^{-1} are given in Sato-Yamazato (1984) and Jurek-Rosinski (1984). As a consequence one find the generators for L_0 from the generators for the class ID of infinitely divisible measures. Further results connected with the class L_0 can be found in the survey article of Yamazato (1984).

The class, S, of stable measures is properly contained in the class L_0. Urbanik (1973) introduced decreasing subclasses L_m of L_0, for $m = 1, 2, \ldots$, such that

$$L_0 \supset L_1 \supset L_2 \supset \ldots \supset L_\infty \supset S, \qquad (0.1)$$

*) on a leave from the University of Wroclaw, Poland.

where $L_\infty := \bigcap_{m>1} L_m$. This approach and the results were extended by Kumar-Schreiber (1979), Sato (1980) and Jurek (1983a) and (1983b). The definitions of the classes L_m, $m = 0, 1, 2, \ldots$ always contains the infinitezimality assumption. Hence all of them are contained in the class ID. The aim of this note is to introduce a class $\mathcal{U}$ such that

$$ID \supset \mathcal{U} \supset L_0, \tag{0.2}$$

thereby extending the sequence of inclusions in (0.1) in the opposite direction. In fact, a family of classes $\mathcal{U}(Q)$ depending on a bounded linear operator Q is defined. The relation between $\mathcal{U}(Q)$ and $L_0(Q)$, and also the generators for the class $\mathcal{U}(Q)$ are examined. Basicly the results extend those of Jurek (1984), where $Q = I$ and the class $\mathcal{U}$ was defined by some non-linear transformations. This is also an example satisfying the hypothesis of Jurek (1984), Section 5. However, here the class $\mathcal{U}(Q)$ is defined as distributions of some random integrals and then identified as a class of limit distributions.

In the last subsection an example associated with the problem of the polar coordinates for an arbitrary one-parameter strongly continuous group of linear operators on a Banach space is given.

1. Let E be a real separable Banach space with a norm $\|\cdot\|$. Let Q be a fixed linear bounded operator on E such that

$$\lim_{t\to\infty} e^{-tQ} = 0 \quad \text{(in the operator topology)}. \tag{1.1}$$

Hence there are $\alpha, \beta \in \mathbb{R}^+$ (positive real numbers) such that

$$\|e^{-tQ}\| \leq \alpha\, e^{-\beta t} \quad \text{for all } t \in \mathbb{R}^+. \tag{1.2}$$

We will say that E-valued random variable (rv) ξ is <u>exp(tQ)-decomposable</u> if for each $t \in \mathbb{R}^+$ there is E-valued rv ξ_t independent of ξ (in symbols: $\xi \perp\!\!\!\perp \xi_t$) such that

$$\xi \overset{d}{=} e^{-tQ}\xi + \xi_t,$$

where $\overset{d}{=}$ means equality in distribution. So, in terms of probability measures on E, we will say that μ is <u>exp(-tQ)-decomposable</u> if for each $t \in \mathbb{R}^+$ there is $\mu_t \in \mathcal{P}(E)$, $\mathcal{P} = \mathcal{P}(E)$ is the set of all probability measures on E, such that

$$\mu = e^{-tQ}\mu * \mu_t, \tag{1.3}$$

where $e^{-tQ}\mu$ is the image of μ by the mapping (operator) e^{-tQ}. Let $L_0(Q)$ be the set of all μ's satisfying (1.3) for all $t \in \mathbb{R}^+$. Note that for $Q = I$-identity operator, $L_0(Q)$ reduces to the Lévy class L_0. Furthermore, $L_0(Q)$ for some Q satisfying (1.1), is the class of full limit distributions of the random variables

$$A_n(\xi_1 + \xi_2 + \cdots + \xi_n) + a_n,$$

where ξ_n are E-valued rv's, A_n are invertible bounded linear operations on E such that

(1) the semigroup generated by $A_n A_m^{-1}$ $(1 \leq m \leq n,\ n = 1,2,\ldots)$ is compact;

(2) the triangular array $A_n \xi_j$, $1 \leq j \leq n$, $n = 1,2,\ldots$ is uniformly infintesimal

cf. Urbanik (1978), Theorem 4.1. See also: Jurek (1983a), Theorem 3.1.

Let $D_E[a,b]$ be a set of E-valued functions, that are right-continuous on $[a,b)$ and have left-hand limits on $(a,b]$, endowed with the Skorohod topology; cf. Billingsley (1968). For an operator A and continuously differentiable real-valued function g, and $D_E[a,b)$-valued rv Y let us put

$$\int_{(a,b]} e^{g(t)A}\, dY(t) := e^{g(t)A}\, Y(t)\Big|_{t=a}^{t=b} - \int_{(a,b]} d\, e^{g(t)A}\, Y(t) =$$
$$= e^{g(t)A}\, Y(t)\Big|_{t=a}^{t=b} - \int_{(a,b]} g'(t) A e^{g(t)A}\, Y(t)\, dt, \tag{1.4}$$

The integral on the right-hand side there exists path-wise, cf. Billingsley (1968), Lemma 14.1. The integrals on $[a,\infty)$ we define as the limit in distribution as $b \to \infty$ in (1.4). In particular, for Q satisfying (1.1) we have

1.1. <u>THEOREM</u>. (Jurek (1983b) or Jacod (1982)). Let Y be a $D_E[0,\infty)$-valued rv with stationary independent increments and such that $Y(0) = 0$ a.s. Then

$$\int_{(0,\infty)} e^{-tQ} dY(t) \quad \text{exists iff} \quad \mathscr{L}(Y(1)) \in ID_{\log},$$

where $\mathscr{L}(\xi)$ denotes the probability distribution of rv ξ, and $ID_{\log}$ is the set of all infinitely divisible measures with finite logarithmic moment.

The above suggest to define the following mapping $I_Q: ID_{\log} \to ID$ by means of random integral

$$I_Q(\nu) := \mathscr{L}\Big(\int_{(0,\infty)} e^{-tQ} dY(t)\Big), \tag{1.5}$$

where Y is $D_E[0,\infty)$-valued rv with stationary independent increments, $Y(0) = 0$ a.s. and $\mathscr{L}(Y(1)) = \nu$. Then we have

1.2 <u>THEOREM</u> (Jurek (1983b) and Jurek-Rosinski (1984)). The mapping I_Q is an algebraic isomorphism between the semigroups $ID_{\log}$ and $L_0(Q)$, i.e. $I_Q : ID_{\log} \to L_0(Q)$ is one-to-one, onto and $I_Q(\nu * \mu) = I_Q(\nu) * I_Q(\mu)$. Moreover, for $\nu_n, \nu \in ID_{\log}$ $I_Q(\nu_n) \Rightarrow I_Q(\nu)$ iff $\nu_n \Rightarrow \nu$ and $\int_E \log(1 + \|x\|)\nu_n(dx) \Rightarrow \int_E \log(1 + \|x\|)\nu(d$

2. Now with above motivation for the study of distributions of random integrals, we introduce the class $U(Q)$ as follows:

$$U(Q) := \{\mathscr{L}(\int_{(0,1]} t^Q dY(t)) : Y \text{ is } D_E[0,1]\text{-valued rv with stationary independent increments, } Y(0) = 0 \text{ a.s.}\} \tag{2.1}$$

Note that (1.4) and (1.2) guarantee the existence of that above integrals. Let us define $J_Q : ID \to U(Q)$ as follows

$$J_Q(\nu) := \mathscr{L}(\int_{(0,1]} t^Q dY(t)) \tag{2.2}$$

where Y is as in the definition of the class $U(Q)$ and $\mathscr{L}(Y(1)) = \nu$. Since in (1.4) the integral on the right-hand side we mean as the Stieltjes-Riemann integral, therefore the left-hand side integral one can approximate by the following partial sums

$$\sum_{j=0}^{n} e^{g(t_j)A}[Y(t_{j+1}) - Y(t_j)] \quad a := t_0 < t_1 < \cdots < t_n := b.$$

Consequently, we get

$$(J_Q(\nu))\hat{}(x^*) = \exp[\int_{(0,1]} \log \hat{\nu}(t^{Q^*}x^*)\, dt], \text{ for } x^* \in E^* \tag{2.3}$$

Suppose that ν has the characteristic functional $\hat{\nu}$ of the form

$$\hat{\nu}(x^*) = \exp\{i\langle x^*,a\rangle - \frac{1}{2}\langle Rx^*,x^*\rangle + \int_{E\setminus\{0\}} [e^{i\langle x^*,x\rangle} - 1 - i\langle x^*,x\rangle 1_B(x)]M(dx)\},$$

where $a \in E$, R is Gaussian covariance operator, M is Lévy measure, 1_B is the indicator of the unit ball in E, and $\langle\cdot,\cdot\rangle$ is a bilinear form between E^* (the dual of E) and E; cf. Araujo-Giné (1980) Theorem 6.2. Then we will write $\nu = [a, R, M]$ or $\nu = \delta(a) * \gamma * c_1$ Poiss (M), where γ is a symmetric Gaussian measure with the covariance operator R. From (2.3) we get that if $\nu = [a, R, M]$ and $J_Q(\nu) =: [a', R', M']$ then

$$a' := \int_0^1 t^Q a\,dt + \int_0^1 \int_{E\setminus\{0\}} t^Q x\,[1_B(t^Q x) - 1_B(x)]M(dx)dt \tag{2.4}$$

$$= (Q+I)^{-1}a + \int_{E\setminus\{0\}} \int_0^1 t^Q x[1_B(t^Q x) - 1_B(x)]dt\, M(dx),$$

$$R' := \int_0^1 t^Q R t^{Q^*}\,dt, \tag{2.5}$$

$$M'(F) := \int_0^1 (t^Q M)(F)\,dt, \quad \text{for all} \quad F \in \mathcal{B}_0 := \mathcal{B}(E\setminus\{0\}). \tag{2.6}$$

Now we are going to establish the algebraic and topological relationship between the sets ID and $U(Q)$.

2.1. THEOREM. The mapping J_Q is a homeomorphism between the topological semigroups ID and $U(Q)$. Moreover, we have

(i) $J_Q(\nu^{*c}) = (J_Q(\nu))^{*c}$ for all $c \in \mathbb{R}^+$:

(ii) $VJ_Q(\nu) = J_Q(V\nu)$ for any bounded linear operator V on E, commuting with Q.

Proof. We prove only that

$$J_Q(\nu_n) \Rightarrow J_Q(\nu) \quad \text{implies } \nu_n \Rightarrow \nu,$$

because the rest of the proof is the same as the proof of Theorem 2.7 in Jurek (1984). To this end, let us choose $D_E[0,\infty)$-valued rv's Y_n, Y with stationary independent increments, $Y_n(0) = 0$ a.s, $Y(0) = 0$ a.s, and $\mathcal{L}(Y_n(1) = \nu_n$, $\mathcal{L}(Y(1) = \nu$, and let us define

$$Z_n(t) := \int_{(0,t]} r^Q\,dY_n(r), Z(t) := \int_{(0,t]} r^Q\,dY(r), \quad \text{for } t \geq 0.$$

Then, $Z_n(\cdot)$ are in $D_E[0,\infty)$, have independent increments and $J_Q(\nu_n) = \mathcal{L}(Z_n(1)) \Rightarrow J_Q(\nu) = \mathcal{L}(Z(1)$. Using (2.3) we get

$$\mathcal{L}(Z_n(t) = \mathcal{L}(\int_{(0,1]} t^Q s^Q\,dY_n(s\cdot t)) = t^Q\,\mathcal{L}(Z_n(1))^{*t} \quad \text{for } t \in \mathbb{R}^+.$$

Consequently, Theorem 1.2 in Jurek (1984) gives

$$\mathcal{L}(Z_n(t_n)) \Rightarrow \mathcal{L}(Z(t)) \quad \text{when } t_n \to t \text{ in } \mathbb{R}^+.$$

Hence, for $t_n > s_n$ and $t_n \to t$, $s_n \to s$ we get

$$\mathcal{L}(Z_n(t_n) - Z_n(s_n)) \Rightarrow \mathcal{L}(Z(t) - X(s)) \quad \text{as } n \to \infty.$$

So, the finite-dimensional distributions of Z_n converge to the ones of Z. Finally this gives

$$\lim_{h\downarrow 0} \limsup_{n\to\infty} \sup_{0\le s<t<s+h\le a} P\{\|Z_n(t) - Z_n(s)\| > \varepsilon\} = 0.$$

Therefore Theorem VI.5.5 in Gihman-Skorohod (1975) with Lindvall (1973) imply $\mathscr{L}(Z_n) \Rightarrow \mathscr{L}(Z)$ in $D_E[0,\infty)$. Applying Continuous Mapping Theorem similarly as in Jurek (1984) we obtain

$$Y_n(1) \stackrel{d}{=} Y_n(2) - Y_n(1) = \int_{(1,2]} r^{-Q}\, dZ_n(r) \Rightarrow \int_{(1,2]} r^{-Q}\, dZ(r) \stackrel{d}{=} Y(1),$$

which completes the proof.

2.2 <u>COROLLARY</u>. $\mathcal{U}(Q)$ is a closed subsemigroup of the semigroup ID.

Next theorem gives some equivalent characterizations of the class $\mathcal{U}(Q)$.

2.3. <u>THEOREM</u>. The following are equivalent:

(i) $\nu \in \mathcal{U}(Q)$.

(ii) $\nu \in$ ID and for every $0<c<1$ there is $\nu_c \in$ ID such that $\nu = c^Q\nu^{*c} * \nu_c$.

(iii) $\nu = [a,R,M]$ and for every $0<c<1$ $R \ge c\,(c^Q R c^{Q*})$ and $M(\cdot) \ge c \cdot (c^Q M)(\cdot)$.

(iv) $\nu = [a,R,M]$ and there is a unique Gaussian covariance operator S and a unique Lévy measure N such that $R = \int_0^1 t^Q S t^{Q*}\, dt$, $M(\cdot) = \int_0^1 t^Q N)(\cdot)\, dt$.

(v) There exists a sequence $(\nu_n) \subseteq$ ID such that $n^{-Q}(\nu_1^{*1} * \nu_2^{*2} * \ldots * \nu_n^{*n})^{*1/n} \Rightarrow \nu$.

<u>Proof</u>. (i) $\Rightarrow$ (ii). For any $0<c<1$ we have

$$\int_{(0,1]} t^Q dY(t) = \int_{(0,c]} t^Q dY(t) + \int_{(c,1]} t^Q dY(t) \stackrel{d}{=} c^Q \int_{(0,1]} t^Q dY(ct) + \int_{(c,1]} t^Q dY(t$$

which gives (ii).

(ii) $\Leftrightarrow$ (iii). It is a consequence of the Lévy-Khintchine representatives of $\nu \in$ ID and the properties of Gaussian covariance operator and the Lévy measures; cf. Jurek (1983b), p. 52.

(iii) $\Rightarrow$ (iv). At first let us note that $R \ge c(c^Q R c^{Q*})$ for $0<c<1$ iff $R \ge e^{-tQ_1} R e^{-tQ_1^*}$ for $0<t<\infty$, where $Q_1 := Q + \frac{1}{2}I$. Consequently from Corollary 2.5 ((c) $\Leftrightarrow$(d) for m = 0) in Jurek (1983b), we get

$$R = \int_0^\infty e^{-tQ_1} S e^{-tQ_1^*} dt = \int_0^1 t^Q S t^{Q^*} dt$$

for a unique Gaussian covariance operator S.

For the second part we need so-called Q-Lévy spectral functions associated with M, cf. Jurek (1982a)

$$L_M(A;r) := -M(\{t^Q u : u \in A \text{ and } t > r\}),$$

where $r \in \mathbb{R}^+$, $A \in \mathcal{B}(S_Q)$, $S_Q := \{x \in E : \|x\|_Q = 1\}$ and $\|\cdot\|_Q$ is an equivalent norm on E given by $\|x\|_Q := \int_0^\infty \|e^{-tQ}x\| dt$, cf. Jurek (1983c). Then it is easy to see that the functions $\mathbb{R}^+ \ni r \mapsto L_M(A;r)$ are concave for every $A \in \mathcal{B}(S_Q)$. Reasoning similarly as in the proof of Theorem 2.2 in Jurek (1984) we get that a Lévy measure M has a form

$$M(F) := \int_{(0,1]} (t^Q N)(F), \quad \text{for } F \in \mathcal{B}_0,$$

where N is some Borel measure on $\mathcal{B}_0$. In fact N is uniquely determined Lévy measure, because the proof of Theorem 1.3 in Jurek (1984) holds true for arbitrary Q not necessarily Q = I.

(iv) $\Rightarrow$ (i). Putting $\mu := [b,S,N]$, where S, N are as in (iv) and

$$b := (Q+I)\{a - \int_{E\setminus\{0\}} \int_0^1 t^Q x[1_B(t^Q x) - 1_B(x)]dt \, N(dx)\}$$

we obtain $\nu = [a,R,M] = J_Q([b,S,N])$, because of (2.4)-(2.6).

(ii) $\Leftrightarrow$ (v). Note that the proof of Theorem 2.5 in Jurek (1984) for Q = I holds true for arbitrary Q.

2.4. <u>COROLLARY</u>. $L_0(Q) \subseteq \mathcal{U}(Q)$.

<u>Proof</u>. Let $\mu \in L_0(Q)$. Then, by (1.3), for all $0<s<1$ there is $\mu_s \in ID$ such that

$$\mu = s^Q\mu * \mu_s = s^Q(\mu^{*s} * \mu^{*1-s}) * \mu_s = s^Q\mu^{*s} * (s^Q\mu^{*1-s} * \mu_s).$$

From Theorem 2.3 ((i) $\Leftrightarrow$(ii)) we get that $\mu \in \mathcal{U}(Q)$, which completes the proof.

Since we have the inclusion $L_0(Q) = \mathcal{I}_Q(ID_{log}) \subseteq J_Q(ID)$ one can naturally ask, when $J_Q(\nu) \in L_0(Q)$. Using the Q-Lévy spectral function L_M and the polar coordinates system given by the norm $\|\cdot\|_Q$, the proof for the case $Q = I$ from Jurek (1984) one can easily extend for arbitrary Q and we get the following theorem.

2.5. THEOREM. $J_Q(\nu) \in L_0(Q)$ for some $\nu \in ID$ iff there is a unique $\mu \in ID_{\log}$ such that $\nu = J_Q(\mu) * \mu$.

Let us recall that an infinitely divisible measure μ is called Q-stable if for each $t > 0$ there is a vector $b(t) \in E$ such that

$$\mu^{*t} = t^Q\mu * \delta(b(t)). \tag{2.7}$$

Hence it is easy to see that Q-stable measures are in $L_0(Q)$, cf. (1.3), and therefore also in $\mathcal{U}(Q)$. Furthermore, if $\mu = [a,S,N]$ then μ is Q-stable (i.e. satisfies (2.7)) if and only if

$$S = QS + SQ^*, \quad N(F) = \int_{S_Q}\int_0^\infty 1_F(t^Q x)t^{-2}\, dt\, m(dx), \tag{2.8}$$

where $S_Q := \{x : \|x\|_Q = 1\}$, $F \in \mathcal{B}_0$, and m is finite Borel measure on S_Q. Let us note that for S and N as in (2.8) the formulas (2.5 and (2.6) imply

$$S' = \int_0^1 t^Q(QS + SQ^*)t^{Q^*}dt = \int_0^1 (dt^Q/dt)St^{Q^*+I}\, dt + \int_0^1 t^Q SQ^* t^{Q^*}\, dt$$

$$= t^Q S t^{Q^*+I}\Big|_{t=0}^{t=1} - \int_0^1 t^Q S(Q^*+I)t^{Q^*}\, dt + \int_0^1 t^Q SQ^* t^{Q^*}\, dt = S - S',$$

$$N'(F) = \int_0^1 \left[\int_{S_Q}\int_0^\infty 1_F((tr)^Q x)r^{-2}\, dr\, m(dx)\right]dt = \frac{1}{2}N(F).$$

Consequently for μ Q-stable we have $J_Q(\mu) = \mu^{*\frac{1}{2}} * \delta(z)$ for some $z \in E$. Conversely we have the following

2.6. THEOREM. If $J_Q\nu = \nu^{*c} * \delta(z)$ for some $c > 0$ and $z \in E$ then $0 < c < 1$ and is $(\frac{c}{1-c})Q$-stable.

Proof. Suppose that $\nu = [a,R,M]$. Then (2.5) and (2.6) give equations

$$cR = \int_0^1 t^Q R t^{Q^*}\, dt \quad \text{and} \quad c\cdot M(\cdot) = \int_0^1 M(t^{-Q}\cdot)\, dt \tag{2.9}$$

Hence, integrating by parts, we get

$$(cQ)R + R(cQ)^* = \int_0^1 \frac{d}{dt}(t^Q)Rt^{Q^*+I}\,dt + \int_0^1 t^Q R t^{Q^*} Q^*\,dt$$

$$= R - \int_0^1 t^Q R(Q^*+I)t^{Q^*}\,dt + \int_0^1 t^Q R t^{Q^*} Q^*\,dt = R - cR = (1-c)R.$$

Expressing the second equality in (2.9) in terms of the Q-Lévy spectral function we obtain the following equation

$$r \int_r^\infty L_M(A;s)/s^2\,ds = c\,L_M(A;r).$$

So, similarly as in the case $Q = I$, we get

$$M(\{t^Q x : x \in A,\ t > r\}) = \gamma(A) r^{-(1-c)/c} \quad \text{for} \quad r \in \mathbb{R}^+ \text{ and } A \in \mathcal{B}(S_Q).$$

Consequently $0<c<1$ and for $\alpha := c/(1-c)$ we have

$$M(F) = \alpha^{-1} \int_{S_Q} \int_0^\infty 1_F(t^Q x) t^{-c^{-1}}\,dt\,\gamma(dx)$$

$$= \int_{S_Q} \int_0^\infty 1_F(s^{\alpha Q} x) s^{-2}\,ds\,\gamma(dx), \quad \text{for } F \in \mathcal{B}_0.$$

This with (2.8) gives that ν is $\frac{c}{1-c}$ Q-stable, which completes the proof.

Now we are going to exploit the continuity of the mapping J_Q (Theorem 2.1) to describe the "generators" of the class $\mathcal{U}(Q)$, i.e. some simple elements form $\mathcal{U}(Q)$ such that their finite convolutions and weak limits give the whole class $\mathcal{U}(Q)$. At first we need the following fact:

2.7. PROPOSITION. A measure $[0,R,0] \in \mathcal{U}(Q)$, or equivalently $R \geq c(c^Q R c^{Q^*})$ for all $0 \leq c \leq 1$ if $QR + RQ^* + R$ is nonnegative operator.

Proof. Note that $R \geq c(c^Q R c^{Q^*})$ for all $0 \leq c \leq 1$ is equivalent to $R \geq e^{-tQ_1} R e^{-tQ_1^*}$ for all $t \geq 0$, where $Q_1 := Q + \frac{1}{2}I$. So, Corollary 2.5 ((a)$\Leftrightarrow$(c) for $m = 0$) in Jurek (1983b) implies that $Q_1 R + RQ_1^* \geq 0$, i.e. $QR + RQ^* + R \geq 0$.

Now we need to know how the measures $c\delta(a)$, $c \in \mathbb{R}^+$, $0 \neq a \in E$ or $[0,0,c\delta(a)]$ are transformed by the integral mapping J_Q. Let $F := \{t^Q u : u \in A,\ t \in B\}$, where $A \in \mathcal{B}(S_Q)$, $B \in \mathcal{B}(\mathbb{R}^+)$, and S_Q is the unit sphere in E, is the new norm $\|\cdot\|_Q$, cf. the proof of the Theorem 2.3. Then (2.6) gives

$$(c\delta(r^Q v))'(F) = c \int_0^1 \delta(r^Q v)(\{t^Q u : u \in A,\ t \in s^{-1}B\})\, ds$$

$$= c\, 1_A(v) \cdot \int 1_{\{0<s\leq 1 \,:\, r\in s^{-1}B\}}\, ds = cr^{-1}\delta(v)(A) \int_0^r 1_B(tx)dt$$

$$= \int_{S_Q} \int_0^r 1_F(t^Q x)\, cr^{-1}\delta(v)(dx)\, dt .$$

Let G_Q be the set of all measures of the form $[x,R,0]$ such $QR + RQ^* + R \geq 0$; cf. Proposition 2.7 or $[x,0,M_{\alpha,\beta,u}]$, where $\alpha,\ \beta \in \mathbb{R}^+$, $u \in S_Q$ and

$$M_{\alpha,\beta,u}(F) = \int_{S_Q} \int_0^\alpha 1_F\ (t^Q x)dt(\beta\delta(u))(dx), \quad \text{for} \quad F \in \mathcal{B}(E\setminus\{0\}).$$

Since all Gaussian measures and the Poisson measures $[x,0,c\delta(a)]$ $c \in \mathbb{R}^+$, $a \in E\setminus\{0\}$ generate the whole class $ID(E)$ and J_Q is continuous isomorphism between ID and $U(Q)$ we obtain the following theorem.

2.8. COROLLARY. The class $U(Q)$ is the smallest closed subsemigroup of $ID(E)$ containing the generators G_Q.

3. Throughout the previous sections we very often exploited the so called "polar coordinates" associated with the group of operators $\{t^Q : t > 0\}$. In Jurek (1983a) are introduced the limit classes $L_m(\mathfrak{U})$ associated with the strongly continuous one-parameter group $\mathfrak{U} := \{U_t : t > 0\}$ satisfying the following "initial" condition

$$\lim_{t\to 0} U_t x = 0 \quad \text{for} \quad x \in E. \tag{3.1}$$

The existence of the polar coordinates in such a general situation is investigated in Jurek (1983c). Note that Kehrer (1983), investigating stable measures with respect to the group $\mathfrak{U}$, assumes also $\lim_{t\to\infty} \|U_t x\| = \infty$, for $x \neq 0$. In this case, Corollary 1 in Jurek (1983c) gives an equivalent norm $\|\cdot\|_{\mathfrak{U}}$ on E such that the mapping

$$S_{\mathfrak{U}} \times \mathbb{R}^+ \ni (x,t) \to U_t x \in E\setminus\{0\}$$

is a homeomorphism, where $S_{\mathfrak{U}} := \{x \in E : \|x\|_{\mathfrak{U}} = 1\}$. For the construction given in Jurek (1983c), it was essential to consider the following "exceptional" set

$$E_{\mathfrak{U}} := \{x \in E : \sup_{t\in\mathbb{R}^+} \|U_t x\| < \infty\}, \tag{3.2}$$

which is of the first Baire category, infinite dimensional linear space, whenever $E \neq \{0\}$, and F_σ-set. The following simple example shows that E may be bigger than $\{0\}$, or that Kehrer assumption is essential.

Let $\phi : \mathbb{R} \longrightarrow \mathbb{R}^+$ be given by $\phi(s) := \min(e^s, 1)$ and

$E := L_1(\phi) = \{f : \mathbb{R} \longrightarrow \mathbb{R} \mid \int_{-\infty}^{\infty} |f(x)|\phi(x)\,dx < \infty\}$ and

$\|f\| := \int_{-\infty}^{\infty} |f(x)|\phi(x)\,dx$. Define for $t \in \mathbb{R}$, $T_t : E \longrightarrow E$ by

$$(T_t f)(x) := f(x-t) \quad \text{for } f \in E \text{ and } x \in \mathbb{R}.$$

Since for $t \leq 0$ we have $\phi(s+t) \leq \phi(s)$ thus

$$\lim_{t\to-\infty} \|T_t f\| = \lim_{t\to-\infty} \int_{-\infty}^{\infty} |f(x)|\phi(x+t)\,dx = 0.$$

On the other hand, for $g(x) := 1_{[a,b]}(x)$ and $t \geq -a$

$$\|T_t g\| = \int_a^b \phi(x+t)\,dx = b-a.$$

So $\lim_{t\to+\infty} \|T_t g\| = b - a > 0$. Putting $U_t := T_{\ln t}$, $t \in \mathbb{R}^+$, we obtain the one-parameter strongly continuous group $\mathcal{U}$ satisfying (3.1) such that all indicator functions of finite intervals are in E ; cf. (3.2). In fact all functions with the support bounded away from $-\infty$ are in E.

REFERENCES

[1] Araujo, A. and Giné, E. (1980). The central limit theorem for real and Banach valued random variables; John Wiley, New York.

[2] Billingsley, P. (1968). Convergence of probability measures; John Wiley, New York.

[3] Gihman, I.I. and Skorohod, A.V. (1975). The theory of stochastic processes. I., Springer Verlag, New York.

[4] Jacod, J. (1982). Sur la non-reversibilite du processus d'Ornstein-Uhlenbeck generalise; Seminare de Probabilities, Université de Rennes, (preprint).

[5] Jurek, Z.J. (1982a). Structure of a class of operator-self-decomposable probability measures; Ann. Probab. 10, 849-856.

[6] Jurek, Z.J. (1982b). How to solve the inequality: $U_t m \leq m$ for $0<t<1$? II. Bull. Acad. Polon. Sci; Serie: Math. 30, 477-483.

[7] Jurek, Z.J. (1983a). Limit distributions and one-parameter groups of linear operators on Banach spaces; J. Multivar. Anal. 13, 578-604.

[8] Jurek, Z.J. (1983b) The classes $L_m(Q)$ of probability measures on Banach spaces; Bull. Pol. Acad.: Math. 31, 51-62.

[9] Jurek, Z.J. (1983c). Polar coordinates in Banach spaces, Bull. Pol. Acad.: Math. 31 or 32 (to appear).

[10] Jurek, Z.J. (1984). Relations between the s-self decomposable and selfdecomposable measures; Ann. Probab. 12 or 13 (to appear).

[11] Jurek, Z.J. and Rosinski, J. (1984). A continuity of some random integral mapping and the uniform integrability; Preprint.

[12] Jurek, Z.J. and Vervaat, W. (1983). An integral representation for selfdecomposable Banach space valued random variables; Z. Wahrscheinlichkeitstherie Verw. Geb. 62, 247-262.

[13] Kehrer, E. (1983). Stabilität von Wahrscheinlichkeitsmassen unter Operatorgruppen auf Banachraumen; University of Tubingen, Germany; Ph.D. Thesis.

[14] Kubik, L. (1962). A characterization of the class L of probability measures; Studia Math. 21, 245-252.

[15] Kumar, A. and Schreiber, B.M. (1979). Representation of certain infinitely divisible probability measures on Banach spaces; J. Multivar. Anal. 9, 288-303.

[16] Lévy, P. (1937). Theorie de l'addition des variable aleatories Paris: Gauthier-Villars (2nd ed. 1954).

[17] Lindvall, T. (1973). Weak convergence of probability measures and random functions in the function space $D[0,\infty)$; J. Appl. Probab. 10, 109-121.

[18] Sato, K. (1980). Class L of multivariate distributions and its subclasses; J. Multivar. Anal. 10, 207-232.

[19] Sato, K. and Yamazato, M. (1984). Operator-selfdecomposable distributions as limit distributions of processes of Ornstein-Uhlenbeck type; Stochastic Process. Appl. 17, 73-100.

[20] Urbanik, K. (1968). A representation of self-decomposable distributions; Bull. Acad. Polon. Sci.; Serie: Math 16, 209-214.

[21] Urbanik, K. (1973). Limit laws for sequences of normed sums satisfying some stability conditions; In: Multivariate Analysis. III (P.R. Krishnaiah, Ed.), pp. 225-237. Academic Press, New York.

[22] Urbanik, K. (1978). Lévy's probability measures on Banach spaces; Studia Math. 63, 283-308.

[23] Weron, A. (1984). Stable processes and measures: A survey; Center for Stochastic processes, Report 65, University of North Carolina at Chapel Hill.

[24] Wolfe, S.J. (1982). On a continuous analogue for the stochastic difference equation $X_n = \rho X_{n-1} + B_n$. Stochastic Process. Appl. 12, 301-312.

[25] Yamazato, M. (1978). Unimodality of infinitely divisible distribution functions of class L; Ann. Probab. 6, 523-531

[26] Yamazato, M. (1984). OL distributions on Euclidean spaces, Theor. Probab. Appl. 29, 3-18.

THE LAW OF THE ITERATED LOGARITHM IN THE ℓ^p SPACES

Michael Klass*
Department of Statistics
University of California
Berkeley, California 94720

and

James Kuelbs**
Department of Mathematics
University of Wisconsin
Madison, Wisconsin 53706

1. Introduction.

Let B denote a real separable Banach space with topological dual B^* and norm $\|\cdot\|$. Throughout $X, X_1, X_2, \cdots$ denote independent, identically distributed, B-valued random variables, and as usual $S_n = X_1 + \cdots + X_n$ for $n \geq 1$. We write Lx to denote the function $\max(1, \log x)$, and write L_2x to denote $L(Lx)$. The classical normalizing constants in the law of the iterated logarithm (LIL) are

$$a_n = \sqrt{2n\, L_2 n}\ .$$

We say X satisfies the bounded LIL (and write $X \in$ BLIL) if

$$\Lambda(X) \equiv \overline{\lim_n}\ \|S_n/a_n\| < \infty \qquad \text{w.p. } 1, \tag{1.1}$$

and satisfies the compact LIL (and write $X \in$ CLIL) if

$$P(\{S_n/a_n\} \text{ is conditionally compact in } B) = 1\,. \tag{1.2}$$

Of course, the CLIL always implies the BLIL, and when [1.2] holds it is well known that the random sequence $\{S_n/a_n\}$ converges to, and clusters throughout, a fixed non-random compact set K with probability one. The set K is determined by the covariance structure of X, and for further information and references on these aspects one can consult [2].

An easy application of the Hewitt-Savage zero-one law shows that $\Lambda(X)$ is always a constant, finite or infinite, with probability one, and in [2] the following result was obtained.

Theorem A. Let X be B-valued such that $E(X) = 0$ and $E(\|X\|^2) < \infty$. Then

$$\max(\sigma(X), \Gamma(X)) \leq \Lambda(X) \leq \sigma(X) + \Gamma(X) \tag{1.3}$$

where

* Supported in part by NSF Grant MCS-8301793.

** Supported in part by NSF Grant MCS-8219742.

$$\sigma(X) = \sup_{\|f\|_{B^*} \leq 1} \{E(f^2(X))\}^{1/2} \qquad [1.4]$$

and

$$\Gamma(X) = \overline{\lim_n} \, E(\|S_n/a_n\|). \qquad [1.5]$$

The inequality [1.3] represents an interesting way to view the LIL, but it is limited in the sense that it has been established only under the classical moment conditions $E(X) = 0$ and $E(\|X\|^2) < \infty$. However, for the LIL in Banach spaces, these classical conditions are neither necessary (for the CLIL) nor sufficient (for the BLIL). The integrability conditions which are necessary in this situation are

$$E(\|X\|^2/L_2\|X\|) < \infty \qquad [1.6]$$

and

$$X \text{ is } WM_0^2 \qquad [1.7]$$

where X is WM_0^2 if $E(f(X)) = 0$ and $E(f^2(X)) < \infty$ for all $f \in B^*$.

The results we prove here involve ℓ^∞, the Banach space of real sequences $x = \{x_i\}$ with norm

$$\|x\| = \sup_{j \geq 1} |x_j|,$$

and the classical ℓ^p spaces, $1 \leq p < 2$, with norm $\|x\| = (\sum_{j \geq 1} |x_j|^p)^{1/p}$. Of course, ℓ^∞ is not separable, but this causes no difficulty in what we prove. In regard to the other ℓ^p spaces, we mention that for $2 \leq p < \infty$ the results in [1] apply as these spaces satisfy the upper Gaussian comparison principle. When $0 < p < 1$ much of what we do when $1 \leq p < 2$ should apply rather directly, but a number of further details need to be checked as these spaces are not Banach spaces. This should be fairly routine, but a bit tedious, so we do not include the details here.

The results we prove are the following.

Theorem 1. Let $X = (\eta_1, \eta_2, \cdots)$ be an ℓ_p valued random variable where $1 \leq p < 2$. Further, assume X is WM_0^2 and $E(\|X\|^p) < \infty$, and let $\{\eta_j' : j \geq 1\}$ be an independent copy of $\{\eta_j : j \geq 1\}$. Then

$$\max(\sigma(X), \Gamma(X)) \leq \Lambda(X) \leq \sigma(X) + \Gamma(X) \qquad [1.8]$$

if either of the following hold:

$$\{|\eta_j|^p - |\eta'_j|^p : j \geq 1\} \text{ is sign-invariant and } E\{[\sum_{j\geq 1} |\eta_j|^{2p}]^{1/p}\} < \infty, \qquad [1.9]$$

or

$$E\{[\sum_{j\geq 1} E(|\eta_j|^p \mid \mathfrak{F}_{j-1})]^{2/p}\} < \infty \text{ and } E[(\sum_{j\geq 1} |\eta_j|^{2p})^{1/p}] < \infty \qquad [1.10]$$

where $\mathfrak{F}_j = \sigma(X_1, \cdots, X_j)$ with $\mathfrak{F}_0 = \{\varphi, \Omega\}$. In addition, if [1.9] or [1.10] hold, then

$$X \in \text{CLIL} \text{ iff } S_n/a_n \xrightarrow{\text{prob}} 0 \text{ iff } \Gamma(X) = 0. \qquad [1.11]$$

Before stating some corollaries and a second theorem we recall that a sequence of real random variables $\{Y_j : j \geq 1\}$ is sign-invariant if for an independent Rademacher sequence $\{\varepsilon_j : j \geq 1\}$ the joint distributions of $\{Y_j : j \geq 1\}$ and $\{\varepsilon_j Y_j : j \geq 1\}$ are identical. We say $\{Y_j : j \geq 1\}$ is m-dependent if the random vectors $(Y_i, \cdots, Y_k)$ and $(Y_{k+n}, \cdots, Y_j)$ are independent whenever $n > m$. In this terminology, an independent sequence is 0-dependent.

Corollary 1. Let $X = (\eta_1, \eta_2, \cdots)$ be ℓ^p-valued with $1 \leq p < 2$ and assume $\{\eta_j : j \geq 1\}$ is m-dependent. Further, assume X is WM_0^2, $E(\|X\|^p) < \infty$, and

$$E\{[\sum_{j\geq 1} |\eta_j|^{2p}]^{1/p}\} < \infty. \qquad [1.12]$$

Then [1.8] and [1.11] hold.

Corollary 2. Let $X = (\eta_1, \eta_2, \cdots)$ be WM_0^2 and ℓ^1-valued where $\{\eta_j : j \geq 1\}$ is m-dependent. Then [1.8] and [1.11] hold.

Theorem 2. Let $X = (\eta_1, \eta_2, \cdots)$ be an ℓ^∞-valued random variable such that $\{\eta_j : j \geq 1\}$ is a sequence of mean zero m-dependent random variables. Further, assume $E(\|X\|^2/L_2\|X\|) < \infty$ and $\sigma^2(X) = \sup_{j\geq 1} E(\eta_j^2)$, and let $\Lambda(X)$ and $\Gamma(X)$ be defined as in [1.1] and [1.5], respectively. Then

$$\max(\sigma(X), \Gamma(X)) \leq \Lambda(X) \leq \sigma(X) + \Gamma(X). \qquad [1.13]$$

Further, under the previous conditions the following implications also hold:

$$\text{If } \lim_j E(\eta_j^2) = 0, \text{ then } \max(\sigma(X), \Gamma(X)) = \Lambda(X). \qquad [1.14]$$

$$\text{If } S_n/a_n \xrightarrow{\text{prob}} 0, \text{ then } \Lambda(X) = \sigma(X). \qquad [1.15]$$

$$\text{If } \lim_j E(\eta_j^2) = 0 \text{ and } S_n/a_n \xrightarrow{\text{prob}} 0, \text{ then } X \in \text{CLIL}. \qquad [1.16]$$

Remarks. (I) In Theorem 2, for $X \in$ CLIL it is not necessary that $\lim_j E(\eta_j^2) = 0$, but it does follow that $S_n/a_n \xrightarrow{\text{prob}} 0$. For example, if $P(\varepsilon = \pm 1) = 1/2$ and

$X = (\varepsilon, \varepsilon, \cdots)$ then $X \in CLIL$ but $E\eta_j^2 = 1$. That $S_n/a_n \xrightarrow{prob} 0$ when $X \in CLIL$ is known from the methods of [4, Theorem 4.1].

(II). Theorem 2 improves Theorem A in the ℓ^∞ setting in the sense that it holds under the integrability conditions necessary for the LIL. The same applies to the result of Corollary 2 in ℓ^1. Of course, the condition that $\{\eta_j : j \geq 1\}$ is m-dependent is highly undesirable.

(III). We will see from the proof that the weak integrability assumptions in Corollary 2 actually imply $E\|X\|^2 < \infty$. Hence in ℓ^1, with $\{\eta_j : j \geq 1\}$ m-dependent, the WM_0^2 assumption is both necessary and sufficient for the LIL.

(IV). The condition $E(|\sum_{j\geq1} |\eta_j|^{2p})^{1/p}) < \infty$ follows from $\sum_{j\geq1} E\eta_j^2 < \infty$ since $1 \leq p < 2$ implies

$$(\sum_{j\geq1} |\eta_j|^{2p})^{1/2p} \leq (\sum_{j\geq1} |\eta_j|^2)^{1/2}.$$

(V). In the proof of Theorem 2 the inequality in [3.24] is perhaps the most important.

2. Proof of Theorem 1 and its corollaries.

To prove [1.8] we will show that under [1.9] or [1.10] we have

$$E(\|X\|^2) < \infty, \tag{2.1}$$

and hence Theorem A gives [1.8]. Further, [4, Theorem 4.1] will then give [1.11].

To prove [2.1] we first assume [1.9]. Now $1 < 2/p \leq 2$, so by the c_r-inequality

$$\begin{aligned}
E(\|X\|^2) &= E\{[\sum_{j\geq1} |\eta_i|^p]^{2/p}\} \\
&\leq 2E\{|\sum_{j\geq1} (|\eta_j|^p - E|\eta_j|^p)|^{2/p} + (\sum_{j\geq1} E|\eta_j|^p)^{2/p}\} \\
&\leq 2[M^{2/p} + E\{|\sum_{j\geq1} (|\eta_j|^p - |\eta_j'|^p)|^{2/p}\}] \\
&\qquad \text{where } M = \sum_{j\geq1} |\eta_j|^p < \infty \text{ since } E\|X\|^p < \infty \\
&= 2[M^{2/p} + E\{|\sum_{j\geq1} \varepsilon_j(|\eta_j|^p - |\eta_j'|^p)|^{2/p}\}] \\
&\qquad \text{since } \{|\eta_j| - |\eta_j'| : j \geq 1\} \text{ is sign-invariant}
\end{aligned} \tag{2.2}$$

$$\leq 4[M^{2/p} + E\{|\sum_{j\geq 1} \varepsilon_j |\eta_j|^p|^{2/p}\}]$$

$$\leq 4[M^{2/p} + E\{[\sum_{j\geq 1} |\eta_j|^{2p}]^{1/p}\}]$$

by Khintchine's inequality when integrating out the Rademacher variables.

Since [1.9] holds, we thus have [2.1] from [2.2], and the theorem is proved under [1.9].

If [1.10] holds we then have

$$E(\|X\|^2) \leq 2E\{|\sum_{j\geq 1} (|\eta_j|^p - E(|\eta_j|^p|\mathfrak{F}_{j-1}))|^{2/p} + |\sum_{j\geq 1} E(|\eta_j|^p|\mathfrak{F}_{j-1})|^{2/p}\}. \qquad [2.3]$$

Now $E\{|\sum_{j\geq 1} E(|\eta_j|^p|\mathfrak{F}_{j-1})|^{2/p}\} < \infty$, so $E(\|X\|^2) < \infty$ if

$$E\{|\sum_{j\geq 1}(|\eta_j|^p - E(|\eta_j|^p|\mathfrak{F}_{j-1}))|^{2/p}\} < \infty. \qquad [2.4]$$

To verify [2.4] we observe that if $\{\varepsilon_j : j \geq 1\}$ is an independent Radamacher sequence and $d_j = |\eta_j|^p - E(|\eta_j|^p|\mathfrak{F}_{j-1})$ $(j \geq 1)$, then $\{\varepsilon_j d_j : j \geq 1\}$ and $\{d_j\}$ are both martingale difference sequences, and hence by [3, Theorem 9] we have finite constants $C_1, C_2, C_3,$ and C_4 such that

$$E\{|\sum_{j\geq 1} d_j|^{2/p}\} \leq C_1 E\{(\sum_{j\geq 1} d_j^2)^{1/p}\}$$

$$\leq C_2 E(|\sum_{j\geq 1} \varepsilon_j d_j|^{2/p})$$

$$\leq C_3[E(|\sum_{j\geq 1} \varepsilon_j|\eta_j|^p|^{2/p}) + E(|\sum_{j\geq 1} \varepsilon_j E(|\eta_j|^p|\mathfrak{F}_{j-1})|^{2/p})] \qquad [2.5]$$

$$\leq C_4\{E[|\sum_{j\geq 1} |\eta_j|^{2p}|^{1/p}] + E[|\sum_{j\geq 1} E(|\eta_j|^p|\mathfrak{F}_{j-1})^2|^{1/p}]\}$$

where the last inequality follows from Khintchine's inequality. Now [1.10] gives $E(|\sum_{j\geq 1} |\eta_j|^{2p}|^{1/p}) < \infty$ and since

$$|\sum_{j\geq 1} E(|\eta_j|^p|\mathfrak{F}_{j-1})^2|^{1/2} \leq \sum_{j\geq 1} E(|\eta_j|^p|\mathfrak{F}_{j-1})$$

we also have

$$E(|\sum_{j\geq 1} E(|\eta_j|^p|\mathfrak{F}_{j-1})^2|^{1/p}) \leq E(|\sum_{j\geq 1} E(|\eta_j|^p|\mathfrak{F}_{j-1})|^{2/p}) < \infty$$

by [1.10]. Hence [2.3] and [2.4] give [2.1], so the theorem is proved.

Proof of Corollary 1: Since $\{\eta_j : j \geq 1\}$ is assumed to be m-dependent we write $X = \sum_{j\geq 1} \eta_j e_j$ where $\{e_j : j \geq 1\}$ is the canonical basis of ℓ^p. We also define

$$Y_k = \sum_{i\geq 0} \eta_{i(m+1)+k}\, e_{i(m+1)+k}$$

for $k = 1, 2, \cdots, m$ and set

$$\eta_j^{(k)} = \begin{cases} \eta_{i(m+1)+k} & \text{if } j = i(m+1)+k, \quad i = 0, 1, \cdots \\ 0 & \text{otherwise}. \end{cases} \tag{2.6}$$

Then

$$Y_k = (\eta_1^{(k)}, \eta_2^{(k)}, \cdots) \qquad (k = 1, 2, \cdots, m)$$

where $\{\eta_j^{(k)} : j \geq 1\}$ is an independent sequence, and hence

$$E(|\eta_j^{(k)}|^p|\mathfrak{F}_{j-1}^{(k)}) = E|\eta_j^{(k)}|^p \qquad (j = 1, 2, \cdots). \tag{2.7}$$

Since $E\|X\|^p < \infty$ and [1.12] holds we thus have [1.10] holding for the sequence $\{\eta_j^{(k)} : j \geq 1\}$, and hence by the proof of Theorem 1 that

$$E\|Y_k\|^2 < \infty \qquad (k = 1, \cdots, m).$$

Now $X = Y_1 + \cdots + Y_m$, so $E\|X\|^2 < \infty$ and Corollary 1 now follows from Theorem A and [4, Theorem 4.1].

Proof of Corollary 2. First recall that the dual space of ℓ^1 is ℓ^∞. Hence, if $\{\eta_j : j \geq 1\}$ is an orthogonal sequence of random variables, the WM_0^2 assumption easily implies that

$$\sum_{j\geq 1} E(\eta_j^2) < \infty.$$

Recalling that $\{\eta_j : j \geq 1\}$ is assumed to be m-dependent we have by the WM_0^2 assumption that

$$\sum_{j\geq 1} E[(\eta_j^{(k)})^2] < \infty \tag{2.8}$$

for $k = 1, \cdots, m$ where $\{\eta_j^{(k)} : j \geq 1\}$ is as in [2.6]. Thus from [2.7] and

$E\|X\|^p < \infty$ we have

$$E\Big(\Big[\sum_{j\geq 1} E(|\eta_j^{(k)}|^p \mid \mathfrak{F}_{j-1}^{(k)})\Big]^{2/p}\Big) < \infty . \qquad [2.9]$$

Combining [2.8] and [2.9] we thus have [1.10] holding for the sequence $\{\eta_j^{(k)} : j \geq 1\}$ (see remark (IV) at the end of section one). Hence by the proof of Theorem 1 we have

$$E\|Y_k\|^2 < \infty$$

for $k = 1, \cdots, m$ and as in the proof of Corollary 1 we have $E\|X\|^2 < \infty$. Thus Corollary 2 follows from Theorem A and [4, Theorem 4.1].

3. Proof of Theorem 2.

That $\Lambda(X) \geq \max(\sigma(X), \Gamma(X))$ is known from [2]. To prove the other half of [1.13] we fix $\varepsilon > 0$ and use the following notation.

We write the sequence $\{X_k : k \geq 1\}$ of independent copies of X as

$$X_k = (\eta_{k1}, \eta_{k2}, \cdots) \qquad (k \geq 1) .$$

Let $\beta > 1$, $n_0 = 0$, and put $n_k = [\beta^k]$ for $k \geq 1$ where $[\cdot]$ denotes the greatest integer function. Let $I(k) = \{n_k+1, \cdots, n_{k+1}\}$ for $k \geq 0$ and set

$$\tau_k = 2\tau n_{k+1} L_2 n_{k+1} = \tau a_{n_{k+1}}^2$$

where $\beta > 1$ and $\tau > 0$ are to be specified later. Then for $j \in I(k)$ let

$$\begin{aligned} u_j &= X_j I(\|X_j\|^2 \leq \tau_k) - E(X_j I(\|X_j\|^2 \leq \tau_k)) \\ w_j &= X_j I(\tau_k < \|X_j\|^2) - E(X_j I(\tau_k < \|X_j\|^2)) \qquad [3.1] \\ &= X_j - u_j , \end{aligned}$$

and set

$$\begin{aligned} U_n &= \sum_{j=1}^{n} u_j \\ W_n &= \sum_{j=1}^{n} w_j . \end{aligned} \qquad [3.2]$$

To prove the right hand side of [1.13] we will first assume $\{\eta_j : j \geq 1\}$ is an independent sequence and $\Gamma(X) < \infty$. Then we will show that for all $\beta > 1$ and $\tau > 0$ we have

$$\overline{\lim_n} \; \| W_n / a_n \| = 0 \qquad \text{w.p. 1,} \qquad [3.3]$$

and for $\beta > 1$, sufficiently close to 1, and $\tau = \varepsilon/32$ we have

$$\overline{\lim_n} \; \| U_n / a_n \| \le \sigma(X) + \Gamma(X) + \varepsilon \qquad \text{w.p. 1.} \qquad [3.4]$$

The proof that [3.3] holds follows because $E(\|X\|^2 / L_2 \|X\|) < \infty$. For details see Lemma 4.1 of [1]. Further, from that lemma we also have

$$\overline{\lim} \; E(\| W_n \| / a_n) = 0, \qquad [3.5]$$

and hence since $S_n = U_n + W_n$ we have

$$\overline{\lim_n} \; E(\| U_n \| / a_n) = \Gamma(X). \qquad [3.6]$$

We prove [3.4] by showing there is a $\beta > 1$, sufficiently close to 1, such that for $\tau = \varepsilon/32$ we have

$$\overline{\lim_r} \; \max_{n \in I(r)} \| U_n / a_n \| \le \sigma(X) + \Gamma(X) + \varepsilon \qquad \text{w.p. 1.} \qquad [3.7]$$

Indeed since $a_n \nearrow \infty$ we have

$$\max_{n \in I(r)} \| U_n / a_n \| \le (a_{n_{r+1}} / a_{n_r}) \max_{n \in I(r)} \| U_n / a_{n_{r+1}} \|, \qquad [3.8]$$

and since $a_{n_{r+1}} / a_{n_r} \sim \sqrt{\beta}$, [3.8] implies

$$\overline{\lim_r} \; \max_{n \in I(r)} \| U_n / a_n \| \le \sqrt{\beta} \; \overline{\lim_r} \; \max_{n \in I(r)} \| U_n \| / a_{n_{r+1}}. \qquad [3.9]$$

Since we are assuming $\{\eta_j : j \ge 1\}$ is an independent sequence it follows that $\{u_j : j \ge 1\}$ is also an independent mean zero sequence. Hence

$$\begin{aligned}
\overline{\lim_r} \; \sup_{n \in I(r)} P(\| U_{n_{r+1}} - U_n \| > \varepsilon a_{n_{r+1}} / 4)
&\le \overline{\lim_r} \; (4/\varepsilon) \sup_{n \in I(r)} E\| U_{n_{r+1}} - U_n \| / a_{n_{r+1}} \\
&\le \overline{\lim_r} \; (4/\varepsilon) \, E\| U_{n_{r+1}} - U_{n_r} \| / a_{n_{r+1}} \\
&= (4/\varepsilon) \, \overline{\lim_r} \; E\| S_{n_{r+1}} - S_{n_r} \| / a_{n_{r+1}} \qquad \text{by [3.5]} \qquad [3.10] \\
&= (4/\varepsilon) \, \overline{\lim_r} \; E\| S_{n_{r+1} - n_r} \| / a_{n_{r+1}} \qquad \text{(by stationarity)} \\
&= (4/\varepsilon) \, \Gamma(X) \; \overline{\lim_r} \; a_{n_{r+1} - n_r} / a_{n_{r+1}}
\end{aligned}$$

$$= (4/\varepsilon)\Gamma(X)\sqrt{1-\frac{1}{\beta}} .$$

Hence for sufficiently large r and $\beta > 1$ sufficiently close to 1, Ottaviani's inequality easily implies that

$$P\left(\max_{n\in I(r)} \|U_n\|/a_{n_{r+1}} > \sigma(X) + \Gamma(X) + \frac{\varepsilon}{2}\right) \le 2P\left(\|U_{n_{r+1}}\| > \left(\sigma(X) + \Gamma(X) + \frac{\varepsilon}{4}\right) a_{n_{r+1}}\right). \qquad [3.11]$$

Thus we need to show

$$\sum_r P\left(\|U_{n_{r+1}}\| > \left(\sigma(X) + \Gamma(X) + \frac{\varepsilon}{4}\right) a_{n_{r+1}}\right) < \infty , \qquad [3.12]$$

since [3.9], [3.11], [3.12], and the Borel-Cantelli lemma then imply that

$$\overline{\lim_r} \max_{n\in I(r)} \|U_n/a_n\| \le \sqrt{\beta}\left(\sigma(X) + \Gamma(X) + \frac{\varepsilon}{2}\right) \qquad \text{w.p. 1.} \qquad [3.13]$$

Taking $\beta > 1$, sufficiently close to one, [3.13] then implies [3.7] since $\sigma(X) < \infty$ and $\Gamma(X) < \infty$ are assumed to hold. Now [3.7] implies [3.4] and this completes the proof. Hence we must establish [3.12].

To establish [3.12] we let $\varphi_j : \ell^\infty \to \mathbb{R}^1$ be the jth coordinate map. Hence $\varphi_j(X) = \eta_j$. We also define

$$B_r = \left\{\|U_{n_r}\| > \left(\Gamma(X) + \frac{\varepsilon}{16}\right) a_{n_r}\right\} \qquad [3.14]$$

and

$$B_{r,j} = \left\{|\varphi_j(U_{n_r})| > \left(\Gamma(X) + \frac{\varepsilon}{16}\right) a_{n_r}\right\}. \qquad [3.15]$$

Then,

$$B_r = \bigcup_{j=1}^{\infty} B_{r,j} , \qquad [3.16]$$

and $\{B_{r,j} : j \ge 1\}$ are independent events.

Now

$$\overline{\lim_r}\, P(B_r) \le \overline{\lim_r} \frac{E\|U_{n_r}\|}{a_{n_r}\Gamma(X)(1+\varepsilon/(16\Gamma(X)))} = \frac{1}{1+\varepsilon/(16\Gamma(X))} = q^* < 1 . \qquad [3.17]$$

Further, since the $\{B_{r,j} : j \ge 1\}$ are independent with [3.17] holding it is easy to show that

$$\frac{\sum_{j\geq 1} P(B_{r,j})}{1+\sum_{j\geq 1} P(B_{r,j})} \leq P(B_r) \leq \sum_{j\geq 1} P(B_{r,j}) . \qquad [3.18]$$

Setting $q = q^* + (1-q^*)/2$ we have $q < 1$ and by combining [3.17] and [3.18] for all r sufficiently large

$$\sum_{j\geq 1} P(B_{r,j}) \leq q(1+\sum_{j\geq 1} P(B_{r,j})) . \qquad [3.19]$$

Hence for all r sufficiently large

$$\sum_{j\geq 1} P(B_{r,j}) \leq \frac{q}{1-q} . \qquad [3.20]$$

Now let

$$T = \inf\{n : |\varphi_j(U_n)| > (\sigma(X) + \tfrac{\varepsilon}{8}) a_{n_r}\}$$

$$N_m = \max_{1\leq k\leq m} |\varphi_j(u_k)| \qquad [3.21]$$

$$M_m = \max_{1\leq n\leq m} |\varphi_j(U_n)| .$$

Then, by an argument due to Hoffmann-Jorgenson, we have

$$\begin{aligned} &P(|\varphi_j(U_{n_r})| > (\sigma(X) + \Gamma(X) + \tfrac{\varepsilon}{4}) a_{n_r}) \\ &\quad = \sum_{k=1}^{n_r} P(T = k, |\varphi_j(U_{n_r})| > (\sigma(X) + \Gamma(X) + \tfrac{\varepsilon}{4}) a_{n_r}) \\ &\quad \leq \sum_{k=1}^{n_r} P(T = k, |\varphi_j(U_{n_r}) - \varphi_j(U_k)| > (\Gamma(X) + \tfrac{\varepsilon}{8} - N_k) a_{n_r}) \qquad [3.22] \\ &\quad \leq \sum_{k=1}^{n_r} [P(T = k; N_k \geq \tfrac{\varepsilon}{16} a_{n_r}) + P(T = k; |\varphi_j(U_{n_r}) - \varphi_j(U_k)| \geq (\Gamma(X) + \tfrac{\varepsilon}{16}) a_{n_r}] \\ &\quad \leq P(N_{n_r} > \varepsilon a_{n_r}/16) + \sum_{k=1}^{n_r} P(T = k) P(|\varphi_j(U_{n_r}) - \varphi_j(U_k)| > (\Gamma(X) + \tfrac{\varepsilon}{16}) a_{n_r}) \end{aligned}$$

by independence.

Now, $\tau = \frac{\varepsilon}{32}$ and $a_n \nearrow \infty$ implies,

$$P(N_{n_r} > \varepsilon a_{n_r}/16) = 0,$$

and since

$$\limsup_r \sup_j \sup_{1 \le k \le n_r} P(|\varphi_j(U_{n_r}) - \varphi_j(U_k)| > (\Gamma(X) + \tfrac{\varepsilon}{16})\, a_{n_r})$$

$$\le \limsup_r \sup_j \frac{E|\varphi_j(U_{n_r})|^2}{(\Gamma(X) + \frac{\varepsilon}{16})^2\, a_{n_r}^2}$$

$$\le \limsup_r \sup_j \frac{n_r E\eta_j^2}{2n_r L_2 n_r (\Gamma(X) + \frac{\varepsilon}{16})^2}$$

$$= 0,$$

we have by Ottaviani's inequality that for all r sufficiently large, uniformly in j, that

$$P(|\varphi_j(U_{n_r})| > (\sigma(X) + \Gamma(X) + \tfrac{\varepsilon}{4})\, a_{n_r})$$

$$\le 2P(M_{n_r} \ge (\sigma(X) + \tfrac{\varepsilon}{8})\, a_{m_r})\, P(|\varphi_j(U_{n_r})| > (\Gamma(X) + \tfrac{\varepsilon}{16})\, a_{n_r})$$

(by Ottaviani on $\varphi_j(U_k)$, $\varphi_j(U_{n_r}) - \varphi_j(U_k)$) [3.23]

$$\le 4P(|\varphi_j(U_{n_r})| > (\sigma(X) + \tfrac{\varepsilon}{8})\, a_{n_r})\, P(|\varphi_j(U_{n_r})| > (\Gamma(X) + \tfrac{\varepsilon}{16})\, a_{n_r}).$$

(by Ottaviani on M_{n_r})

Hence for all r sufficiently large by [3.20] and [3.23] we have

$$\sum_{j \ge 1} P(|\varphi_j(U_{n_r})| > (\sigma(X) + \Gamma(X) + \tfrac{\varepsilon}{4})\, a_{n_r})$$

$$\le 4 \max_j P(|\varphi_j(U_{n_r})| > (\sigma(X) + \tfrac{\varepsilon}{8})\, a_{n_r}) \sum_{j \ge 1} P(B_{r,j})$$

$$\le \frac{4q}{1-q} \max_j P(|\varphi_j(U_{n_r})| > (\sigma(X) + \tfrac{\varepsilon}{8})\, a_{n_r}) \qquad [3.24]$$

$$\le \frac{4q}{1-q} \{\max_j [2CP(|G_j| > (\sigma(X) + \tfrac{\varepsilon}{64})\sqrt{2L_2 n_r}) + C \sum_{k=1}^{n_r} \frac{E|\varphi_j(u_k)|^{2+\alpha}}{a_{n_r}^{2+\alpha}}]\}$$

where $G_j = N(0, \sigma_j^2)$ with $\sigma_j \le \sigma(X)$ and C is a finite constant by the argument in [1, pp. 113-114]. Hence by [3.24] and the definition of the sup-norm on ℓ^∞

$$\sum_r P(\|U_{n_r}\| > (\sigma(X) + \Gamma(X) + \frac{\varepsilon}{4})a_{n_r})$$

$$\leq \sum_r \sum_j P(|\varphi_j(U_{n_r})| > (\sigma(X) + \Gamma(X) + \frac{\varepsilon}{4})a_{n_r}) \tag{3.25}$$

$$\leq \frac{8Cq}{1-q} \sum_r [\int_{(1+\delta)\sqrt{2L_2 n_r}}^{\infty} \frac{e^{-u^2/2}}{\sqrt{2\pi}} du + \sum_{k=1}^{n_r} \frac{E\|u_k\|^{2+\alpha}}{a_{n_r}^{2+\alpha}}]$$

where $\delta = \varepsilon/(64\ \sigma(X)) > 0$. Applying a standard argument as in [1, p. 113-114] we have $\sum_r \sum_{k=1}^{n_r} E\|u_k\|^{2+\alpha}/a_{n_r}^{2+\alpha} < \infty$, and since $\delta > 0$, $\sum_r \int_{(1+\delta)\sqrt{2L_2 n_r}} e^{-u^2/2}\, du/\sqrt{2\pi} < \infty$ which gives [3.12]. Hence the right side of [1.13] holds when $\{\eta_j : j \geq 1\}$ is an independent sequence and $\Gamma(X) < \infty$.

If $\Gamma(X) = \infty$, the right hand side of [1.13] is obvious, so it suffices to assume $\Gamma(X) < \infty$ and $\{\eta_j : j \geq 1\}$ m-dependent.

When $\{\eta_j : j \geq 1\}$ is m-dependent we proceed as in Corollary 1 defining the independent sequences

$$\{\eta_j^{(k)} : j \geq 1\}$$

for $k = 1, \cdots, m$ and related variables

$$Y_k = (\eta_1^{(k)}, \eta_2^{(k)}, \cdots) \qquad (k = 1, \cdots, m).$$

Then $X = Y_1 + \cdots + Y_m$ and the definition of the sup-norm implies that $\|X\| = \sup_{1 \leq k \leq m} \|Y_k\|$. Applying the previously established result we have

$$\overline{\lim_n}\ \|S_n(Y_k)\|/a_n \leq \sigma(Y_k) + \Gamma(Y_k) \qquad \text{w.p. } 1$$

for each $k = 1, \cdots, m$ where $S_n(Y)$ denotes the sum of n independent copies of Y, $\sigma(Y_k) = \sup_j E(\eta_j^{(k)})^2$, and $\Gamma(Y_k) = \overline{\lim_n} \frac{E\|S_n(Y_k)\|}{a_n}$. Since $S_n(X) = S_n(Y_1) + \cdots + S_n(Y_m)$ with $\|S_n(X)\| = \sup_{1 \leq k \leq m} \|S_n(Y_k)\|$ the result is now immediate since

$$\sup_{1 \leq k \leq m} \sigma(Y_k) = \sigma(X)$$

and

$$\sup_{1 \leq k \leq m} \Gamma(Y_k) = \Gamma(X).$$

To establish [1.14] is easy once we know [1.13] holds. That is, consider the mappings Π_N and Q_N defined for $x = \{x_i\} \in \ell^\infty$ by

$$\Pi_N(x) = (x_1, \cdots, x_N, 0, 0, \cdots)$$

and

$$Q_N(x) = (0, \cdots, 0, x_{N+1}, x_{N+2}, \cdots) .$$

Then it is well known that

$$\Gamma(Q_N(X)) = \Gamma(X)$$

and applying [1.13] to the random vector $Q_N(X)$ we get

$$\overline{\lim_n} \| S_n(Q_N(X))/a_n \| \leq \sigma(Q_N(X)) + \Gamma(X) . \qquad [3.26]$$

Since

$$\| \frac{S_n}{a_n}(X) \| = \max\{ \frac{\| S_n(\Pi_N(X)) \|}{a_n} , \frac{\| S_n(Q_N(X)) \|}{a_n} \},$$

we have by the finite dimensional LIL applied to $\Pi_N(X)$ and [3.26] that

$$\overline{\lim_n} \| \frac{S_n}{a_n}(X) \| \leq \max\{ \sigma(\Pi_N(X)), \Gamma(X) + \sigma(Q_N(X)) \} . \qquad [3.27]$$

Since N is arbitrary with $\lim_N \sigma(Q_N(X)) = 0$ and $\lim_N \sigma(\Pi_N(X)) = \sigma(X)$, [3.27] gives [1.14].

If $S_n/a_n \xrightarrow{\text{prob}} 0$, then it is well known that $\Gamma(X) = 0$ and hence [1.15] holds immediately from [1.13].

To verify [1.16] is now trivial since $S_n/a_n \xrightarrow{\text{prob}} 0$ implies $\Lambda(Q_N(X)) = \sigma(Q_N(X))$ for each N. Since $\lim_j E\eta_j^2 = 0$ we have $\lim_N \sigma(Q_N(X)) = 0$ and thus for each $\varepsilon > 0$ there is an N such that with probability one

$$\overline{\lim_n} \| \frac{S_N(Q_N(X))}{a_n} \| = \overline{\lim_n} \| \frac{Q_N(S_n)}{a_n} \| = \sigma(Q_N(X)) < \varepsilon .$$

Since $\varepsilon > 0$ is arbitrary standard arguments now imply that $\{S_n/a_n\}$ is conditionally compact with probability one, and hence Theorem 2 is proved.

REFERENCES

[1] Acosta, A. de and Kuelbs, J. (1983). Some results on the cluster set $C(\{S_n/a_n\})$ and the LIL. Ann. Probability 11, pp. 102-122.

[2] Acosta, A. de, Kuelbs, J. and Ledoux M. An inequality for the law of the iterated logarithm. Lecture Notes in Mathematics, 990, pp. 1-29.

[3] Burkholder, D. L. (1966). Martingale transforms, Ann. Math. Statist. 37, pp. 1494-1504.

[4] Kuelbs, J. (1977). Kolmogorov's law of the iterated logarithm for Banach space valued random variables. Illinois J. Math. 21, pp. 784-800.

A SQUARE ROOT LAW FOR DIFFUSING PARTICLES

Raoul LePage
Department of Statistics and Probability
Michigan State University, E. Lansing, MI 48823
Bertram M. Schreiber
Department of Mathematics
Wayne State University, Detroit, MI 48202

Abstract. In an earlier paper, we proved that a plot of log n independent Brownian motions in dimension d = 1, for times in [0,n], is nearly certain to give the appearance of a shaded region with square root boundaries, when subjected to the rescaling of the functional iterated logarithm law. Here we prove that for every finite dimension d the same conclusion holds if just one point is plotted from each of log n Brownian paths having variance parameter equal to 1, provided these points are selected uniformly in the time interval [0,n] and independently of the paths.

Introduction. Our earlier result may be visualized by thinking of a smoke stack which, at time zero, emits a large number log n of burning particles whose gaseous trajectories are observed over the time interval [0,n] as they drift in horizontally moving air. Our result suggested a plume which, when viewed from the side and appropriately rescaled, densely occupies a region with square root boundaries. In effect, the log n independent $\mathbb{R}^1$-valued Brownian motions over times $[0,n]$, when rescaled, fill the space between the square root boundaries before they spill much outside them.

Our new result suggests the same type of plume if the particles leave no trails but are emitted at random times which are uniform over $[0,n]$. Moreover, the new result extends to an arbitrary finite dimension d, when plotting one randomly selected point from each of log n independent $\mathbb{R}^d$-valued Brownian motions (having variance parameter 1) over time $[0,n]$. For convenience we prove the equivalent result for selecting one random point from each of $(\log n)^d$ independent $\mathbb{R}^d$-valued Brownian motions, having variance parameter d, over time $[0,n]$.

Formulation. We use ℓx to denote $\log_e x$. The evaluations of independent $\mathbb{R}^d$-valued Brownian motions at any fixed times are independent normal random vectors. Fix an arbitrary $c > 0$ and finite dimension d. For each integer $n \geq 3$ consider the 'plume' given by

$$\{ (K_i, \sqrt{K_i} Y_i) : 1 \leq i \leq c_n \}, \quad c_n = [c(\ell n)^d],$$

where Y_i are iid $N(0, I_{d \times d})$ and the times K_1, K_2, ... are iid uniform on the integers 1, ..., n, and independent of the Y_i. We compare this plume with the 'iterated logarithm region',

$$\{ (t,y) : e \leq t \leq n, \|y\|_d \leq \sqrt{2dt\ell\ell t} \} \subset [e,n] \times \mathbb{R}^d .$$

Our result makes the comparison after rescaling time by n, and displacement $\| \ \|_d$ by $a_n = \sqrt{2n\ell\ell n}$. Let

$$T_{in} = K_i/n \ ; \ Y^*_{in} = \sqrt{K_i} Y_i / a_n ,$$

and define the 'rescaled plume'

$$\mathcal{P}_n = \{ \pi_{in} : 1 \le i \le c_n \},$$

where $\pi_{in} = (T_{in}, Y^*_{in})$, $n \ge 3$, $1 \le i \le n$. Observe that the same rescaling applied to the iterated logarithm boundaries gives $\sqrt{2dnt\ell\ell nt}/a_n = \sqrt{dt} + o(1)$, for $e/n \le t \le 1$. Define the 'limit region'

$$\mathcal{R} = \{ (t,y): 0 \le t \le 1, \|y\|_d \le \sqrt{dt} \}.$$

We prove that $|\mathcal{P}_n - \mathcal{R}|$ converges to zero in probability, where

$$|\mathcal{P}_n - \mathcal{R}| = |\mathcal{P}_n - \mathcal{R}|_1 + |\mathcal{P}_n - \mathcal{R}|_0,$$

$$|\mathcal{P}_n - \mathcal{R}|_1 = \sup\{ \varepsilon > 0: \exists i \le c_n, \text{ such that } \| Y^*_{in}\|_d > \sqrt{dT_{in}} + \varepsilon \},$$

$$|\mathcal{P}_n - \mathcal{R}|_0 = \sup\{ \varepsilon > 0: \exists (t,y) \in \mathcal{R}, \text{ such that } \|(t,y) - \pi_{in}\|_{d+1} > \varepsilon, \forall i \le c_n \}.$$

<u>Outer Law</u>: $|\mathcal{P}_n - \mathcal{R}|_1$ converges in probability to zero.

Proof. For each $\varepsilon > 0$ define events A_n by,

$$A_n = \{ \exists i \le c_n \text{ such that } \|Y^*_{in}\|_d > \sqrt{dK_i/n} + \varepsilon \}.$$

Note that A_n is the event $|\mathcal{P}_n - \mathcal{R}|_1 > \varepsilon$ and,

$$P(A_n^c) = \prod_{i \le c_n} P(\|Y^*_{in}\|_d \le \sqrt{dT_{in}} + \varepsilon).$$

$$= \prod_{i \le c_n} P(X_d^2 \le 2\,\ell\ell n\,(\sqrt{d} + \varepsilon/\sqrt{K_i/n})^2)$$

$$\ge \left(1 - \gamma_d \int_{2\ell\ell n\,(\sqrt{d}+\varepsilon)^2}^{\infty} x^{\frac{d}{2}-1} e^{-x/2}\,dx\right)^{c_n}, \quad \gamma_d = 2^{-\frac{d}{2}}\Gamma(\tfrac{d}{2}).$$

As $n \to \infty$, with $u = \ell\ell n$, the last line above is asymptotically,

$$\sim \exp - \left\{ c e^{du} \gamma_d \int_{2u(\sqrt{d}+\varepsilon)^2}^{\infty} x^{\frac{d}{2}-1} e^{-x/2} dx \right\}$$

$$\sim \exp - \left\{ \frac{c}{d}\gamma_d (2u(\sqrt{d} + \varepsilon)^2)^{\frac{d}{2}-1} 2((\sqrt{d} + \varepsilon)^2 e^{-((\sqrt{d} + \varepsilon)^2 - d)u} \right\}$$

$\to 1$ as $n \to \infty$.

Therefore $P(A_n) \to 0$ as $n \to \infty$. □

Remark. For d = 1 this is weaker than our corresponding outer law for paths.

<u>Inner Law:</u> $|P_n - R|_0$ converges in probability to zero.

Proof. We will cover the event $\{|\mathcal{P}_n - \mathcal{R}|_0 > 2\varepsilon+\sqrt{d\varepsilon}\}$ by a bounded (with n) number of events, then show that the probability of each of these events tends to zero as $n \to \infty$, for each fixed $\varepsilon > 0$. Fix $\varepsilon = 2^{-J}$, J a positive integer. Let $A_{n\varepsilon} = \{ |\mathcal{P}_n - \mathcal{R}|_0 > 2\varepsilon+\sqrt{d\varepsilon} \}$. Then $A_{n\varepsilon} \subset \bigcup_{k=2}^{2^J} B_{nk}$ where,

$$B_{nk} = \{ \exists(t,y)\varepsilon\mathcal{R} \text{ with } t\in\Delta_k \text{ such that } \forall i \text{ with } T_{in}\in\Delta_k,$$
$$\|(t,y) - \pi_{in}\|_{d+1} \geq \varepsilon + \sqrt{d\varepsilon} \},$$

where

$$\Delta_k = \Delta(k,\varepsilon) = [(k-1)\varepsilon, k\varepsilon], \ k = 1, 2, \ldots, [\varepsilon^{-1}]+1.$$

We next prove $B_{nk} \subset C_{nk}$ where

$$C_{nk} = \{ \exists y \varepsilon \mathbb{R}^d \text{ with } \|y\|_d \leq r_{k-1} \text{ and}$$
$$\|y - Y^*_{in}\|_d \geq \varepsilon, \ \forall i \text{ with } T_{in}\in\Delta_k \},$$

with $r_k = \sqrt{d\varepsilon k}$, $k = 1,2,\ldots,\varepsilon^{-1}$. Suppose $\varepsilon<1$. Let (t,y) and k be as in the definition of B. If $\|y\|_d \leq r_{k-1}$ then

$$\|y-Y^*_{in}\| \geq \|(t,y)-\pi_{in}\|_{d+1} - |t-T_{in}| > \varepsilon+\sqrt{d\varepsilon}-\varepsilon > \varepsilon,$$

satisfying the condition for C_{nk}. If, on the other hand, $\|y\|_d > r_{k-1}$ let $y_0 = y\, r_{k-1}/\|y\|_d$. On $B_{nk} \cap C^c_{nk}$

$$\varepsilon + \sqrt{d\varepsilon} < \|(t,y)-\pi_{in}\|_{d+1}$$
$$\le \|(t,y)-(t,y_0)\|_{d+1} + \|(t,y_0)-\pi_{in}\|_{d+1}$$
$$< \|y\|_d - r_{k-1} + \varepsilon \le \sqrt{d\varepsilon} + \varepsilon.$$

This contradiction establishes $B_{nk} \subset C_{nk}$.

To deal with C_{nk}, consider a covering of the sphere $\|y\|_d = r_{k-1}$ by finitely many (d-1)-dimensional regions of identical shape and size and having maximal diameter less than $\varepsilon/2$ in every direction. For example these regions, which may overlap, may be spherical caps. We can move one into another by rotation of the sphere. Now shrink these regions, in steps of size $\delta_k = (\varepsilon/2)r_{k-1}$, toward the origin in $\mathbb{R}^d$. The result is a covering of the ball $\|y\|_d \le r_{k-1}$ by finitely many regions $G(j,\theta)$ concentrically placed about the origin, and generated by a finite number $M_{\varepsilon d}$ of rotations from the $2/\varepsilon$ regions along a given ray. Each of the regions $G(j,\theta)$ can be identified by its distance $j\delta_k$, $j=1, 2, \ldots, 2/\varepsilon$, on a ray from the origin, and the radial projection of its center to a point θ on the unit sphere in $\mathbb{R}^d$. By spherical symmetry of Y_1,

$$P(C_{nk}) \le M_{\varepsilon d} \sum_{j=2}^{2^J} P(C_{nkj}),$$

where

$$C_{nkj} = \{ \|Y^*_{in}\|_d \notin (r_{k-1}/2)\Delta_j, \ \forall i \text{ with } T_{in} \in \Delta_k \}.$$

It is enough to prove that for each $k \ge 2$, $j \ge 2$, $P(C_{nkj}) \to 0$ as $n \to \infty$. By conditioning on the σ-field $\mathcal{F}$ generated by $\mathbb{K}_i$, $i \ge 1$

$$P(C_{nkj}) = E\, P(\chi^2_d \notin b^2_n[(j-1)^2, j^2], \ \forall i \text{ with } T_{in} \in \Delta_k | \mathcal{F}),$$

where

$$b_n^2 = (a_n^2/K_i)(\varepsilon/2)^2 r_{k-1}^2$$

$$\leq 2d(\ell\ell n)(\varepsilon/2)^2, \text{ when } T_{in} \in \Delta_k.$$

Also, $b_n^2(j-1)^2 \to \infty$ as $n \to \infty$, so for $j \geq 2$ it is eventually greater than the mode of χ_d^2, beyond which the χ_d^2 density is decreasing. For n sufficiently large therefore, since $j \leq 2/\varepsilon$,

$$P(C_{nkj}) \leq E \min[1, (1 - b_n^2(j^2-(j-1)^2)\gamma_d(2d\ell\ell n)^{\frac{d}{2}-1}(\ell n)^{-d})^N]$$

where $N = N_{kn} = \#(\{ i \leq c_n : T_{in} \in \Delta_k \})$, so $N/(\ell n)^d \to \varepsilon$ almost surely as $n \to \infty$. The last line above is

$$\leq E \min[1, \exp - \{ b_n^2(j^2-(j-1)^2)\gamma_d(2d\ell\ell n)^{\frac{d}{2}-1}(\ell n)^{-d}N \}].$$

Since $K_i \leq n$ the definition of b_n implies that exponent above is

$$\geq \; 2r_{k-1}^2(\varepsilon/2)^2(j^2-(j-1)^2)\gamma_d d^{-1}(2d\ell\ell n)^{d/2}(\ell n)^{-d}N$$

$$\to \infty \text{ a.s. when } n \to \infty .$$

We have proved $P(|\mathcal{F}_n - \mathcal{R}|_0 > 2\varepsilon+\sqrt{d}\varepsilon) \to 0$ as $n \to \infty$. □

Combining the inner and outer laws we obtain the main result.

<u>Theorem:</u> $|\mathcal{F}_n - \mathcal{R}|$ converges in probability to zero as $n \to \infty$.

Work in progress will extend these results to infinite dimensions.

REFERENCES

[1] LePage, R. and Schreiber, B. M. An iterated logarithm law for families of Brownian paths. To appear, Z. W. Verw. Geb. (1984).

[2] Richardson, L. F. Some measurements of atmospheric turbulance. Phil. Trans. Roy. Soc. London, A. Vol. 221 (1921), p.1.

[3] Taylor, G. I. Diffusion by continuous movements. Proc. London Math. Soc. 20 (1921), 196-211.

STOCHASTIC PROCESSES WITH SAMPLE PATHS IN EXPONENTIAL ORLICZ SPACES

Michael B. Marcus*
Texas A&M University
College Station, TX USA

and

Gilles Pisier*
Université Paris VI
Paris, France

I. Introduction

Let $(T,\mathcal{T},\mu)$ be a measure space and μ a probability measure. Let $\{X(t),\ t \in T\}$ be a real or complex valued stochastic process defined on some probability space. We are concerned with the following question: When is

$$\int_T \exp\left|\frac{X(t)}{\alpha}\right|^q \mu(dt) < \infty \tag{1.1}$$

for some $0 < q \leq \infty$ and some $\alpha > 0$? We will consider this question for a variety of stochastic processes with particular emphasis on stationary Gaussian processes and, more generally, on strongly stationary p-stable processes with $1 < p \leq 2$. Nevertheless, in order to clarify the significance of the question that we have raised, in this introduction we will discuss our results only as they apply to Gaussian processes. Therefore, let us assume that $\{X(t),\ t \in T\}$ is a Gaussian process with $\beta = \sup_{t\in T} E|X(t)|^2 < \infty$. Then

$$E\exp\left|\frac{X(t)}{\alpha}\right|^2 < \infty \tag{1.2}$$

as long as $\alpha > (2\beta)^{1/2}$, and so by Fubini's Theorem (1.1) is always finite for these processes for $q \leq 2$. Clearly, (1.2) is not finite for any power greater than 2 (unless $X(t) \equiv 0,\ \forall t \in T$) but this doesn't mean that (1.1) can not be finite for

*Supported in part by a grant from the National Science Foundation.

$q > 2$. If $\sup_{t\in T} |X(t)| < \infty$ a.s., (1.1) is finite for all α and $q > 0$. However, we shall see that (1.1) can also be finite for $q > 2$ when $\{X(t),\ t \in T\}$ is not bounded. In fact one can consider a classification of unbounded Gaussian processes according to whether (1.1) holds for some q and α.

We only obtain fragmentary results in this paper, nevertheless they are intriguing because they seem to be generalizations of well known results on continuity and boundedness of stochastic processes. For example Theorem 3.1 contains the following sufficient condition for (1.1) in the case of Gaussian processes: Let $\{X(t),\ t \in T\}$ be a Gaussian process. Define

$$d = d(s,t) = \left(E|X(s) - X(t)|^2\right)^{1/2} \ ; \quad s,\ t \in T,$$

and assume that $\hat{d} = \sup_{s,t\in T} d(s,t) < \infty$. Let $N(T,d;\varepsilon)$ denote the minimum number of open balls in the d metric or pseudo-metric with centers in T that covers T. Then for $2 \leq q \leq \infty$

$$(1.3) \qquad \int_0^{\hat{d}} \left(\log N(T,d;\varepsilon)\right)^{1/2 - 1/q} d\varepsilon < \infty$$

is a sufficient condition for (1.1) for some $\alpha < \infty$. When $q = \infty$ this is Dudley's sufficient condition for boundedness of Gaussian processes. (Theorem 3.1 also extends Dudley's continuity condition.) When $q = 2$ (1.1) is trivially true. Otherwise this is a new result.

In Theorem 4.5, when dealing with Gaussian random Fourier series on a compact Abelian group G in which the group characters are contained in a Sidon set we show that for $2 \leq q < \infty$

$$(1.4) \qquad \sup_{\varepsilon>0} \varepsilon\left(\log N(G,d;\varepsilon)\right)^{1/2 - 1/q} < \infty$$

is necessary and sufficient for (1.1) for some $\alpha < \infty$. The reader will immediately note that (1.4) is less restrictive than (1.3). This shows that (1.3) is not best possible even for stationary Gaussian processes. On the other hand we show in Theorem 3.5 that if (1.1) holds for all probability measures μ on T then necessarily (1.4) holds.

The appropriate way to study the problem stated at (1.1) is to consider the Orlicz space formed with the function $\psi_q(z) = (\exp|z|^q)-1$, $z \in \mathbb{C}$ and the measure μ, which we will denote by $L^{\psi_q}(d\mu)$. For the convenience of the reader we will give some well known inequalities involving this space in Section 2 and present an important relationship between L^{ψ_q} and $\ell_{p,\infty}$, where $\frac{1}{p}+\frac{1}{q}=1$ and $p > 1$. In Section 3 we give the extension of Dudley's theorem alluded to above. In Theorem 3.3 we apply this result to the processes (on a commutative group) considered in [4] and [5]. In Section 4 we show how, in many results, $\ell_{p,\infty}$ $(p > 1)$ plays a similar role for processes almost surely in L^{ψ_q} that ℓ_1 does for processes with almost surely continuous sample paths. We also give, in Theorem 4.5, necessary and sufficient conditions for p-stable random Fourier series on compact Abelian groups, in which the group characters lie in a Sidon set, to have sample paths almost surely in $L^{\psi_{q'}}$, where $\frac{1}{p}+\frac{1}{q}=1$ and $2 \le q < q' < \infty$. At the very end of the paper we point out that our results can be used to study almost sure properties of occupation time distributions and local times of the stochastic processes that we have considered.

II. Preliminaries

Let $(T,\mathcal{T},\mu)$ be a measure space where μ is a probability measure. Let $\phi\colon R^+ \to R^+$ be an increasing convex function with $\phi(0) = 0$. Relative to the measure space $(T,\mathcal{T},\mu)$ we denote by $L^\phi(d\mu)$ the so called "Orlicz space" formed by all measurable functions $f\colon T \to \mathbb{C}$ for which there is a $c > 0$ such that

$$\int_T \phi\left(\left|\frac{f}{c}\right|\right) d\mu(t) < \infty .$$

We equip this space with the norm

$$\|f\|_\phi = \inf\left\{c > 0\colon \int_T \phi\left(\left|\frac{f}{c}\right|\right)d\mu(t) \le 1\right\} . \tag{2.1}$$

To be more specific let us denote by ψ_q the function

$$\psi_q(x) = \exp|x|^q - 1, \quad 1 \le q \le \infty .$$

We will be concerned with the Orlicz space $L^{\psi_q}(d\mu)$. As is customary we admit the value $q = \infty$. In this case, by definition,

$$\|f\|_{L^{\psi_\infty}(d\mu)} = \sup_{t\in T} |f(t)| = \|f\|_\infty .$$

Let I be an index set. For any family $\{\alpha_i\}_{i\in I}$ of complex numbers tending to zero at infinity we can define a sequence $\{\alpha_n^*\}_{n\in \mathbb{N}}$, (N denotes the integers), which is the non-increasing rearrangement of $\{|\alpha_i|\}_{i\in I}$. For $0 < p < \infty$ we denote by $\ell_{p,\infty}(I)$ the space of all families $\{\alpha_i\}_{i\in I}$ of complex numbers such that $\sup_{n\geq 1} n^{1/p} \alpha_n^* < \infty$, $0 < p < \infty$, and we define

$$\|\{\alpha_i\}_{i\in I}\|_{p,\infty} = \sup_{n\geq 1} n^{1/p} \alpha_n^* . \tag{2.2}$$

It is well known that for $p > 1$, the functional $\|\cdot\|_{p,\infty}$ is equivalent to a norm on $\ell_{p,\infty}(I)$ with which $\ell_{p,\infty}(I)$ is a Banach space.

The two "norms" (2.1) and (2.2) are intimately related in the study of random series. Let $\{\varepsilon_i\}_{i\in I}$ be a Rademacher sequence on some probability space, i.e. a sequence of i.i.d random variables satisfying $P(\varepsilon_i = 1) = P(\varepsilon_i = -1) = 1/2$. It is easy to check that for $p < 2$, $\ell_{p,\infty}(I) \subset \ell_2(I)$. Therefore if $\{\alpha_i\}_{i\in I}$ is in $\ell_{p,\infty}(I)$ the series $S = \sum_{i\in I} \varepsilon_i \alpha_i$ converges a.s.. We shall consider S to be defined on the probability space (Ω, F, P). A proof of the next lemma can be found in either [8], Proposition 2.2 or [5], Lemma 3.1.

<u>Lemma 2.1</u>: If $\{\alpha_i\}_{i\in I}$ belongs to $\ell_{p,\infty}(I)$, $1 < p < 2$, then S belongs to $L^{\psi_q}(dP)$ and we have

$$k_p^{-1}\|\{\alpha_i\}_{i\in I}\|_{p,\infty} \leq \|S\|_{L^{\psi_q}(dP)} \leq k_p\|\{\alpha_i\}_{i\in I}\|_{p,\infty} , \tag{2.3}$$

where k_p is a constant depending only on p.

Let $\{X(t),\ t \in T\}$ be a stochastic process on the probability space (Ω, F, P) and let $\omega \in \Omega$. Recall that $(T, \mathcal{T}, \mu)$ is also a probability space. We will consider $X(t,\omega)$ as defined on the product probability space $(\Omega \otimes T,\ F \times \mathcal{T},\ \mu \otimes P)$. In what follows we use the notation

$$\|X(t,\omega)\|_{L^1(d\mu)(L^{\psi_q}(dP))} = \int_T \|X(t,\omega)\|_{L^{\psi_q}(dP)}\, d\mu(t)$$

and

$$\|X(t,\omega)\|_{L^\infty(dP)(L^{\psi_q}(d\mu))} = \sup_{\omega\in\Omega} \|X(t,\omega)\|_{L^{\psi_q}(d\mu)}$$

and similarly when $d\mu$ and dP are interchanged. The following inequalities are well known.

Lemma 2.2: Let $(T,\mathcal{T},\mu)$ be a measure space where μ is a probability measure. Let $\{X(t),\ t \in T\}$ be a stochastic process on $(\Omega,\mathcal{F},P)$. Then for $1 \le q \le \infty$

$$(2.4)\qquad \frac{\log 2}{1+\log 2}\max\left\{\|X(t,\omega)\|_{L^1(d\mu)(L^{\psi_q}(dP))},\ \|X(t,\omega)\|_{L^1(dP)(L^{\psi_q}(d\mu))}\right\}$$

$$\le \|X(t,\omega)\|_{L^{\psi_q}(d\mu\otimes dP)}$$

$$\le \inf\left\{\|X(t,\omega)\|_{L^\infty(d\mu)(L^{\psi_q}(dP))},\ \|X(t,\omega)\|_{L^\infty(dP)(L^{\psi_q}(d\mu))}\right\}.$$

Proof: Let

$$c(\omega) = \|X(t,\omega)\|_{L^{\psi_q}(d\mu)}.$$

Then for $\lambda \ge 1$, by Jensen's inequality and the definition of the Orlicz space norm

$$P(c(\omega) > \lambda\delta) \le P\left[\int \exp\left|\frac{X(t,\omega)}{\lambda\delta}\right|^q d\mu(t) \ge 2\right]$$

$$\le P\left[\int \exp\left|\frac{X(t,\omega)}{\delta}\right|^q d\mu(t) \ge 2^{\lambda^q}\right]$$

$$\le \iint \exp\left|\frac{X(t,\omega)}{\delta}\right|^q d\mu(t)dP\cdot 2^{-\lambda^q},$$

where at the last step we use Chebyshev's inequality. Now set

$$\delta = \|X(t,\omega)\|_{L^{\psi_q}(d\mu\otimes dP)},$$

and observe that

(2.5) $$P(c(\omega) > \lambda\delta) \le 2\cdot 2^{-\lambda^q} .$$

Therefore, by (2.5)

$$E\left(\frac{c(\omega)}{\delta}\right) = 1 + \int_1^\infty P\left(\frac{c(\omega)}{\delta} > \lambda\right)d\lambda$$

$$\le 1 + (\ln 2)^{-1} ,$$

which is what we obtain when $q = 1$. This gives the inequality on the left in (2.4) for the term $\|X(t,\omega)\|_{L^1(dP)(L^{\psi_q}(d\mu))}$. However the other inequality is exactly the same since both P and μ are probability measures.

The upper bound is easier. We have

$$\int \exp \left|\frac{X(t,\omega)}{c(\omega)}\right|^q d\mu(t) \le 2 ,$$

which implies that

$$\iint \exp\left|\frac{X(t,\omega)}{\sup_\omega c(\omega)}\right|^q d\mu(t)dP \le 2 .$$

Therefore

$$\delta \le \sup_\omega c(\omega)$$

and similarly

$$\delta \le \sup_t c(t) ,$$

where $c(t) = \|X(t,\omega)\|_{L^{\psi_q}(dP)}$.

Lemma 2.2 enables us to obtain some inequalities for processes that will interest us later on.

<u>Lemma 2.3</u>: Let $\{G(t), t \in T\}$ be a real valued mean zero Gaussian process defined on (Ω,F,P) such that $\sup_{t\in T} \left(E|G(t)|^2\right)^{1/2} = \sigma < \infty$. Let

(2.6) $$Y(t) = \sum_{k=1}^\infty \varepsilon_k f_k(t), \quad t \in T$$

where $\{\varepsilon_k\}_{k=1}^{\infty}$ is a Rademacher sequence and $\{f_k(t)\}_{k=1}^{\infty}$ are complex valued functions on T such that $\sup_{t\varepsilon T} \|\{f_k(t)\}\|_{p,\infty} < \infty$. Then

(2.7) $$E\|G(t,\omega)\|_{L^{\psi_2}(d\mu)} \leq (8/3)^{1/2} \frac{1+\log 2}{\log 2} \sigma \equiv \beta\sigma$$

and furthermore, for $\lambda \geq 1$

(2.8) $$P\left(\|G(t,\omega)\|_{L^{\psi_2}(d\mu)} > \lambda\right) \leq 2 \exp\left(-\log 2 \left|\frac{\lambda}{\beta\sigma}\right|^2\right) .$$

Similarly for $2 < q < \infty$,

(2.9) $$E\left(\|Y(t,\omega)\|_{L^{\psi_q}(d\mu)}\right) \leq C_q \sup_{t\varepsilon T} \|\{f_k(t)\}\|_{p,\infty} \equiv \alpha,$$

where C_q is a constant depending only on q and $\frac{1}{p} + \frac{1}{q} = 1$. Furthermore for $\lambda \geq 1$

(2.10) $$P\left(\|Y(t,\omega)\|_{L^{\psi_q}(d\mu)} > \lambda\right) \leq 2 \exp\left(-\log 2 \left|\frac{\lambda}{\alpha}\right|^q\right) .$$

Proof: By (2.4)

$$E\|G(t,\omega)\|_{L^{\psi_2}(d\mu)} < \frac{1+\log 2}{\log 2} \sup_{t\varepsilon T} \|G(t,\omega)\|_{L^{\psi_2}(dP)} .$$

Since $G(t)$ is a Gaussian process it is elementary to compute $\|G(t,\omega)\|_{L^{\psi_2}(dP)}$ and we get (2.7). The probability estimate (2.8) follows from (2.4) and (2.5). To obtain (2.9) we again use (2.4) to get

$$E\|Y(t,\omega)\|_{L^{\psi_q}(d\mu)} < \frac{1+\log 2}{\log 2} \sup_{t\varepsilon T} \|Y(t,\omega)\|_{L^{\psi_q}(dP)}$$

and then use (2.3). The probablity estimate also follows from (2.3), (2.4) and (2.5).

The next series of Lemmas is directed towards obtaining an extension of Dudley's continuity condition for stochastic processes to also show that processes have sample paths in $L^{\psi_q}(d\mu)$. The first is a well known interpolation inequality.

Lemma 2.4: Let $(T,\mathcal{T},\mu)$ be as above and let $f: T \to \mathbb{C}$. Then for $0 < q \leq q' \leq \infty$,

(2.11)
$$\|f\|_{L^{\psi_{q'}}(d\mu)} \leq \|f\|^{q/q'}_{L^{\psi_q}(d\mu)} \|f\|_\infty^{1-q/q'} .$$

Proof: This follows immediately from the obvious inequality

$$\int \exp\left|\frac{f(t)}{c}\right|^{q'} d\mu(t) \leq \int \exp\left|\frac{\|f\|_\infty^{1-q/q'}|f(t)|^{q/q'}}{c}\right|^{q'} d\mu(t) .$$

Lemma 2.5: Let $\{Z_i\}_{i=1}^N$ be real or complex valued random variables on $(\Omega, ,P)$ satisfying

(2.12)
$$\|Z_i\|_{L^{\psi_q}(dP)} \leq d, \quad 1 \leq i \leq N .$$

Then, for $1 \leq q < \infty$,

(2.13)
$$E \sup_{1\leq i\leq N} |Z_i| \leq d(\log N)^{1/q}\left(1 + \frac{2}{q \log N}\right), \quad N \geq 2 .$$

Proof: By the definition of the Orlicz space norm and Chebyshev's inequality we see that (2.12) implies that for $1 \leq i \leq N$

$$P\left[\frac{|Z_i|}{d} > u\right] \leq 2e^{-|u|^q}, \quad \forall u > 0 ,$$

Therefore

$$P\left[\sup_{1\leq i\leq N} |Z_i| > u\right] \leq 2Ne^{-\left|\frac{u}{d}\right|^q}, \quad \forall u > 0 ,$$

and

$$E \sup_{1\leq i\leq N} |Z_i| \leq d(\log N)^{1/q} + N \int_{d(\log N)^{1/q}}^{\infty} e^{-\left|\frac{u}{d}\right|^q} du$$

which gives (2.13).

Lemma 2.6: Let $T = \{1, 2, \ldots, N\}$ and let $\{Z(t), t \in T\}$ be a stochstic process on $(\Omega,\mathcal{F},P)$ satisfying

(2.14) $$\|Z(t)\|_{L^{\psi_q}(dP)} \leq d, \quad \forall\, t \in T .$$

Then for $1 \leq q \leq q' \leq \infty$ we have

(2.15) $$E\|Z(t,\omega)\|_{L^{\psi_{q'}}(d\mu)} \leq C_q\, d(\log N)^{1/q - 1/q'}$$

where C_q is a constant depending only on q.

Proof: By Lemma 2.4 and convexity we have

$$E\|Z(t,\omega)\|_{L^{\psi_{q'}}(d\mu)} \leq \left(E\|Z(t,\omega)\|_{L^{\psi_q}(d\mu)}\right)^{q/q'} \left(E \sup_{t\varepsilon T} |Z(t)|\right)^{1-q/q'},$$

which by Lemmas 2.2 and 2.5

$$\leq \left(\frac{1 + \log 2}{\log 2} \sup_{t\varepsilon T} \|Z(t,\omega)\|_{L^{\psi_q}(dP)}\right)^{q/q'} \left(\left(1 + \frac{2}{q \log N}\right) d(\log N)^{1/q}\right)^{1 - q/q'}.$$

Using (2.14) we get (2.15).

We will also use the following simple observation. In what follows we will let $P(T)$ denote the probability measures on $(T,\mathcal{T})$.

Lemma 2.7: Suppose that $f: T \to C$ is such that

(2.16) $$\|f\|_{L^{\psi_q}(d\mu)} < \infty, \quad \forall\, \mu \in P(T) .$$

Then there exists a constant C such that

(2.17) $$\|f\|_{L^{\psi_q}(d\mu)} \leq C, \quad \forall\, \mu \in P(T) .$$

Proof: If (2.17) is false there exist probability measures $\mu_k \in P(T)$ such that $\|f\|_{L^{\psi_q}(d\mu_k)} \geq 2^{k+1}$ which implies that

$$\int \exp\left|\frac{X(t)}{2^k}\right|^q d\mu_k > 2 .$$

Therefore by convexity, for $j \leq k$

$$2^{2^{(k-j)q}} \leq \left(\int \exp\left|\frac{X(t)}{2^k}\right|^q d\mu_k \right)^{2^{(k-j)q}} \tag{2.18}$$

$$\leq \int \exp\left|\frac{X(t)}{2^j}\right|^q d\mu_k .$$

Using (2.18) we see that

$$\int \exp\left|\frac{X(t)}{2^j}\right|^q d\left(\sum_{k=1}^{\infty} \frac{\mu_k}{2^k}\right) \geq \int \exp\left|\frac{X(t)}{2^j}\right|^q d\left(\sum_{k=j}^{\infty} \frac{\mu_k}{2^k}\right)$$

$$\geq \sum_{k=j}^{\infty} 2^{-k}\, 2^{2^{(k-j)q}} = \infty .$$

This implies that $\|f\|_{L^{\psi_q}(d\nu)} = \infty$ for $\nu = \sum_{k=1}^{\infty} \frac{\mu_k}{2^k}$ which contradicts (2.16).

III. Extensions of Dudley's Theorem

We first show that Dudley's sufficient condition for continuity of sample paths of Gaussian processes, and its many generalizations, can be extended to also provide conditions for these sample paths to lie in exponential Orlicz spaces.

Theorem 3.1: Let $(T,\mathcal{T},\mu)$ be a compact topological measure space with μ a probability measure. Let $\{\tilde{X}(t),\ t \in T\}$ be a real valued stochastic process defined on $(\Omega,\mathcal{F},P)$ satisfying

$$\|\tilde{X}(s) - \tilde{X}(t)\|_{L^{\psi_q}(dP)} \leq d(s,t), \quad \forall\ s,\ t \in T \tag{3.1}$$

for some $1 \leq q < \infty$ and continuous metric or pseudo-metric $d(s,t)$ on $T \otimes T$. Let $\hat{d} = \sup_{s,t\in T} d(s,t)$ and assume that

$$J_{q,q'}(T,d) = \int_0^{\hat{d}} \left(\log N(T,d;\varepsilon)\right)^{1/q - 1/q'} d\varepsilon < \infty \tag{3.2}$$

where $q \leq q' \leq \infty$ and $N(T,d;\varepsilon)$ denotes the minimal number of open balls of radius ε in the d metric or pseudometric, with centers in T, that covers T. Then there exists a version $\{X(t),\ t \in T\}$ of the process satisfying

(3.3) $$E\|X\|_{L^{\psi_{q'}}(d\mu)} \leq C_{q,q'}\{E|X(t_0)| + \hat{d} + J_{q,q'}(T,d)\}$$

for any $t_0 \in T$, where $C_{q,q'}$ is a constant depending only on q and q'. Furthermore if we consider the space $\overline{T} = \{(s,t) \in T \otimes T: d(s,t) \leq \varepsilon\}$ with the induced topology and measure $d\nu = (d\mu(t) \otimes d\mu(s))(\int\int_{\overline{T}} d\mu(t) \otimes d\mu(s))^{-1}$ we get

(3.4) $$E\|X(s) - X(t)\|_{L^{\psi_{q'}}(d\nu)} \leq C'_{q,q'}\{\varepsilon + \int_0^{\varepsilon} (\log N(T,d;\varepsilon))^{1/q - 1/q'} d\varepsilon\}$$

where $C'_{q,q'}$ is a constant depending only on q and q'.

<u>Proof</u>: There are $N(T,d;\delta_n) \equiv N(\delta_n)$ balls of radius δ_n that cover T. From this set of balls extract disjoint sets $\{A_i^n\}_{i=1}^{N(\delta_n)}$ such that $T = \bigcup_{n=1}^{N(\delta_n)} A_i^n$ and from each A_i^n, $i = 1, \ldots, N(\delta_n)$ select a point t_i^n. For a sequence $\{\delta_n\}_{n=0}^{\infty}$ with $\delta_0 = \hat{d}$ and $\delta_n \downarrow 0$ define for each $n \geq 0$

$$X_0(t) = \tilde{X}(t_0)I_{[t\varepsilon T]}$$
$$\vdots$$
$$X_n(t) = \sum_{i=1}^{N(\delta_n)} I_{[t\varepsilon A_i^n]} \tilde{X}(t_i^n) .$$

Let $\Delta^n(t) = X_n(t) - X_{n-1}(t)$ and consider

(3.5) $$X(t) = \tilde{X}(t_0)I_{[t\varepsilon T]} + \sum_{n=1}^{\infty} \Delta^n(t) .$$

It is clear that $\{X(t), t \in T\}$ and $\{\tilde{X}(t), t \in T\}$ agree on the cylinder sets of R^T. Upon closer examination we see that

(3.6) $$\Delta^n(t) = \sum_{i=1}^{N(\delta_n)} \sum_{j=1}^{N(\delta_n)-1} I_{[t\varepsilon A_i^n \cap A_j^{n-1}]} (\tilde{X}(t_i^n) - \tilde{X}(t_j^{n-1})).$$

That is, $\Delta^n(t)$ is a stochastic process on a finite sample space having no more than $N^2(\delta_n)$ elements. Furthermore

$$(3.7) \qquad \|\tilde{X}(t_i^n) - \tilde{X}(t_j^{n-1})\|_{L^{\psi_q}(d\mu)} \leq 2\delta_{n-1}$$

since if there exists an $s \in A_i^n \quad A_j^{n-1}$ we have

$$\|\tilde{X}(t_i^n) - \tilde{X}(t_j^{n-1})\|_{L^{\psi_q}(dP)} \leq \|\tilde{X}(t_i^n) - \tilde{X}(s)\|_{L^{\psi_q}(dP)} + \|\tilde{X}(t_j^{n-1}) - \tilde{X}(s)\|_{L^{\psi_q}(dP)}$$

$$\leq \delta_n + \delta_{n-1} \, .$$

Therefore, using (3.5) - (3.7) and lemma 2.6 we have

$$(3.8) \qquad E\|X\|_{L^{\psi_q}(d\mu)} \leq \tilde{C}_{q,q'}\Big(E|X(t_0)| + \sum_{n=1}^{\infty} \delta_n (\log N(\delta_n))^{1/q - 1/q'}\Big) ,$$

where $\tilde{C}_{q,q'}$ is a constant depending only on q and q'. If we take $\delta_n = 2^{-n}$ and bound the sum in (3.8) by an integral expression we get (3.3). The inequality (3.4) follows immediately from (3.3) if we note that $N(\overline{T},d;\varepsilon) \leq N(T \otimes T, d;\varepsilon) \leq N^2(T,d;\varepsilon/2)$ which is shown in Lemma 1.3, Chapter II, [4]. This completes the proof of Theorem 3.1.

Remark 3.2: If $\{\tilde{X}(t), t \in T\}$ is a mean zero Gaussian or subgaussian process then

$$\|\tilde{X}(s) - \tilde{X}(t)\|_{L^{\psi_2}(dP)} \leq C\big(E|\tilde{X}(s) - \tilde{X}(t)|^2\big)^{1/2}$$

for some constant C. With this choice of norm and with $q' = \infty$ Theorem 3.1 gives Dudley's Theorem. However, as is well known, Dudley's approach can be easily adapted to give all the results of Theorem 3.1 in the case $q' = \infty$. (See e.g. Lemma 3.2 [5].) More complete extensions of Dudley's Theorem, in the sense of the different norms considered in (3.11) and the corresponding entropy integral as in (3.2) exist [2], [9], [11], but only for $q' = \infty$. One can see from the proof of Theorem 3.1 above that in all these cases some kind of interpolation can be used to obtain conditions for the sample paths to be in weaker Banach spaces than $\mathbb{C}(T)$.

For instance Theorem 3.1 is actually valid as stated for $0 < q \leq q' \leq \infty$.

Theorem 3.1 can be applied to strongly stationary p-stable processes as defined in [5]. For the convenience of the reader we will repeat the definition here and since later we will also consider these processes in greater generality we will give the more general definition.

Let T be a set. We will denote by $R^{(T)}$ (resp. $C^{(T)}$) the space of all finitely supported families $(\alpha(t))_{t\varepsilon T}$ of real (resp. complex) numbers. Let $0 < p \leq 2$. We will say that a real (resp. complex) stochatic process $\{X(t), t \varepsilon T\}$ is p-stable if there exists a positive measure m on R^T (resp. C^T) equipped with the cylindrical σ-algebra such that $\forall \alpha \varepsilon R^{(T)}$

$$(3.9) \qquad E \exp i \sum_{t\varepsilon T} \alpha(t)X(t) = \exp - \int \Big| \sum_{t\varepsilon T} \alpha(t)\beta(t)\Big|^p dm(\beta) ,$$

(resp. $\alpha \varepsilon C^{(T)}$)

$$(3.10) \qquad E \exp i \operatorname{Re}\left[\sum_{t\varepsilon T} \overline{\alpha(t)}X(t) \right] = \exp - \int \Big| \sum_{t\varepsilon T} \overline{\alpha(t)}\beta(t)\Big|^p dm(\beta).$$

When T is a locally compact Abelian group G with dual group Γ. A real (resp. complex) valued p-stable process $\{X(t), t\varepsilon G\}$ will be called <u>strongly stationary</u> if it admits a representation as in (1.12) (resp. (1.13)) where the measure m is a finite positive Radon measure supported on Γ. Strongly stationary p-stable processes can be represented as random Fourier series and random Fourier transforms. See Section I, [5] for further details. Finally note that 2-stable processes are Gaussian processes.

In both the real and imaginary case we associate with the processes $\{X(t), t \varepsilon T\}$ the metric or pseudo-metric

$$(3.11) \qquad d_X(s,t) = \left(\int |\beta(s) - \beta(t)|^p m(d\beta)\right)^{1/p} .$$

<u>Theorem 3.3</u>: Let G be a locally compact Abelian group with dual group Γ and for $1 < p \leq 2$ let $\{\tilde{X}(t), t \varepsilon G\}$ be a strongly stationary p-stable stochastic process with associated metric or pseudo-metric d_X as defined in (3.9), (3.10) and (3.11). Let $K \subset G$ be a compact neighborhood of the unit element of G and let μ

be a probability measure supported on $(K,\mathcal{K})$ where $\mathcal{K}$ is the induced topology. (If G is compact we can take $K = G$ and μ can be taken to be Haar measure but we are not restricted to Haar measure.) Let $\frac{1}{p} + \frac{1}{q} = 1$ and assume that

$$J_{q,q'}(K,d_X) < \infty, \quad 2 \leq q \leq q' \leq \infty . \tag{3.12}$$

Then there exists a version $\{X(t), t \in G\}$ of $\{\tilde{X}(t), t \in G\}$ satisfying

$$E\|X\|_{L^{\phi_{q'}}(d\mu)} \leq C_{q,q'}(K)\{|m|^{1/p} + J_{q,q'}(K,d_X)\} \tag{3.13}$$

where $|m| = m(R^T)$ (resp. $m(\mathbb{C}^T)$) in the real (resp. complex) case and $C_{q,q'}(K)$ is a constant depending only on q,q' and K. Furthermore if we consider the set $\overline{K} = \{(s,t) \in K \otimes K: d_X(s,t) \leq \varepsilon\}$ and measure $d\nu = (d\mu(t) \otimes d\mu(s))(\int\int_{\overline{K}} d\mu(t) \otimes d\mu(s))^{-1}$ we have

$$E\|X(s) - X(t)\|_{L^{\phi_{q'}}(d\nu)} \leq C'_{q,q'}(K)\{\varepsilon + \int_0^{\varepsilon} (\log N(K,d_X;\varepsilon))^{1/q - 1/q'} d\varepsilon\} \tag{3.14}$$

where $C'_{q,q'}(K)$ is a constant depending only on q, q' and K.

<u>Proof</u>: Theorem 3.3 follows from Theorem 3.1 in exactly the same manner as the sufficiency part of Theorem A follows from Lemma 3.2 in [5].

<u>Remark 3.4</u>: The case $p = 2$ as defined in (3.9) - (3.11) is a repetition of the Gaussian case. However, by following the proof of the sufficiency part of Theorem 1.1 Chapter I, [4] one sees that (3.13) and (3.14) with $p = q = 2$ are also valid for the more general class of random Fourier series considered in (1.10) and (1.11) of Chapter I, [4].

We also note that the measure ν that appears in (3.4) and (3.14) need not be a product measure but can be any probability measure on $\overline{T}$ or $\overline{K}$ respectively.

Finally, let $L_0^{\phi_q}$ be the closure in L^{ϕ_q} of the space of simple functions. As stated Theorems 3.1 and 3.3 allow us to view the process $\{X(t), t \in T\}$ as a random variable with values in the Banach space $L^{\psi_{q'}}(d\mu)$. However, it is easy to see from the proofs of these Theorems, that we actually obtain a random variable with values in the smaller space $L_0^{\psi_{q'}}(d\mu)$ which is separable. Following the methods of [4], (3.14) (of this paper) enables us to prove a Central Limit Theorem for the processes considered in Theorem 1.1, Chapter IV, [4]. Also by Lemma 2.2 [7], one can prove a Central Limit Theorem in $L^{\psi_{q'}}(d\mu)$ for the processes considered in Theorem 3.1 as long as $0 < q \leq 2$.

When $q' = \infty$ (3.13) and (3.14) is best possible since $J_{q,\infty}(K,d_X) < \infty$ is a necessary and sufficient condition for a strongly stationary p-stable process to have a version with continuous sample paths. We shall see in Section 4 that for $q < q' < \infty$, $J_{q,q'}(K,d_X) < \infty$ is not necessary for $\{X(t,\omega),\ t \in K\}$ to have a version in $L^{\psi_q}(d\mu)$ a.s. Nevertheless the next theorem, a partial converse of Theorem 3.3, shows that Theorem 3.3 is "not too bad". Note that in this we do not restrict ourselves to strongly stationary processes but consider p-stable processes in general.

<u>Theorem 3.5</u>: Let $\{X(t),\ t \in T\}$ be a p-stable process on (Ω,F,P), $1 < p \leq 2$, with associated metric d_X as defined in (3.9), (3.10) and (3.11). Let $\frac{1}{p} + \frac{1}{q} = 1$ and $q < q' \leq \infty$. If

(3.15) $$\|X(t,\omega)\|_{L^{\psi_{q'}}(d\mu)} < \infty \quad \text{a.s} \quad \forall\ \mu \in P(T)\ ,$$

then there exists a constant C such that

(3.16) $$\sup_{\varepsilon>0} \varepsilon\big(\log N(T,d_X;\varepsilon)\big)^{1/q - 1/q'} < C\ .$$

($P(T)$ is defined just before Lemma 2.7).

<u>Proof:</u> By a result of Fernique, Landau and Shepp in the case $p = 2$ (see eg. Theorem 3.8, Chapter II, [4]) or by a result of de Acosta [1] in the case $1 < p < 2$, (3.15) implies that

(3.17) $$E\|X(t,\omega)\|_{L^{\psi_{q'}}(d\mu)} < \infty \ , \quad \forall \mu \in P(T) \ .$$

Given $\varepsilon > 0$ let $M(\varepsilon) = M(T,d_X;\varepsilon)$ denote the maximum number of disjoint balls of radius ε in the metric or pseudo-metric d_X with centers in T. For one such maximal packing let $\{t_i\}_{i=1}^{M(\varepsilon)}$ denote the centers of the balls. Consider the discrete uniform probability measure ν on T that is supported on $\{t_i\}_{i=1}^{M(\varepsilon)}$. Since

$$\int \exp\left|\frac{X(t)}{c}\right|^{q'} d\nu \geq M^{-1} \exp\Big(\sup_{1\leq i\leq M(\varepsilon)} \left|\frac{X(t_i)}{c}\right|^{q'}\Big)$$

we see that

(3.18) $$\|X(t,\omega)\|_{L^{\psi_{q'}}(d\nu)} \geq \sup_{1\leq i\leq M(\varepsilon)} |X(t_i,\omega)|(\log 2M(\varepsilon))^{-1/q'} \ .$$

Also, by Corollary 2.7, [5] we have

(3.19) $$E \sup_{1\leq i\leq M(\varepsilon)} |X(t_i,\omega)| \geq d_{q,q'} \inf_{1\leq i\neq j\leq M(\varepsilon)} E|X(t_i) - X(t_j)|(\log M(\varepsilon))^{1/q} \ ,$$

for some constant $d_{q,q'}$ depending only on q and q'. It is well known that $E|X(t_i) - X(t_j)|$ and $d_X(t_i,t_j)$ are equivalent metrics or pseudo-metrics on $\{t_i\}_{i=1}^{M(\varepsilon)}$. (To see this, in the real case, note that by (3.9), $\xi = (X(t_i) - (X_j))(\int|\beta(t_i) - \beta(t_j)|^p \, dm(\beta))^{-1/p}$ satisfies $E \exp i\lambda\xi = \exp - |\lambda|^p$ and for $1 < p \leq 2$, $E|\xi|$ exists. The complex case is similar.) Therefore, by this last remark, (3.18) and (3.19) we see that

(3.20) $$E\|X(t,\omega)\|_{L^{\psi_{q'}}(d\nu)} \geq d_{q,q'} \inf_{1\leq i\neq j\leq M(\varepsilon)} d_X(t_i,t_j)(\log M(\varepsilon))^{1/q - 1/q'} \ ,$$

where $d_{q,q'}$ is a constant depending only on q and q'. Finally we recall that $d_X(t_i,t_j) \geq \varepsilon$ for $1 \leq i \neq j \leq M(\varepsilon)$ and that, by definition, $M(\varepsilon) \geq N(T,d_X;2\varepsilon)$. Using these two observations in (3.19) along with Lemma 2.7 we get (3.16).

§4. Weaker entropy conditions for L^{ψ_q}.

In this section we show that the sufficient conditions in Theorems 3.1 and 3.3 are not best possible. However in the case of random Fourier series on Sidon sets, we do obtain a condition which is necessary and sufficient. We begin by giving some elementary conditions on processes expressed as series of independent terms to have sample paths in L^{ϕ_q}.

Lemma 4.1: (i) Let $\{a_k\}_{k=1}^{\infty}$ be real numbers satisfying $\|\{a_k\}_{k=1}^{\infty}\|_{p,\infty} < \infty$. Let $\{\eta_k\}_{k=1}^{\infty}$ be i.i.d. symmetric real valued random variables with $E|\eta_1|^p < \infty$ and let $\{\phi_k(t)\}_{k=1}^{\infty}$ be real valued functions on T satisfying $\sup_{1\le k\le\infty} \sup_{t\in T} |\phi_k(t)| < \infty$. Then for $(T,\mathcal{T},\mu)$ as in Theorem 3.1, $\frac{1}{p}+\frac{1}{q}=1$ and $q \geq 2$ we have

$$(4.1)\qquad E\Big\|\sum_{k=1}^{\infty} a_k\eta_k\phi_k(t)\Big\|_{L^{\phi_q}(d\mu)} \le C_q'\|\{a_k\}_{k=1}^{\infty}\|_{p,\infty}(E|\eta_1|^p)^{1/p} \sup_{1\le k\le\infty}\sup_{t\in T}|\phi_k(t)|$$

where C_q' is a constant depending only on q.

(ii) Let $\{a_k\}_{k=1}^{\infty}$, as above, satisfy $\|\{a_k\}_{k=1}^{\infty}\|_p < \infty$ and let $\{\xi_k\}_{k=1}^{\infty}$ be i.i.d. symmetric real valued p-stable random variables i.e.,

$$(4.2)\qquad E\exp i\lambda\xi_1 = \exp -|\lambda|^p.$$

Then, for $(T,\mathcal{T},\mu)$ as in Theorem 3.1, $\frac{1}{p}+\frac{1}{q}=1$ and $q > 2$ we have

$$(4.3)\qquad E\Big\|\sum_{k=1}^{\infty} a_k\xi_k\phi_k(t)\Big\|_{L^{\phi_q}(d\mu)} \le C_q\Big(\sum_k |a_k|^p\Big)^{1/p} \sup_{1\le k\le\infty}\sup_{t\in T}|\phi_k(t)|$$

where C_q is a constant depending only on q.

Proof: These inequalities follow immediately from (2.9) and some inequalities in [5]. Let $\{\varepsilon_k\}_{k=1}^{\infty}$ be a Rademacher sequence independent of $\{\eta_k\}_{k=1}^{\infty}$ and let $\{\varepsilon_k\}_{k=1}^{\infty}$ be defined on $(\Omega,\mathcal{F},P)$ and $\{\eta_k\}_{k=1}^{\infty}$ be defined on $(\Omega_1,\mathcal{F}_1,P_1)$. We consider the series $\sum_{k=1}^{\infty} a_k\varepsilon_k\eta_k\phi_k(t)$ to be defined on the product probability space $(\Omega\times\Omega_1, \mathcal{F}\times\mathcal{F}_1, P\times P_1)$. Clearly this series is equal in distribution to

$$(4.4)\qquad X(t;\omega,\omega_1) = \sum_{k=1}^{\infty} a_k\varepsilon_k(\omega)\eta_k(\omega_1)\phi_k(t), \quad t\in T.$$

Let E_ε denote expectation with respect to $(\Omega,\mathcal{F},P)$, E_1 denote expectation with respect to (Ω_1, F_1, P_1) and let $\omega_1 \in \Omega_1$. Then by (2.9) we have

$$(4.5) \qquad E_\varepsilon \|X(t,\omega_1,\cdot)\|_{L^{\phi_q}(d\mu)} \leq C_q \|\{a_k \eta_k\}_{k=1}^\infty\|_{p,\infty} \sup_{1 \leq k \leq \infty} \sup_{t \in T} |\phi_k(t)|.$$

The inequality in (4.1) follows from (4.5) since by (3.22) of [5]

$$(4.5a) \qquad E_1 \|\{a_k \eta_k\}_{k=1}^\infty\|_{p,\infty} \leq \|\{a_k\}_{k=1}^\infty\|_{p,\infty} E_1 \|\{k^{-1/p} \eta_k\}_{k=1}^\infty\|_{p,\infty}$$

$$\leq h_q \|\{a_k\}_{k=1}^\infty\|_{p,\infty} (E|\eta_1|^p)^{1/p}$$

where h_q is a constant depending only on q.

By exactly the same method we see that

$$(4.6) \qquad E\|\sum_{k=1}^\infty a_k \xi_k \phi_k(t)\|_{L^{\phi_q}(d\mu)} \leq C_q \, E\|\{a_k \xi_k\}_{k=1}^\infty\|_{p,\infty} \sup_{1 \leq k \leq \infty} \sup_{t \in T} |\phi_k(t)|.$$

We use (4.6) to obtain (4.3) since by Corollary 3.8, [5]

$$E\|\{a_k \xi_k\}_{k=1}^\infty\|_{p,\infty} \leq h\left(\sum_{k=1}^\infty |a_k|^p\right)^{1/p}$$

where h is a constant. This completes the proof of Lemma 4.1.

We have the following corollary for p-stable series.

<u>Corollary 4.2</u>: Let $\{a_k\}_{k=1}^\infty$ be real numbers and let $\{\phi_k(t)\}_{k=1}^\infty$ be real valued functions on T satisfying $\sup_{1 \leq k \leq \infty} \sup_{t \in T} |\phi_k(t)| = 1$. Consider the series $\sum_{k=1}^\infty a_k \xi_k \phi_k(t)$ as defined in (ii) of Lemma 4.1 with $1 < p \leq 2$, $\frac{1}{p} + \frac{1}{q} = 1$, $\frac{1}{p'} + \frac{1}{q'} =$ and $q < q' < \infty$. Then for $(T, \mathcal{T}, \mu)$ as in Theorem 3.1 we have

$$(4.7) \qquad E\|\sum_{k=1}^\infty a_k \xi_k \phi_k(t)\|_{L^{\phi_q}(d\mu)} \leq C_q \left(\sum_{k=1}^\infty |a_k|^p\right)^{1/p}$$

and

$$(4.8) \qquad E\|\sum_{k=1}^\infty a_k \xi_k \phi_k(t)\|_{L^{\phi_{q'}}(d\mu)} \leq C_{q'} \|\{a_k\}_{k=1}^\infty\|_{p',\infty},$$

where C_q and $C_{q'}$ are constants depending only on q and q'.

Proof: The statement in (4.7) is merely (4.3). The statement in (4.8) follows from (4.1) since if ξ_1 is p-stable, $E|\xi_1|^{p'} < \infty$.

A reinterpretation of (4.7) leads to the following interesting statement.

Corollary 4.3: Let $(T,\mathcal{T},\mu)$ be as in Theorem 3.1. The map from $L^\infty(T)$ to $L^{\psi_q}(d\mu)$ is stable type p for $1 < p < 2$.

Proof: By definition we need to show that for all $x_1, \ldots, x_n \in L^\infty(T)$; $\xi_1, \ldots, \xi_n$ i.i.d. and satisfying (4.2) and all $n \in \mathbb{N}$ there exists a constant C such that

$$E\|\sum_{k=1}^{n} \xi_k x_k\|_{L^{\psi_q}(d\mu)} \leq C\left(\sum_{k=1}^{n} \|x_k\|^p_{L^\infty(T)}\right)^{1/p} .$$

But this follows immediately from (4.7) since we can write $x_k = \|x_k\|_{L^\infty(T)}(x_k/\|x_k\|_{L^\infty(T)})$.

Finally we note that we can extend (4.7) to the larger class of p-stable processes defined in (3.9) and (3.10).

Corollary 4.4: Let $\{\tilde{X}(t),\ t \in T\}$ be a p-stable stochastic process on $R^T(\mathbb{C}^T)$ determined by (3.9) ((3.10)), $1 < p < 2$. Assume further that the measure m is supported on the functions $\{\beta(t),\ t \in T\}$ which satisfy $\sup_{t\in T}|\beta(t)| \leq 1$. Let $(T,\mathcal{T},\mu)$ be as given in Theorem 3.1. Then $\{\tilde{X}(t),\ t \in T\}$ has a version $\{X(t),\ t \in T\}$ in $L^{\psi_q}(d\mu)$ a.s. and furthermore

$$E\|X(t)\|_{L^{\psi_q}(d\mu)} \leq C'_q , \qquad \frac{1}{p} + \frac{1}{q} = 1,$$

where C'_q is a constant depending only on q.

Proof: The proof follows from a representation of $\{\tilde{X}(t),\ t \in T\}$ that is given in [5]. Here we will sketch the proof in the real case using the terminology of [5].

Note that for the complex case we can simply use this proof for the real and imaginary parts separately. We can find a version of $\{\tilde{X}(t),\ t \in T\}$ expressed as the series

$$X(t) = \sum_{k=1}^{\infty} \varepsilon_k (\Gamma_k)^{-1/p} \beta_k(t), \quad t \in T$$

where $\Gamma_k = X_1 + \dots + X_k$ for $\{X_k\}_{k=1}^{\infty}$ i.i.d. random variables with $P(X_1 > \lambda) = e^{-\lambda}$, $\{\varepsilon_k\}_{k=1}^{\infty}$ is a Rademacher sequence and $\{\beta_k\}_{k=1}^{\infty}$ is an i.i.d. sequence of R^T valued random variable where the distribution of β_1 is given by m. Furthermore $\{\varepsilon_k\}_{k=1}^{\infty}$, $\{\Gamma_k\}_{k=1}^{\infty}$ and $\{\beta_k\}_{k=1}^{\infty}$ are independent of each other. Following the proof of Lemma 4.1 we get

$$E\|X(t)\|_{L^{\phi_q}(d\mu)} < C_q\, E\|\{\Gamma_j^{-1/p}\}_{j=1}^{\infty}\|_{p,\infty}$$

$$< C_q\, E \sup_{1<j<\infty} \left(\frac{j}{\Gamma_j}\right)^{1/p} < C_q',$$

where at the last step we use (1.25) of [5].

We now consider random Fourier series where the group characters are contained in a Sidon set. Let G be a compact Abelian group with dual group Γ and let μ denote Haar measure on G normalized so that $\mu(G) = 1$. Let $\Lambda = \{\gamma_k\colon k \in \mathbb{N}\} \subset \Gamma$ be a Sidon set. Λ is called a Sidon set if for all sequences $\{a_k\}_{k=1}^{\infty} \in \mathbb{C}^{\mathbb{N}}$, there exists a constant $S(\Lambda)$, depending on Λ but not on $\{a_k\}_{k=1}^{\infty}$, such that

(4.9) $$\sup_{t\in G} \sum_{k=1}^{\infty} |a_k\gamma_k(t)| \geq S(\Lambda) \sum_{k=1}^{\infty} |a_k|.$$

Let $\{\xi_k\}_{k=1}^{\infty}$ be an i.i.d symmetric real valued p-stable sequence of random variables as defined in (4.2). We will be concerned with the series

(4.10) $$Y(t) = \sum_{k=1}^{\infty} a_k\xi_k\gamma_k(t), \quad t \in G$$

where, to repeat, $\gamma_k \in \Lambda$, $k = 1, \dots, \infty$. Associated with this series we consider the metric or pseudometric

$$\delta_p(s,t) = \Big(\sum_{k=1}^{\infty} |a_k|^p |\gamma_k(s) - \gamma_k(t)|^p\Big)^{1/p} . \tag{4.11}$$

The next result applies to these series.

<u>Theorem 4.5.</u> Let $\{Y(t), t \in G\}$ be a p-stable random Fourier series as defined in (4.10), i.e. the group characters $\{\gamma_k\}_{k=1}^{\infty}$ are contained in a Sidon set, and let δ_p be given by (4.11). Let $\frac{1}{p} + \frac{1}{q} = \frac{1}{p'} + \frac{1}{q'} = 1$ and let $1 < p' < p \le 2$. Then we have

$$S(\Lambda)^{-1} b_{p'} \|\{a_k\}_{k=1}^{\infty}\|_{p',\infty} \le E\|Y\|_{L^{\psi_{q'}}(d\mu)} \le S(\Lambda) B_{p,p'} \|\{a_k\}_{k=1}^{\infty}\|_{p',\infty} \tag{4.12}$$

and

$$\begin{aligned} S(\Lambda)^{-1} c_{p,p'} \|\{a_k\}_{k=1}^{\infty}\|_{p',\infty} &\le \sup_{\varepsilon>0} \varepsilon\big(\log N(G, \delta_p;\varepsilon)\big)^{1/q - 1/q'} \\ &\le C_{p,p'} \|\{a_k\}_{k=1}^{\infty}\|_{p',\infty} \end{aligned} \tag{4.13}$$

where $S(\Lambda)$ is given in (4.9) and $b_{p'}$, $B_{p,p'}$, $c_{p,p'}$ and $C_{p,p'}$ are constants depending only on p and p'.

<u>Proof:</u> Without loss of generality, we assume that $|a_k|$ is non-increasing in k. The inequalities in (4.12) follow from Proposition 2.7 [8], in which it is shown that

$$(S(\Lambda)A_{p'})^{-1} \|\{a_k\}_{k=1}^{\infty}\|_{p',\infty} \le \Big\|\sum_{k=1}^{\infty} a_k\gamma_k\Big\|_{L^{\psi_{q'}}(d\mu)} \le S(\Lambda)B_{p'}\|\{a_k\}_{k=1}^{\infty}\|_{p',\infty}. \tag{4.14}$$

Thus we have

$$(S(\Lambda)A_{p'})^{-1} E\|\{\alpha_k\xi_k\}_{k=1}^{\infty}\|_{p',\infty} \le E\|Y\|_{L^{\psi_{q'}}(d\mu)} \le S(\Lambda)B_{p'}E\|\{a_k\xi_k\}_{k=1}^{\infty}\|_{p',\infty}. \tag{4.15}$$

Using (4.15) and (4.5a) we get the upper bound in (4.12). (Of course $E|\xi_1|^{p'} < \infty$, $\forall\ p' < p$.) To obtain the lower bound in (4.12) we note that

$$\begin{aligned} E\|\{a_k\xi_k\}_{k=1}^{\infty}\|_{p',\infty} &\ge h_{p'}\|\{a_k E\ \xi_k\}_{k=1}^{\infty}\|_{p',\infty} \\ &= h_{p'} E\ \xi_k\ \|\{a_k\}_{k=1}^{\infty}\|_{p',\infty}. \end{aligned}$$

These inequalities are possible, because, as is well known, $\| \|_{p',\infty}$ is equivalent to a norm on the subspace of R^N where it is finite.

By (4.1) and the fact that $E|\xi_1|^{p'} < \infty$ we see that for <u>any</u> probability measure ν on G

$$E\|Y\|_{L^{\psi_{q'}}(d\nu)} \le C'_{p,p'}\|\{a_k\}_{k=1}^{\infty}\|_{p',\infty} . \tag{4.16}$$

If $\{a_k\}_{k=1}^{\infty}$ and $\{\gamma_k\}_{k=1}^{\infty}$ are real then $\{Y(t),\ t \in G\}$ is a p-stable process and Theorem 3.5, in particular (3.20) and the lines following it give the upper bound in (4.13). Exactly the same proof works in the complex case however we first need to clarify some minor technical points which we will do in Remark 4.7 below.

To obtain the lower bound in (4.13) let us assume that $\|\{a_k\}_{k=1}^{\infty}\|_{p',\infty} = 1 + \delta$ for some $\delta > 0$. Then there exists an n for which $|a_n| \ge n^{-1/p'}$. Therefore

$$\delta_p(s,t) \ge n^{-1/p'}\Big(\sum_{k=1}^{n} |\gamma_k(s) - \gamma_k(t)|^p\Big)^{1/p} \equiv \tilde{\delta}_p$$

and

$$N(G, \delta_p; \varepsilon) \ge N\big((G, \tilde{\delta}_p; \varepsilon) . \tag{4.17}$$

Let

$$m(\varepsilon) = \mu\Big[u \in G:\ n^{-1/p'}\Big(\sum_{k=1}^{n} |1 - \gamma_k(u)|^p\Big)^{1/p} < \varepsilon\ \Big] .$$

It is well known that

$$N(G, \tilde{\delta}_p; \varepsilon) \ge \frac{1}{m(\varepsilon)} \tag{4.18}$$

(see, e.g. Lemma 1.1, Chapter II, [4]).

Let us note that

$$|1 - \gamma_k(u)|^p = 2^p\left|\frac{1 - \operatorname{Re}\gamma_k(u)}{2}\right|^{p/2} \ge 2^{p-1}(1 - \operatorname{Re}\gamma_k(u))$$

since $\left|\frac{1 - \operatorname{Re}\gamma_k(u)}{2}\right| \le 1$ and $p/2 \le 1$. Thus

$$(4.19)\qquad m(\varepsilon) \le \mu[u \in G: (\sum_{k=1}^{n} (1 - \mathrm{Re}\,\gamma_k(u)))^{1/p} < \varepsilon n^{1/p'} 2^{1/p - 1}]$$

$$= \mu[u \in G: \sum_{k=1}^{n} \mathrm{Re}\,\gamma_k(u) > n - (\varepsilon n^{1/p'} 2^{1/p - 1})^p].$$

Let $\bar{\varepsilon} = (2^{2/p - 1} n^{1/p' - 1/p})^{-1}$. Substituting this in (4.19) we get

$$m(\bar{\varepsilon}) \le \mu[u \in G: \sum_{k=1}^{n} \mathrm{Re}\,\gamma_k(u) > n/2]$$

$$= \mu[u \in G: \frac{\sum_{k=1}^{n} \mathrm{Re}\,\gamma_k(u)}{\lambda n^{1/p'}} > \frac{n^{1/q'}}{2\lambda}]$$

$$\le \exp(- \frac{n}{|2\lambda|^{q'}}) \int_G \exp\left|\frac{\sum_{k=1}^{n} \gamma_k(u)}{\lambda n^{1/p'}}\right|^{q'} d\mu .$$

Finally by (4.14) with $\lambda = S(\Lambda)B_{p'}$ we have

$$(4.20)\qquad m(\bar{\varepsilon}) \le 2 \exp(- \frac{n}{|2S(\Lambda)B_{p'}|^{q'}}) \equiv 2 \exp(- \frac{n}{\beta_{p'}}) .$$

Using (4.17), (4.18) and (4.19) we see that

$$\bar{\varepsilon}(\log N(G, \delta_p; \bar{\varepsilon}))^{1/q - 1/q'} \ge 2^{1 - 2/p}(\frac{1}{\beta_{p'}} - \frac{\log 2}{n})^{1/q - 1/q'}$$

and so we see that there exists a constant $\tilde{c}_{p,p'}$ such that

$$(4.21)\qquad \sup_{\varepsilon>0} \varepsilon(\log N(G, \delta_p; \varepsilon))^{1/q - 1/q'} \ge \tilde{c}_{p,p'}(S(\Lambda))^{-1}$$

if $n \ge n_0(p,p')$ for n_0 sufficiently large depending on p and p'. However, it is obvious that the left side of (4.21) is greater than zero for each $n < n_0(p,p')$ and so we get the left side of (4.13) when $\|\{a_k\}_{k=1}^{\infty}\|_{p',\infty} = 1 + \delta$ for any $\delta > 0$. The actual statement in (4.13) follows by homogeneity.

For emphasis we reformulate Theorem 4.5 as follows:

Corollary 4.6: Let $\{Y(t),\ t \in G\}$ and δ_p be as in Theorem 4.6. Then the following are equivalent:

(i) $\|\{a_k\}\|_{p',\infty} < \infty$,

(ii) $E\|Y\|_{L^{\Phi_{q'}}(d\mu)} < \infty$,

(iii) $\sup_{\varepsilon > 0} \varepsilon(\log N(G, \delta_p; \varepsilon))^{1/q - 1/q'} < \infty$.

Remark 4.7: If $\{a_k\}_{k=1}^{\infty}$ and $\{\gamma_k\}_{k=1}^{\infty}$ are real then $Y(t)$ in (4.14) is a strongly stationary p-stable process as defined in (3.9). (In this case Λ is the support of m). However $Y(t)$ is not a strongly stationary p-stable process when $\{\alpha_k\}_{k=1}^{\infty}$ and $\{\gamma_k\}_{k=1}^{\infty}$ are complex. For it to be so we would have to replace ξ_k by $\tilde{\xi}_k$ where $E \exp i \operatorname{Re}(\overline{z\tilde{\xi}_1}) = \exp - |z|^p$. Nevertheless, Theorem 4.5 and Corollary 4.6 can be used to show that (4.16), (4.17) and (i), (ii), and (iii) of Corollary 4.6 also hold for

$$\tilde{Y}(t) = \sum_{k=1}^{\infty} a_k \tilde{\xi}_k \gamma_k(t), \qquad t \in G \tag{4.22}$$

with $\{\gamma_k\}_{k=1}^{\infty} \subset \Lambda$ and δ_p as given in (4.11). (Note that it follows from (4.16) that

$$E\|\tilde{Y}\|_{L^{\Phi_{q'}}(d\nu)} < C''_{p,p'} \|\{a_k\}_{k=1}^{\infty}\|_{p',\infty}$$

where we use the same notation as in (4.16). This inequality enables us to obtain the upper bound in (4.13) in the complex case.)

The reader will note that (iii) is a necessary and sufficient condition for $\{Y(t),\ t \in G\}$, as considered in Theorem 4.5, or for $\{\tilde{Y}(t),\ t \in G\}$ to be in $L^{\Phi_{q'}}(d\mu)$. Furthermore for $\varepsilon \leq \hat{d}$

$$J_{q,q'}(G,\delta_p) \geq \int_0^{\varepsilon} (\log N(G, \delta_p; u))^{1/q - 1/q'} du$$

$$\geq \varepsilon(\log N(G, \delta_p; \varepsilon))^{1/q - 1/q'} .$$

Indeed it is easy to see that $J_{q,q'}(G,\delta_p)$ is not comparable with $\sup_{\varepsilon>0} \varepsilon(\log N(G,\delta_p;\varepsilon))^{1/q - 1/q'}$ when $q < q' \leq \infty$. For example let $G = [0,2\pi]$ and $\gamma_k = e^{i2^k t}$, $k = 1,\ldots,\infty$, with δ_p as given in (4.11). If $|a_k|$ is regularly varying and non-increasing one can easily show that

$$J_{q,q'}(G,\delta_p) \geq \sum_{n=1}^{\infty} \frac{|a_n|}{n^{1/q'}} .$$

(This fact is also obtainable from the much more difficult inequality (1.11), Chapter VII, [4].) Considering the right side of (4.13) we see that there are stochastic processes for which $J_{q,q'}(G,\delta_p) = \infty$ and $\sup_{\varepsilon>0} \varepsilon(\log N(G, \delta_p; \varepsilon))^{1/q - 1/q'} < \infty$. This shows that the bounds in (3.13) and (3.14) are not best possible except, as is well known, when $q' = \infty$.

The above remarks lead to the following question. Let $\{X(t), t \in G\}$ be a strongly stationary p-stable process as defined in (3.9) and (3.10) and let d_X be as defined in (3.11). Let $K \subset G$ be a compact neighborhood of the unit element of G and let μ be Haar measure on G normalized so that $\mu(K) = 1$. Then, for $\frac{1}{p} + \frac{1}{q} = 1$ and $2 < q < q' < \infty$, is

(4.23) $$\sup_{\varepsilon>0} \varepsilon(\log N(K, d_X; \varepsilon))^{1/q - 1/q'} < \infty$$

a necessary and sufficient condition for

(4.24) $$E\|X\|_{L^{\psi_{q'}}(d\mu)} < \infty \; ?$$

We have neither been able to answer this question in the negative nor to give any other examples than the random Fourier series with group characters in a Sidon set for which the answer is positive.

The metric entropy condition that appears in Theorem 4.5 also gives a measure of the size of $\ell_{p',\infty}$ blocks. For a sequence $\{a_k\}_{k=1}^{\infty}$ of non-negative real numbers we define a block

$$B_{\{a_k\}} = \{ x = (x_1, x_2, \ldots) : |x_k| \leq a_k \}$$

where $x \in \mathbb{C}^{\mathbb{N}}$. $B_{\{a_k\}}$ is called an $\ell_{p',\infty}$ block if $\{a_k\}_{k=1}^{\infty} \in \ell_{p',\infty}$. For $x, y \in B_{\{\alpha_k\}}$ let σ_p denote the ordinary ℓ_p metric on $B_{\{a_k\}}$, i.e.

$$\sigma_p(x,y) = \Big(\sum_{k=1}^{\infty} |x_k - y_k|^p \Big)^{1/p} .$$

<u>Theorem 4.8</u>: Let $B_{\{a_k\}}$ be an $\ell_{p',\infty}$ block with $p > p'$. Then

$$(4.25) \qquad c_{p,p'} \|\{a_k\}_{k=1}^{\infty}\|_{p',\infty} \leq \sup_{\varepsilon>0} \varepsilon \big(\log N(B_{\{\alpha_k\}}, \sigma_p;\varepsilon)\big)^{\frac{1}{p'} - \frac{1}{p}} \leq C_{p,p'} \|\{\alpha_k\}_{k=1}^{\infty}\|_{p',\infty}$$

where $N(B_{\{a_k\}}, \sigma_p;\varepsilon)$, the metric entropy of $B_{\{a_k\}}$ with respect to σ_p, is defined in the statement of Theorem 3.1 and $c_{p,p'}$ and $C_{p,p'}$ are constants depending only on p and p'.

Proof: This follows easily by applying Theorem 4.5 to the canonical Sidon set formed by the sequence of coordinate functions on $T^{\mathbb{N}}$.

We conclude this paper by considering the classical random Fourier series and employing a technique used by Salem and Zygmund in [10].

<u>Theorem 4.9</u>: Let $\{\xi_k\}_{k=1}^{\infty}$ be i.i.d symmetric p-stable random variables as defined in (4.2) and $\{a_k\}_{k=1}^{\infty} \in \mathbb{C}^N$ and consider

$$(4.26) \qquad W(t) = \sum_{k=1}^{\infty} a_k \xi_k e^{i2\pi kt}, \quad t \in [0,1]$$

and

$$\sigma_p(u) = \Big(\sum_{k=1}^{\infty} |a_k|^p |e^{i2\pi ku} - 1|^p \Big)^{1/p}, \quad u \in [0,2\pi] .$$

Then, for $\frac{1}{p} + \frac{1}{q} = \frac{1}{p'} + \frac{1}{q'} = 1$ and $2 \leq q < q' < \infty$ we have

$$(4.27) \qquad E\|W\|_{L^{\Phi_{q'}}(d\mu)} \geq C_{p,p'} \sup_n n^{-1/q'} \sum_{k=0}^{n-1} \Big(\sum_{j=2^k}^{2^{k+1}-1} |a_j|^p \Big)^{1/p}$$

where dt denotes integration with respect to Lebesgue measure.

In particular, if $b_k = \left(\sum_{j=2^k}^{2^{k+1}-1} |a_j|^p \right)^{1/p}$ is non-increasing, then

$$E\|W\|_{L^{\psi_{q'}}(dt)} \geq C_{p,p'} \|\{b_{k-1}\}_{k=1}^{\infty}\|_{p',\infty} . \tag{4.28}$$

Here $C_{p,p'}$ and $D_{p,p'}$ are constants depending only on p and p'.

Proof: Let $p_n(t) = \sum_{k=1}^{n} c_k e^{i2\pi kt}$, $t \in [0,1]$, $\{c_k\}_{k=1}^{n} \in \mathbb{C}^N$. Then by Bernstein's Theorem ([3], pg. 41)

$$\lambda[t: p_n(t) > \tfrac{1}{2} \|p_n\|_\infty] \geq \frac{c}{n}$$

where λ is Lebesgue measure and c is a constant independent of n. Therefore

$$\int_0^{2\pi} \exp\left| \frac{\sum_{k=1}^{n} a_k \xi_k e^{i2\pi kt}}{\alpha} \right|^{q'} dt \geq \frac{c}{n} \exp\left| \frac{\left\| \sum_{k=1}^{n} a_k \xi_k e^{i2\pi kt} \right\|_\infty}{2\alpha} \right|^{q'}$$

and consequently

$$\left\| \sum_{k=1}^{2^n-1} a_k \xi_k e^{i2\pi kt} \right\|_{L^{\psi_{q'}}(dt)} \geq C_{q'} n^{-1/q'} \left\| \sum_{k=1}^{2^n-1} a_k \xi_k e^{i2\pi kt} \right\|_\infty . \tag{4.29}$$

Therefore, using (4.29) and Levy's inequality we see that

$$\begin{aligned} E\|W\|_{L^{\psi_{q'}}(dt)} &\geq \frac{1}{2} E\left\| \sum_{k=1}^{2^n-1} a_k \xi_k e^{i2\pi kt} \right\|_{L^{\psi_{q'}}(dt)} \\ &\geq \frac{1}{2} C_{q'} n^{-1/q'} E\left\| \sum_{k=1}^{2^n-1} a_k \xi_k e^{i2\pi kt} \right\|_\infty \\ &\geq C_{p,p'} n^{-1/q'} \sum_{k=0}^{n-1} \left(\sum_{j=2^k}^{2^{k+1}-1} |a_j|^p \right)^{1/p} \end{aligned} \tag{4.30}$$

where, at the last step we use a classical result of Salem and Zygmund which is given in Section 2, [6]. We get (4.27) from (4.30). The inequality in (4.28) follows immediately from (4.27) when b_k is non-increasing.

Remark 4.10: In the last line of (4.30) we used an inequality which also implies that

(4.31)
$$E\left\|\sum_{k=1}^{\infty} a_k\xi_k e^{i2\pi kt}\right\|_{\infty} \geq C_p \sum_{k=0}^{\infty}\left(\sum_{j=2^k}^{2^{k+1}-1} |a_j|^p\right)^{1/p}$$
$$= C_p \|\{b_{k-1}\}_{k=1}^{\infty}\|_1, \quad 1 < p \leq 2,$$

where $\{b_k\}_{k=0}^{\infty}$ is as defined in Theorem 4.9. Comparing (4.28) and (4.31) we see that $\ell_{p',\infty}$ seems to play the same role with respect to $L^{\psi_{q'}}(dt)$ that ℓ_1 does with respect to $L_{\infty}(G) = L^{\psi_{\infty}}(dt)$. This same relationship between ℓ_1 and $\ell_{p',\infty}$ is present in (4.8) since, obviously we also have, in the notation of Corollary 4.2,

$$E\|\sum_{k=1}^{\infty} a_k\xi_k\phi_k(t)\|_{\infty} \leq C_{q'}\|\{a_k\}_{k=1}^{\infty}\|_1 .$$

Let us also note that (4.31) is valid without any conditions on the $\{b_k\}_{k=0}^{\infty}$ but the inequality can not be reversed, even with a different constant. (Although the reverse inequality with an appropriate constant holds if $\{b_k\}_{k=0}^{\infty}$ is non-increasing.) The correct necessary and sufficient condition for the left side of (4.31) to be finite is $J_{q,\infty}(G,\sigma_p) < \infty$ and this only agrees with $\|\{b_{k-1}\}_{k=0}^{\infty}\|_1 < \infty$ when the $\{a_k\}_{k=1}^{\infty}$ are smooth enough. A propos to (4.28) if we consider $\{a_k\}_{k=10}^{\infty}$ of the form

$$a_k = [k^{\alpha}(\log k)^{\beta}(\log\log k)^{\gamma}]^{-1}$$

for real numbers α, β, γ then it is not too difficult to check that for $2 \leq q < q' < \infty$

$$\|\{b_{k-1}\}_{k=1}^{\infty}\|_{p',\infty} < \infty \quad \text{iff} \quad \sup_{\varepsilon>0} \varepsilon\big(\log N(C, \sigma_p; \varepsilon)\big)^{1/q - 1/q'} < \infty .$$

In other words, in so far as we have determined, (4.28) is consistent with a yes answer to the question posed in (4.23) and (4.24).

In conclusion let us mention that these results apply to the study of occupation time distributions and local times of stochastic processes. By a change of variables we can write

$$(4.32) \qquad \int_0^1 \exp\left|\frac{X(t,\omega)}{\alpha}\right|^q dt = \left|\int_0^\infty \exp\left|\frac{u}{\alpha}\right|^q d\lambda(t \in [0,1]: X(t,\omega) > u)\right|$$

where λ is Lebesgue measure. Furthermore, if $\lambda[t \in [0,1]: X(t,\omega) > u]$ is absolutely continuous with respect to Lebesgue measure on the real line we denote its Radon Nikodym derivative by $L(u,\omega)$ ($L(u,\omega)$ is also called the local time of $X(t,\omega)$ with respect to $[0,1]$) and have that (4.32) also

$$= \left|\int_0^\infty \exp\left|\frac{u}{\alpha}\right|^q L(u,\omega)du\right|.$$

Thus the results obtained in this paper concerning the first integral in (4.32) give information about the exponential moments of the occupation time distribution and local time of the relevant stochastic processes. It is significant that these are "almost sure" results. This gives a new slant on much of the current work in this field in which the results are given in "distribution", not "almost surely" (see, e.g. Simeon Berman's paper in this Volume and the references cited therein.)

of variables we can write

$$(4.32) \qquad \int_0^1 \exp\left|\frac{X(t,\omega)}{\alpha}\right|^q dt = \int_0^\infty \exp\left|\frac{u}{\alpha}\right|^q d\lambda(t \in [0,1]: X(t,\omega) > u)$$

where λ is Lebesgue measure. Furthermore, if $\lambda[t \in [0,1]: X(t,\omega) > u]$ is absolutely continuous with respect to Lebesgue measure on the real line we denote its Radon Nikodym derivative by $L(u,\omega)$ ($L(u,\omega)$ is also called the local time of $X(t,\omega)$ with respect to $[0,1]$) and have that (4.32) also

$$= \left|\int_0^\infty \exp\left|\frac{u}{\alpha}\right|^q L(u,\omega)du\right|.$$

Thus the results obtained in this paper concerning the first integral in (4.32) give information about the exponential moments of the occupation time distribution and local time of the relevant stochastic processes. It is significant that these are "almost sure" results. This gives a new slant on much of the current work in this field in which the results are given in "distribution", not "almost surely" (see, e.g. Simeon Berman's paper in this Volume and the references cited therein.)

References

1. DeAcosta, A., Stable measures and semi-norms, Ann. Probability, 3 (1975), 865-875.

2. Fernique, X., Régularité de Fonctions Aléatoires non gaussiennes, Ecole d'Eté de St. Flour, Lecture Notes in Math., 976 (1983), 7-69, Springer Verlag, New York.

3. Kahane, J. P., Some random series of functions, (1968), D. C. Heath, Lexington, Mass. USA

4. Marcus, M. B. and Pisier, G., Random Fourier series with applications to Harmonic Analysis, Ann. Math. Studies, Vol. 101 (1981), Princeton Univ. Press, Princeton, N.J.

5. Marcus, M. B. and Pisier, G., Characterisations of almost surely continuous p-stable random Fourier series and strongly stationary processes, Acta Mathematica, Vol. 152 (1984), 245-301.

6. Marcus, M. B. and Pisier, G., Some results on the continuity of stable processes and the domain of attraction of continuous stable processes, Ann. Inst. Henri Poincaré, Vol. 20, (1984), 177-199.

7. Morrow, G. On a Central Limit Theorem Motivated by Some Random Fourier Series with Dependent Coefficients, preprint.

8. Pisier, G., De nouvelles caracterisations des ensembles de Sidon, Mathematical Analysis and Applications, (19811), 686-726, in the series Advances in Math., Supplementary Studies, 7B, Academic Press, New York, N.Y.

9. Pisier, G., Some applications of the metric entropy condition to harmonic analysis, in Banach spaces, Harmonic Analysis and Probability, Proceeding 80-81. Lecture Notes in Math., 995 (1983), 123-154, Springer-Verlag, New York.

10. Salem, R. and Zygmund, A., Some properties of trigonometric systems whose terms have random signs. Acta Mathematica, Vol. 91 (1954), 245-301.

11. Weber, M., Analyse Infinitesimal de Fonctions Aléatoires, Ecole d'Eté de St. Flour, Lecture Notes in Math., 976 (1983), 384-464, Springer Verlag, New York.

A Skorohod - like representation in infinite dimensions

Terry R. McConnell
Department of Mathematics
Syracuse University
Syracuse, NY 13210

Abstract: Every discrete parameter Banach space-valued martingale may be embedded in a continuous parameter martingale which has continuous sample paths and is adapted to the σ-fields generated by one-dimensional Brownian motion.

1. Introduction and Statement of Results. Let $B(t)$ be standard Brownian motion defined on $(\Omega,\mathcal{F},P)$. Skorohod [14, p. 180] proved that any probability measure ν on $\mathbf{R}$ having mean zero and finite variance may be realized as the distribution of $B(\tau)$, where τ is a stopping time of finite expectation. For example, suppose ν is concentrated on at most two points, $a \leq 0 \leq b$. The probabilities assigned to these points are uniquely determined by the condition that ν have mean zero. Let τ be the time of the first exit of $B(t)$ from $[a,b]$. Then $B(\tau)$ has a distribution which is concentrated on $\{a,b\}$, has mean zero, and hence agrees with ν. In the general case we may represent ν as a mixture of such two-point distributions and use for τ a mixture (randomization) of the two-point stopping times.

This is essentially Skorohod's original approach. It suffers from two shortcomings. The first, primarily esthetic, problem is that the values of τ depend not only on the Brownian path, but also on additional independent information. The second problem is that the method does not apply to measures on $\mathbf{R}^n$ for $n \geq 2$. Subsequent to Skorohod's work there have been a number of ingenious constructions of suitable intrinsically defined stopping times, (see, e.g., [7], [12], [13], [6], and [1]) and a number of partial results in higher dimensions ([10], [3], and [8]).

In dimensions $n \geq 2$ it is generally impossible to represent a given probability measure as the distribution of stopped n-dimensional Brownian motion $B(t)$. Indeed, since $B(t)$ does not hit points, it is impossible to represent any measure which assigns positive probability to any point. However, if one is allowed to select in a non-anticipating way an arbitrary amount of dependence among the components of $B(t)$, then a Skorohod-type representation is always possible, even in infinite dimensions.

Theorem 1. Let B be a Banach space and X a strongly measurable B-valued random variable defined on a probability space $(\Sigma, \mathcal{H}, \mu)$. Suppose also that X satisfies $EX = 0$ and $E\,|X|_B^p < \infty$ for some $1 \le p \le \infty$. Then there is a B-valued martingale $M(t)$ relative to the completed σ-fields generated by one-dimensional standard Brownian motion, such that $M(t)$ has continuous sample paths and converges almost surely and in L^p to a random variable $M(\infty)$ having the same distribution as X.

This is a natural generalization of Skorohod's theorem: Consider the case $B = \mathbf{R}$ and $p = 2$. Every continuous square integrable martingale $M(t)$ may be represented as $B(\tau(t))$ where $\tau(t)$ is a nondecreasing, continuously indexed family of Brownian stopping times (see e.g. [9, Theorem 7.2]). The convergence of $B(\tau(t))$ in L^2 entails the convergence of $\tau(t)$ in L^1 to a stopping time τ which may then be used as the stopping time in Skorohod's theorem.

The proof of Theorem 1 has the following finite dimensional corollary which has also been obtained by Bass [2] using different methods.

Corollary 1. Let ν be a probability measure on $\mathbf{R}^n$ having mean zero and finite second moments. There exists a square integrable, nonanticipating functional, $F(t)$, of n-dimensional Brownian motion, $B(t)$, taking values in the space of $n \times n$ matrices, such that the Itô integral $\int_0^\infty F(t)dB(t)$ has probability distribution given by ν.

One of the most useful ramifications of Skorohod's result - the embedding of discrete parameter martingales in Brownian motion - also has an analogue in higher dimensions.

Theorem 2. Let $f_1, f_2, \ldots,$ be a martingale with values in a Banach space B. There is a martingale $M(t)$ relative to the Brownian σ-fields with continuous sample paths, and there exist stopping times $\tau_0 \le \tau_1 \le \ldots$ such that the sequences f_i and $M(\tau_i)$ have the same distribution.

This result is particularly useful as a method of controlling the jumps of martingales (see [4]).

The proofs of Theorems 1 and 2 and of Corollary 1 are givenproofs rely on methods for martingales ("splicing" and "spreading") which the author learned from D.L. Burkholder. Related constructions are used in [5] and [11].

2. Proofs. We may assume without loss of generality that the Banach space B in which all random variables under consideration take their values is separable. Fix for the remainder of the discussion a countable dense subset $y_1, y_2, \ldots$ of B.

We will prove Theorem 1 in a somewhat stronger form needed for the proof of Theorem 2. Let $B(t)$ be standard Brownian motion defined on a probability space $(\Omega, \mathcal{F}, P)$. Let $(\Sigma, \mathcal{H})$ be a separable measurable space and X an $\mathcal{H}$-measurable random variable which generates $\mathcal{H}$. Let M_p denote the polish space of probability measures μ on $\mathcal{H}$ such that

$$\int_\Sigma |X|_B^p \, d\mu < \infty \quad \text{and} \quad \int_\Sigma X \, d\mu = 0 ,$$

with metric d given by

$$d(\mu,\nu) = \sup_{A \in \mathcal{H}} \left| \int_A |X|_B^p \, d\mu - \int_A |X|_B^p \, d\nu \right| + ||\mu - \nu|| \tag{2.1}$$

We will show that for each $\mu \in M_p$ there is a sequence $x_0, x_1, x_2, \ldots$ of vectors in B and a sequence of Brownian stopping times such that:

(2.2) $f_n = \sum_{i=0}^{n} x_i [B(\tau_{i+1}) - B(\tau_i)]$ converges a.s. and in L_B^p to a random variable $f(\mu)$ having the same distribution as X has under μ;

(2.3) the sequence τ_i converges to an a.s. finite stopping time $\tau(\mu)$; and

(2.4) the maps $\mu \to \tau(\mu)$ and $\mu \to f(\mu)$ are Borel measurable.

We remark here that (2.3) follows directly from (2.2) if and only if B is 2-convexifiable. (See Corollary 3.1 and Proposition [illegible].4 of [11].) For general Banach spaces some care must be exercised in the construction in order to ensure that (2.2) and (2.3) both hold.

The construction breaks into the following four steps.

Step 1. We express X as the sum of a martingale difference sequence X_j, with each X_j a simple function.

Step 2. On a new probability space we construct a "spread out" version of the martingale of Step 1. More precisely we construct a martingale g_n of the form $g_n = \sum_{k=0}^{n} x_k d_k$, where the x_k belong to B, the d_k form a real-valued martingale difference sequence, and g_n converges to a random variable g_∞ having the same distribution as X.

<u>Step 3</u>. We apply real-valued Skorohod embedding to the d_k.

<u>Step 4</u>. We verify (2.2) - (2.4).

Before carrying out these steps in detail, let us indicate how Theorem 1 follows. Define $M(t) = M(t,\mu)$ by

$$M(t) = \sum_{i=0}^{n-1} x_i [B(\tau_{i+1}) - B(\tau_i)] + x_n [B(t) - B(\tau_n)]$$

on $\{\tau_n \le t < \tau_{n+1}\}$ and $M(t) = M_{\tau(\mu)}$ on $\{\tau(\mu) \le t\}$.

Since M(t) has continuous paths and $M(\infty)$ has the same distribution as X there remains only to check that M(t) converges to $M(\infty)$ a.s. and in L^p_B (although $M(\infty)$ is L^p_B-bounded we cannot deduce this convergence directly since B is not assumed to have the Radon-Nikodym property.) Choose a subsequence n_k such that

$$E|M(\tau_{n_k}) - M(\tau_{n_{k-1}})|_B^p \le 2^{-k}.$$

This is possible since we have L^p and almost sure convergence along a subsequence of the τ_n. Let

$$D_k = \sup_{\tau_{n_k} \le t \le \tau_{n_{k+1}}} |M(t) - M(\tau_{n_k})|.$$

By Doob's inequality we have $P(D_k > \varepsilon) \le \varepsilon^{-p} 2^{-k}$ for each $\varepsilon > 0$. It then follows from the Borel-Cantelli Lemma that D_k converges to zero almost surely, hence M(t) converges almost surely.

<u>Step 1</u>. First define Z_n by $Z_n(\sigma) = y_k$ where k is the smallest index such that $|x(\sigma) - y_k| \le 1/(n+1)$. Then the Z_n converge to X uniformly. Let $A_{n,k} = \{\sigma: |X(\sigma) - y_k| \le 1/(n+1)\}$. Choose $N = N(n,\mu)$ as small as possible

so that $\mu(\Sigma \setminus \bigcup_{k=1}^{N} A_{n,k}) \le 1/n$. Then $Y_n = Z_n 1_{\bigcup_{k=1}^{N} A_{n,k}}$ are simple random variables which converge to X μ-almost everywhere and in L^p. Finally, define the martingale difference sequence X_n by $X_0 = Y_0$ and

(2.5) $$\sum_{i=0}^{n} X_i = Y_n - E(\Delta Y_n | Y_0 , Y_1 \dots Y_{n-1}), \; n \ge 1,$$

where $\Delta Y = Y_n - Y_{n-1}$. The convergence of the martingale defined by the X_n to X follows from the convergence of Y_n to X. For future reference, note that all values assumed by X_n and the probabilities of these values depend measurably on μ.

Step 2. We begin with the following lemma.

Lemma 2.1. Let Z be a simple random variable defined on $(\Sigma, \mathcal{H}, \mu)$ assuming values $x_j \in B$, $j = 1,2,\dots,m$. Then there is a probability space $(S, \mathfrak{C}, Q)$ and a martingale difference sequence $u_0, u_1, \dots, u_{m-1}$ defined on this space such that $u_0 = EZ$,

(2.6) $$\sum_{j=0}^{m-1} u_j \overset{d}{=} Z,$$

and

(2.7) each u_j assumes at most 2 nonzero values.

PROOF: We may assume without loss of generality that $EZ = 0$ and $m = 2^N$ for some integer N. We proceed by induction on N. For $N = 1$ take $(S, \mathfrak{C}, Q) = (\Sigma, \mathcal{H}, \mu)$ and $u_1 = Z$.

Assume the result proved for $m = 2^N$ and let Z be a random variable on $(\Sigma, \mathcal{H}, \mu)$ assuming 2^{N+1} values. Let Σ_1 be the event that Z assumes one of the values $x_1, x_2, \dots, x_{2^N}$. Let $Z_1 = Z_{|\Sigma}$, $\mathcal{H}_1 = \{A \cap \Sigma_1 : A \in \mathcal{H}\}$, and define a probability measure μ_1 on Σ_1 by $\mu_1(A \cap \Sigma_1) = \mu(A|\Sigma_1)$. Define similarly a random variable Z_2 on $(\Sigma_2, \mathcal{H}_2, \mu_2)$ by using $x_{2^N+1}, \dots, x_{2^{N+1}}$ in the place of $x_1, x_2, \dots, x_{2^N}$.

By the inductive hypothesis there are two martingale difference sequences u^i_j, $i = 1,2$; $j = 0,1,2,\dots,2^N - 1$, defined on $(S_i, \mathfrak{C}_i, Q_i)$ such that $u^i_0 = EZ_i$

and both (2.6) and (2.7) hold with Z replaced by Z_i.

Let $(S_0, \mathcal{E}_0, Q_0)$ be a probability space on which there is defined a random variable α such that $Q_0(\alpha=1) = 1-Q_0(\alpha=0) = \mu(\Sigma_1)$. Let $(S, \mathcal{E}, Q)$ be the product probability space $S_0 \times S_1 \times S_2$ equipped with the product σ-field $\mathcal{E}$ and product measure Q. Define random variables $u_0, u_1, \ldots, u_{M-1}$ on $(S, \mathcal{E}, Q)$ by $u_0 = 0 = EZ$,

$$u_1(s_0,s_1,s_2) = \alpha(s_0)u_0^1(s_1) + (1-\alpha(s_0))u_0^2(s_2)$$

and

$$u_j(s_0,s_1,s_2) = \begin{cases} \alpha(s_0)u_{j-1}^1(s_1), & 2 \le j \le 2^N \\ (1-\alpha(s_0))u^2_{j-2^N}(s_2), & 2^N < j \le 2^{N+1}-1 . \end{cases}$$

It is easy to verify that $u_0, u_1, \ldots, u_{2^{N+1}-1}$ satisfy (2.6) and (2.7).

Let X_n be the martingale difference sequence constructed in Step 1. Denote by m the number of values assumed by X_n and put $\ell_n = (\prod_{i=0}^{n} m_i)$. We will next use Lemma 2.1 to construct a new martingale difference Δ_k defined on a large probability space. This sequence is constructed in blocks of length ℓ_n and has the properties that each Δ_k assumes at most 2 nonzero values and that the subsequence defined by

$$\sum_{j=0}^{n} \sum_{i=k_{j-1}}^{k_j-1} \Delta_i ,$$

where $k_j = \sum_{\gamma=0}^{j} \ell_\gamma$ and $k_{-1} = 0$, has the same distribution as the sequence of nth partial sums of the X_i. We shall describe the construction of the nth such block in detail: The partial sum $\sum_{j=0}^{n-1} X_j$ is a simple random variable which assumes ℓ_{n-1} values, possibly not all distinct. For each of these values, s_i, let Z_i be a random variable defined on some probability space with distribution equal to the conditional distribution of X_n given that $\sum_{j=0}^{n-1} X_j$ equal

. For each i construct using Lemma 2.1 with $Z = Z_i$ martingale difference quences $u_0^i, u_1^i, \dots, u_{m-1}^i$ on probability spaces $(S_i^{(n)}, \mathfrak{E}_i^{(n)}, Q_i^{(n)})$. We tended these sequences so as to be defined and independent for distinct i the product space

$$(S^{(n)}, \mathfrak{E}^{(n)}, Q^{(n)}) = \prod_{i=1}^{m_n} (S_i^{(n)}, \mathfrak{E}_i^{(n)}, Q_i^{(n)}),$$

d again extend similarly to the product space

$$\prod_{n=0}^{\infty} (S^{(n)}, \mathfrak{E}^{(n)}, Q^{(n)}) .$$

is on the latter space that the entire sequence Δ_n is defined.

To finish the construction of the nth block, assuming all previous blocks ready constructed, let

$$A_i = \{ \sum_{j<k_{n-1}} \Delta_j = s_i \}$$

d define

$$\Delta_{k_{n-1} + m_n(i-1) + \gamma} = 1_{A_i} \cdot u_\gamma^i, \; i = 1,\dots,\ell_{n-1}; \; \gamma = 0,\dots,m_n - 1.$$

By the argument used above to prove convergence of M(t) we have that the rtial sums $\sum_{j=1}^{N} \Delta_j$ converge a.s. and in L^p. It thus follows that we have

$$\sup_N E|\Delta_N|_B^p \le c(\mu) < \infty.$$

nce each Δ_N assumes values in a one-dimensional subspace of B (this is a y point) we may write $\Delta_n = x_n d_n$ where $x_n \in B$ with $|x_n|_B = 2^n$, and the d_n rm a real-valued martingale difference sequence each assuming at most 2 non-ro values and satisfying

.8) $$E|d_n|^p \le 2^{-np} c(\mu).$$

ep 3. By the one-dimensional Skorohod embedding there are stopping times of one-dimensional standard Brownian motion B(t) such that the sequence $\tau_{n+1}) - B(\tau_n)$ has the same distribution as the sequence d_n.

ep 4. The B-valued martingale

$$f_n = \sum_{j=0}^{n} x_j [B(\tau_{j+1}) - B(\tau_j)]$$

converges a.s. and in L to a random variable having the same distribution as X by construction. There remains only to check that (2.3) and (2.4) hold. For (2.3) we have by [4] and (2.8) the estimate

$$P\{\tau_{n+1} - \tau_n \geq \varepsilon\} \leq \sqrt{2}\, \varepsilon^{-1/2}\, 2^{-n} c(\mu)^{1/p}$$

for every $\varepsilon > 0$, and the convergence of τ_n follows from the Borel-Cantelli Lemma.

As for (2.4), first note that each value x_j depends measurably on μ. Moreover, if we agree, say, to use Root's method [12] to construct each τ_n, then the τ_n depend measurably on μ in the topology of convergence in probability. The desired measurability is an easy consequence of these observations. This completes Step 4 and the proof of Theorem 1.

Proof of Corollary 1. (Sketch) By Steps 1-3 of the proof of Theorem 1 we represent ν as the distribution of the limit of $\sum_{j=0}^{N} x_j d_j$, where $x_j \in \mathbf{R}^n$ and d_j is a real-valued martingale difference sequence. Let Π_j denote the orthogonal projection onto the one-dimensional subspace of $\mathbf{R}^n$ spanned by x_j. The strong Markov property of Brownian motion implies that for any stopping time τ of n-dimensional Brownian motion, B(t), the process $\Pi_j(B(t+\tau) - B(\tau))$ is a standard one-dimensional Brownian motion. Using this we may construct inductively stopping times τ such that the sequence $\Pi_j(B(t+\tau_j) - B(\tau_j))$ has the same distribution as d_j. The desired non-anticipating functional is then given by

$$F(t) = \sum_{j=0}^{\infty} |x_j| \Pi_j 1_{[\tau_j, \tau_{j+1}]}(t).$$

Proof of Theorem 2. We shall consider only the case of a two step martingale f_0, f_1. It will be clear how to extend the construction so as to embed martingales of arbitrary finite or infinite length.

Let B(t) be standard Brownian motion defined on $(\Omega, \mathcal{F}, P)$. Let $\mathcal{F}_t$ denote the usual completed σ-fields of B(t). By Theorem 1 there exists a stopping time τ_0 and a continuous martingale $M_0(t)$ with respect to $\mathcal{F}_t$ such that $M_0(\tau_0)$ has the same distribution as f_0. Let $\{\mu_x\}_{x \in B}$ be a regular version of the conditional probabilities $P\{f_1 \in \cdot \mid f_0 = x\}$.

We shall now use the construction in the proof of Theorem 1 in which the measurable space $(\Sigma, \mathcal{H})$ is given by B with its family of Borel sets and the random variable X is the identity function on B. Recall the family M_1. We may assume without loss of generality that $\mu_x \in M_1$ for every choice of $x \in B$. Since the map $x \to \mu_x(A)$ is Borel for each $A \in \mathcal{H}$, it is easy to check that the map $x \to \mu_x$ is measurable from B into M_1. Let $\tau(\mu_x)$ and $M_{(x)}(t)$ be the stopping time and continuous martingale constructed in Theorem 1 with $\mu = \mu_x$. Taking into account the measurability of those constructions and the strong Markov property of B(t) it is then easy to see that

$$\tau_1 = \tau(\mu_{M_0(\tau_0)})\circ\theta_{\tau_0} + \tau_0$$

is an $\mathcal{F}_t$-stopping time and that the new process M(t) defined by

$$M(t) = \begin{cases} M_0(t) & , \quad t \le \tau_0 \\ M_{(M_0(\tau_0))}(t-\tau_0) & , \quad \tau_0 \le t \le \tau_1 \end{cases}$$

is a continuous martingale with respect to $\mathcal{F}_t$ such that we have equality of the distributions of (f_0, f_1) and $(M(\tau_0), M(\tau_1))$. The proof of Theorem 2 is complete.

Acknowledgement

The author wishes to thank the auditors of the graduate topics course in stochastic processes at Cornell University where this material was presented. This work was done when the author was employed by Cornell University.

REFERENCES

[1] Azéma, J. and Yor, M., Une solution simple au probléme de Skorohod, Sem. Probab. XIII. Lecture notes in Math., Springer (1979).

[2] Bass, R.F., Skorohod imbedding via stochastic integrals, Sem. Probab. XVII, Lecture notes in Math., Springer (1983).

[3] Baxter, J.R. and Chacon, R.V., Potentials of stopped distributions, Illinois J. of Math., 18 (1974), 649-656.

[4] Burkholder, D.L., A sharp inequality for martingale transforms, Ann. Probability 7 (1979), 858-863.

[5] Burkholder, D.L., A geometrical characterization of Banach spaces in which martingale difference sequences are unconditional, Ann. Probability 9 (1981), 997-1001.

[6] Chacon, R.V. and Walsh, J.B., One-dimensional potential embedding, Sem. Probab. X, Lecture notes in Math., Springer (1976).

[7] Dubins, L.E., On a problem of Skorohod, Ann. Math. Statist. 39 (1968), 2094-2097.

[8] Falkner, N., On Skorohod embedding in n-dimensional Brownian motion by means of natural stopping times, Sem. Probab. XIV, Lecture notes in Math., Springer (1980).

[9] Ikeda, N. and Watanabe, S., Stochastic differential equations and diffusion processes, North Holland, Amsterdam, 1981.

[10] Kiefer, J., Skorohod embedding of Multivariate RV's, and the Sample DF, Z. Wahrscheinlichkeitstheorie verw. Geb. 24 (1972), 1-35.

[11] Pisier, G., Martingales with values in uniformly convex spaces, Israel J. Math. 20 326-350.

[12] Root, D.H., The existence of certain stopping times of Brownian motion, Ann. Math. Statist. 40 (1969), 715-718.

[13] Rost, H., The stopping distributions of a Markov process, Invent. Math. 14 (1971), 1-16.

[14] Skorohod, A., Studies in the Theory of Random Processes, Addison Wesley, Reading, 1965.

MOMENT INEQUALITIES FOR REAL AND VECTOR p-STABLE STOCHASTIC INTEGRALS[1]

J. Rosinski[2] *and W.A. Woyczynski*
University of North Carolina at Chapel Hill
and
Case Western Reserve University

1. Introduction.

In the present paper we obtain moment inequalities for single and double stochastic integrals with respect to p-stable motion (Section 3). The proofs are based on our own work on the structure of single and multiple p-stable integrals (cf. [12], [11] and [13]) which is summarized in some detail in Section 2, and on the work of R.F. Bass and M. Cranston [1] on inequalities for moments of exit times of a p-stable motion. Their results, as stated in [1], do not apply directly to the situation in which we want to use them, in particular, because one dimensional processes are explicitly excluded there. So, we offer the needed variation of their result in Section 3 and, for the sake of completeness, provide its full proof in the Appendix.

In Section 4 we propose an extension of the theory of stochastic integration with respect to a p-stable motion, to the case when the latter takes values in a Banach space.

2. Single and double p-stable integrals.

Let (Ω,F,P) be a probability space and let $(F_t)_{t>0}$ be a right continuous, increasing family of P-complete sub-σ-fields of F. Let $0 < p < 2$. We will denote by $(M(t))_{t>0}$ an (F_t)-p-stable motion i.e. an (F_t)-adapted process with $M(0)=0$, sample paths a.s. in $D[0,\infty)$ and

$$E\{\exp [i\lambda(M(t) - M(s))]|F_s\} = \exp [-(t-s)|\lambda|^p]$$

for every $0 \leq s \leq t$, and $\lambda \in \mathbb{R}$. For a simple (F_t)-adapted process F such that

$$F(t,\omega) = \begin{cases} \phi_0(\omega) & \text{for } t=0 \\ \phi_i(\omega) & \text{for } t_i < t \leq t_{i+1},\ i=0,1,\ldots, \end{cases}$$

[1] Research supported by AFOSR Grant No. F49620 82 C 0009
[2] On leave from Wroclaw University

the stochastic integral is defined as usual:

$$\int_0^t F(s,\omega)dM(s,\omega) = \sum_{i=0}^{n-1} \phi_i(\omega)(M(t_{i+1},\omega)-M(t_i,\omega))+\phi_n(\omega)(M(t,\omega)-M(t_n,\omega)),$$

if $t_n \leq t \leq t_{n+1}$, $n=0,1,2,\ldots$. Clearly, the above integral is a process with sample paths a.s. in $D[0,\infty)$.

DEFINITION 2.1. An (F_t)-adapted measurable process $F=(F(t,\omega))_{t\geq 0}$ is said to be M-integrable if there exists a sequence (F_n) of simple (F_t)-adapted processes such that for each $T > 0$:

(i) $F_n \to F$ in measure $dPdt$ on $\Omega \times [0,T]$ as $n \to \infty$,

(ii) $\int_0^t F_n dM$ converge a.s. uniformly in $t \in [0,T]$ as $n \to \infty$,

and the limiting process in (ii) does not depend on the choice of a sequence (F_n) satisfying conditions (i) and (ii). This limit process (with sample paths a.s. in $D[0,\infty)$) will be denoted by $\int_0^t F(s)dM(s)$, $t \geq 0$.

THEOREM 2.1. ([12]) *The process* F *is* M-*integrable if and only if* $F \in L^p_{a.s.}$ *i.e. if*

$$P\{\int_0^T |F(t,\omega)|^p dt < \infty\} = 1$$

for each $T > 0$.

The sufficiency in the above theorem (which also follows from a general result of O. Kallenberg [5]) is obtained by means of a pathwise construction which parallels a known Brownian integral construction and which depends on the following inequality for simple processes F: there exists a constant $c = c(p) > 0$ such that for each $T > 0$

$$c^{-1}||F||^p_{p,T} \leq \sup_{\lambda>0} \lambda^p P\{\sup_{t\leq T} |\int_0^t F dM| > \lambda\} \leq c||F||^p_{p,T},$$

where

$$||F||^p_{p,T} = E\int_0^T |F(s,\omega)|^p ds.$$

The inequality implies that the mapping $F \to \int F dM$ extends to an isomorphic embedding of $L^p(L^p)$ into a Lorentz space $\Lambda^p(L^\infty)$. The upper estimate was obtained by E. Giné and M.B. Marcus in [4].

The proof of necessity uses device of the inner clock for p-stable stochastic integrals the usefulness thereof is established by the following:

THEOREM 2.2 ([12]). *Let* $F \in L^p_{a.s.}$ *be such that*

$$\tau(u) \overset{df}{=} \int_0^u |F|^p dt \to \infty \quad \text{a.s.}$$

as $u \to \infty$. *Then, if*

$$\tau^{-1}(t) = \inf\{u: \tau(u) > t\} \text{ and } A_t = F_{\tau^{-1}(t)},$$

then the time-changed stochastic integral

$$\tilde{M}(t) = \int_0^{\tau^{-1}(t)} F(s)dM(s)$$

is an (A_t)*-p-stable motion.*

The above theorem can also be used to establish properties of integrals which are "pathwise inherited" from the properties of p-stable motion itself. For example, the above result immediately yields the following corollary to the classical Khinchine's result on the local behavior of processes with stationary and independent increments:

THEOREM 2.3 ([12]). *Let* F *be as in Theorem* 2.2 *and suppose that* $\phi: (0,\infty) \to \mathbb{R}^+$ *is such that* $t^{1/p}\phi(t)$ *is increasing and* $\lim_{t\to 0}\phi(t)=\infty$. *Then*

$$\int_0^t F(s)dM(s) = o(\tau^{1/p}(t)\phi(\tau(t))) \quad \text{a.s.}$$

as $t \to 0$, *if and only if*

$$\int_0^1 t^{-1}\phi^{-p}(t)\, dt < \infty .$$

Theorem 2.1 implies that the necessary and sufficient condition for existence of the double integral

$$\int_0^T \left(\int_0^t f(s,t)dM(s)\right)dM(t) \tag{2.1}$$

is that

$$(2.2)\qquad P\{\int_0^T |F(t)|^p dt < \infty\} = 1,$$

where

$$F(t) = \int_0^t f(s,t)dM(s), \qquad t \in [0,T] .$$

The condition (2.2) is equivalent to the property that the integral operator

$$L^{p'}[0,T] \ni \phi \to \int_0^T f(s,t)\chi_\Delta(s,t)\phi(t)\, dt \in L^p[0,T],$$

where $1/p'+1/p=1$, and $\Delta = \{(s,t): 0 \leq s < t \leq T\}$, is Θ_p-radonifying (or, by Kwapien-Maurey Theory, completely summing) (cf. [7], [3]). The above equivalence follows, in particular, from the following result which gives a natural necessary condition for f to satisfy (2.2). Although this result may have been known in the folklore, we were unable to locate a published proof of it and decided to provide our own proof below. Proof of Thm 2.4 relies on Prop. 2.1.

PROPOSITION 2.1. *Let T be a measurable space, μ be a σ-finite measure on T, and let*

$$X(t) = \int_0^1 f(t,s)dM(s) , \qquad t \in T,$$

be a p-*stable process, where* $f: T \times [0,1] \to \mathbb{R}$ *is a jointly measurable function. Then if*

$$\int_T |X(t)|^p \mu(dt) < \infty \qquad a.s.$$

then

$$\int_0^1 \int_T |f(t,s)|^p \mu(dt)\, ds < \infty .$$

Proof. Observe that $\forall q < p \; \exists \; C \; \forall \; t_1,\dots,t_n \in T$

$$(2.3)\qquad \left(\int_0^1 \frac{1}{n}\sum_{i=1}^n |f(t_i,s)|^p ds\right)^{1/p} \leq C\left(E\left(\frac{1}{n}\sum_{i=1}^n |X(t_i)|^p\right)^{q/p}\right)^{1/q} .$$

Note, that (2.3) is just a special case of the "stable-cotype-p" inequality (valid in an arbitrary Banach space E, cf. e.g. [7], Cor. 7.3.5):

$$\left(\int_0^1 ||\vec{f}(s)||_E^p \, ds\right)^{1/p} \le C\left(E\left|\left| \int_0^1 \vec{f}(s)dM(s)\right|\right|_E^q\right)^{1/q},$$

where $\vec{f}$ is taken to be as follows:

(2.4) $$\vec{f}\colon [0,1] \ni s \to \sum_{i=1}^{n} I_{[\frac{i-1}{n},\frac{i}{n}]}(u)f(t_i,s) \in L_p[(0,1),du] \equiv E.$$

Assume initially that $\mu(T) = 1$, and define (on a new probability space (Ω_U, P_U)) a sequence of i.i.d. random variables $U_n\colon \Omega_U \to T$, $n=1,2,\ldots$, such that $L(U_n) = \mu$. Then, by (2.3) for each $\omega_U \in \Omega_U$

(2.5)
$$\begin{aligned} &C\left(E_X\left[\frac{1}{n}\sum_{i=1}^{n}|X(U_i)|^p\right]^{q/p}\right)^{1/q} \\ &\ge \left(\int_0^1 \frac{1}{n}\sum_{i=1}^{n}|f(U_i,s)|^p ds\right)^{1/p} \\ &= \left(\frac{1}{n}\sum_{i=1}^{n}||f(U_i,.)||^p_{L^p((0,1),ds)}\right)^{1/p}. \end{aligned}$$

Now, by the Kolmogorov's Law of Large Numbers, for each ω such that $\int|X(t,\omega)|^p d\mu(t) < \infty$ we get that

$$\frac{1}{n}\sum_{i=1}^{n} |X(U_i,\omega)|^p \underset{P_U\text{-a.s.}}{\to} E_U\, X(U_1,\omega)|^p = \int_T |X(t,\omega)|^p d\mu(t) < \infty$$

as $n \to \infty$. By Fubini's Theorem, for P_U-almost all ω_U's

$$\frac{1}{n}\sum_{i=1}^{n}|X(U_i(\omega_U),.)|^p \to \int |X(t)|^p d\mu(t) \qquad P\text{-a.s.}$$

Since for any p-stable random vectors $Y, Y_1, Y_2, \ldots$ with values in a Banach space E, and any $q < p$, $||Y_n||_E \to ||Y||_E$ in P if and only if $E||Y_n||_E^q \to E||Y||_E^q$, (cf. [7], Prop. 7.3.11) we can use the same idea as in (2.4) to obtain that P_U-a.s.

$$E_X\left(\frac{1}{n}\sum_{1}^{n}|X(U_i)|^p\right)^{q/p} \to E_X\left(\int_T |X(t)|^p d\mu(t)\right)^{q/p}$$

as $n \to \infty$. By (2.5)

$$\sup_n \frac{1}{n}\sum_{1}^{n}||f(U_{i,.})||^p_{L^p(ds)} < \infty, \qquad P_U\text{-a.s.},$$

and by Kolmogorov's Law of Large Numbers

$$\int_{T}\int_{0}^{1} |f(t,s)|^p ds\mu(dt) = E_U \| f(U_i,.)\|^p_{L^p(ds)} < \infty ,$$

which completes the proof in the case $\mu(T) = 1$.

Notice now, that letting $n \to \infty$ in (2.5) one immediately obtains the inequality

$$\left(\int_T\int_0^1 |f(t,s)|^p ds\mu(dt)\right)^{1/p} \leq C\left(E\left(\int_T |X(t)|^p d\mu(t)\right)^{q/p}\right)^{1/q}$$
$$= C\left(E\|X(.)\|^q_{L^p(T)}\right)^{1/q} < \infty$$

from which the extension to σ-finite μ's follows. Q.E.D.

The necessary condition for a.s. p-integrability of sample paths of p-stable processes established by Proposition 2.1 is, however, not sufficient and the following result gives full analytic description of kernels f which have the property (2.2):

THEOREM 2.4. [12]. *Let* $1 \leq p < 2$ *and let the parameter set* T *be a separable metric space equipped with a* σ*-finite Borel measure* μ*. For a measurable symmetric* p*-stable process*

$$X(t) = \int_0^1 f(s,t)dM(s) , \quad t \in T,$$

we have that

$$P\{\int_T |X(t)|^p \mu(dt) < \infty\} = 1,$$

if and only if

$$A_p(f) \overset{df}{=}$$

$$\int_T\int_0^1 |f(s,t)|^p [1+\log_+ \frac{|f(s,t)|^p \int_T\int_0^1 |f(s,t)|^p \mu(dt)ds}{\int_0^1 |f(s,t)|^p ds \int_T |f(s,t)|^p \mu(dt)}]ds\mu(dt) < \infty .$$

The proof depends on the following two facts:

(i) $X(t)$ has sample paths a.s. in $L^p(T,\mu)$ if and only if the series $\Sigma j^{-1/p}\ r_j V_j(t)$ converges a.s. in $L^p(T,\mu)$, where (r_j) are Rademacher r.v.'s and V_j are independent copies of a process $V(t)$, $t \in T$, which has sample paths a.s. in a sphere of $L^p(T,\mu)$, and which has finite dimensional distributions completely determined by f (cf [12] but the idea really goes back to [8], Remark 3.15);

(ii) for i.i.d. symmetric r.v.'s $X, X_1, X_2, \ldots$

$$c^{-1}E|X|^p(1+\log_+\frac{|X|}{E|X|^p}) \le E|\sum_{j=1}^{\infty} j^{-1/p}X_j|^p \le cE|X|^p(1 + \log_+\frac{|X|^p}{E|X|^p}),$$

where $0 < p < q$, and $c = c(p)$ is a numerical constant (cf [3]).

COROLLARY 2.1 ([12]). *Let* $1 \le p < 2$. *The double integral*

$$\int_\Delta\int f(s,t)dM(s)dM(t) \quad ,$$

$\Delta = \{(s,t): 0 \le s < t \le T\}$, *exists if and only if* $A_p(fI_\Delta) < \infty$.

For f constant on rectangles the double integral (2.1) becomes a random quadratic form

$$Q^2(a) = \sum_{i<j} a(i,j)\, M_i M_j \;,$$

where $M_1, M_2, \ldots$ are i.i.d. p-stable random variables, and one immediately obtains from Corollary 2.1 the following

COROLLARY 2.2 ([3]). *Let* $1 \le p < 2$. $Q^2(a)$ *converges a.s. if and only if*

$$\sum_{i<j}|a(i,j)|^p[1+\log_+\frac{|a(i,j)|^p}{\sum_{l=1}^{i-1}|a(l,j)|^p \sum_{l=j+1}^{\infty}|a(i,l)|^p}] < \infty .$$

Although we don't have at this point a good theory of n-tuple p-stable integrals for $n \ge 3$, the theorem below, concerning general multilinear random forms may be considered as a step towards such a theory.

THEOREM 2.5 ([13]). *Let* $0 < p < 2$. *Let* $X, X_1, X_2, \ldots$ *be i.i.d. with symmetric distributions such that*

$$\lim_{x\to\infty} x^p P\{|X| > x\} = c > 0.$$

If

$$N_p^{(k)}(a) \overset{df}{=} \sum_{i_1<i_2<\ldots<i_k} |a(i_1,\ldots,i_k)|^p(1+\log_+^{k-1}|a(i_1,\ldots,i_k)|^{-1}) < \infty$$

then the sequence

$$Q_n^{(k)}(a) = \sum_{i_1<i_2<\ldots<i_k\leq n} a(i_1,\ldots,i_k)X_{i_1}\ldots X_{i_k}, \quad n=1,2,\ldots,$$

converges unconditionally (i.e. $Q_n^{(k)}(\varepsilon a)$ *converges for all* $\varepsilon: \mathbb{N}^k \to \{-1,1\}$) *in* L^q *for every* $q<p$ *to a* $Q^{(k)}(a)$ *which, for all* $x>0$ *satisfies the following inequality*

$$P\{|Q^{(k)}(a)|>x\} \leq D_{k,p}x^{-p}(1 + \log_+^{k-1}x)N_p^{(k)}(a),$$

where $D_{k,p}$ *is a constant.*

The proof relies on the tail estimation for $Q_n^{(k)}(a)$ which uses the fact that

$$\lim_{x\to\infty} x^p(p \log x)^{1-k}P\{|X_1\cdot X_2\cdot\ldots\cdot X_k|>x\} = c^k/(k-1)!\ .$$

For further results in this direction see a recent paper by W. Krakowiak and J. Szulga [6].

3. Moment inequalities for exit times of stable processes and for p-stable stochastic integrals.

In this section we present a version of R.F. Bass and M. Cranston's [1] inequalities for moments of exit times of a stable process X in the case when X takes values in a separable Banach space E. As corollaries we also obtain moment inequalities for single and double stochastic integrals with respect to p-stable motion.

Recall that a non-zero E-valued stochastic process $X(t)$, $t \geq 0$, is said to be a *symmetric* p-*stable Lévy process*, $0 < p < 2$, if

(i) X has independent and stationary increments,
(ii) $X(t) - X(s) \overset{D}{=} |t - s|^{1/p}X(1)$ for every $t,s \geq 0$,
(iii) $X(.,\omega) \in D_E[0,\infty)$ and $X(0) = 0$ a.s.

The characteristic functional of $X(t)$ can be written in the form

$$(3.1)\qquad E\exp[ix^*X(t)] = \exp[-t\int_U |x^*x|^p m(dx)],$$

where m is a unique, finite, symmetric (i.e. $m(-B) = m(B)$ for every Borel set $B \subset U$) positive measure on the unit sphere U of E. Such an m is called the spectral measure of X. The distribution of X(t) is infinitely divisible without the Gaussian component and with Lévy measure represented in polar coordinates as $tc_p m(dx)dr/r^{1+p}$, $(r,x) \in (0,\infty) \times U$, where $c_p > 0$ depends only on p.

Let X(t), $t \geq 0$, by a symmetric p-stable Lévy process in E and let $\{A_t\}_{t\geq 0}$ be a right continuous filtration such that X(t) is A_t-measurable and $\sigma(X(u)-X(t))$ is independent of A_t for every $u>t\geq 0$.

We will say that a continuous function $\Phi:[0,\infty) \to [0,\infty)$ *grows more slowly than* λ^p, $p > 0$, if there exist constants c, α_0 and $q < p$ such that

$$\Phi(\alpha\lambda) \leq c\alpha^q\Phi(\lambda)$$

for all $\lambda > 0$ and all $\alpha \geq \alpha_0$.
The proof of the following version of a theorem of R.F. Bass and M. Cranston [1] is supplied in the Appendix.

Theorem 3.1. *If* Φ *grows more slowly than* λ^p, $0 < p < 2$, *then there exist positive constants* c_1 *and* c_2 *depending only on* p, c, α_0 *and* q *such that for every finite* $\{A_t\}$*-stopping time* T

$$c_1 E\Phi(\tau^{1/p}) \leq E\Phi(X^*(\tau-)) \leq E\,(X^*(\tau)) \leq c_2 E\Phi(\tau^{1/p}),$$

where $X^*(\tau-) = \sup_{t<\tau} ||X(t)||$ *and* $X^*(\tau) = \sup_{t\leq\tau} ||X(t)||$.

Define now, a p-stable process starting at $x \in E$ by means of the formula

$$Y_x(t) = x + X(t), \qquad t \geq 0,\ x \in E.$$

Then, with the help of Theorem 3.1, one can obtain the following

COROLLARY 3.1. *If* Φ *grows more slowly than* λ^p, $0 < p < 2$, *then there exist positive constants* d_1 *and* d_2 *depending only on* p, c, α_0 *and* q *such that for every finite* $\{A_t\}$*-stopping time* τ *and* $x \in E$

$$d_1 E\Phi((||x||^p+\tau)^{1/p}) \le E\Phi(Y_x^*(\tau-)) \le E\Phi(Y_x^*(\tau)) \le d_2 E\Phi((||x||^p+\tau)^{1/p}).$$

Proof. An elementary application of the triangle inequality yields that

$$\frac{1}{3}(X^*(\tau) + ||x||) \le Y_x^*(\tau) \le X^*(\tau) + ||x||.$$

Similar inequality holds for $X^*(\tau-)$ and $Y_x^*(\tau-)$. Hence, with $L \sim R$ standing for the inequalities $C^{-1}L \le R \le CL$, where C is a positive constant depending perhaps on p,c,α_0, and q, we have

$$\Phi(Y_x^*(\tau)) \sim \Phi(X^*(\tau) + ||x||) = \Phi_{||x||}(X^*(\tau))$$

where $\Phi_u(\lambda) = \Phi(\lambda + u)$, and an analogous result obtains for $X^*(\tau-)$ and $Y_x^*(\tau-)$. Since $\Phi_u(\alpha\lambda) \le c\alpha^q \Phi_u(\lambda)$ for all $\lambda > 0$, $u \ge 0$ and $\alpha \ge \alpha_0 \vee 1$, Theorem 3.1 gives that

$$E\Phi_{||x||}(\tau^{1/p}) \sim E\Phi_{||x||}(X^*(\tau-)) \sim E\Phi_{||x||}(X^*(\tau))$$

which concludes the proof. Q.E.D.

We shall apply now the above theorem to obtain moment estimates for stochastic integrals.

THEOREM 3.2. *Let* $M(t)$ *be a real* (F_t)-*p-stable motion and let* $F \in L^p_{a.s.}$. *If* Φ *grows more slowly than* λ^p *then there exist positive constants* c_1 *and* c_2 *depending only on* p,c,α_0 *and* q *such that for each* $u > 0$

$$c_1 E\Phi((\int_0^u |F|^p dt)^{1/p}) \le E\Phi(\sup_{t<u} |\int_0^t F(s)dM(s)|)$$
$$\le E\Phi(\sup_{t\le u} |\int_0^t F(s)dM(s)|) \le c_2 E\Phi((\int_0^u |F(t)|^p dt)^{1/p})$$

Proof. Since u is fixed here, we can always extend F in such a way that

$$\tau(u) = \int_0^u |F|^p dt \to \infty \qquad \text{a.s.}$$

Therefore, by Theorem 2.2

$$X(t) = \int_0^{\tau^{-1}(t)} F(s)dM(s)$$

is an (A_t)-p-stable motion, where

$$A_t = F_{\tau^{-1}(t)}$$

and

$$\tau^{-1}(t) = \inf\{u: \tau(u) > t\}.$$

Applying Theorem 3.1 to $X(t)$ and $\tau = \tau(u)$ one immediately obtains our result. Q.E.D.

Taking $f(s,t)$ such that $A_p(f) < \infty$ and substituting in the above Theorem

$$F(t) = \int_0^t f(s,t)dM(s)$$

one immediately obtains from Corollary 2.1 the following result:

COROLLARY 3.3. *If Φ grows more slowly than λ^p then there exist positive constants c_1 and c_2 depending only on p,c,α_0 and q such that for each $u > 0$*

$$c_1 E\Phi\left\{\left(\int_0^u \left|\int_0^t f(s,t)dM(s)\right|^p dt\right)^{1/p}\right\}$$

$$\leq E\Phi\left\{\sup_{v\leq u}\left|\int_0^v\int_0^t f(s,t)dM(s)dM(t)\right|\right\}$$

$$\leq c_2 E\Phi\left\{\left(\int_0^u \left|\int_0^t f(s,t)dM(s)\right|^p dt\right)^{1/p}\right\}.$$

The following theorem summarizes recent results concerning moment inequalities for double p-stable stochastic integrals:

THEOREM 3.3. *Let $1 < q < p \leq 2$. Then there exist positive constants c_1, c_2 and c_3 depending only on p,q, such that*

$$c_1(A_p(f))^{q/p}$$

$$\leq c_2 E\left[\int_0^u \left|\int_0^t f(s,t)dM(s)\right|^p dt\right]^{q/p}$$

$$\leq E\left|\int_0^u \int_0^t f(s,t)dM(s)dM(t)\right|^q$$

$$\leq E \sup_{v\leq u} \left|\int_0^v \int_0^t f(s,t)dM(s)dM(t)\right|^q$$

$$\leq c_3(A_p(f))^{q/p}.$$

The proof of the two sided estimate between the first and second quantities has been recently obtained by J. Rosinski [10], between the second and third quantities by T. McConnell and M. Taqqu [9].

In this situation Theorem 3.3 follows directly from Corollary 3.3.

4. Intëgration with respect to a vector-valued p-stable motion.

Let X be a symmetric p-stable Lévy process, $0 < p < 2$, with values in a separable Banach space E (see Section 3), and let $\{F_t\}_{t\geq 0}$ be a right continuous filtration such that $X(t)$ is F_t-measurable and $\sigma(X(u)-X(t))$ is independent of F_t for every $u>t\geq 0$.

THEOREM 4.1. *For each real process* $F \in L^p_{a.s.}$ *there exists an* E*-valued process* $Y(t)$ *(denoted* $\int_0^t F(s)dX(s)$*) with sample paths in* $D_E[0,\infty)$ *such that for each* $x^* \in E^*$ *and* $t \in \mathbb{R}^+$ *we have that*

$$x^*Y(t) = \int_0^t F(s)d(x^*X(s)) \qquad \text{a.s.}$$

Proof. Without loss of generality we can assume that

$$\tau(t) = \int_0^t |F(s)|^p ds \to \infty$$

a.s. as $t \to \infty$. For each $x^* \in E^*$ the process

$$a(x^*)x^*X(t),\ t \geq 0, \text{ where } a(x^*) = \left(\int_U |x^*x|^p m(dx)\right)^{-1/p} < \infty$$

and $a(x^*)=0$ otherwise, is a real p-stable motion (see (3.1)). By Theorem 2.2, for any fixed $x^* \in E^*$, the real processes

$$Z_{x^*}(t) = \int_0^{\tau^{-1}(t)} F(s)d(x^*X(s)),\ t \geq 0, \text{ and } x^*X(t),\ t \geq 0,$$

have the same finite dimensional distributions. Moreover, $Z_{x^*}(t)$ is A_t-measurable and the increments $Z_{x^*}(t+h) - Z_{x^*}(t)$, $h \geq 0$, are independent of A_t.

Observe now, that for any fixed $t \geq 0$, $Z_{x^*}(t)$, $x^* \in E^*$, is a linear process on E^*, equidistributed with the linear decomposable process $\underset{\sim}{x}^*X(t)$, $x^* \in E^*$. Therefore, there exists an E-valued random vector $\tilde{X}(t)$ such that for each $x^* \in E^*$, $x^*\tilde{X}(t) = Z_{x^*}(t)$ a.s. Also, by the above remarks, the process $\tilde{X}(t)$ is (A_t)-adapted and the increments $\tilde{X}(t+h)-\tilde{X}(t)$, $h \geq 0$, are independent of A_t. Therefore $\tilde{X}(t)$, $t \geq 0$, has the same finite dimensional distributions as $X(t)$, $t \geq 0$, and we can select a modification of $\tilde{X}$ (also denoted by $\tilde{X}$) with all sample paths in $D_E[0,\infty)$. Hence $Y(t)$, $t \geq 0$, defined by the formula

$$Y(t) = \tilde{X}(\tau(t)),$$

has sample paths in $D_E[0,\infty)$ and satisfies, for any $x^* \in E^*$ and $t \geq 0$, the formula

$$x^*Y(t) = \int_0^t F(s)d(x^*X(s)) \qquad \text{a.s.}$$

Q.E.D.

Remark. The above construction, with obvious modifications, works for an E-valued Brownian motion as well.

The following result, besides providing moment estimates for the integral $\int F(s)dX(s)$, shows that the latter exists also in the strong sense. It follows immediately from the construction given in the proof of Theorem 4.1 and from Theorem 3.1.

THEOREM 4.2. *If Φ grows more slowly than λ^p, $0 < p < 2$, then there exist positive constants c_1 and c_2 depending only on p, c, α_0 and q such that*

$$c_1 E\Phi\left(\left(\int_0^t |F(s)|^p ds\right)^{\frac{1}{p}}\right) \leq E\Phi\left(\sup_{u \leq t} \left\|\int_0^u F(s)dX(s)\right\|\right) \leq c_2 E\Phi\left(\left(\int_0^t |F(s)|^p ds\right)^{\frac{1}{p}}\right),$$

Appendix

Proof of Theorem 3.1. (cf. [1]). Clearly, it suffices to prove that

(A.1) $$c_1 E(\phi^{1/p}) \leq E\Phi(X^*(\tau-)),$$

and

(A.2) $$E\Phi(X^*(\tau)) \leq c_2 E\Phi(\tau^{1/p}).$$

To obtain (A.1), it is enough to show that for $\beta>1$, $\delta>0$ and $\lambda>0$

$$P[t^{1/p} > \beta\lambda,\ X^*(\tau-) \leq \delta\lambda] \leq c(\beta,\delta)P[\tau^{1/p}>\lambda],$$

where $c(\beta,\lambda) \to 0$ as either $\beta \to \infty$ or $\delta \to 0$ (see Burkholder (1973), Lemma 7.1; the assumption $\Phi(0)=0$ is not necessary in this case).

Setting $a = \lambda^p$ and $b = (\beta\lambda)^p$ one obtains

$$\begin{aligned} P[\tau^{1/p}>\beta\lambda,\ X^*(\tau-)\leq\delta\lambda] &= P[\tau>b,\ X^*(\tau-) \leq \delta\lambda]\leq P[\tau>a, ||X(b)-X(a)||\leq 2\delta\lambda] \\ &= P[\tau>a]P[||X(b)-X(a)||\leq 2\delta\lambda] \\ &= P[\tau^{1/p}>\lambda]P[||X(1)||\leq \frac{2\delta}{(\beta^p - 1)^{1/p}}], \end{aligned}$$

which proves (A.1).

To obtain (A.2), we define an $\{A_t\}$-stopping time

$$\sigma = \inf\{t>0: ||X(t \wedge \tau)|| > \lambda\} .$$

Then we have that

$$\begin{aligned} &P[X^*(\tau) > \beta\lambda,\ \tau^{1/p} \leq \delta\lambda] = \\ &= P[X^*(\tau)>\beta\lambda, \tau^{1/p}\leq\delta\lambda, ||X(\sigma)||<\tfrac{\beta\lambda}{2}]+P[||X^*(\tau)||>\beta\lambda, \tau^{1/p}\leq\delta\lambda, ||X(\sigma)||\geq\tfrac{\beta\lambda}{2}] \\ &= I+J. \end{aligned}$$

Put $a = (\delta\lambda)^p$. For $\beta > 2$ we have

$$I = P[X^*(\tau) > \beta\lambda, \tau^{1/p} \leq \sigma\lambda, ||X(\sigma)|| < \frac{\beta\lambda}{2}, \sigma < \tau]$$

$$\leq P[\sup_{t \leq a} ||X(\sigma + t) - X(\sigma)|| > \frac{\beta\lambda}{2}, \sigma < \tau]$$

$$= P[X^*(a) > \frac{\beta\lambda}{2}] \; P\,[\sigma < \tau]$$

$$\leq 2P\,[||X(1)|| > \frac{\beta}{2\delta}] \cdot P[X^*(\tau) > \lambda]$$

$$\leq 2 \left(\frac{2\delta}{\beta}\right)^p \Lambda_p\,(X)\; P[X^*(\tau) > \lambda],$$

where
$$\Lambda_p(X) = \sup_{\lambda > 0} \lambda^p P\,[||X(1)|| > \lambda] < \infty.$$

Next, we obtain an estimate for J. Note that if n is the Lévy measure for X(1) then

$$n(B_R^c) = \int_{(R,\infty)\times U} c_p m(dx)dr/r^{1+p} = c_p m(U)p^{-1}R^{-p} = CR^{-p},$$

where B_R is the ball in E with radius R and center at 0. Let

$$Y(t) = \sum_{s \leq t} I(||\Delta X(s)|| > R),$$

where $\Delta X(s)$ is the jump process associated with X(s). Then Y(t) is a Poisson process with parameter $n(B_R^C)$, so that $Y(t)-t\, n\,(B_R^c)$ is a martingale (with respect $\{A_t\}$). By the optional sampling theorem, for every bounded $\{A_t\}$ stopping time τ

$$E \sum_{s \leq \tau} I(||\Delta X(s)|| > R) = EY(\tau) = n(B_R^c)E\tau$$

Let $\sigma_1 = \sigma \wedge \tau \wedge a$. If $s<\sigma_1 \leq \sigma \wedge \tau$ then $||\Delta X(s)|| \leq 2\lambda$ by definition of σ. Hence, if $R > 2\lambda$ then

$$P[||\Delta X(\sigma_1)|| > R] = E \sum_{s \leq \sigma} I(||\Delta X(s)|| > R) = n(B_R^c)\; E\; \sigma_1.$$

Since $\sigma_1 \leq \tau$ we have $X^*(\tau) \geq ||X(\sigma_1)||$, and, consequently, for $\beta>\sigma$ we obtain that

$$J = P[X^*(\tau) > \beta\lambda, \tau \leq a, ||X(\sigma)|| > \frac{\beta\lambda}{2}, \sigma \leq \tau]$$

$$\leq P[||\Delta X(\sigma_1)|| > (\frac{\beta}{2} - 1)\lambda] = n(B^c_{(\frac{\beta}{2}-1)\lambda})E\sigma_1$$

$$= C(\tfrac{\beta}{2} - 1)^{-p}\lambda^{-p}E\sigma_1 = 3^p(\tfrac{\beta}{2} - 1)^{-p}n(B^C_{3\lambda})\ E\ \sigma_1$$

$$= 3^p(\tfrac{\beta}{2} - 1)^{-p}P[||\Delta X(\sigma_1)|| > 3\lambda] \leq 3^p(\tfrac{\beta}{2} - 1)^{-p}P[||X(\sigma_1)|| > \lambda]$$

$$\leq 3^p(\tfrac{\beta}{2} - 1)^{-p}P[X^*(\tau) > \lambda].$$

Putting together estimates for I and J we get that

$$P[X^*(\tau)>\beta\lambda] \leq P[X^*(\tau) > \beta\lambda, \tau^{1/p}\leq\delta\lambda] + P[\tau^{1/p} > \delta\lambda]$$

$$\leq c(\beta,\delta,p)P[X^*(\tau)>\lambda] + P[\tau^{1/p} > \sigma\lambda],$$

where

$$c(\beta,\delta,p) = \beta^{-p}[2^{p+1}\sigma^p\Lambda_p(X) + 3^p(\tfrac{1}{2} - \tfrac{1}{\beta})^{-p}].$$

Therefore

$$E\Phi(\beta^{-1}X^*(\tau)) = \Phi(0) + \int_0^\infty P[X^*(\tau) > \beta\lambda]d\Phi(\lambda) \leq$$

$$\leq \Phi(0) + c(\beta,\delta,p)\int_0^\infty P[X^*(\tau) >\lambda]d\Phi(\lambda) + \int_0^\infty P[\tau^{1/p}>\delta\lambda]d\Phi(\lambda)$$

$$\leq c(\beta,\delta,p)\ E\Phi(X^*(\tau)) + E\Phi(\delta^{-1}\tau^{1/p}).$$

If $\beta > \alpha_0$ and $\delta < \alpha_0^{-1}$ then

$$E\Phi((X^*(\tau)) = E\Phi(\beta\beta^{-1}X^*(\tau)) \leq c\beta^qE\Phi(\beta^{-1}X^*(\tau)),$$

and

$$E\Phi(\delta^{-1}\tau^{1/p}) \leq c\delta^{-q}E\Phi(\tau^{1/p}).$$

Finally, we obtain the inequality

$$[c^{-1}\beta^{-q}-c(\beta,\delta,p)]\ E\Phi(X^*(\tau)) \leq c\delta^{-q}E\Phi(\tau^{1/p})$$

which proves (A.2) since the constant on the left hand side can be made positive by taking β large enough and δ small enough (remember that $q < p$).

REFERENCES

[1] R.F. BASS and M. CRANSTON, Exit times for symmetric stable processes in $\mathbb{R}^n$, Annals of Probability 11 (1983), 578-588.

[2] D. BURKHOLDER, Distribution function inequalities for martingales, Annals of Probability 1 (1973), 19-42.

[3] S. CAMBANIS, J. ROSIŃSKI and W.A. WOYCZYŃSKI, Convergence of quadratic forms in p-stable random variables and Θ_p-radonifying operators, Annals of Probability 13 (1985) (to appear)

[4] E. GINÉ and M.B. MARCUS, The central limit theorem for stochastic integrals with respect to Lévy processes, Annals of Probability 3 (1983), 58-77.

[5] O. KALLENBERG, On the existence and path properties of stochastic integrals, Annals of Probability 3 (1975), 262-280.

[6] W. KRAKOWIAK and J. SZULGA, Random multilinear forms, Wroclaw University, Preprint, 1984.

[7] W. LINDE, Infinitely divisible and stable measures on Banach spaces, Teubner, Leipzig 1983.

[8] M.B. MARCUS and G. PISIER, Characterization of almost surely continuous p-stable random Fourier series and strongly stationary processes, Acta Mathematica (1984), 245-301.

[9] T.R. McCONNELL and M.S. TAQQU, Double integration with respect to symmetric stable processes, Cornell University, Dept. of Operations Research, Technical Report # 618, 1984.

[10] J. ROSIŃSKI, On stochastic integral representation of stable processes with sample paths in Banach spaces, UNC, Center for Stochastic Processes, Technical Report # 88, 1985.

[11] J. ROSIŃSKI and W.A. WOYCZYŃSKI, Products of random measures, multilinear random forms and multiple stochastic integrals, Proc. Conf. Measure Theory, Oberwolfach 1983, Springer's Lecture Notes in Mathematics (1984), 22 pp.

[12] J. ROSIŃSKI and W.A. WOYCZYŃSKI, On Itô stochastic integration with respect to p-stable motion: inner clock, integrability of sample paths, double and multiple integrals, Annals of Probability 13 (1985) (to appear).

[13] J. ROSIŃSKI and W.A. WOYCZYŃSKI, Multilinear forms in Pareto-like random variables and product random measures, Colloquium Mathematicum, S. Hartman Festschrift (to appear).

[14] J. SZULGA and W.A. WOYCZYŃSKI, Existence of a double random integral with respect to stable measure, Journal of Multivariate Analysis 13 (1983), 194-201.

A NOTE ON THE CONVERGENCE TO GAUSSIAN LAWS OF SUMS OF STATIONARY ϕ-MIXING TRIANGULAR ARRAYS

Jorge D. Samur
Universidad Nacional de La Plata

1. Introduction.

This paper deals with sufficient conditions for the convergence to Gaussian laws of the row sums of stationary, ϕ-mixing triangular arrays of Banach space valued random vectors (see the definitions below). Some results of [9] are improved in the sense that we remove a certain dependence restriction about contiguous random vectors (the hypothesis $\phi(1)<1$). This is carried out by using an inequality of N. Herrndorf ([5,Lemma 3.1]) and a version for the dependent case of an inequality contained in a result of [4] (see Lemma 2.2 below) in place of some analogous inequalities previously used in [9].

For the case of stationary sequences of random vectors we give results which extend to the ϕ-mixing case and in the Hilbert space setting part of a classical criterion of the independent case for the convergence to the normal law. Corollary 3.4 below generalizes a result of Z.Y. Lin ([7]) for m-dependent real random variables and also includés a well-known theorem of I.A. Ibragimov and its extension given in [8] (see also [9,Remark on page 405]).

Now we recall some definitions. By a triangular array $\{X_{nj}\}$ we mean a family $\{X_{nj}: j=1,\ldots,j_n, n\in N\}$ (N is the set of nonzero natural numbers) of B-valued random vectors (r.v.'s) defined on a common probability space $(\Omega,\mathcal{A},P)$; here and throughout the paper, B denotes a real separable Banach space with norm $\|\cdot\|$ and B' will denote its dual space. We will assume that $j_n\to\infty$ as $n\to\infty$ and we shall write $S_{nk}=\sum_{j=1}^{k}X_{nj}$ if $k=1,\ldots,j_n$, $S_n=S_{nj_n}$.

If $\mathcal{M},\mathcal{N}$ are two sub-σ-algebras of $\mathcal{A}$ we will consider the coefficient

$$\phi(\mathcal{M},\mathcal{N}) = \sup\{ |\frac{P(E\cap F)}{P(E)} - P(F)| : E\in\mathcal{M}, F\in\mathcal{N}, P(E)>0\}.$$

Given a triangular array $\{X_{nj}:j=1,\ldots,j_n, n\in N\}$ let $\mathcal{M}_{hk}^{(n)}=\sigma(\{X_{nj}:j=h,\ldots,k\}$ (the σ-algebra generated by the indicated set of r.v.'s) for $n\in N$ and $1\le h\le k\le j_n$ and define

$$\phi(k) = \sup_{n\in N, j_n>k} \max_{1\le h\le j_n-k} \phi(\mathcal{M}_{1h}^{(n)}, \mathcal{M}_{h+k,j_n}^{(n)})$$

($k\in N$). Note that $\phi(1)\le 1$ and that $\{\phi(k)\}$ is nonincreasing. It is said that $\{X_{nj}\}$ is <u>φ-mixing</u> if $\phi(k)\downarrow 0$ as $k\to\infty$; we will say that $\{X_{nj}\}$ is <u>stationary</u> (has <u>stationary sums</u>) if $\mathcal{L}(X_{n1},\ldots,X_{nh})=\mathcal{L}(X_{n,k+1},\ldots,X_{n,k+h})$ ($\mathcal{L}(X_{n1}+\ldots+X_{nh})=\mathcal{L}(X_{n,k+1}+\ldots+X_{n,k+h})$, respectively) for $1\le h\le j_n, 1\le k\le j_n-h$, $n\in N$ (if Z is a random vector, $\mathcal{L}(Z)$ denotes its law). We have similar definitions for a sequence and for a finite set $\{X_1,\ldots,X_n\}$ of B-valued r.v.'s; in the last case, we shall write $S_k=\sum_{j=1}^k X_j$ for $k=1,\ldots,n$.

If X is a B-valued r.v. and $\delta>0$ we write $X_\delta = XI_{\{\|X\|\le\delta\}}$ (I_A is the indicator function of the set A); for a triangular array $\{X_{nj}\}$ and $\delta>0$ we write $S_{n;\delta}=\sum_{j=1}^{j_n} X_{nj\delta}$. If γ is a centered Gaussian measure on B, Φ_γ denotes its covariance. We denote by $\to_w$ the weak convergence of probability measures and by $\to_P$ the convergence in probability of r.v.'s.

2. Two inequalities and some consequences.

The first inequality is due to N. Herrndorf ([5, Lemma 3.1]).

<u>2.1 Lemma</u>. Let $\{X_1,\ldots,X_n\}$ be a set of B-valued r.v.'s. Suppose $q\in N$, $q+1\le n$ and let $a>0$. Then

$$(1-\phi(q)-\max_{q\le k\le n} P(\|S_n-S_k\|>a))P(\max_{1\le k\le n}\|S_k\|>3a)$$
$$\le P(\|S_n\|>a)+P((q-1)\max_{1\le j\le n}\|X_j\|>a).$$

The following inequality is obtained by modifying an argument in the proof of Lemma 2, page 383 of [4]. A less general version is contained in [9, Proposition 2.4].

<u>2.2 Lemma</u>. Let $\{X_1,\ldots,X_n\}$ be a set of B-valued r.v.'s such that $\|X_j\|\le M$ a.s. $(j=1,\ldots,n)$. Suppose $q\in N$, $q+1\le n$ and let $t>0$. Then for every $\ell\in N$ we have

$$P(\max_{1\le k\le n}\|S_k\|>\ell(t+qM))$$
$$\le(\phi(q)+P(\max_{1\le k\le n}\|S_k\|>t/2))^{\ell-1}P(\max_{1\le k\le n}\|S_k\|>t+qM).$$

<u>Proof</u>. If $s>0$ define $E^{(s)}=\{\max_{k\le n}\|S_k\|>s\}$, $E_1^{(s)}=\{\|S_1\|>s\}$, $E_i^{(s)}=\{\max_{k\le i-1}\|S_k\|\le s,\ \|S_i\|>s\}$ for $i=2,\ldots,n$. Now assume that $s+qM<u$. Then we have that $E^{(u)}=\bigcup_{i=1}^n E_i^{(s)}\cap E^{(u)}$, $E_i^{(s)}\cap E^{(u)}\subset E_i^{(s)}\cap\{\max_{i+q\le k\le n}\|S_k-S_{i-1+q}\|>u-s-qM\}$ if $i=1,\ldots,n-q$ and $E_i^{(s)}\cap E^{(u)}=\emptyset$ if $i=n-q+1,\ldots,n$ (for example, assume $i\in\{1,\ldots,n-q\}$ and that $E_i^{(s)}\cap E^{(u)}$ occurs; then $\|S_{k_0}\|>u$ for some $k_0\ge i+q$ and, moreover, $\|S_{k_0}-S_{i-1+q}\|\ge\|S_{k_0}\|-\|S_{i-1}\|-\|S_{i-1+q}-S_{i-1}\|>u-s-qM$). Therefore

$$P(E^{(u)})=\sum_{i=1}^n P(E_i^{(s)}\cap E^{(u)})$$
$$\le\sum_{i=1}^{n-q}(\phi(q)+P(\max_{i+q\le k\le n}\|S_k-S_{i-1+q}\|>u-s-qM))P(E_i^{(s)})$$
$$\le(\phi(q)+P(\max_{k\le n}\|S_k\|>(u-s-qM)/2))P(E^{(s)}).$$

From this we conclude that, given $t>0$, we have

$$P(E^{(\ell(t+qM))})\le(\phi(q)+P(\max_{k\le n}\|S_k\|>t/2))P(E^{((\ell-1)(t+qM))})$$

for every integer $\ell\ge 2$. This implies the desired conclusion.

Combining this two inequalities we can now improve some results of [9].

<u>2.3 Lemma</u>. Let $\{X_1,\ldots,X_n\}$ be a set of B-valued r.v.'s such that $\|X_j\|\le M$ a.s. $(j=1,\ldots,n)$. Suppose $q\in N$, $q+1\le n$, $\phi(q)<\alpha<1$ and let $t>0$. Then if $\max_{q\le k\le n}P(\|S_n-S_k\|>t/6)\le 1-\alpha$ we have for every $\ell\in N$

$$P(\max_{1\le k\le n}\|S_k\|>\ell(t+qM))$$
$$\le(\phi(q)+(\alpha-\phi(q))^{-1}\{P(\|S_n\|>t/6)+P((q-1)\max_{1\le j\le n}\|X_j\|>t/6)\})^{\ell-1}\times$$

$\times(\alpha-\phi(q))^{-1}\{P(\|S_n\|>t/6)+P((q-1)\max_{1\le j\le n}\|X_j\|>t/6)\}.$

Remark. Using this result in place of Proposition 2.4 of [9] we conclude that the hypothesis $\phi(1)<1$ can be omitted in Proposition 3.5 (for this purpose, it will be convenient to use the particular case of Lemma 2.3 in which $(q-1)M\le t/6$) and in the "if" part of Theorem 4.1 of that paper. But this is not the case for its "only if" part and, consequently, for Theorem 4.2 of [9], as the following example shows. Let $Y_j, j\in N$, $\eta_j, j\ge 0$ be independent symmetric real random variables such that $Y_j, j\in N$, are identically distributed with $E(Y_1^2)=1$ and $\eta_j, j\ge 0$, are identically distributed with $\lim_{x\to\infty} x^2 P(|\eta_1|>x)=\infty$. Define $X_j=Y_j+\eta_j-\eta_{j-1}$ for $j\in N$ and $\{X_{nj}\}=\{n^{-1/2}X_j : j=1,\dots,n, n\in N\}$; then $\{X_j\}$ is stationary and 1-dependent. We have that $\mathcal{L}(S_n)\to_w N(0,1)$ and $\{X_{nj}\}$ satisfies the condition (*) considered in [9] (see Remark 2 on page 395 there) but, given $\varepsilon>0$,

$$nP(|X_{n1}|>\varepsilon)\ge nP(|\eta_1|>2\varepsilon n^{1/2})P(|Y_1|\le(\varepsilon/2)n^{1/2})P(|\eta_0|\le(\varepsilon/2)n^{1/2})$$

which goes to ∞ as $n\to\infty$. This example also shows that the hypothesis $\phi(1)<1$ can not be omitted in Theorem 3.4 of [9].

For the proof of the following proposition use the preceding lemma and remark and see the proofs of Theorems 4.1 and 4.2 of [9].

2.4 Proposition. Let $\{X_{nj}\}$ be a ϕ-mixing triangular array with stationary sums which satisfies

$$(*)\quad \{r_n\}\subset N,\ r_n\le j_n,\ r_n/j_n\to 0 \text{ imply } \sum_{j=1}^{r_n} X_{nj}\to_P 0.$$

Suppose that $\mathcal{L}(S_n)\to_w \nu$ and that $j_n P(\|X_{n1}\|>\varepsilon)\to 0$ for every $\varepsilon>0$. Then ν is Gaussian. Moreover, if $\{X_{nj}\}$ is stationary and $\nu=\delta_z * \gamma$, where $z\in B$ and γ is a centered Gaussian measure, then for every $\delta>0$

(a) $\lim_n Ef^2(S_{n;\delta}-ES_{n;\delta})=\Phi_\gamma(f,f)$ for each $f\in B'$,

(b) $\mathcal{L}(S_n-ES_{n;\delta})\to_w\gamma$, $S_n-S_{n;\delta}\to_P 0$, $\mathcal{L}(S_{n;\delta}-ES_{n;\delta})\to_w\gamma$ and $ES_{n;\delta}\to z$ in B.

By this result we conclude that the hypothesis $\phi(1)<1$ can be removed from Theorem 4.4 of [9] and its corollaries (see the proofs there).

3. The Hilbert space case.

From now on, we will assume that B is a separable Hilbert space. We write $d_k(x)=\inf\{\|x-y\|:y\in F_k\}$, F_k being the subspace spanned by $\{e_1,\ldots,e_k\}$, where $\{e_i:i\in N\}$ is a fixed (but arbitrary) orthonormal basis of B, when B is infinite-dimensional; if the dimension of B is finite we have an orthonormal basis $\{e_1,\ldots,e_n\}$ ($n\in N$) and we put $d_k=0$ for $k\geq n$.

Given a stationary triangular array $\{X_{nj}:j=1,\ldots,j_n,n\in N\}$, $\delta>0$ and $f\in B'$ we write

$$V_n(\delta,f)=j_nEf^2(X_{n1\delta}-EX_{n1\delta})$$
$$+2j_n\sum_{j=1}^{j_n-1}E(f(X_{n1\delta}-EX_{n1\delta})f(X_{n,j+1,\delta}-EX_{n,j+1,\delta})).$$

As observed at the end of the preceding section, we have the following improvement of Corollary 4.5 of [9].

3.1 Theorem. Let $\{X_{nj}\}$ be a stationary, ϕ-mixing triangular array with $\sum_{j=1}^{\infty}\phi^{1/2}(j)<\infty$. Assume

(1) for every $\varepsilon>0$, $j_nP(\|X_{n1}\|>\varepsilon)\to 0$,

(2) there exist a sequentially w*-dense subset W of B' and $\delta>0$ such that for every $f\in W$

$$C_{\delta,f}=\sup_n j_nEf^2(X_{n1\delta}-EX_{n1\delta})<\infty$$

and the limit

$$\Phi(f)=\lim_n V_n(\delta,f)\text{ exists,}$$

(3) there exists $\beta>0$ such that

$$\lim_k \overline{\lim}_n j_nEd_k^2(X_{n1\beta}-EX_{n1\beta})=0.$$

Then there exists a centered Gaussian measure γ such that $\Phi_\gamma(f,f)=\Phi(f)$ for every $f\in W$ and $\mathcal{L}(S_n-ES_{n;\tau})\to_w\gamma$ for every $\tau>0$.

3.2 Corollary. Assume that $\{X_{nj}\}$ satisfies the hypotheses of Theorem 3.1

with (2) replaced by

(2') there exist a sequentially w*-dense subset W of B' and $\delta>0$ such that for every $f\in W$ and each $j\in N$ the limit

$$\Phi_j(f)=\lim_n j_n E(f(X_{n1\delta}-EX_{n1\delta})f(X_{nj\delta}-EX_{nj\delta})) \text{ exists.}$$

Then for every $f\in W$ the sum

$$\Phi(f)=\Phi_1(f)+2\sum_{j=2}^{\infty}\Phi_j(f)$$

converges and there exists a centered Gaussian measure γ such that $\Phi_\gamma(f,f)=\Phi(f)$ for every $f\in W$ and $\mathcal{L}(S_n-ES_{n;\tau})\to_w\gamma$ for each $\tau>0$.

Proof. We will show that (2) holds. Fix $f\in W$ and observe first that $C_{\delta,f}$ is finite because $\Phi_1(f)$ exists. To see that the sum defining $\Phi(f)$ converges absolutely note that

$$\begin{aligned}\sum_{j=2}^{\infty}|\Phi_j(f)| &\le \underline{\lim}_n \sum_{j=2}^{\infty}|j_n E(f(X_{n1\delta}-EX_{n1\delta})f(X_{nj\delta}-EX_{nj\delta}))| \\ &\le \underline{\lim}_n \sum_{j=2}^{\infty} j_n 2\phi^{1/2}(j-1)Ef^2(X_{n1\delta}-EX_{n1\delta}) \\ &\le 2C_{\delta,f}\sum_{j=1}^{\infty}\phi^{1/2}(j)<\infty\end{aligned}$$

(we have used a well-known inequality of I.A. Ibragimov, quoted, for example, in [9,Proposition 2.5]). Let $\varepsilon>0$. Choose $j_0\in N$ such that $2(\sum_{j>j_0}\phi^{1/2}(j))C_{\delta,f}\le\varepsilon$ and $\sum_{j>j_0}|\Phi_j(f)|\le\varepsilon$; then, breaking the sums which define $V_n(\delta,f)$ and $\Phi(f)$ at j_0 and using the definition of $\Phi_j(f)$, we obtain that $\overline{\lim}_n\ |V_n(\delta,f)-\Phi(f)|\le 2\varepsilon$. This shows that $\lim_n V_n(\delta,f)=\Phi(f)$.

Now we turn to triangular arrays arising from a single sequence by normalization.

3.3 Corollary. Let $\{X_j : j\in N\}$ be a stationary, ϕ-mixing sequence with $\sum_{j=1}^{\infty}\phi^{1/2}(j)<\infty$. Then the sum

$$V(x,f)=Ef^2(X_{1,x}-EX_{1,x})+2\sum_{j=1}^{\infty}E(f(X_{1,x}-EX_{1,x})f(X_{j+1,x}-EX_{j+1,x}))$$

converges for each $x>0$. Suppose that $E\|X_{1,x}\|^2>0$ for some $x>0$ and assume

(i) $\displaystyle\lim_{x\to\infty}\frac{x^2P(\|X_1\|>x)}{E\|X_{1,x}\|^2}=0$,

(ii) there exists a sequentially w*-dense subset W of B' such that for every $f \in W$ the limit

$$\Phi(f) = \lim_{x\to\infty} \frac{V(x,f)}{E\|X_{1,x}\|^2} \text{ exists,}$$

(iii) $\lim_k \overline{\lim_{x\to\infty}} \frac{Ed_k^2(X_{1,x})}{E\|X_{1,x}\|^2} = 0.$

Then $E\|X_1\|<\infty$ and there exist a sequence $\{a_n\}$ with $a_n>0$, $a_n\to\infty$ and a centered Gaussian measure γ such that $\Phi_\gamma(f,f)=\Phi(f)$ for each $f\in W$ and $\mathcal{L}(a_n^{-1}(X_1+\ldots+X_n-nEX_1))\to_w \gamma$.

Proof. First, observe that by an inequality previously used we have

$$|E(f(X_{1,x}-EX_{1,x})f(X_{j+1,x}-EX_{j+1,x}))| \le 2\phi^{1/2}(j)Ef^2(X_{1,x}-EX_{1,x}) \tag{3.1}$$

for each $x>0$. Thus the series defining $V(x,f)$ converges absolutely. Hypothesis (i) says that the function $U(x)= E\|X_{1,x}\|^2$ is slowly varying at ∞ (that is, $\lim_{x\to\infty} U(tx)(U(x))^{-1}=1$ for each $t>0$ –see, for example, [1,Chapter 2,Corollary 6.16]); then there exist a sequence $\{a_n\}$ with $a_n>0$, $a_n\to\infty$ such that $na_n^{-2}U(\delta a_n)\to 1$ for every $\delta>0$ (see [1,page 87]; if $a^2= E\|X_1\|^2<\infty$, $a>0$, we can take $a_n=an^{1/2}$). We will apply Theorem 3.1 to the triangular array $\{X_{nj}\}=\{a_n^{-1}X_j:j=1,\ldots,n,n\in N\}$. Our hypotheses (i) and (iii) imply that $\{X_{nj}\}$ satisfies (1) and (3). In order to see that (2) holds, take $\delta>0$ and $f\in W$; by the choice of a_n, $C_{\delta,f}$ is finite. On the other hand, we have

$$V_n(\delta,f)= na_n^{-2}U(\delta a_n)\{V(\delta a_n,f)(U(\delta a_n))^{-1} -2(U(\delta a_n))^{-1}\textstyle\sum_{j=n}^{\infty}E(f(X_{1,\delta a_n}-EX_{1,\delta a_n})f(X_{j+1,\delta a_n}-EX_{j+1,\delta a_n}))\}$$

and, by (3.1),

$$|U(x)^{-1}\textstyle\sum_{j=n}^{\infty}E(f(X_{1,x}-EX_{1,x})f(X_{j+1,x}-EX_{j+1,x}))|\le 2\sum_{j=n}^{\infty}\phi^{1/2}(j);$$

using hypothesis (ii) we conclude that (2) is satisfied with our Φ (if $a^2 = E\|X_1\|^2 < \infty$ we have, in fact, $\Phi(f) = a^{-2}\{Ef^2(X_1 - EX_1) + 2\sum_{j=1}^{\infty} E(f(X_1 - EX_1) \times \times f(X_{j+1} - EX_{j+1}))\}$ for each $f \in B'$). Then there exists the desired Gaussian measure γ such that $\mathcal{L}(a_n^{-1}(X_1 + \ldots + X_n - nEX_{n1,1})) \to_w \gamma$. The final conclusion follows from the fact that, as a consequence of (i), $E\|X_1\| < \infty$ and $na_n^{-1} E(\|X_1\| I_{\{\|X_1\| > a_n\}}) \to 0$ as $n \to \infty$.

Let X_1 be a B-valued r.v.. Recall that if $E\|X_1\|^2 = \infty$ then $(E\|X_{1,x}\|)^2 = o(E\|X_{1,x}\|^2)$ as $x \to \infty$; for each $f \in B'$ write

$$m^2(f) = \lim_{x \to \infty} \frac{(Ef(X_{1,x}))^2}{E\|X_{1,x}\|^2} = \begin{cases} \dfrac{(Ef(X_1))^2}{E\|X_1\|^2} & \text{if } 0 < E\|X_1\|^2 < \infty \\ 0 & \text{if } E\|X_1\|^2 = \infty . \end{cases}$$

<u>3.4 Corollary</u>. Assume that $\{X_j\}$ satisfies the hypothesis of Corollary 3.3 with (ii) replaced by

(ii') there exists a sequentially w*-dense subset W of B' such that for every $f \in W$ and each $j \in N$ the limit

$$\Phi_j^0(f) = \lim_{x \to \infty} \frac{E(f(X_{1,x}) f(X_{j,x}))}{E\|X_{1,x}\|^2} \quad \text{exists.}$$

Then $E\|X_1\| < \infty$, the sum

$$\Phi(f) = (\Phi_1^0(f) - m^2(f)) + 2\sum_{j=2}^{\infty} (\Phi_j^0(f) - m^2(f))$$

converges for every $f \in W$ and there exist a sequence $\{a_n\}$ with $a_n > 0$, $a_n \to \infty$ and a centered Gaussian measure γ such that $\Phi_\gamma(f,f) = \Phi(f)$ for each $f \in W$ and $\mathcal{L}(a_n^{-1}(X_1 + \ldots + X_n - nEX_1)) \to_w \gamma$.

<u>Proof</u>. For each $f \in W$ and every $j \in N$ we have

$$\lim_{x \to \infty} (E\|X_{1,x}\|^2)^{-1} E(f(X_{1,x} - EX_{1,x}) f(X_{j,x} - EX_{j,x})) = \Phi_j^0(f) - m^2(f).$$

Arguing as in the proof of Corollary 3.2 we can show that the series defining the Φ of our statement converges absolutely and that $\lim_{x \to \infty}$

$(E\|X_{1,x}\|^2)^{-1}V(x,f)=\Phi(f)$ for each $f\in W$. Now we can apply Corollary 3.3.

Examples. Fix $\alpha\in(0,1)$ and write $\beta=1-\alpha$. Let $I=\{0\}\cup N$, $(p_{ij})_{i,j\in I}$ be the stochastic matrix defined by

$$p_{ij}=\begin{cases}\beta & \text{if } j=0\\ \alpha & \text{if } j=i+1\\ 0 & \text{otherwise}\end{cases}$$

$(i,j\in I)$ and π be the probability measure defined on (the class of all subsets of) I by $\pi=\sum_{i\in I}\beta\alpha^i\delta_i$ where δ_i denotes the unit point mass at $i\in I$. Let $\{Y_k:k\in N\}$ be the canonical version of the Markov chain with state space I, stationary transition probabilities (p_{ij}) and initial distribution π (see examples (2.m) -"success runs"- and (7,f) in [3, Chapter XV]). Now take a function $h:I\to B$ and define $X_k=h(Y_k)$ $(k\in N)$. It follows that $\{X_k:k\in N\}$ is a ϕ-mixing stationary sequence with $\phi(k)=O(\alpha^k)$ (see Section 1.2 of [6]; the coefficient of ergodicity of the matrix (p_{ij}) is β).

(a) Let $B=\mathbb{R}$ and take $h(i)=(\alpha^{-1/2})^i$ $(i\in I)$. We have that $U(x)=E\|X_{1,x}\|^2\sim\beta(\log\alpha^{-1/2})^{-1}\log x$ (the ratio of both sides tends to 1 as $x\to\infty$) which shows that U varies slowly at ∞; this is equivalent to (i). Noting that the n-step transition probabilities are

$$p_{ij}^{(n)}=\begin{cases}\beta\alpha^j & \text{if } j=0,1,\dots,n-1\\ \alpha^n & \text{if } j=i+n\\ 0 & \text{otherwise}\end{cases} \tag{3.2}$$

$(i,j\in I, n\in N)$ we can see that $E(X_{1,x}X_{k,x})\sim(\alpha^{1/2})^{k-1}\beta(\log\alpha^{-1/2})^{-1}\log x$ for each $k\in N$. Now observe that in order to verify condition (ii') in the real valued case it is sufficient to consider for each $k\in N$ the limit

$$\Phi_k^0=\lim_{x\to\infty}\frac{E(X_{1,x}X_{k,x})}{E\|X_{1,x}\|^2}$$

which is $(\alpha^{1/2})^{k-1}$ in the present case. Then the conclusion of Corollary 3.4 holds if $\{a_n\}$ is a sequence such that $na_n^{-2}U(a_n)\to1$ as $n\to\infty$ (see the proof of Corollary 3.3); we can take $a_n=(2^{-1}\beta(\log\alpha^{-1/2})^{-1}n\log n)^{1/2}$.

Since $EX_1=1+\alpha^{1/2}$ we conclude that

$$\mathcal{L}(a_n^{-1}((\alpha^{-1/2})^{Y_1}+\ldots+(\alpha^{-1/2})^{Y_n}-n(1+\alpha^{1/2}))) \to_w N(0,\Phi)$$

with $\Phi=\Phi_1^0+2\sum_{k=2}^{\infty}\Phi_k^0=(1+\alpha^{1/2})(1-\alpha^{1/2})^{-1}$. Let us remark that the same result is valid if π is replaced by any probability measure π' on I (see the argument at the end of the following example).

(b) Now we give a more curious example. Let $p_1,p_2,\ldots$ be the prime numbers arranged in increasing order and let $\mathcal{P}_s=\{i\in N: i=p_s^m \text{ for some } m\in N\}$ for each $s\in N$. Take $B=\ell^2$, the Hilbert space of square-summable sequences $\{x_s\}_{s\in N}$ of real numbers, and let $\{e_s: s\in N\}$ be its canonical orthonormal basis. Define $h: I\to\ell^2$ by

$$h(i)=\begin{cases}(\alpha^{-1/2})^i(\alpha^{1/2})^{p_s}\, e_s & \text{if } i\in\mathcal{P}_s\\ 0 & \text{otherwise}\end{cases}$$

If $x\geq 1$ we have

$$U(x)=E\|X_{1,x}\|^2=\beta\sum_{s=1}^{\infty}\alpha^{p_s}+\beta\sum_{s:p_s\leq v^{-1}(x)}\alpha^{p_s}(z(s,x)-1)$$

where $v(t)=(\alpha^{-1/2})^{t^2-t}$ $(t\geq 1)$ and

$$z(s,x)=\operatorname{card}\{m\in N: (\alpha^{-1/2})^{p_s^m-p_s}\leq x\}$$

$$=[(\log p_s)^{-1}\log(p_s+(\log\alpha^{-1/2})^{-1}\log x)]$$

($s\in N$, $x\geq 1$; $[\cdot]$ denotes the integer part of a real number). Since $v^{-1}(x)=o(\log x)$ we obtain that $\log(v^{-1}(x)+(\log\alpha^{-1/2})^{-1}\log x)\sim\log(\log x)$, which implies that

$$U(x)\sim a\beta\log(\log x) \text{ where } a=\sum_{s=1}^{\infty}\alpha^{p_s}(\log p_s)^{-1}.$$

This shows that (i) holds. On the other hand, observe that if $r\in N$ and $x\geq 1$ we have

$$Ed_r^2(X_{1,x})=\beta\sum_{s=r+1}^{\infty}\alpha^{p_s}+\beta\sum_{s:s\geq r+1,\,p_s\leq v^{-1}(x)}\alpha^{p_s}(z(s,x)-1).$$

Arguing as above we can conclude that

$$\overline{\lim_{x\to\infty}} \frac{Ed_r^2(X_{1,x})}{E\|X_{1,x}\|^2} \le a^{-1}\sum_{s=r+1}^{\infty} \alpha^{p_s}(\log p_s)^{-1}$$

for each $r\in N$. Then (iii) is satisfied.

Now take W as the set of all $f\in(\ell^2)'$ which correspond, by the Riesz representation, to an element $\{f_s\}_{s\in N}$ of ℓ^2 such that $f_s=0$ except for at most a finite number of f_s's. Fix $f\in W$ associated to $\{f_s\}_{s\in N}\in\ell^2$. If $x\geq 1$ we have

$$Ef^2(X_{1,x}) = \beta\sum_{s=1}^{\infty} f_s^2\alpha^{p_s} + \beta\sum_{s:p_s\le v^{-1}(x)} f_s^2\alpha^{p_s}(z(s,x)-1).$$

This implies that $\Phi_1^0(f)=a^{-1}\sum_{s=1}^{\infty} f_s^2\alpha^{p_s}(\log p_s)^{-1}$. Now take an integer $k\geq 2$. Using (3.2) we obtain that

$$\begin{aligned} E(f(X_{1,x})f(X_{k,x})) &= \sum f_s f_{s'}(\alpha^{1/2})^{p_s^m}(\alpha^{1/2})^{p_{s'}^{m'}}\beta^2(\alpha^{1/2})^{p_s}(\alpha^{1/2})^{p_{s'}} \\ &\quad + \sum f_s f_{s'}(\alpha^{1/2})^{p_s}(\alpha^{1/2})^{p_{s'}}\beta\alpha^{(k-1)/2} \\ &= c_{k,x} + c'_{k,x} \qquad \text{(say)} \end{aligned}$$

where the first sum extends over all $s,s',m,m'\in N$ such that $(\alpha^{-1/2})^{p_s^m-p_s}\le x$, $(\alpha^{-1/2})^{p_{s'}^{m'}-p_{s'}}\le x$, $p_{s'}^{m'}\le k-2$ and the second one extends over all s,s', $m,m'\in N$ such that $(\alpha^{-1/2})^{p_s^m-p_s}\le x$, $(\alpha^{-1/2})^{p_{s'}^{m'}-p_{s'}}\le x$, $p_s^m+(k-1)=p_{s'}^{m'}$. Since

$$|c_{k,x}| \le \beta^2\Big(\sum_{n=1}^{\infty}(\alpha^{1/2})^n\Big)^2\Big(\sum_{s=1}^{\infty}|f_s|(\alpha^{1/2})^{p_s}\Big)^2<\infty$$

we have $c_{k,x}=o(U(x))$; on the other hand,

$$|c'_{k,x}| \le \beta\alpha^{(k-1)/2}\sum_{s,s'\in N}|f_s|(\alpha^{1/2})^{p_s}|f_{s'}|(\alpha^{1/2})^{p_{s'}}r_{p_s,p_{s'},k-1}(z(s,x))$$

where

$$r_{p,q,b}(n) = \operatorname{card}\{m\in N: m\le n \text{ and } p^m+b=q^{m'} \text{ for some } m'\in N\}$$

(p,q are primes, $b\in N$, $n\in N$). Since $f\in W$, we will have that $c'_{k,x}=o(U(x))$

and hence that $\Phi_k^0(f)=0$ if we show that $r_{p,q,b}(n)=o(n)$ for every pair of primes p,q and each $b\in N$. When p=q it is easy to see that $r_{p,p,b}(n)\leq 1$ for every $n\in N$ and each $b\in N$. Suppose $p\neq q$ and fix $b\in N$. Assume that $r_{p,q,b}(n)\to\infty$ as $n\to\infty$ and let $m_1<m_2<\ldots<m_k<\ldots$ be all the numbers $m\in N$ for which there exists $m'\in N$ such that $p^m+b=q^{m'}$; given $k\in N$, let m_k' be the exponent associated in that way to m_k. Writing $d_k=p^{m_{k+1}}-p^{m_k}=q^{m_{k+1}'}-q^{m_k'}$ $(k\in N)$ we see that p^{m_k} and $q^{m'_k}$ both divide d_k; since they are relatively primes, their product divides d_k. So $d_k>(p^{m_k})^2$ for each $k\in N$. This implie that $p^{m_{k+1}}>(p^{m_1})^{2^k}$ for every $k\in N$. Then, by the definition of $r_{p,q,b}(n)=r(n)$, we have that $2^{r(n)-1}m_1\log p\leq m_{r(n)}\log p\leq n\log p$ $(n\in N)$. Therefore $r_{p,q,b}(n)\leq(\log 2)^{-1}\log n+1$ $(n\in N)$, which gives the desired result.

We have verified that (ii') holds with the above indicated values of $\Phi_k^0(f)$ $(f\in W,\ k\in N)$. Now observe that if $a_n=(a\beta n\log(\log n))^{1/2}$ then $na_n^{-2}U(a_n)\to 1$ and that $EX_1=\{x_s\}_{s\in N}$ with $x_s=(\alpha^{1/2})^{p_s}\sum_{m=1}^{\infty}(\alpha^{1/2})^{p_s^m}$. We can formulate the conclusion which we obtain from Corollary 3.4 as follows:

$$\mathcal{L}(\{a_n^{-1}(\alpha^{1/2})^{p_s}(\sum_{1\leq k\leq n,Y_k\in\mathcal{P}_s}(\alpha^{-1/2})^{Y_k}-n\sum_{m=1}^{\infty}(\alpha^{1/2})^{p_s^m})\}_{s\in N})\to_w\gamma$$

where γ is the centered Gaussian measure on ℓ^2 with covariance

$$\Phi_\gamma(f,f)=a^{-1}\sum_{s=1}^{\infty}f^2(e_s)\alpha^{p_s}(\log p_s)^{-1}\qquad(f\in(\ell^2)').$$

Now we show that in this result π can be replaced by an arbitrary probability measure (initial distribution) π' on I. Denote by E_π and $E_{\pi'}$ the corresponding expectation operators. It is sufficient to show that $E_{\pi'}u(S_n)-E_\pi u(S_n)\to 0$ as $n\to\infty$ for each bounded uniformly continuous function $u:B\to\mathbb{R}$, where $S_n=a_n^{-1}\sum_{j=1}^n(X_j-E_\pi X_1)$. Let u be such a function. If $n,r\in N$, $n>r$, write $Z_{n,r}=a_n^{-1}\sum_{j=r+1}^n(X_j-E_\pi X_1)$. Given $\varepsilon>0$, since we can apply Lemma 7.2 of [2,Chapter V], we can choose $r\in N$ such that $|E_{\pi'}u(Z_{n,r})-E_\pi u(Z_{n,r})|\leq\varepsilon$ for every integer $n>r$. Therefore we have for those n that

$$|E_{\pi'}u(S_n)-E_{\pi}u(S_n)| \leq \epsilon + E_{\pi'}|u(S_n)-u(Z_{n,r})| + E_{\pi}|u(Z_{n,r})-u(S_n)|.$$

Since $u(S_n)-u(Z_{n,r})$ tends pointwise to zero as $n\to\infty$, we conclude that $\overline{\lim}_n\ |E_{\pi'}u(S_n)-E_{\pi}u(S_n)|\leq\epsilon$. This implies our assertion.

Acknowledgments. I thank Professor A. de Acosta for several useful conversations and N. Coleff, N. Bucari and A. Maltz for their comments on a question about primes involved in example (b).

References

[1] A. Araujo and E. Giné (1980). The central limit theorem for real and Banach valued random variables. Wiley, New York.

[2] J.L. Doob (1953). Stochastic processes. Wiley, New York.

[3] W. Feller (1968). An introduction to probability theory and its applications I, 3rd ed. Wiley, New York.

[4] I.I. Gihman and A.V. Skorohod (1974). The theory of stochastic processes I. Springer-Verlag, New York.

[5] N. Herrndorf (1983). The invariance principle for ϕ-mixing sequences. Z. Wahrsch. verw. Gebiete 63, 97-108.

[6] M. Iosifescu and R. Theodorescu (1969). Random processes and learning. Springer-Verlag, Berlin.

[7] Z.Y. Lin (1981). Limit theorem for a class of sequences of weakly dependent random variables. Chinese Ann. Math. 2 n°2, 181-185 (MR 82h:60042).

[8] V.V. Mal'tsev and E.I. Ostrovskii (1982). Central limit theorem for stationary processes in Hilbert space. Theor. Probab. Appl. 27 n°2, 357-359.

[9] J.D. Samur (1984). Convergence of sums of mixing triangular arrays of random vectors with stationary rows. Ann. Probab. 12, 390-426.

Departamento de Matemática
Facultad de Ciencias Exactas, UNLP
Casilla de correo 172
1900 La Plata
Argentina

MAX-INFINITE DIVISIBILITY AND MAX-STABILITY IN INFINITE DIMENSIONS

Pirooz Vatan
Mathematics Department
Tufts University
Medford, MA 02155

This paper extends the notions of max-stability and max-infinite divisibility to infinite dimensional spaces. The duality with the notions of stability and infinite divisibility is stressed. It is shown that this duality, in which, for example, the operations of sum and maximum, convolution and multiplication, and the characteristic functionals and distribution functions are dual pairs, extends to infinite dimensions. In particular max-infinitely divisible and max-stable laws on $\mathbb{R}^\infty$, ℓ^∞, and c_0 are characterized; characterizations of certain classes of max-stable laws on the ℓ^p spaces are mentioned; max-stable processes are represented by a "stochastic maxima" which is dual to LePage's (1980) and Kuelbs' (1973) representations of stable processes by stochastic series and integrals; necessary and sufficient conditions for continuity in probability of max-stable processes are found.

1. Introduction

1.1 Notation. The following notation is in force throughout this paper. $\mathbb{R}^\infty$ is the usual space of real sequences $\{a_n\}_{n\geq 1}$ with product σ-algebra $\mathcal{B}(\mathbb{R}^\infty)$. Similarly, $[-\infty,\infty)^\infty$ is the countable product space $[-\infty,\infty)\times[-\infty,\infty)\times\ldots$ with product σ-algebra. If E is any subset of $[-\infty,\infty)^\infty$, $\mathcal{B}(E)$ will denote the Borel subsets of E with respect to the product topology. All operations and relations involving elements of $\mathbb{R}^n$ or $\mathbb{R}^\infty$ are component-wise. Note that "$\geq$" does not mean "greater than or equal to" if $n\geq 2$. For any elements $x=(x_1,\ldots,x_n)$ and $y=(y_1,\ldots,y_n)$ in $\mathbb{R}^n$, let $[x,y]:=[x_1,y_1]\times\ldots\times[x_n,y_n]$. The intervals $[x,y)$, $(x,y]$ and (x,y) are defined in an analogous manner. Given $m\in\{0,1,2,\ldots,\infty\}$ and $y\in(-\infty,\infty]^m$, let $A_y:=\prod_{i\geq 1} A_i$, where Π denotes the cartesian product of sets and $A_i:=[-\infty,y_i]$ if $y_i\neq\infty$, and $A_i:=[-\infty,\infty)$ if $y_i=\infty$.

1.2 Background and motivation. The existence of the so-called extremal distributions on the real line is a well-known fact (see e.g., Gnedenko (1943)). They consist of three distinct families of types:

$$\phi(x) := \begin{cases} \exp(-x^{-\alpha}), & \text{if } x > 0 \\ 0, & \text{if } x \leq 0 \end{cases} \qquad \alpha > 0$$

$$\psi(x) := \begin{cases} 1, & \text{if } x > 0 \\ \exp(-(-x^{\alpha}), & \text{if } x \leq 0 \end{cases} \qquad \alpha > 0 \qquad [1.1]$$

$$\Lambda(x) := \exp(-\exp(-x)), \qquad x \in \mathbb{R},$$

These laws arise as the only possible distributional limits of affinely scaled maxima of i.i.d. random variables, i.e. weak limits of expressions of the form

$$\left(\max_{1 \leq i \leq n} X_i + a_n\right) / b_n$$

as $n \to \infty$, where $a_n > 0$, $b_n \in \mathbb{R}$, and $X_1, X_2, \ldots$ is an i.i.d. sequence of random variables.

The following characterizing property of the extremal distributions suggests the other name by which they are often called, namely max-stable (MS) laws: $\forall n \; \exists a_n > 0$, $b_n \in \mathbb{R}$, such that $\forall x \in \mathbb{R}$

$$F^n(x) = F(a_n X + b_n) ,$$

i.e. the type of a random variable X with extremal distribution F remains the same under the operation of taking the maximum of n independent copies of X.

The apparent similarity of the above to the case of stable distributions which arise as the distributional limits of affinely scaled sums of i.i.d. random variables is not accidental. It turns out that most results in one theory have appropriate duals in the other.

The theory of stable laws has been adequately generalized to separable Banach spaces. These generalizations, as well as the one-dimensional theory, can be based on the analysis and characterization of a wider class of laws, the infinitely divisible probability distributions. Similarly, higher dinensional MS laws arise naturally as the limits of partial maxima obtained from multidimensional data which is normed (affinely scaled) component-wise.

The dual of infinite divisibility, max-infinite divisibility, is defined as follows: A law with distribution function F is max-infinitely divisible (MID) if and only if for all $t > 0$, $F^{1/t}$ is also a distribution function (d.f.). In dimension one all probability laws are MID. This is probably why initial efforts to determine MS laws in $\mathbb{R}^n$ did not use the (non-trivial) notion of max-infinite divisibility. These efforts, which ignored the, at least formal, duality that exists between the two theories, were not terribly successful.

Grenander (1963) appears to be the first study of max-infinite divisibility (in $\mathbb{R}^1$) via semi-group methods. This led Balkema and Resnick (1977) to define and characterize all MID laws on $\mathbb{R}^2$. Their results involve notions dual to those present in the theory of infinitely divisible laws, such as Levy measure, triangular arrays, and independent increment stable processes. However, as least two of their statements and three of their proofs do not generalize to higher dimensions in a straightforward way. Furthermore, they fail to provide a 1-1 correspondence between a MID law and its "exponent measure," the dual of a Levy measure.

DeHaan and Resnick (1977) use Balkema and Resnick's result to derive, at least in $\mathbb{R}^2$, a representation for MS laws that is quite similar to the stable case. Pickands (1976) also obtains a similar representation. However, besides the fact that de Haan and Resnick's representation in $\mathbb{R}^n$ depends on the (unproven) extension of Balkema and Resnick's theorem (and hence does not provide a 1-1 correspondence between "m-spectral measures" and MS laws), the formula expressing the d.f. does not lend itself to immediate extension to an infinite-dimensional space. In contrast Kuelbs' representation for the characteristic function (the dual of a d.f.) of a stable law in a separable Hilbert space is formally the same in all dimensions (see Kuelbs (1973))

This paper extends Balkema and Resnick's main theorem to $\mathbb{R}^n$, $n \geq 1$ and $\mathbb{R}^\infty := \mathbb{R}\times\mathbb{R}\times\mathbb{R}\times\ldots$. Actually our result, Theorem 2.3, is stronger since it provides the 1-1 correspondence mentioned above. In addition, MID laws on ℓ^∞ and c_0, as subsets of $\mathbb{R}^\infty$, are also characterized in Theorem 2.10. Using these characterizations, Theorem 3.9 provides a representation of MS laws on $\mathbb{R}^n$ as well as some infinite-dimensional spaces. It appears that the dual of the separable Hilbert space ℓ^2 is the space ℓ^∞ of bounded real sequences considered as a subspace of $\mathbb{R}^\infty$. In fact, our characterization is most natural for the spaces ℓ^∞ and c_0. We also mention similar, but weaker and less natural, results for MS laws on the spaces ℓ^p, $p \in (0,\infty)$, of p-summable real sequences (Theorem 3.15). The weakness of these results is probably due to the fact that the defining condition of ℓ^p is in terms of summability and not in terms of properties more natural for partial maxima.

In another direction, LePage, Woodroofe, and Zinn (1981) show that a symmetric stable random variable can be obtained by summing the (randomly) weighted inverses of points of a Poisson random measure on the real line. More precisely, suppose $\{\Gamma_k\}_{k \geq 1}$ are points of a standard Poisson random measure P on $(0,\infty)$ and $\{f_j\}_{j \geq 1}$ is a Rademacher sequence independent of P. Then for any $0 < \alpha < 2$, the sum

$$\sum_{k \geq 1} f_k \Gamma_k^{-1/\alpha}$$

is a symmetric stable random variable of order α. This is a special case of a series representation for infinitely divisible probability measures in terms of the points of P (see Ferguson and Klass (1972) and LePage (1980)).

A similar result holds in higher dimensions. In fact LePage (1980) uses Kuelbs' characterization of the characteristic functionals of stable laws on a separable Hilbert space H to obtain a series representation for stable laws in H. There, the coefficients f_j are certain random vectors in H. This representation is both elegant and useful in many ways.

We obtain the duals of these representations for MID and MS laws on $\mathbb{R}^n$, $\mathbb{R}^\infty$, and ℓ^∞. Theorems 4.2 and 4.4 show that any MS law on any of these spaces can be represented by the maximum of weighted inverses of points of a Poisson random measure. This series representation can be generalized to MS continuous time processes which are continuous in probability (Theorem 4.5). Moreover, Theorem 4.7 provides a condition which is equivalent to continuity in probability of MS processes. DeHaan (1984) has independently derived similar but slightly weaker versions of Theorems 3.10 and 4.5.

There are numerous other parallels between the theories of stable and MS laws. We try to demonstrate as many of these as possible, hoping that this will eventually lead to a unified treatment of both subjects based on, say, common algebraic properties of addition and the maximum operation such as the semi-group structure. This in turn might lead to development of other theories with new operations replacing addition and maxima. Possible candidates in this direction are the "norm functions":

$$f_n(x,y) := (x^n + y^n)^{1/n}, \qquad n = \pm 1, \pm 2, \pm 3, \ldots .$$

Note that addition, maximum and minimum are special and limiting cases of the f_n.

2. Max-infinitely divisible laws. As mentioned before, Balkema and Resnick (1977) define max-infinite divisibility for laws on $\mathbb{R}^2$ and characterize their d.f.'s via a representation theorem. They make no comments on the possibility of extending these results to $\mathbb{R}^n$, $n > 2$. The fact that all laws on $\mathbb{R}^1$ are MID suggests that the proofs are not dimension-free (in fact their construction of the exponent measure fails in $\mathbb{R}^1$). Therefore, a generalization should be carried out carefully. Their main result is extended to all Euclidean spaces in this section. The statement to be proved is slightly stronger due to a uniqueness clause. Also, the notion of a MID law on $\mathbb{R}$ is defined and an appropriate representation is obtained. Additional information is provided in the special cases of laws concentrated on ℓ^∞ and c_0 as subsets of $\mathbb{R}^\infty$.

Definition 2.1. A law P on $\mathbb{R}^n$, $n \geq 1$, with d.f. F is called max-infinitely divisible (MID) iff for any integer $m \geq 1$, $F^{1/m}$ is a d.f. on $\mathbb{R}^n$. A law P on $\mathbb{R}^\infty$ is called max-infinitely divisible iff all its finite-dimensional marginals are MID. A random variable taking values in $\mathbb{R}^n$ (or $\mathbb{R}^\infty$) is called max-infiritely divisible iff its law is MID.

Definition 2.2. The λ-support of a measure P on $[-\infty,\infty)^n$, $n \geq 1$, is the smallest set in $[-\infty,\infty)^n$ of the form $[b_1,\infty) \times \ldots \times [b_n,\infty)$ with $(b_1,\ldots,b_n) \in [-\infty,\infty)^n$ which contains the support of P. The λ-support of a law P on $\mathbb{R}^\infty$ is defined to be $Q := [q_1,\infty) \times [q_2,\infty) \times \ldots$ iff the λ-supports of all of the finite-dimensional marginals of P correspond to the finite-dimensional projections of Q.

The statement and proof of the following theorem are very closely based on Balkema and Resnick's Theorem 3. The proof is slightly more complicated. The main differences in the statement and proof are:

1) Our representation is unique, giving a 1-1 correspondence between the exponent measures and MID laws.
2) The exponent measures are not σ-finite. (Actually they are σ-finite on $[-\infty,\infty)^n$ minus one point.)
3) Our construction of the exponent measure has a different λ-support. We think that this construction is more natural.

Theorem 2.3. Fix $n \in \{1,2,\ldots,\infty\}$ and $Q := [q_1,\infty) \times \ldots \times [q_n,\infty)$ for some $(q_1,\ldots,q_n) \in [-\infty,\infty)^n$. Then there is a 1-1 correspondence between

A) MID laws P on $\mathbb{R}^n$ with λ-support Q

and

B) Positive measures ν on $[-\infty,\infty)^n$ with λ-support Q, satisfying

a) $\nu(\{q\}) = \infty$

b) $\nu(A_y^c) < \infty$ for all $y \in \{x \in (q_1,\infty] \times \ldots \times (q_n,\infty]: x_i \neq \infty$ for only finitely many $i\}$

c) if $q_i = -\infty$ then $\nu(\prod_{j=1}^{n} B_j) = \infty$ where $B_j := [-\infty,\infty)$ if $j \neq i$ and $B_i := \mathbb{R}$.

The correspondence is given by

$$P(A_y) = e^{-\nu(A_y^c)} \quad \text{for all} \quad y \in (q_1,\infty] \times \ldots \times (q_n,\infty] \ . \qquad [2.4]$$

ν is called the <u>exponent measure</u> of P.

<u>Proof</u>. <u>Case $n < \infty$</u>. Let ν be as in part (B) above. Proceeding as in Balkema and Resnick (1977), define $H: \mathbb{R}^n \to [0,\infty]$ by

$$H(x) := \nu(A_x^c) \quad \text{for all} \quad x \in \mathbb{R}^n \ .$$

Construct a Poisson random measure N on $\mathbb{R} \times (Q - \{q\})$, where $q := (q_1,\ldots,q_n)$, with intensity measure $\lambda \times \nu'$, where ν' is the restriction of ν to $Q - \{q\}$ and λ is the Lebesque measure on $\mathbb{R}$. ν' is σ-finite by (b). Let $\{(T_k,X_k)\}_{k \geq 1}$ represent the sample points of the Poisson process and define a $[-\infty,\infty]^n$-valued process $\{Y(t)\}_{t\in(0,\infty)}$ by

$$Y(t) := \begin{cases} q, & \text{if no } T_k \leq t \\ \sup\{X_k: T_k \leq t\}, & \text{otherwise} \end{cases} .$$

Observe that for all t, $Y(t) \geq q$. Furthermore, for all $y \in Q$ and $t > 0$,

$$\begin{aligned} P(Y(t) \leq y) &= P(N((0,t] \times (Q - A_y)) = 0) \qquad [2.5] \\ &= e^{-t\nu'(Q - A_y)} \\ &= e^{-t\nu(A_y^c)} \\ &= e^{-tH(y)} \ . \end{aligned}$$

By (b) and (c) we can assume that $Y(t)$ is real-valued. By (a) for $y \notin Q$, $H(y) = \infty$. Also, for $y \notin Q$, $P(Y(t) \leq y) = 0$. Therefore, formula (2.5) holds for all $y \in \mathbb{R}^n$. Consequently, $e^{-\nu(A_y^c)}$ is the d.f. of the real-valued random variable $Y(1)$. It is clear that $Y(1)$ is MID.

Now suppose that ν_1 and ν_2 give rise to the same MID distribution F. Then $\nu_1(A_y^c)$ and $\nu_2(A_y^c)$ are the same for all y in $\mathbb{R}^n$. Furthermore, they are finite for y in $(q_1,\infty) \times (q_2,\infty) \times \ldots$. This implies that ν_1 and ν_2 agree on Q. But, by definition, ν_1 and ν_2 are zero outside of Q. Therefore $\nu_1 = \nu_2$ everywhere. Thus we have shown that two different measures ν_1 and ν_2, satisfying the properties in (B) give rise to two different MID distributions.

Now assume that P is a MID distribution on $\mathbb{R}^n$ with d.f. F. For $k = 1,2,\ldots$ and $x \in \mathbb{R}^n$ let $H_k(x) := k(1 - F^{1/k}(x))$. Note that whenever $F > 0$, $F^{1/k} \to 1$ as $k \to \infty$. A Taylor expansion for $\log(x)$ allows the deduction that

$$H_k(x) \to H(x) := -\log F(x), \quad \text{for all } x \in \mathbb{R}^n .$$

If ν_k is the finite measure on $\mathbb{R}^n$ whose d.f. is $kF^{1/k}$, then $H_k(y) = \nu_k(A_y^c)$ for all y in $\mathbb{R}^n$. It remains to establish that the limit H also corresponds to a measure ν, which however is not finite. Also needed is verification that ν satisfies the conditions in (B). Let Q' be the smallest set of the form $[p_1,\infty) \times \ldots \times [p_n,\infty)$ containing all the points in $[-\infty,\infty)^n$ at which H is finite, where $(p_1,\ldots,p_n)$ is in $[-\infty,\infty)^n$. Then $Q' = Q$, the λ-support of P. H is decreasing in all arguments since F is. Since $F^{1/k}$ is a d.f., we have

$$F^{1/k}(\min(a,b)) \geq F^{1/k}(a) + F^{1/k}(b) - 1 \qquad \text{for all } a,b \in \mathbb{R}^n ,$$

and therefore

$$H^k(\min(a,b)) \leq H_k(a) + H_k(b) \qquad \text{for all } a,b \in \mathbb{R}^n .$$

Hence H satisfies the same condition. From these facts it follows that H is finite for $x > q$ and infinite outside Q.

We now define a measure ν_0 on the Borel subsets of $Q - \{q\}$. Let C be the semi-ring of rectangles of the form $A := \prod_{i=1}^{n} A_i$ where each A_i has one of the following forms: $(a_i,b_i]$ or $[q_i,b_i]$ with $q < a < b$. Define ν_0 on C by

$$\nu_0(\prod_{i=1}^{n} A_i) := \Delta_A H := \sum_x \operatorname{sgn}_A(x)H(x)$$

where the sum extends over those vertices $x = (x_1,\dots,x_n)$ of A such that $x_i = a_i$ or $x_i = b_i$ (and not q_i) for all $i = 1,\dots,n$ and $\mathrm{sgn}_A(x)$ is +1 or -1 according as the number of i, $1 \le i \le n$, satisfying $x_i = a_i$ is odd or even.

First we must show that $\Delta_A H$ is non-negative. Note that for all k and A as above

$$\Delta_A H_k := \sum_x \mathrm{sgn}_A(x) H_k(x) = \nu_k\Big(\prod_{i=1}^{n} B_i\Big) \ge 0 ,$$

where each B_i has one of the forms $(-\infty, b_i]$ or $(a_i, b_i]$ according to whether A_i has the form $[q_i, b_i]$ or $(a_i, b_i]$. Now use $\Delta_A H = \lim_{k\to\infty} \Delta_A H$ to obtain $\Delta_A H \ge 0$. The facts that $\lim_{x\to\infty} H(x,\dots,x) = 0$ and that H is lower semi-continuous together with a slight modification of Theorem 12.5 of Billingsley (1979) (see Theorem 2.8) give that ν_0 extends uniquely to the σ-algebra generated by these rectangles which is precisely the Borel subsets of $Q - \{q\}$. Finally, define the measure ν on the Borel subsets of $[-\infty,\infty)^n$ to be equal to ν_0 on $Q - \{q\}$, equal to ∞ at $\{q\}$ and equal to 0 outside Q. Condition (a) and the fact that the λ-support of ν is Q are now immediate. That ν satisfies (2.4) follows from the defining relationship of ν on rectangles and the fact that ν is 0 outside Q. Finally condition (c) is obtained by noting that

$$\begin{aligned}\nu\Big(\prod_{j=1}^{n} B_j\Big) &= \lim_{b\to\infty}\lim_{a\to-\infty} \nu([q_1,b]\times\dots\times[q_{i-1},b]\times[a,b]\times[q_{i+1},b]\times\dots\times[q_n,b]) \\ &= \lim_{b\to\infty} H(b,\dots,b) - \lim_{a\to-\infty}\lim_{b\to\infty} H(b,\dots,b,a,b,\dots,b) \\ &= \infty - 0 = \infty .\end{aligned}$$

Since ν determines the d.f. via (2.4), there can be only one MID law P corresponding to an exponent measure ν, thereby completing the proof of the theorem for $n < \infty$.

<u>Case $n = \infty$</u>. First note that the marginals of an MID distribution P are MID. Also the λ-supports of these marginals correspond to the appropriate projections of the λ-support of P. Now if P is a MID law on $\mathbb{R}^\infty$ with λ-support Q then the family $\{P_G\}_G$ of finite dimensional marginals of P gives rise to a consistent family of measures $\{\nu_G\}_G$ on the spaces $[-\infty,\infty)^G$. This family in turn gives rise to a measure ν on the space $[-\infty,\infty)^\infty$ (see Theorem 2.9 below). By definition of the λ-support for laws on $\mathbb{R}^\infty$, we see that ν has

the right λ-support. Equation (2.4) holds because it holds for every finite dimensional y and P is countably additive. Similar reasoning verifies conditions (a) - (c). Uniqueness of ν follows from the uniqueness of its finite-dimensional marginals.

Conversely, given an exponent measure ν on $\mathbb{R}^\infty$ satisfying (B), construct a family of MID laws P_G corresponding to the finite-dimensional marginals of ν. This family is consistent by inspection. Therefore it gives rise to a MID law P on $\mathbb{R}^\infty$ with λ-support Q. Again uniqueness of P is deduced from the fact that ν determines the finite-dimensional marginals of P. Q.E.D.

Remark 2.6. Condition (a) of Theorem 2.3 is necessary for two reasons. On $\mathbb{R}^1$ it is required if P has an atom at the infimum of its λ-support. Moreover, it's necessary if we require that the exponent measures of the marginals correspond to marginals of the exponent measure. The existence of some y satisfying condition (b) is necessary to insure that P have no mass on the subset of $[-\infty,\infty]^n$ consisting of points with at least one coordinate $= \infty$. That this condition should hold for all y insures the uniqueness of ν. Condition (c) seems to be the least natural of the three. It ensures that P has no mass on the subset of $[-\infty,\infty)^n$ consisting of points with at least one coordinate $= -\infty$. Note that $-\infty$ plays the same role for maxima that 0 plays for addition: they are both identity elements. For this reason it appears natural to expand our definition of MID random variables to those taking values in $[-\infty,\infty)^n$. In that case Theorem 2.3 still holds if we only remove condition (c). If we further expand our definition to $[-\infty,\infty]^n$-valued random variables then Theorem 2.3 holds if condition (c) is removed and ν is allowed to be a measure on $[-\infty,\infty]^n$.

Remark 2.7. It is possible to show that the support of a MID law is equal to the closure under maxima of the support of its exponent measure. See Balkema and Resnick (1977).

The following theorem guarantees the existence of an exponent measure corresponding to a certain "cumulative distribution function." The proof is, almost word for word, the same as Theorem 12.5 of Billingsley (1979) which considers cumulative d.f.s corresponding to measures assigning finite values to bounded rectangles. It has to be modified so that it applies to measures which are possibly infinite in a neighborhood of a fixed point (see Vatan (1984a)).

Fix an integer k and a point $q \in [-\infty,\infty)^k$ and let $Q := [q_1,\infty) \times \ldots \times [q_k,\infty)$. For $q_i \leq a_i \leq b_i$ define $I(a_i,b_i)$ to be $(a_i,b_i]$ if $a_i > q_i$ and $[a_i,b_i]$ if $a_i = q_i$. Let C be the semi-ring of rectangles of the form $A := \prod_{i=1}^{k} A_i$, where each A_i is of the form $I(a_i,b_i)$ for some $(q_1,\ldots,q_k) \leq (a_1,\ldots,a_k) \leq (b_1,\ldots,b_k)$ such that there is at least one $a_i > q_i$.

Theorem 2.8. Suppose that $H: (q_1,\infty) \times \ldots \times (q_k,\infty) \to \mathbb{R}$ is semi-continuous from below and satisfies $\Delta_A H \geq 0$ for rectangles $A \in C$ where $\Delta_A H$ is defined as in the proof of Theorem 2.3. Then there exists a unique exponent measure ν on $\mathbb{R}^k$ satisfying

$$\nu(A) = \Delta_A H .$$

The next theorem shows the existence of an exponent measure on $\mathbb{R}^\infty$ given a consistent family $\{\nu_G\}$ of exponent measures on finite dimensional spaces $\mathbb{R}^G$ such that ν_G is induced by the standard projection from $\mathbb{R}^\infty$ to $\mathbb{R}^G$. This is an extension of Kolmogorov's existence theorem for countable indices. The usual proof of Kolmogorov's theorem does not apply immediately since the ν_G are neither finite nor σ-finite. A modification of that proof which works in this more general setting can be found in Vatan (1984a).

Theorem 2.9. Fix $q := (q_1,q_2,\ldots) \in [-\infty,\infty) \times [-\infty,\infty) \times \ldots$ and let $Q := [q_1,\infty) \times [q_2,\infty) \times \ldots$. Suppose that $\{\nu_G$: G a finite subset of $\{1,2,\ldots\}\}$ is a consistent family of exponent measures, i.e. each ν_G satisfies the conditions of Theorem 2.3 and for all finite subsets $F = \{f_1,\ldots,f_m\}$ and $G = \{g_1,\ldots,g_n\}$, $\nu_G = \nu_F \circ f_{FG}^{-1}$, where f_{FG}: $[-\infty,\infty)^F \to [-\infty,\infty)^G$ is the standard projection. Then there exists a unique exponent measure ν on Q such that, for all finite subsets $F = \{f_1,\ldots,f_m\}$ of $\{1,2,\ldots,\}$, $\nu_F = \nu_G \circ f_F^{-1}$, where $f_F: Q \to [q_{f_1},\infty) \times \ldots \times [q_{f_m},\infty)$ is the standard projection.

Theorem 2.3 extends to other ℓ^∞ spaces as well as c_0.

Theorem 2.10. Assume that P is a MID law on $[0,\infty)^\infty$ with exponent measure ν. Then

a) P is concentrated on ℓ^∞ iff ν is.

b) P is concentrated on c_0 iff ν is.

Proof.

a) $$\nu(\{x: \|x\|_\infty = \infty\}) = \nu\Big(\bigcap_{n\geq 1} \bigcup_{m\geq 1} \{x: \max_{1\leq i\leq m} x_i > n\}\Big)$$
$$= \lim_{n\to\infty} \lim_{m\to\infty} \nu\{x: \max_{1\leq i\leq m} x_i > n\}$$
$$= - \lim_{n\to\infty} \lim_{m\to\infty} \log P\{x: \max_{1\leq i\leq m} x_i \leq n\}$$
$$= - \log P\Big(\bigcup_{n\geq 1} \bigcap_{m\geq 1} \{x: \max_{1\leq i\leq m} x_i \leq n\}\Big)$$
$$= - \log P\{x: \max_{i\geq 1} x_i < \infty\}$$
$$= - \log P\{x: \|x\|_\infty < \infty\} .$$

b) $$\nu(\{x: \lim_{i\to\infty} x_i \neq 0\}) = \nu\Big(\bigcup_{m\geq 1} \bigcap_{n\geq 1} \{x: \max_{i\geq n} \geq 1/m\}\Big)$$
$$= \lim_{m\to\infty} \lim_{n\to\infty} \nu(\{x: \max_{i\geq n} x_i \geq 1/m\})$$
$$= \lim_{m\to\infty} \lim_{n\to\infty} \nu(A^c_{(\infty,\ldots,\infty,1/m,1/m,\ldots)})$$
$$= - \lim_{m\to\infty} \lim_{n\to\infty} \log P(A^c_{(\infty,\ldots,\infty,1/m,1/m,\ldots)})$$
$$= - \log P\Big(\bigcap_{m\geq 1} \bigcup_{n\geq 1} \{x: \max_{i\geq n} x_i < 1/m\}\Big)$$
$$= - \log P(\{x: \lim_{i\to\infty} x_i = 0\}).$$

Q.E.D.

3. Max-stable laws. An important subclass of MID laws is the set of max-stable (MS) laws. A law P on $\mathbb{R}^n$ with d.f. F is MS if all powers F^k of F are distributions of the same type as F, i.e. F^k can be obtained by scaling and translating F. MS laws on the real line were first characterized by Fisher and Tippett (1928). Grenander (1943) and von Mises (1936) developed the theory further by characterizing max-domains of attraction of such laws.

Results on representations of multivariate MS distributions have appeared in Finkelstein (1953), Geffroy (1958/1959), Tiago deOliveira (1958), and Sibuya (1960). All of these, however, consider only the bivariate case. These representations seem ad hoc and not easy to use. Galambos (1978) also contains representations of MS laws on $\mathbb{R}^n$. The first satisfactory treatment of the bivariate case was done by deHaan and Resnick (1977) where they characterize all such laws in a way very similar to the characterization of symmetric stable laws on $\mathbb{R}^n$. This result is based on Balkema and Resnick's representation of MID laws on

$\mathbb{R}^2$ and it generalizes to $\mathbb{R}^n$ easily. However, their formula does not lend itself to immediate extension to infinite-dimensional spaces.

The spectral representation for a symmetric stable law on a separable Banach space B shows that the Levy measure factors as a product of a radial measure and a finite symmetric positive measure (the spectral measure) on the unit ball of B (see, e.g., Araujo and Gine (1980)). This section establishes a similar representation for a max-stable law P on the spaces c_0 and ℓ^p, $p \in (0,\infty]$, regarded as subsets of R^∞. The result is analogous, namely the exponent measure ν of P factors as a product of a radial measure and a finite positive measure (the "m-spectral measure") on the positive quadrant of the appropriate unit ball.

A few preliminaries are needed.

Definition 3.1. Let $X_1, X_2, \ldots$ be i.i.d random vectors in $\mathbb{R}^n$ with law P and d.f. F. Let Z_m, $m = 1,2,\ldots$ be the partial maxima of the X_i, i.e. $Z_m := \max_{1 \le i \le m} X_i$. Then P and F are called max-stable (MS) iff there exists a law Q with d.f. G, and constant vectors a_m and b_m in $\mathbb{R}^n$, $m = 1,2,\ldots$ with $a_m > 0$, such that $L((Z_m - b_m)/a_m) \xrightarrow[w]{} Q$ or, equivalently, $F^m(a_m x + b_m) \xrightarrow[w]{} G(x)$. A law P on $\mathbb{R}^\infty$ is called max-stable (MS) iff all its finite dimensional marginals are MS.

We collect a few simple facts concerning MS laws on $\mathbb{R}^n$. See, e.g., Galambos (1978). Let H be a d.f. on $\mathbb{R}^n$ with one-dimensional marginals H_i, $i = 1,2,\ldots,n$.

Fact 3.2. H is MS iff for all $s > 0$ there exist constant vectors A_s and B_s in $\mathbb{R}^n$ such that $A_s > 0$ and

$$H^s(A_s x + B_s) = H(x), \qquad \text{for all } x \text{ in } \mathbb{R}^n .$$

In particular, H being MS implies H is MID.

Fact 3.3. If H is MS then, for all $x = (x_1,\ldots,x_n)$ in $\mathbb{R}^n$,

$$H(x) \geq \prod_{i=1}^{n} H_i(x_i) .$$

Fact 3.4. H is MS on $\mathbb{R}$ iff it is degenerate or has one of the extremal types given in (1.1).

Fact 3.5. All marginals of a MS law on $\mathbb{R}^n$ are MS on $\mathbb{R}$. However, max-stability of the one-dimensional marginals of a law P does not imply max-stability of P.

Fact 3.6. Let H and F be MS on $\mathbb{R}^n$ and $\mathbb{R}$, respectively. Then the d.f.

$$G(x) := H(x_1,\dots,x_{i-1},H_i^{-1}(F(x_i)),x_{i+1},\dots,x_n)$$

is MS on $\mathbb{R}^n$ with F as its ith one-dimensional marginal. Here H_i^{-1} is any left inverse of H_i.

It is convenient to restrict attention to a class of MS laws which are particularly easy to handle.

Definition 3.7. A MS law on $\mathbb{R}^n$ or $\mathbb{R}^\infty$, is called simple (SMS) iff its one-dimensional marginals have d.f. s of the form $\phi_1(\lambda x)$ for some $\lambda > 0$.

No generality is lost in restricting attention to the SMS laws because Fact 3.6 allows MS laws on $\mathbb{R}^n$, or $\mathbb{R}^\infty$, to be converted to SMS laws on $\mathbb{R}^n$, or $\mathbb{R}^\infty$, respectively.

It is clear that the λ-support of a SMS law on $\mathbb{R}^n$ is $[0,\infty)^n$, $n = 1,2,3,\dots,\infty$.

Fact 3.8. By 3.2, H is SMS iff

$$H^s(sx) = H(x), \qquad \text{for all } x \in \mathbb{R}^n, \quad s > 0 .$$

Now we are ready to establish the existence of a spectral representation for SMS laws on the subsets ℓ^∞, ℓ^p, $p \in (0,\infty)$, and c_0 of $\mathbb{R}^\infty$. Theorem 2.10 showed that the exponent measures for MID laws on ℓ^∞ and c_0 are themselves concentrated on ℓ^∞ and c_0. This is also true for SMS laws on ℓ^p, $p \in (0,\infty)$. Furthermore, the exponent measure ν of a SMS law on any ℓ^p or c_0 decomposes into the product $\mu \times \sigma$ of a radial measure μ and a finite positive measure σ concentrated on the positive quadrant of the unit sphere. σ will be called an m-spectral measure.

To proceed we first generalize and strengthen a finite dimensional result due to deHaan and Resnick (1977). The strengthening requires the uniqueness in Theorem 2.3. It also demonstrates that there is nothing special about the ℓ^2 sphere as the domain of the m-spectral measure. This is important because ℓ^∞ becomes significant in the infinite-dimensional extension.

Theorem 3.9. Let $n \in \{1,2,\ldots,\infty\}$. There is a 1-1 correspondence between

a) SMS laws P on $\mathbb{R}^n$, concentrated on $\ell^\infty(c_0)$ if $n = \infty$

and

b) finite positive measures on the positive unit sphere S where S can be chosen as $S^+_{n,p} := \{x \in \mathbb{R}^n : x \geq 0,\ \|x\| = 1\}$ for any $p \in (0,\infty]$ if $n < \infty$ and $S^+_\infty := S^+_{\infty,\infty} :=$ $\{x \in \ell^\infty(c_0) : x \geq 0,\ \|x\|_\infty = 1\}$ if $n = \infty$.

The correspondence is given by

$$-\log p(A_y) = \int_S \max_i \theta_i/y_i \, d\sigma(\theta), \quad \text{for all } y \in (0,\infty]^n. \qquad [3.10]$$

Proof. We consider only the case $n = \infty$, since the finite-dimensional proof is identical with this case upon replacing ∞ by n, S^+_∞ by $S^+_{n,p}$, ℓ^∞ by $\mathbb{R}^n$, $[0,\infty)^\infty$ by $[0,\infty)^n$ and utilizing Theorem 2.3 instead of Theorem 2.10. Suppose P is SMS on $\mathbb{R}^\infty$. Then it is MID and therefore has an exponent measure ν such that

$$-\log P(A_y) = \nu(A_y^c), \text{ for all } y \in (0,\infty]^\infty. \qquad [3.11]$$

By Fact 3.8 $\nu(A_y^c) = s\nu(sA_y^c)$ for all $y \in (0,\infty]^\infty$. Since $\{A_y^c\}_{y\in(0,\infty]^\infty}$ form a determining class,

$$\nu(B) = s\nu(sB), \qquad \text{for all } B \in \mathcal{B}([0,\infty)^\infty). \qquad [3.12]$$

By Theorem 2.10, ν is concentrated on $\ell^\infty \cap [0,\infty)^\infty$. Now define a "polar coordinate transformation"

$$T: \ell^\infty \cap [0,\infty)^\infty - \{0\} \to (0,\infty) \times S^+_\infty$$

$$T(x) := (r(x),\theta(x)), \text{ where}$$

$$\theta(x) := (\theta_1(x),\theta_2(x),\ldots), \quad \text{with } \theta_i(x) := x_i/\|x\|_\infty, \quad i \geq 1$$

$$r(x) := \|x\|_\infty .$$

T is a Borel equivalence, i.e. T is one-to-one and onto and both T and T^{-1} are measurable with respect to the product σ-algebra of $(0,\infty) \times S^+_\infty$. Define $\mu := \nu \circ T^{-1}$. Let μ_1 be the measure on $(0,\infty)$

with density $1/x^2$ and let σ be the measure on S_∞^+ with

$$\sigma(B) := \nu\{x \in \ell^\infty \cap [0,\infty)^\infty: r(x) > 1,\ \theta(x) \in B\}, \text{ for all } B \in \mathcal{B}(S_\infty^+).$$

It will be shown that μ can be written as the product of μ_1 and σ. For any $r > 0$, and $B \in \mathcal{B}(S_\infty^+)$,

$$\begin{aligned}\mu((r,\infty)\times B) &= \nu(\{x \in \ell^\infty \cap [0,\infty)^\infty: r(x) > r,\ \theta(x) \in B\}) \\ &= \nu(r\cdot\{x \in \ell^\infty \cap [0,\infty)^\infty: r(x) > 1, \theta(x) \in B\}) \\ &= 1/r\cdot\nu(\{x \in \ell^\infty \cap [0,\infty)^\infty: r(x) > 1, \theta(x) \in B\}) \quad \text{by (3.12)} \\ &= \mu_1(r,\infty)\times\sigma(B).\end{aligned}$$

So we have

$$\begin{aligned}\nu(A_y^c) &= \int_{A_y^c} d\nu = \int_{TA_y^c} d\mu_1 d\sigma \qquad [3.13] \\ &= \int_{\{(r,\theta):\ r\theta_i \le y_i,\ i=1,2,\dots\}^c} d\mu_1(r)d\sigma(\theta) \\ &= \int_{\{(r,\theta):\ r > \min_{i\ge 1} y_i/\theta_i\}} d\mu_1(r)d\sigma(\theta) \\ &= \int_{S_\infty^+}\int_{\min_{i\ge 1} y_i/\theta_i}^{\infty} 1/r^2 dr\ d\sigma(\theta) \\ &= \int_{S_\infty^+} \max_{i\ge 1} \theta_i/y_i d\sigma(\theta).\end{aligned}$$

Combining (3.13) with (3.12) gives

$$-\log P(A_y) = \int_{S_\infty^+} \max_{i\ge 1} \theta_i/y_i\ d\sigma(\theta), \qquad \text{for all } y \in (0,\infty]^\infty.$$

To establish the finiteness of σ, use (3.12) and the fact that P is concentrated on ℓ^∞, to obtain

$$\begin{aligned}\sigma(S_\infty^+) &= \nu(\{x: r(x) > 1\}) \le \nu(\{x: x < (1,1,1,\dots)\}^c) \\ &= -\log P(\{x: x < (1,1,1,\dots)\}) < \infty.\end{aligned}$$

For the converse, let σ be a finite positive measure on S_∞^+ as in the theorem. Construct ν via the transformation T. Then ν satisfies the conditions of Theorem 2.3 and therefore gives rise to a MID law P on $\mathbb{R}$ which is SMS by construction.

To show P is concentrated on ℓ^∞ it suffices to show the existence of an m such that $P(A_{(m,m,\ldots)}) > 0$. Indeed,

$$P(\ell^\infty) = \lim_{m\to\infty} P(A_{(m,m,\ldots)})$$

$$= \lim_{m\to\infty} P(m\cdot A_{(1,1,\ldots)})$$

$$= \lim_{m\to\infty} P^{1/m}(A_{(1,1,\ldots)}) \qquad \text{by Fact 3.8.}$$

Hence, if $P(A_{(m,m,\ldots)}) > 0$ for some m, it follows that $P(\ell^\infty) = 1$. The following calculation shows that this is true for any $m > 0$.

$$-\log P(A_{(m,m,\ldots)}) = \int_{S_\infty^+} \max_{i\geq 1} \theta_i/m \, d\sigma(\theta) \qquad [3.14]$$

$$= 1/m \int_{S_\infty^+} \max_{i\geq 1} \theta_i \, d\sigma(\theta)$$

$$= \sigma(S_\infty^+)/m < \infty .$$

This completes the proof of the theorem for ℓ^∞.

The c_0 case follows from the ℓ^∞ case, Theorem 2.10, and the fact that $\mu_1 \times \sigma$ is concentrated on c_0 iff σ is, where μ_1 and σ are as before. Q.E.D.

Note that there exist infinite measures σ on S_∞^+ which give rise to SMS laws on $\mathbb{R}^\infty$ via (3.10). For example, the measure which assigns 1 to those vertices of S_∞^+ which lie on the coordinate axes of $\mathbb{R}^\infty$, and is 0 everywhere else.

For completeness, our extensions to ℓ^p will be listed, although the proofs will appear elsewhere (Vatan (1984b)).

<u>Theorem 3.15</u>. Fix $p \in (0,\infty]$. Suppose P is a SMS law on $\mathbb{R}^\infty$ which is concentrated on ℓ^p, and such that there exists $y \in \ell^p$ with $P(A_y) > 0$. Then there exists a unique positive finite measure σ on $S_p^+ := \{x \in \ell^p\colon x \geq 0, \|x\|_p = 1\}$ such that

$$-\log P(A_y) = \int_{S_\infty^+} \max_{i\geq 1} \theta_i/y_i \, d\sigma(\theta), \qquad \text{for all } y \in (0,\infty]^\infty .$$

It is possible to apply the previous results to arbitrary SMS laws on $\mathbb{R}^\infty$ via the next result.

Proposition 3.16. Suppose $\{X_n\}_{n\geq 1}$ is SMS on $\mathbb{R}^\infty$, $p \in (0,\infty]$, and $y = (y_1, y_2, \ldots) \in \ell^p$ with $y_i > 0$, for all $i \geq 1$. Then there exists a sequence $k := \{k_n\}_{n\geq 1}$ of positive numbers such that $\{k_n X_n\}_{n\geq 1}$ is SMS on ℓ^p and $P(kX \leq y) > 0$.

The following proposition gives a condition, in terms of the scale factors of the components of a SMS sequence $\{X_n\}_{n\geq 1}$, for $\{X_n\}_{n\geq 1}$ to be in ℓ^p a.s.

Proposition 3.17. Suppose $\{X_n\}_{n\geq 1}$ is a SMS sequence and let $\lambda_n := -\log P(X_n \leq 1)$. If there is a $y \in \mathbb{R}^\infty$, $y > 0$ such that $\sum_{n\geq 1} \lambda_n / y_n < \infty$ then $P(X \leq y) > 0$. In particular, if $y \in \ell^p$ then $X \in \ell^p$ a.s. and Theorem 3.15 can be applied.

It is clear that the above condition can only be satisfied for $y \in \ell^p$ if $\lim_{n\to\infty} \lambda_n = 0$. Here is a sufficient condition.

Corollary 3.18. If $\{\lambda_n\}_{n\geq 1} \in \ell^q$, $q \in (0,1)$ and $\{X_n\}_{n\geq 1}$ is a SMS sequence with $\lambda_n = -\log P(X_n \leq 1)$, $n = 1,2,\ldots$, then the condition of Proposition 3.17 is satisfied with $y_n := \lambda_n^{1-q}$, $n = 1,2,\ldots$ and $y \in \ell^{q/(1-q)}$.

The next theorem gives necessary and sufficient conditions for an independent SMS sequence to be a.s. in ℓ^p, $p \in [1,\infty]$. A useful corollary is that the converse of Theorem 3.15, at least for $p = 1$, is not true, i.e. there exists a finite measure on the positive unit sphere of ℓ^1 which gives rise to a SMS sequence not concentrated on ℓ^1.

Theorem 3.19. Suppose $\{X_n\}_{n\geq 1}$ is an independent SMS sequence with $P(X_n \leq x) = \phi_1(\lambda_n x) := \exp(-\lambda_n/x)1_{x>0}$, $n = 1,2,\ldots$. Then

a) $\{X_n\}_{n\geq 1} \in \ell^1$ a.s. if and only if

$$\sum_{n\geq 1} \lambda_n \log(1/\lambda_n) < \infty \quad \text{and} \quad \lim_{n\to\infty} \lambda_n = 0 . \qquad [3.20]$$

b) $\{X_n\}_{n\geq 1} \in \ell^p$ a.s., $p \in (1,\infty]$ if and only if $\sum_{n\geq 1} \lambda_n < \infty$.

Corollary 3.21. If $\{X_n\}_{n\geq 1}$ is an independent MS sequence in ℓ^∞ with scale factors λ_n then $\{X_n\}_{n\geq 1} \in \ell^p$ for all $p > 1$.

The following example shows that condition (3.20) in Theorem 3.19 is not equivalent to $\sum_{n\geq 1} \lambda_n < \infty$ and therefore, the converse of Theorem 3.15 does not hold for ℓ^1. It also shows that Theorem 3.9 does not extend, in its full generality, to ℓ^1.

Example 3.22. Let $\lambda_n := (n \log^2 n)^{-1}$, $n = 2,3,\ldots$, and $\lambda_1 := 1$. Then $\sum_{n\geq 1} \lambda_n < \infty$ but condition (3.20) does not hold.

4. The series representation. In Section 3 the d.f.s of MS laws on ℓ^∞ and c_0 were characterized in a manner completely analogous to the finite dimensional case. As mentioned earlier, this is similar to Kuelbs (1973) where he obtains a representation of symmetric stable laws on a separable Hilbert space in terms of their characteristic functions. His representation is also completely analogous to the finite dimensional case. In the same paper Kuelbs utilizes this generalization to find a representation for a symmetric stable sequence of random variables (i.e. a sequence with stable finite dimensional marginals) in terms of stochastic integrals originally defined by Schilder (1970).

In attempting to find a similar representation for simple max-stable laws on $\mathbb{R}^n$ based on Schilder's method one finds that his Lemmas 3.1, 3.2, Theorem 2.1, and Corollary 2.1 remain true if one replaces "symmetric stable" with "simple max-stable," addition with the maximum operation, and "stable independent increment processes" with "extremal processes." This leads to replacing integration with respect to a measure by its counterpart, some kind of limiting maximum operation with respect to a measure. Instead of developing a theory in this direction, we note that Theorem 2.3 of Section 2 already provides a representation of a MID law in terms of the points of a Poisson random measure. The dual of this representation would be summing the points of the dual process. This in turn could be interpreted as a stochastic integral of the identity function with respect to the random measure.

In fact, as mentioned in Section 1, LePage (1980) has obtained a representation for symmetric stable laws on a separable Hilbert space in terms of sums of weighted inverses of the points of a Poisson random measure. This is, in fact, the sum whose dual we are seeking.

The objective of this section is to obtain a "series" ("spectral" or "stochastic maxima") representation for these laws on finite and infinite dimensional spaces. Three facets of our proof are analogous to the method employed by Kuelbs and Schilder: (1) We construct a measurable map T from the unit interval to the positive unit sphere in the appropriate dimension. (2) We construct a measure γ on $[0,1]$ which induces the m-spectral measure σ of the simple max-stable law under T, i.e. $\sigma = \gamma \circ T^{-1}$. (3) We transform a general MS sequence into a MS sequence almost surely in ℓ^{∞} via multiplication by an appropriate sequence of positive reals.

Kuelbs (1973) also directly extends his stochastic integral representation for symmetric stable sequences to those continuous time symmetric stable processes which are continuous in probability. It is also possible to extend our result in this direction in a very analogous manner. As in Kuelbs, the essence of the proof is in obtaining a criterion for continuity in probability of SMS processes. This is the final result of this section which provides a characterization of continuity in probability of a continuous time SMS process in terms of a condition on the bi-variate m-spectral measures. First we need a standard measure theoretic result.

Theorem 4.1. Suppose γ is a Polish space and ν is any finite Borel measure ν on Y concentrated on a Borel subset A of Y. Let ν' be the restriction of ν to A. Then there exists a finite Borel measure γ on $[0,1]$ and a map $f: [0,1] \to A$ such that $\nu' = \gamma \circ f^{-1}$.

Theorem 4.2. Suppose X is a SMS random vector in $\mathbb{R}^n$, $1 \leq n \leq \infty$ which is concentrated on ℓ^{∞} if $n = \infty$. Then there exists a finite measure γ on $[0,1]$ and a Borel measurable map $f := (f_1,\ldots,f_n)$: $[0,1] \to S^+_{n,\infty}$ such that

$$L(X) = L(\max_{k \geq 1} (1/\Gamma_k) \times f(\theta_k)) , \qquad [4.3]$$

where $\{(\Gamma_k,\theta_k)\}_{k \geq 1}$ are the points in a Poisson random measure on $(0,\infty) \times [0,1]$ with intensity measure $\lambda \times \gamma$ and λ is the Lebesgue measure on $(0,\infty)$. Furthermore, the m-spectral measure on X is induced by γ via f. Conversely, for any such f and γ the right hand side in (4.3) is a SMS random vector in $\mathbb{R}^n$, concentrated on ℓ^{∞} if $n = \infty$, with m-spectral measure $\gamma \circ f^{-1}$.

Proof. First consider $n < \infty$. Let σ be the m-spectral measure of the SMS law of X, so that

$$-\log P(A_y) = \int_{S^+_{n,\infty}} \max_{i\geq 1} (\theta_i/y_i)\,d\sigma(\theta), \qquad \text{for all} \quad y \in [-\infty,\infty)^n$$

where $S^+_{n,\infty} := \{x \in \mathbb{R}^n:\ x \geq 0,\ \|x\|_\infty = 1\}$. The existence of a map $f = (f_1,\dots,f_n)\colon [0,1] \to S^+_{n,\infty}$ and a finite measure γ on $[0,1]$ such that $\sigma = \gamma \circ f^{-1}$ is guaranteed by Theorem 4.1, since $S^+_{n,\infty}$ is a complete separable metric space. Let $f_i(\theta)$ be the components of $f(\theta)$, $i = 1,2,\dots,n$, and let $\{(\Gamma_k,\theta_k)\}_{k\geq 1}$ be the points of the Poisson random measure as in theorem. Then the calculation below establishes both parts of the theorem. Applying the definition of the Poisson random measure,

$$P(\max_{k\geq 1} (1/\Gamma_k) f_i(\theta_k) \leq y_i,\ i = 1,\dots,n)$$

$$= P(N(\{(r,\theta):\ r > \min_{1\leq i\leq n} y_i/f_i(\theta)\}) = 0)$$

$$= e^{-\lambda\times\gamma(\{(r,\theta):\ r > \min_{1\leq i\leq n} y_i/f_i(\theta)\})}$$

$$= e^{-\int_0^1 \int_{\min_{1\leq i\leq n} y_i/f_i(\theta)}^{\infty} 1/R^2\, dR\, d\gamma(\theta)}$$

$$= e^{-\int_0^1 \max_{1\leq i\leq n} f_i(\theta)/y_i\ d\gamma(\theta)}$$

$$= e^{-\int_{S^+_{n,\infty}} \max_{1\leq i\leq n} \theta_i/y_i\ d\sigma(\theta)}, \quad \text{since} \quad \sigma = \gamma\circ f^{-1}.$$

If $n = \infty$, then by Theorem 3.10 we know that X has an m-spectral measure σ on S^+_∞. Since S^+_∞ is a Borel subset of the Polish space $\mathbb{R}^\infty$, Theorem 4.1 can be applied. The rest of the proof is identical with the proof of the previous theorem upon obvious modifications.

We now obtain a representation for general SMS sequences.

<u>Theorem 4.4</u>. Suppose $X := (X_1,X_2,\dots)$ is a SMS random sequence. Then there exists a finite measure γ on $[0,1]$ and a Borel measurable map $f := (f_1,f_2,\dots)\colon [0,1] \to \mathbb{R}^\infty$ such that each f_i is a bounded positive function and

$$\mathcal{L}((X_1,X_2,\dots)) = \mathcal{L}(\max_{k\geq 1}(1/\Gamma_k)(f_1(\theta_k),f_2(\theta_k),\dots))$$

where (Γ_k,θ_k) are as in the previous theorem.

Proof. Using Theorem 3.16, multiply X by an appropriate sequence to obtain a SMS sequence Y concentrated on ℓ^∞. Then apply Theorem 4.3 to Y. Q.E.D.

Theorem 4.4 can be extended to continuous time SMS processes (i.e. those with SMS finite dimensional marginals) which are continuous in probability.

Theorem 4.5. Let $\{X_t: t \in T\}$ be a simple max-stable process. If T is a topological space with a countable dense subset T_0 and $\{x_t: t \in T\}$ is continuous in probability on $T - T_0$ then there is a finite measure γ on $[0,1]$ and a family of bounded positive functions $\{f_t: t \in T\}$ such that the finite dimensional marginals of $\{X_t: t \in T\}$ are the same as those of the process defined by

$$\{\max_{k\geq 1} (1/\Gamma_k) f_t(\theta_k): t \in T\} \qquad [4.6]$$

where, as before, $\{(\Gamma_k, \theta_k)\}_{k\geq 1}$ are the points of the Poisson random measure with intensity $\lambda\times\gamma$.

Our proof of Theorem 4.5, which will appear elsewhere, requires a characterization of continuity in probability for continuous time SMS processes. Our next objective is to give one such characterization for a large class of such processes in terms of the behaviour of their bivariate m-spectral measures. First consider the case of identically distributed components. Note that the bivariate m-spectral measure of a MS random vector in $\mathbb{R}^2$ can be taken on the interval $[0,\pi/2]$ due to Theorem 3.9 with $p = 2$ and the fact that a measure on the positive unit sphere in $\mathbb{R}^2$ can be interpreted as a measure on $[0,\pi/2]$.

Theorem 4.7. Suppose $\{Y_t: t \in T\}$ is a SMS process with i.i.d. components, where T is an interval in $\mathbb{R}$. Let $S_{s,t}$ be the bivariate m-spectral measure of (Y_s, Y_t), $s,t \in T$, represented over $[0,\pi/2]$. If all $S_{s,t}$, $s \neq t$, have continuous densities with respect to Lebesgue measure, then $\{T_t: t \in T\}$ is continuous in probability if and only if for any $t \in T$, $\phi \in [0,\pi/2)$, and $\psi \in (\pi/4,\pi/2]$, $S_{s,t}([0,\pi]) \to 0$, and $S_{s,t}([\psi,\pi/2]) \to 0$, as $s \to t$.

Proof. Let $f_{s,t}$ be a density for $S_{s,t}$, $s,t \in T$. Let $G_{s,t}$ be the d.f. of (Y_s, Y_t), $s,t \in T$. Since $f_{s,t}$ is continuous, $G_{s,t}$ has a density (assuming w.l.o.g. that $Y_t \sim \Phi_1(x)$, $\forall t$) $g_{s,t}$. In polar coordinates one has

$$G_{s,t}(r,\theta) = e^{-A_{s,t}(\theta)/r}$$

$$g_{s,t}(r,\theta) = e^{-A_{s,t}(\theta)/r}(f_{s,t}(\theta)/r^3 + B_{s,t}(\theta)C_{s,t}(\theta)/r^4) \qquad \text{where}$$

$$A_{s,t}(\theta) := (1/\cos\theta)\int_0^\theta f_{s,t}(\psi)\cos\psi d\psi + (1/\sin\theta)\int_\theta^{\pi/2} f_{s,t}(\psi)\sin\psi d\psi$$

$$B_{s,t}(\theta) := (1/\cos^2\theta)\int_0^\theta f_{s,t}(\psi)\cos\psi d\psi$$

$$C_{s,t}(\theta) := (1/\sin^2\theta)\int_\theta^{\pi/2} f_{s,t}(\psi)\sin\psi d\psi .$$

$$I_1(s,t,\varepsilon) := \iint_{x_1 - x_2 > \varepsilon} g(x_1,x_2)dx_1dx_2$$

$$I_2(s,t,\varepsilon) := \iint_{x_2 - x_1 > \varepsilon} g(x_1,x_2)dx_1dx_2 .$$

$$I(s,t,\varepsilon) := I_1(s,t,\varepsilon) + I_2(s,t,\varepsilon) .$$

Continuity in probability is equivalent to $\forall\varepsilon > 0$, $\lim_{s\to t} I(s,t,\varepsilon) = 0$.

Sufficiency: Letting $\delta := \varepsilon 2^{-1/2}$

$$I_1(s,t,\varepsilon) = \int_0^{\pi/4}\int_{\gamma(\theta)}^\infty e^{-A_{s,t}(\theta)/r} f_{s,t}(\theta)/r^3 \, rdrd\theta$$

$$+ \int_0^{\pi/4}\int_{\gamma(\theta)}^\infty e^{-A_{s,t}(\theta)/r} B_{s,t}(\theta)C_{s,t}(\theta)/r^4 \, rdrd\theta$$

$$:= I_{1,1}(s,t,\delta) + I_{1,2}(s,t,\delta) ,$$

where $\gamma(\theta) := \delta/\sin(\pi/4-\theta)$.

Step 1. First note that

$$I_{1,1}(s,t,\delta) = \int_0^{\pi/4} f_{s,t}(\theta)D_{s,t}(\theta)d\theta, \qquad \text{where}$$

$$D_{s,t}(\theta) := (1 - e^{-A_{s,t}(\theta)/\gamma(\theta)})/A_{s,t}(\theta) .$$

But for any $\theta_1,\theta_2 \in (0,\pi/4)$, $|D_{s,t}|$ is bounded uniformly in s,t over $[\theta_1,\theta_2]$ (cf. (g) in Lemma 4.15 below). Therefore

$$\lim_{s\to t} I_{1,1}(s,t,\delta) = 0$$

$$\Longleftrightarrow \tag{4.8}$$

$$\lim_{s\to t} S_{s,t}([\theta_1,\theta_2]) = 0, \text{ for all } \theta_1,\theta_2 \in (0,\pi/4),\ t \in T\ .$$

Step 2. Next, consider $\lim_{s\to t} I_{1,2}(s,t,\delta)$. Upon integration by parts and using the fact that $A_{s,t}(\theta) > 0$,

$$I_{1,2}(s,t,\delta) = \int_0^{\pi/4} B_{s,t}(\theta)C_{s,t}(\theta)/A^2_{s,t}(\theta)d\theta$$

$$- \int_0^{\pi/2} B_{s,t}(\theta)C_{s,t}(\theta)\sin(\pi/4-\theta)/(\delta A_{s,t}(\theta))e^{-A_{s,t}(\theta)\sin(\pi/4-\theta)/\delta}d\theta$$

$$- \int_0^{\pi/2} B_{s,t}(\theta)C_{s,t}(\theta)/A^2_{s,t}(\theta)e^{-A_{s,t}(\theta)\sin(\pi/4-\theta)/\delta}d\theta$$

$$=: I_{1,2,1}(s,t) - I_{1,2,2}(s,t,\delta) - I_{1,2,3}(s,t,\delta)\ .$$

We shall prove that if $S_{s,t}([0,\psi]) \to 0$ as $s \to t$, for all $\psi \in [0,\pi/4)$ then each of these integrals goes to 0, thereby establishing that $\lim_{s\to t} I_{2,2}(s,t,\delta) = 0$. Since this condition includes the condition for $\lim_{s\to t} I_{1,1}(s,t,\delta) = 0$, we will have proved that $\lim_{s\to t} I_1(s,t,\varepsilon) = 0$.

For $I_{1,2,1}$ use the leftmost bounds in (c) and (e) of Lemma 4.12 below to get that the denominator of the integrand is uniformly bounded away from 0. Then apply the Lebesque Dominated Convergence Theorem.

Since $I_{1,2,2}(s,t,\delta) \le I_{1,2,1}(s,t)$ for all s,t and δ we can also conclude that $\lim_{s\to t} I_{1,2,2}(s,t,\delta) = 0$.

Finally consider $I_{1,2,3}$. The integrand is bounded above by an L^1 function. (On $[\pi/8,\pi/2]$ it is bounded by $2^{1/2}/(\alpha \sin\pi/8)$ and on $[0,\pi/8]$ it is bounded by the function $2^{1/2}\exp(-\delta_{\pi/8}\sin(\pi/8)/(\delta\sin\theta))/(\alpha\sin\theta)$ (cf. (f) of Lemma 4.12 below.) Now the assumption $\lim_{s\to t} S_{s,t}([0,\psi]) > 0$ for all $\psi \in [0,\pi/4)$ implies that $\lim_{s\to t} I_{1,2,3}(s,t,\delta) = 0$. Hence it also implies that $\lim_{s\to t} I_1(s,t,\varepsilon) = 0$. By symmetry the condition $\lim_{s\to t} S_{s,t}([\psi,\pi/2]) = 0$ for all $\psi \in (\pi/4,\pi/2]$ implies that $\lim_{s\to t} I_2(s,t,\varepsilon) = 0$.

Necessity. We already have proved necessity of the condition

$$\lim_{s\to t} S_{s,t}([\theta_1,\theta_2]) = 0, \qquad \text{for all } \theta_1,\theta_2 \in (0,\pi/4)\ .$$

So assume this condition. Note that convergence in probability as $s \to t$ implies convergence in probability and hence in distribution of (X_s, X_t) to (X_t, X_t) as $s \to t$. But this is true if and only if

$$\lim_{s \to t} A_{s,t}(\theta) = A_{t,t}(\theta) = \begin{cases} 1/\sin\theta, & \text{if } \theta < \pi/4 \\ 1/\cos\theta, & \text{if } \theta > \pi/4 \end{cases} . \qquad [4.9]$$

Additionally, for all $\theta \in [0, \pi/4)$,

$$\lim_{s \to t} \left(\int_0^{\pi/2} f_{s,t}(\psi) \sin\psi d\psi\right) / \sin\theta = 1/\sin\theta . \qquad [4.10]$$

Now if there exists $\psi_1 \in [0, \pi/4)$ such that $\lim_{s \to t} S_{s,t}([0,\psi_1]) \neq 0$, then

$$\lim_{s \to t} \left(\int_0^{\psi_1} f_{s,t}(\psi) \cos\psi d\psi\right) / \cos\psi_1 \neq 0 . \qquad [4.11]$$

But (4.10) and (4.11) contradict (4.9). Hence, $\lim_{s \to t} S_{s,t}([0,\psi]) = 0$, for all $\psi \in [0, \pi/4)$ is necessary. Q.E.D.

The following lemma is a simple exercise in estimating integrals.

Lemma 4.12. With notation the same as above and $\sigma = S([0, \pi/2])$, we have:

(a) $2^{1/2} \leq \sigma \leq 2$ and each value in this interval is possible.

(b) For all $\theta \in [0, \pi/2]$,

$$S((\theta, \pi/2]) \geq \sigma 2^{1/2} \sin(\pi/4 - \theta)/(1 + 2^{1/2} \sin(\pi/4 - \theta))$$

$$\geq 2 \sin(\pi/4 - \theta)/(1 + 2^{1/2} \sin(\pi/4 - \theta)) =: \varepsilon_\theta .$$

The next inequality is relevant for the numerator of $A(\theta)$.

(c) $2^{1/2} \sin 2\theta \leq \sigma/2 \sin 2\theta$

$$\leq \sin\theta \int_0^\theta \cos\psi S(d\psi) + \cos\theta \int_\theta^{\pi/2} \sin\psi S(d\psi) \leq \sigma \leq 2,$$

for all $\theta \in [0, \pi/2]$.

(d) $A(\theta) \geq \sigma \geq 2^{1/2}$, for all $\theta \in [0, \pi/2]$.

The inequalities in (c) can be improved for θ near 0.

(e) For all $\theta \in [0, \psi]$, where $\psi \in [0, \pi/4]$

$$\delta_\psi \cos\psi \leq \delta_\psi \cos\theta \leq \sin\theta \int_0^\theta \cos\psi S(d\psi) + \cos\theta \int_\theta^{\pi/2} \sin\psi S(d\psi)$$

where $\delta_\psi := \sin\psi\cdot\varepsilon_\psi := 2\sin\psi\sin(\pi/4-\psi)/(1+2^{1/2}\sin(\pi/4-\psi)) > 0$ if $\psi \neq 0$ or $\pi/4$, and ε_ψ is defined as in (b).

(f) For all $\theta \in [0,\psi]$, where $\psi \in (0,\pi/4)$

$$\gamma_\psi/\sin\theta \geq A(\theta) \geq \delta_\psi/\sin\theta, \quad \text{where } \gamma_\psi := 2/\cos\psi .$$

(g) For all $\theta \in [\theta_1,\theta_2]$ and $\theta_1,\theta_2 \in (0,\pi/4)$ there exists $C_{\theta_1,\theta_2} > 0$ such that

$$0 < C_{\theta_1,\theta_2} \leq 1/A(\theta)(1 - e^{-A(\theta)\sin(\pi/4-\theta)/\delta}) \leq 2^{-1/2} .$$

Remark 4.13.

(a) Theorem 4.7 is applicable even when the components are not identically distributed.

(b) It is easy to show that the limit in probability of a SMS sequence is itself a SMS random variable.

(c) It is also easy to show that a SMS sequence $X, X_1, X_2, \ldots$ converges in probability to X iff the bivariant m-spectral measures S_n of (X_n, X) converge to the m-spectral measure of (X,X). This in turn is equivalent to the convergence in L^1 of f_n to f where the functions f, f_n correspond to X, X_n, respectively, as defined in Theorem 4.4.

(d) Theorems 4.7 and 4.4 can be combined with remark (c) above to give a proof of Theorem 4.5. (First apply 4.4 to the SMS sequence $\{X_q\}_{q\in Q\cap T}$, where Q is the set of rationals.)

Acknowledgement. The author wishes to express his grateful thanks to M. G. Hahn for her encouragement and advice during her supervision of this work.

REFERENCES

Araujo, A. and Gine, E. (1980). The Central Limit Theorem for Real and Banach Valued Random Variables. Wiley, New York.

Balkema, A. and Resnick, S. I. (1977). Max-infinite divisibility. J. Appl. Prob. 14, 309-309.

Cambanis, S. and Miller, G. (1981). Linear Problems in p^{th} Order and Stable Processes. SIAM J. Appl. Math. 41 43-69.

deHaan, L. (1984). A Spectral Representation for Max-Stable Processes. Ann. of Prob. 12, 1194-1204.

deHaan, L. and Resnick, S. I. (1977). Limit Theory for Multivariate Sample Extremes. Z. Wahrscheinlichkeitstheorie verw. Gebiete, 40, 317-337.

Ferguson, T. and Klass, M. (1972). A representation of independent increments processes without Gaussian components. Ann. Math. Statist., 43, 5, 1634-1643.

Galambos, J. (1978). The Asymptotic Theory of Extreme Order Statistics. New York, Wiley.

Gnedenko, B.V. (1943). Sur la distribution limite du terme maximum d'une serie aleatoire. Ann. Math. 44, 423-453.

Grenander, V. (1963). Probabilities on Algebraic Structures. Almquist and Wiskall.

Gumbel, E. J. (1958). Statistics of Extremes. Columbia University Press.

Kuelbs, J. (1973). A representation theorem for symmetric stable processes and stable meansures on H. Z. Wahrscheinlichkeithstheorie verw. Gebiete, 26, 259-271.

LePage, R. (1980). Multidimensional infinitely divisible variables and processes. Unpublished manuscript.

LePage, R., Woodroofe, M., and Zinn, J. (1981). Convergence to a stable distribution via order statistics. Ann. of Prob. 9, 624-632.

Pickands, J. (1976). Multivariate extreme value distributions. Preprint. University of Pennsylvania.

Schilder, M. (1970). Some structure theorems for the symmetric stable laws. Ann. of Math. Statist. 41, 412-421.

Vatan, P. (1984a). Max-stable and max-infinitely divisible laws on infinite dimensional spaces. Ph.D. Thesis, Mathematics Department, M.I.T. Cambridge, MA.

Vatan, P. (1984b). Some results on max-stable and max-infinitely divisible processes. To appear.

A MAXIMAL LAW OF THE ITERATED LOGARITHM FOR OPERATOR-NORMALIZED STOCHASTICALLY COMPACT PARTIAL SUMS OF I.I.D. RANDOM VECTORS

Daniel Charles Weiner
Department of Mathematics
University of Wisconsin-Madison
Madison WI 53706/USA

1. Introduction

In this paper we will prove a finite-dimensional version of Chung's (1948) celebrated Maximal (usually known as "other") Law of the Iterated Logarithm for the Partial Sums of a sequence of independent random variables. We will be extending the i.i.d. version due to Jain and Pruitt (1973) to the case of operator-normalization of the partial sums: Where they assumed weak convergence of the (appropriately) constant-normalized i.i.d. sums, we assume stochastic compactness (tightness with nondegenerate limits) along with a suitable "nonasymmetry" condition for the operator-normalized sums of i.i.d. $\mathbb{R}^d$-valued random vectors.

We will also sketch a related extension of a Law of the Iterated Logarithm for Stable Summands, due to Chover (1966), to the case of Operator-Stable Summands. These two strong limit theorems, along with another operator L.I.L. to be found in Weiner (1984a) (for variables attracted to a standard Gaussian law, a bounded L.I.L. with cluster set is obtained), begin to illustrate how general the operator method is for passing from weak limit theorems to strong via classical techniques despite the fact that, when operator-normalization becomes necessary, different directions (i.e., projections) for the r.v.'s can have grossly different tail behaviors and hence vastly different growth rates when summed.

2. The Maximal L.I.L.: Background

Let $X_1, X_2, \ldots,$ be independent, real-valued random variables, with $S_n = X_1 + \cdots X_n$. Under conditions on the third moments $E|X_j|^3$, Chung (1948) proved a result on the minimal growth of the maximum partial sum,

$$\liminf_{n\to\infty} \left(\frac{LLn}{n}\right)^{1/2} \max_{k\leq n} |S_k| = \frac{\pi}{\sqrt{8}}, \text{ a.s.,} \qquad [*]$$

where LLn stands for log log n.

When the X's are identically distributed, his condition amounts to $E|X_1|^3 < \infty$. Since then various authors have attempted to weaken the moment condition on X_1 in the i.i.d. case, until in 1973, Jain and Pruitt were able to prove a result similar to (*) whenever X_1 is in the domain of attraction of a strictly-stable, not completely asymmetric random variable. Thus when the distribution $L(S_n/d_n)$ converges weakly to a strictly stable distribution whose density is everywhere positive, they proved there is a finite, positive constant c with

$$\liminf_{n\to\infty} \frac{\max_{k\leq n} |S_k|}{d([n/LLn])} = c, \quad \text{a.s.,} \qquad [**]$$

where [x] denotes the greatest integer in x, and $d(m) = d_m$, $m \geq 1$.

The constant c was identified only in 1982, by Jain, using the techniques of large deviation theory, where it was shown that c is dependent only on the limiting stable law, not on $L(X_1)$, so that [**] is actually an invariance principle.

We will prove a suitable analogue for (**) in finite dimensions, where instead of normalizing S_n in the Central Limit Theorem by constants d_n, we normalize S_n by linear operators T_n, using the techniques developed by Hahn and Klass (1982) and Griffin (1983). Moreover, instead of assuming $\{L(T_nS_n)\}$ is convergent, we merely assume that it is tight with all subsequential limits fully supported. The corresponding "nonasymmetry" condition and precise assymptions will be discussed in section 3, and the theorem will be stated and proved in section 4. The main difficulties in proving the analogue of [**] for this new situation are (1) the variation of $\{T_n\}$ - that is, behavior of $\{T_nT^{-1}_{[an]}\}$, where a is a constant, and (2) what partial maximum to consider. (1) was solved in the remark following lemma 2 in Hudson, Veeh and Weiner (1984), and (2) is overcome by normalizing by the correct operator $T_{[n/LLn]}$ first, then taking norms, and then finding the partial maxima of these pre-normalized sums. That this process works is due to the main reason for the use and utility of operators in the Central Limit Theorem - different directions having different growth rates are normlaized differently, so the sums ought to grow the same in all directions, and thus have a reasonable minimal growth rate.

We remark that in our theorem the value of the lim inf constant, and whether or not it depends on $L(X_1)$ or just the limit laws, are not determined, but inasmuch as the constant is finite and positive, the correct rate has been achieved. At this time to the author's knowledge no work in operator large deviations theory has appeared, so that Jain's 1982 results are as yet inaccessible to the operator method.

3. Notation and Preliminary Facts

Let $X, X_1, X_2, \ldots,$ be independent, identically distributed (i.i.d.) random vectors taking values in $\mathbb{R}^d$, whose common distribution law $L(X)$ is full, i.e., not supported on any $(d-1)$-dimensional hyperplane. Put $S_n = X_1 + X_2 + \cdots + X_n$. Now Griffin (1983) and Hahn and Klass (1982) have determined a necessary and sufficient condition that there exist linear operators $\{T_n\}$ on $\mathbb{R}^d$ and centerings $\{b_n\} \subset \mathbb{R}^d$ such that $\{L(T_nS_n-b_n): n \geq 1\}$ is stochastically compact, i.e., tight with only full subsequential limits. Their condition is

$$\overline{\lim_{t\to\infty}} \sup_{\|\theta\|=1} \frac{t^2P(|\theta(X)| \geq t)}{E(\theta^2(X) \wedge t^2)} < 1 \tag{1}$$

where $\|\cdot\|$ is the Euclidean norm on $\mathbb{R}^d$, $\theta(X) = (\theta, X)$ is the Euclidean inner product of θ and X, and $s \wedge t$ denotes the minimum of s and t. When [1] holds, suitable operators and centerings are constructed. All we need to know here is that T_n is diagonal with respect to an orthonormal basis depending on n, and enjoys the property that

$$\overline{\lim_{n}}\, nE\|T_nX\|^2 I(\|T_nX\| \leq C) < \infty, \tag{2}$$

for every $C > 0$, where I denotes indicator function.

Assuming that $\{L(T_nS_n-b_n)\}$ is stochastically compact, let L denote its limit set. Now L is stochastically compact, and members of L have C^∞-densities (Griffin (1983)), and of course are infinitely divisible. Although L depends on the choice of operators used to normalize the sums, it is easy to see using lemma 2ff. in Hudson, Veeh and Weiner (1984) that any other choice of normalizers will modify L only by a compact set of invertible operators and centerings, and thus for our limit theorem a different choice can only change the (unknown) value of the lim inf constant, not its finite positivity.

We are assuming that it is not necessary to use centerings to achieve stochastic compactness, i.e., that we may take $b_n \equiv 0$. When $E\|X\| < \infty$, this is no restriction, as one merely replaces X by $X-EX$, but when $E\|X\| = \infty$ this represents an additional assumption.

It is well-known that extremely asymmetric distributions have a different order of growth for their partial maxima on the line (see, e.g., Jain and Pruitt (1973)), so it is clear some kind of "nonasymmetry" assumption has to be made. Letting $\{\mu^t : t \geq 0\}$ denote the weakly continuous semi-group associated with $\mu \in L$, such that $\mu^0 = \delta(0) =$ point mass at 0, and $\mu^1 = \mu$, we assume

$$\delta = \inf_{t \geq 1} \inf_{\mu \in L} \min_{\alpha} \mu^t(\mathbb{R}^d(\alpha)) > 0, \qquad [3]$$

where with respect to one fixed orthonormal basis $\{e_1, \ldots, e_d\}$ one writes $x \in \mathbb{R}^d$ as $x = (x_1, x_2, \ldots, x_d)$, and given $\alpha = (\alpha_1, \ldots, \alpha_d) \in \{-1,1\}^d$, one puts $\mathbb{R}^d(\alpha) = \{x = (x_1, \ldots, x_d) : \alpha_i x_i \geq 0, i = 1, \ldots, d\}$. Thus assumption (3) essentially insures no loss of mass in any "orthant", regardless of high convolution powers. It is satisfied by strictly stable variables on the line whose densities are everywhere positive, and in the case of strictly operator-stable r.v.'s (cf. Sharpe (1969)) it suffices to assume an everywhere positive density when the exponent is diagonalizable over the reals. Other examples in the operator-stable case are available, but being overly technical they are not included here.

We have need of some additional properties of X and L.

<u>Lemma 1</u>. $\lim_{\lambda \to 0} \sup_{\mu \in L} \mu\{x : \|x\| \leq \lambda\} = 0.$

<u>Proof</u>: If not, choose $\varepsilon > 0$, $\{\mu_n\} \subset L$, $\lambda_n \searrow 0$ with $\mu_n\{x : \|x\| \leq \lambda_n\} > \varepsilon$, for every $n \geq 1$. Since L is stochastically compact, take $n_k \nearrow \infty$ with $\{\mu_{n_k}\}$ convergent, say, to $\mu \in L$. Since μ has a density, we know for every $t > 0$ that $\mu\{x : \|x\| = t\} = 0$, so that $\mu_{n_k}\{x : \|x\| \leq t\} \to \mu\{x : \|x\| \leq t\}$ as $k \to \infty$. Since μ has no atom at 0 (due to the existence of a density for μ), choose $t > 0$ so that $\mu\{x : \|x\| \leq t\} < \varepsilon/2$. Find N so that $k > N$ implies $\lambda_{n_k} < t$ and $\mu_{n_k}\{x : \|x\| \leq t\} < \mu\{x : \|x\| \leq t\} + \varepsilon/2 < \varepsilon$. Then by assumption $\varepsilon < \mu_{n_k}\{x : \|x\| \leq \lambda_{n_k}\} \leq \mu_{n_k}\{x : \|x\| \leq t\} < \varepsilon$, for $k > N$, a contradiction.

For the last lemma let $\mathbb{R}^d(\alpha, a)$ denote the set $\{x = (x_1, \ldots, x_d) : 0 \leq \alpha_i x_i \leq a, i = 1, \ldots, d\}$, where $a > 0$.

<u>Lemma 2</u>. Given $\lambda < 1$ and $\rho \in \mathbb{N}$, these exists $t > 0$ such that

$$\liminf_{n \to \infty} \min_{\alpha} P(T_n S_{n\rho} \in \mathbb{R}^d(\alpha, t)) \geq \lambda\delta, \qquad [4]$$

where $\delta > 0$ is defined as in (3).

<u>Proof</u>: Fix $0 < \lambda < 1$, and $\rho \in \mathbb{N}$. We first claim

$\liminf_{t\to\infty} \inf_{\mu\in L} \min_{\alpha} \mu^{\rho}(\mathbb{R}^d(\alpha,t)) \geq \delta$. But an easy contradiction argument like that in lemma 1, using the compactness of L and the fact that $\bigcup_{t>0} \mathbb{R}^d(\alpha,t) = \mathbb{R}^d(\alpha)$, establishes the claim. So find t so large that $\inf_{\mu\in L} \min_{\alpha} \mu^{\rho}(\mathbb{R}^d(\alpha,t)) > \lambda\delta$. We now claim (4) holds, because given $n_k \uparrow \infty$ with $\mu_{n_k} \to \mu \in L$, we have $P(T_{n_k} S_{n_k \rho} \in \mathbb{R}^d(\alpha,t)) \to \mu^{\rho}(\mathbb{R}^d(\alpha,t)) > \lambda\delta$, for each $\alpha \in \{-1,1\}^d$. Thus another contradiction argument using the compactness of L establishes (4), and the lemma is proved.

The following is useful in applying Ottaviani's inequality (see, e.g., Chung (1968), p. 120), which replaces Lévy's inequality in some nonsymmetric situations.

Lemma 3. Given $m \in \mathbb{N}$, the set $A = \{L(T_n S_j): j \leq mn, n \geq 1\}$ is tight, and thus for each $\lambda > 0$,

$$\lim_{\lambda\to\infty} \max_{j\leq mn} P(\|T_n S_j\| \geq \lambda) = 0. \qquad [5]$$

Proof: We note (5) is just the definition of tightness. Given any subsequence $\{L(T_{n(1)} S_{j(1)})\}$ from A, since $0 < j(1)/n(1) \leq m$, we choose a further subsequence $\{n(2)\} \subset \{n(1)\}$ obeying $j(2)/n(2)$ $\beta \in [0,m]$. Then, by stochastic compactness of $\{L(T_n S_n): n \geq 1\}$, we choose a further subsequence $\{n_3\} \subset \{n_2\}$ so that $L(T_{n(3)} S_{n(3)}) \xrightarrow{w} \mu \in L$. Then, by Proposition 2.1(a) of de Acosta (1982), we have $j(3)/n(3) \to \beta$ implies $L(T_{n(3)} S_{j(3)}) \xrightarrow{w} \mu^{\beta}$. Thus A is tight, although the possibility $\beta = 0$ excludes the possibility that A is stochastically compact.

4. Main Theorem and Proof

Here is our main result.

Theorem. Assume $\{L(T_n S_n): n \geq 1\}$ is stochastically compact, and assume the nonasymmetry condition (3). Then there exists a positive, finite constant c, such that

$$\liminf_{n\to\infty} \max_{k\leq n} \|T_{[\frac{n}{LLn}]} S_k\| = c, \quad \text{a.s.}$$

Proof: It is proved in Griffin (1983) that $\lim_{n\to\infty} \|T_n\| = 0$, so the usual zero-one law implies that there exists a constant $c \in [0,\infty]$ with $\liminf_{n\to\infty} \max_{k\leq n} \|\Lambda_n S_k\| = c$, a.s., where Λ_n denotes $T_{[n/LLn]}$.

We must, therefore, only show that $\liminf_n \max_{k\leq n} \|\Lambda_n S_k\|$ is finite and positive, a.s. For convenience we divide the proof into lemmas.

Lemma 4. There exists $d_1 > 0$ with $\liminf_{n\to\infty} \max_{k\leq n} \|\Lambda_n S_k\| \geq d_1$, a.s.

Proof: Let M_n denote $\max_{k\leq n} \|\Lambda_n S_k\|$, and let $m_n = [n/LLn]$. Then for $\lambda > 0$,

$$P(M_n \leq \lambda) \leq \prod_{i=1}^{[LLn]} P(\|\Lambda_n(S_{im_n} - S_{(i-1)m_n})\| \leq 2\lambda)$$

$$= P(\|\Lambda_n S_{m_n}\| \leq 2\lambda)^{[LLn]}$$

$$= P(\|T_{m_n} S_{m_n}\| \leq 2\lambda)^{[LLn]} .$$

Thus $\overline{\lim}_{n\to\infty} \frac{1}{LLn} \log P(M_n \leq \lambda) \leq \overline{\lim}_{n\to\infty} \log P(\|T_n S_n\| \leq 2\lambda)$

$\leq \log \sup_{\mu\in L} \mu\{x:\|x\| \leq 2\lambda\} \to -\infty$, as $\lambda \to 0$,

by lemma 1. Thus given $\alpha > 1$, there is a $\lambda > 0$ with $\log \sup_{\mu\in L} \mu\{x:\|x\| \leq 2\lambda\} < -\alpha$.

So, letting $n_k = 2^k$, let K be so large that $k \geq K$ implies $\frac{1}{LLn_k} \log P(M_{n_k} \leq \lambda) \leq -\alpha$. Then

$$\sum_{k\geq K} P(M_{n_k} \leq \lambda) \leq \sum_{k\geq K} \exp(-\alpha LLn_k)$$

$$= \sum_{k\geq K} \frac{1}{(k \log 2)^{\alpha}} < \infty,$$

since $\alpha > 1$. Thus, the Borel-Cantelli lemma gives $\liminf_{k\to\infty} M_{n_k} \geq \lambda$, a.s.

To fill in the other times n with $2^k < n < 2^{k+1}$, we quote the regularity result in the remark following lemma 2 in Hudson, Veeh and Weiner (1984), which implies $\lim_{k\to\infty} \max_{n_k<n\leq n_{k+1}} \|\Lambda_{n_k} \Lambda_n^{-1}\| = B < \infty$.

Thus, if $n_k < n \leq n_{k+1}$, we have (almost surely), for sufficiently large n,

$$M_n = \max_{j\leq n} \|\Lambda_n S_j\|$$

$$\geq \max_{j\leq n_k} \|\Lambda_n S_j\|$$

$$\geq \max_{j\leq n_k} \frac{\|\Lambda_{n_k} S_j\|}{\|\Lambda_{n_k}\Lambda_n^{-1}\|}$$

$$\geq \max_{j\leq n_k} \|\Lambda_{n_k} S_j\| / \max_{n_k<m<n_{k+1}} \|\Lambda_{n_k}\Lambda_m^{-1}\|$$

$$\geq M_{n_k}/2B \geq \lambda/2B$$

so that for d_1 in lemma 4 we may take $\lambda/2B$, and the lemma is proved.

Lemma 5. There exists C such that for some $\xi > 1$ and all sufficiently large n we have $P(M_n \leq C) \geq (\log n)^{-\xi}$.

Proof: Let $F_n = \sigma(X_1, \ldots, X_n)$ denote the σ-field generated by the first n X's, let $\varepsilon > 0$, $\lambda > 0$, and choose an integer $\rho > -\log(\frac{1}{2}\delta)$, where $\delta > 0$ was defined in equation (3). As the notation for the subsequences becomes involved, we suppress some of the dependence on n in the following: Let $v = [n/LLn]$, $m = \rho v$, $M = [n/m]$, $A_m(j) = \max_{1\leq i\leq m} \|\Lambda_n(S_{jm+i} - S_{jm})\|$,

$E_j = [\|\Lambda_n S_{(j+1)m}\| \leq \varepsilon,\quad A_m(j) \leq \lambda]$. Note $M \sim 1/(\rho LLn)$.

Now $[M_n \leq \varepsilon + \lambda] \supset \bigcap_{j=0}^{M} E_j$. Moreover, $E_j \in F_{Mm}$ for $j < M$, and $P(E_m | F_{Mm}) = P(E_m | S_{Mm})$, a.s.

Thus, letting $I(F)$ denote the indicator function of the set F we have

$$\begin{aligned} P(M_n \leq \lambda + \varepsilon) &\geq P(\bigcap_{j=0}^{M} E_j) \\ &= E(E(I(\bigcap_{j=0}^{M} E_j | F_{Mm}))) \\ &= E(\{\prod_{j=0}^{M-1} I(E_j)\} E(I(E_M) | F_{Mm})) \\ &= E(\{\prod_{j=0}^{M-1} I(E_j)\} E(I(E_M) | S_{Mm})). \end{aligned} \qquad [6]$$

Suppose that we can produce $b > 0$ so that independently of j, for all sufficiently large n, we have $E(I(E_j) | S_{jm}) \geq b$, a.s., or

E_{j-1}. Then (6) yields $P(M_n \le \lambda + \varepsilon) \ge b^M P(E_0)$, and it is easy to see that $\liminf_{n\to\infty} P(E_0) > 0$, using Ottaviani's inequality, lemma 3, and the fact that $\{L(\Lambda_n S_m): n \ge 1\}$ is stochastically compact, since $m \sim \rho n/LLn$.

To produce b, we note $P(E_j | S_{jm}) = P(\|\Lambda_n S_{(j+1)m}\| \le \varepsilon | S_{jm}) P(A_m(j) \le \lambda)$, because the event $[A_m(j) \le \lambda]$ is independent of $\sigma(S_{jm})$. Now another application of Ottaviani's inequality gives (note the left side is independent of j by i.i.d.)

$$P(A_m(j) > \lambda) \le \frac{1}{1-c_n} P(\|\Lambda_n S_m\| > \tfrac{1}{2}\lambda)$$

where $c_n = \max_{i\le m} P(\|\Lambda_n S_i\| > \frac{1}{2}\lambda) \to 0$ uniformly in n, as $\lambda \to \infty$, again by lemma 3. Thus uniformly in j,n we have $P(A_m(j) \le \lambda) \to 1$ as $\lambda \to \infty$.

Turning to the evaluation of $P(\|\Lambda_n S_{(j+1)m}\| \le \varepsilon | S_{jm})$ over the set E_{j-1}, let us assume without loss of generality that our norm $\|\cdot\|$ was chosen to be the norm $\|\sum_{i=1}^{d} \beta_i e_i\| = \max_{i\le d} |\beta_i|$, where $\{e_1, \ldots, e_d\}$, used in equation (3), was an orthonormal basis of $\mathbb{R}^d$. Because all norms are equivalent, only the value of c in the theorem will be affected.

If $\omega \in E_{j-1}$, then $\|\Lambda_n S_{jm}(\omega)\| \le \varepsilon$, so $|(e_i, \Lambda_n S_{jm}(\omega))| \le \varepsilon$, $i = 1,\ldots,d$. Hence for some element $\alpha = (\alpha_1, \ldots, \alpha_d) \in \{-1,1\}^d$, we have

$$0 \le \alpha_i (e_i, \Lambda_n S_{jm}(\omega)) \le \varepsilon, \quad i = 1,\ldots,d. \tag{7}$$

To be precise, let $\alpha_i = \operatorname{sgn}(e_i, \Lambda_n S_{jm}(\omega))$. Suppose ω also satisfies $-\varepsilon \le \alpha_i (e_i, \Lambda_n (S_{(j+1)m}(\omega) - S_{jm}(\omega))) \le 0$, $i = 1,\ldots,d$. Adding to equation (7), we obtain

$$-\varepsilon \le (e_i, \Lambda_n S_{(j+1)m}(\omega)) \le \varepsilon, \quad i = 1,\ldots,d,$$

or, by choice of norm, $\|\Lambda_n S_{(j+1)m}(\omega)\| \le \varepsilon$. Thus we have shown that, a.s. on E_{j-1},

$$P(\|\Lambda_n S_{(j+1)m}\| \le \varepsilon)$$

$$\ge \min_{\alpha\in\{-1,1\}^d} P(-\varepsilon \le \alpha_i (e_i, \Lambda_n (S_{(j+1)m} - S_{jm})) \le 0, \quad i = 1,\ldots,d),$$

and of course by lemma 2, the right side, which is $\min P(\Lambda_n S_m \in \mathbb{R}^d(\alpha,\varepsilon))$, is as close to $\delta > 0$ as we please, by choosing $\varepsilon > 0$ large enough.

Thus, any fraction of δ will serve as our b, provided n is large enough, and so to finish the proof of the lemma, we choose $1 > \lambda > \frac{1}{2}$ and then find that b may be taken at least $\lambda\delta$. Hence for $C = \varepsilon + \lambda$ as large as indicated above, we find $P(M_n \leq C) \geq b^M(\text{constant}) \geq (\text{constant})(\lambda\delta)^M \geq (\text{constant})(\lambda\delta)^{LLn/\rho} = (\log n)^{-\xi}$, where $\xi = -\frac{1}{\rho}\log(\lambda\delta) < 1$, by choice of ρ.

<u>Lemma 6</u>. There exists $d_2 < \infty$ with $\liminf_{n\to\infty} \max_{k\leq n} \|\Lambda_n S_k\| \leq d_2$, a.s.

<u>Proof</u>: Choose C and $\xi < 1$, as in lemma 5. Now the remark following lemma 2 in Hudson, Veeh and Weiner (1984) shows that if $n_k = \exp(k^\gamma)$, where $1 < \gamma < 1/\xi$, then $n_{k-1}/n_k \to 0$ and thus $\{\Lambda_{n_k}\Lambda^{-1}_{n_k-n_{k-1}}\}$ is relatively compact with inverses also relatively compact.

Note that

$$M_{n_k} \leq \max_{1\leq j\leq n_{k-1}} \|\Lambda_{n_k} S_j\| + \max_{n_{k-1}<j\leq n_k} \|\Lambda_{n_k}(S_j - S_{n_{k-1}})\|$$

$$\leq \max_{1\leq j\leq n_{k-1}} \|\Lambda_{n_k} S_j\| + \|\Lambda_{n_k}\Lambda^{-1}_{n_k-n_{k-1}}\| \max_{n_{k-1}<j<n_k} \|\Lambda_{n_k-n_{k-1}}(S_j - S_{n_{k-1}})\|.$$

Because of the identical distributions, lemma 5 implies

$$P\Big(\max_{n_{k-1}<j<n_k} \|\Lambda_{n_k-n_{k-1}} S_j\| \leq C\Big)$$

$$= P\Big(\max_{1\leq j\leq n_k-n_{k-1}} \|\Lambda_{n_k-n_{k-1}} S_j\| \leq C\Big) = P(M_{n_k-n_{k-1}} \leq C)$$

$$\geq (\log(n_k - n_{k-1}))^{-\xi},$$

and the sum over k of the right member diverges since $1 < \gamma < \xi^{-1}$. But, $\{S_j - S_{n_{k-1}} : n_{k-1} < j \leq n_k\}_{k=1}^{\infty}$ is an independent sequence of blocks so the independence part of the Borel-Cantelli lemma gives $\liminf_{k\to\infty} \max_{n_{k-1}<j\leq n_k} \|\Lambda_{n_k-n_{k-1}}(S_j - S_{n_{k-1}})\| \leq C$. It remains to show only that $\max_{1\leq j\leq n_{k-1}} \|\Lambda_{n_k} S_j\|$ is almost surely bounded by a uniform constant, because $\|\Lambda_{n_k}\Lambda^{-1}_{n_k-n_{k-1}}\|$ is uniformly bounded, as we have seen. But we claim, in fact, that $\max_{j\leq n_{k-1}} \|\Lambda_{n_k} S_j\| \to 0$, a.s., as $k \to \infty$. We appeal once more to Ottaviani's inequality, which is useful here

because $\max_{j\le n_{k-1}} P(\|\Lambda_{n_k}(S_{n_{k-1}}-S_j)\| > \lambda) = \max_{j\le n_{k-1}} P(\|\Lambda_{n_k} S_{n_{k-1}-j}\| > \lambda) \le$

$\max_{j\le n_{k-1}} P(\|\Lambda_{n_k}\Lambda_{n_{k-1}}^{-1}\| \, \|\Lambda_{n_k} S_{n_{k-1}-j}\| > \lambda) \to 0$ as $k \to \infty$, for each

$\lambda > 0$, by lemma 3 and the consequence of the remark after lemma 2 in Hudson, Veeh and Weiner (1984) that $n_{k-1}/n_k \to 0$ forces $\|\Lambda_{n_k}\Lambda_{n_{k-1}}^{-1}\| \to 0$.

Thus we must show only that $P(\|\Lambda_{n_k} S_{n_{k-1}}\| > \lambda)$ is summable over k, for any $\lambda > 0$. For $j \le n_{k-1}$, put $Y_j = X_j I(\|\Lambda_{n_k} X_j\| \le \lambda)$. Now via the Central Limit Theorem, lemma 3 implies $\{nET_n XI(\|T_n X\| \le \lambda)\}$ is bounded for each λ. Thus $\{\frac{n_k}{LLn_k} E\Lambda_{n_k} XI(\|\Lambda_{n_k} X\| \le \lambda)\}$ is bounded, but then $n_{k-1}E\Lambda_{n_k} XI(\|\Lambda_{n_k} X\| \le \lambda) \to 0$ since $\frac{n_{k-1}LLn_k}{n_k} \to 0$. Thus $E\Lambda_{n_k} \sum_{j=1}^{n_{k-1}} Y_j \to 0$, and hence we only have to show that

$P(\|\Lambda_{n_k} S_{n_{k-1}} - E\Lambda_{n_k} \sum_{j=1}^{n_{k-1}} Y_j\| > \lambda)$ is summable over k. But,

$$P(\|\Lambda_{n_k} S_{n_{k-1}} - E\Lambda_{n_k} \sum_{j=1}^{n_{k-1}} Y_j\| > \lambda)$$

$$\le P(\|\Lambda_{n_k} \sum_{j=1}^{n_{k-1}} Y_j - E\Lambda_{n_k} \sum_{j=1}^{n_{k-1}} Y_j\| > \lambda)$$

$$+ n_{k-1}P(\|\Lambda_{n_k} X\| > \lambda)$$

$$\le \frac{1}{\lambda^2} E\|\Lambda_{n_k} \sum_{j=1}^{n_{k-1}} Y_j - E\Lambda_{n_k} \sum_{j=1}^{n_{k-1}} Y_j\|^2$$

$$+ n_{k-1} \frac{LLn_k}{n_k} \frac{n_k}{LLn_k} P(\|T_{[\frac{n_k}{LLn_k}]} X\| > \lambda) \quad .$$

Now, again by the Central Limit Theorem, $\overline{\lim_n}\, nP(\|T_n X\| > \lambda) < \infty$, for for each $\lambda > 0$, so the last term on the right side is dominated by a constant times $\frac{LLn_k}{n_k} n_{k-1}$, which is summable (recalling that

$n_k = \exp(k^\gamma)$, where $\gamma > 1$).

Whereas, since all norms are equivalent we may assume $\|\cdot\|$ is Euclidean, and then

$$E\|\Lambda_{n_k}\sum_{j=1}^{n_{k-1}} Y_j - E\Lambda_{n_k}\sum_{j=1}^{n_{k-1}} Y_j\|^2$$

$$= n_{k-1}E\|\Lambda_{n_k}Y_1 - E\Lambda_{n_k}Y_1\|^2$$

$$\leq n_{k-1}E\|\Lambda_{n_k}X\|^2 I(\|\Lambda_{n_k}X\| \leq \lambda)$$

$$= O(n_{k-1}LLn_k/n_k),$$

which is again summable over k, where we have used the property (2) of $\{T_n\}$, and recalled that $\Lambda_n = T_{[n/LLn]}$. Thus an application of the Borel-Cantelli lemma completes the proof of the upper bound, and hence that of the theorem.

5. A Related Operator Strong Limit Theorem

In 1966 J. Chover proved a Law of the Iterated Logarithm for symmet ric stable (non Gaussian) summands with real values. The proof, a combination of classical L.I.L. techniques (Borel-Cantelli lemma, almost independent subsequences, Levy's inequality, etc.) and a nonlinear norma ization specially suited to the well-known polynomial decay for the larg large-value probabilities of a stable random variable (as opposed to the exponential decay for Gaussian random variables) extends easily to Banach space valued symmetric stable summands, as these have the same tail probabilities. Thus if $(B, \|\cdot\|)$ is a Banach space and $S_n = X_1 + X_2 + \cdots + X_n$, where $\{X_i\}$ are i.i.d. α-stable symmetric B-valued random variables $(0 < \alpha < 2)$, Chover's result is

$$\limsup_{n\to\infty} \left\| \frac{S_n}{n^{1/\alpha}} \right\|^{\frac{1}{LLn}} = e^{1/\alpha}, \quad \text{a.s.,} \qquad [8]$$

where LLn stands for $\log\log n$. He also proved that the cluster set of the random sequence $\{\|S_n/n^{1/\alpha}\|^{1/LLn} : n > 1\}$ almost surely contains the interval $[1, e^{1/\alpha}]$.

In finite dimensions the concept of operator stability was investi gated by M. Sharpe (1969). Briefly, if the distribution $L(X)$ is fu (i.e., not supported on any proper hyperplane) then we say X is operator-stable if and only if there exists an invertible operator A (called an exponent) such that for each n, if $S_1, \ldots, X_n$ are i.i.d.-$L(X)$ with $S_n = X_1 + \cdots + X_n$, there is a vector b_n with $L(X) = L(n^{-A}S_n - b_n)$. Here $n^{-A} = \sum_{k=0}^{\infty} (A \log n)^k/k!$ is defined in in the usual way. Operator stable laws are precisely the set of possible limit laws of affine-normalized i.i.d. sums of full r.v.'s in finite dimensions. We remark that A above need not be unique, but we refer the reader to Sharpe's paper for further details.

We will prove the analogue of [8] for strictly operator-stable non-Gaussian summands in a finite-dimensional vector space V. Strictness means the shifts described above can all be taken to be 0. Sharpe proved that unless 1 is an eigenvalue of an exponent, there is a shift of the operator-stable law rendering it strictly operator-stable. (The case of 1 being an eigenvalue is related to the Cauchy variables on the line which are not strictly stable unless symmetric.) Non-Gaussian simply means that $\frac{1}{2}I$ is not an exponent for the law, where I is the identity operator on V.

We note that the norm $\|\cdot\|$ on V is irrelevant, since all such norms on V are equivalent, and the nature of the power in our version of [8] shows that the constants of equivalence will be absorbed in the limit.

The main tool in the proof of the Theorem is this consequence of the stability property of $L(X)$ due to the Central Limit Theorem: The result of Hudson (1980), asserting that operator-stable laws have continuous Lebesgue densities, implies

$$0 < \lim_{n\to\infty} nP(\|n^{-A}X\| \geq C) < \infty, \qquad [9]$$

for each $C > 0$. Here the non-Gaussian property of $L(X)$ gives a nonzero value for the limit, and is responsible for the proof of the lower bound in the Theorem.

Here is our result:

<u>Theorem</u>. Let $X, X_1, X_2, \ldots,$ be i.i.d., strictly operator-stable non-Gaussian V-valued random vectors, with $S_n = X_1 + \cdots + X_n$. Then for any exponent A for $L(X)$ we have

$$\limsup_{n\to\infty} \|(n \log n)^{-A}S_n\|^{\frac{1}{LLn}} = 1, \quad \text{a.s.}, \qquad [10]$$

and for every $0 < \lambda < 1$,

$$1 \text{ is a cluster point of } \{\|n^{-A}(\log n)^{-\lambda A} S_n\|^{\frac{1}{LLn}} : n \geq 1\}, \text{ a.s.} \quad [11]$$

Sketch of Proof: The upper bound is obtained by using the Borel-Cantelli lemma along the subsequence $n_k = 2^k$, and by using Ottaviani's maximal inequality to reduce to consideration of this subsequence. Here strict stability is required.

The lower bound is obtained by using the independence part of the Borel-Cantelli lemma on the independent differences $\{(n_k \log n_k)^{-A}(S_{n_k} - S_{n_{k-1}})\}$, where now $n_k = k^k$ (so that $\frac{n_{k-1}}{n_k} \sim LLn_k/e \log n_k$); then the upper bound result is used to complete the argument for the lower bound in the standard way, as in Chover's proof. We remark that in Chover's paper, the recommendation $n_k = k^k$ was accidently deleted in the lower bound argument.

Finally, the clustering statement on the norms, [11] is proved using the independence part of the Borel-Cantelli lemma on

$$P(\|n_k^{-A} (\log n_k)^{-\lambda A}(S_{n_k} - S_{n_{k-1}})\|^{\frac{1}{LLn_k}} \in (1-\varepsilon, 1+\varepsilon)) \sim$$

$$P(\|n_k^{-A} (\log n_k)^{-\lambda A}(S_{n_k} - S_{n_{k-1}})\|^{\frac{1}{LLn_k}} > 1 - \varepsilon),$$ where now $n_k = \exp(k^{1/\lambda})$; the upper bound result is quoted again to complete the clustering proof just as the lower bound argument was completed in this way, only now the property $n_{k-1}/n_k \leq \exp(-\frac{\delta}{\lambda}(\log n_k)^{1-\lambda})$, for any $0 < \delta < 1$, of the subsequence $\{n_k\}$ is used.

We remark that the main ingredients of the proof are, of course, th stability property $L(n^{-A}S_n) = L(X)$, and the tail estimate [9].

References

de Acosta, A. (1982). Invariance Principles in Probability for Triangular Arrays of B-valued Random Vectors and Some Applications. Ann. Probability 10, 346-373.

Chover, J. (1966). A Law of the Iterated Logarithm for Stable Summand Proc. Am. Math. Society 17, 441-443.

Chung, K. (1948). On the Maximum Partial Sums of Sequences of Independent Random Variables. Trans. Am. Math. Society 64, 205-233.

Chung, K. (1968). A Course in Probability Theory. Academic, New York

Griffin, P. (1983). Matrix Normalized Sums of Independent, Identically Distributed Random Vectors. Preprint.

Hahn, M., and Klass, M. (1982). Affine Normability of Partial Sums of I.I.D. Random Vectors: A Characterization. Preprint.

Hudson, W. (1980). Operator-Stable Distributions and Stable Marginals. J. Multivariate Anal. 10, 26-37.

Hudson, W., Veeh, J. and Weiner, D. (1984). Moments of Distributions Attracted to Operator Stable Laws. To appear in J. Multivariate Anal.

Jain, N. (1982). A Donsker-Varadhan Type of Invariance Principle. Z. Wahrsch. verw. Gebiete 59, 117-132.

Jain, N., and Pruitt, W. (1973). Maxima of Partial Sums of Independent Random Variables. Z. Wahrsch. verw. Gebiete 27, 141-151.

Sharpe, M. (1969). Operator-Stable Probability Distributions on Vector Groups. Trans. Am. Math. Society 136, 51-65.

Weiner, D. (1984a). A Law of the Iterated Logarithm for Distributions in the Generalized Domain of Attraction of a Nondegenerate Gaussian Law. Submitted for Publication.

Weiner, D. (1984b). Ph.D. Thesis, University of Wisconsin-Madison.

STABLE MEASURES AND PROCESSES IN STATISTICAL PHYSICS[1]

Aleksander Weron[2]
Center for Stochastic Processes
Department of Statistics
University of North Carolina
Chapel Hill, NC 27514

and

Karina Weron[2]
Department of Physics and Astronomy
Louisiana State University
Baton Rouge, LA 70803

Abstract. It is shown how α-stable distributions arise in statistical physics. A probabilistic proof of Khalfin's formula for decaying quantum systems is given. Also ergodic properties of symmetric α-stable flows in classical statistical mechanics are discussed.

1. Introduction.

Exactly sixty years ago Lévy (1924) has initiated the theory of stable distributions. His theory was completed by Gnedenko and Kolmogorov (1954), who mentioned *"it is probable that the scope of applied problems in which stable distributions play an essential role will become in due course rather wide"*. The double anniversary motivated us to present some applications of stable distributions and processes to statistical physics. In passing, let us remark that in probability books only reference to Holtsmark (1915) work on the gravitational field of stars (3/2-stable distribution) is made. The only exception is a very recent book of Zolotarev (1983).

In recent years, inverse power long tails have become more evident in the analysis of physical phenomena and therefore stable distributions provide useful models. In the rest of this section we mention a number of works in this area. In section 2 we show how completely asymmetric α-stable distributions arise in quantum statistical physics and we sketch a probabilistic proof of Khalfin's formula for non-decay probability function. In section 3 we discuss how recent results of Cambanis, Hardin and Weron (1984) on ergodic properties of stationary symmetric α-stable processes can be used

---

[1]Supported in part by AFOSR Grant No. F49620 82 C 0009.
[2]On leave from Technical University of Wroclaw, 50-370 Wroclaw, Poland.

in classical statistical mechanics.

Let us recall that a probability distribution μ on $(-\infty,\infty)$ is α-stable if its characteristic function $\hat{\mu}(t) = \phi(t)$ is given by

$$\operatorname{Log} \phi(t) = \begin{cases} i\gamma t - (\sigma|t|^{\alpha}) \{1 - i\beta \operatorname{sign}(t) \tan(\pi\alpha/2)\} & \text{if } \alpha \neq 1 \\ i\gamma t - \sigma|t| - i\beta(2/\pi)\sigma t \log|t| & \text{if } \alpha = 1, \end{cases} \tag{1}$$

where α,β,γ and σ are real constants with $\sigma \geq 0$, $0 < \alpha \leq 2$ and $|\beta| \leq 1$. See, Gnedenko and Kolmogorov (1954). Here α is the characteristic exponent, γ and σ determine location and scale. The coefficient β indicates whether the α-stable distribution is symmetric ($\beta=0$) or completely asymmetric ($|\beta| = 1$). Only in the case $0<\alpha<1$ the α-stable densities with $|\beta| = 1$ are one-sided i.e., their support is $[0,+\infty)$ for $\beta = 1$ and $(-\infty,0]$ for $\beta = -1$.

For a comprehensive survey of the recent works on α-stable processes and their relation via "correspondence principle" with α-stable measures on vector spaces cf. Weron (1984).

There are many physical phenomena which exhibit both space and time long tails and thus seem to violate the requirement of Gaussian distribution as a limit in the traditional central limit theorem. However, since these physical systems are stable in the sense a Gaussian is but without second moments, one suspects the use of stable distributions which have long tails to be relevant in the physics of these phenomena. A clear physical basis is required to justify the use of stable distribution in much the same way Khinchin (1949) gave a physical justification for the use of Gaussian distributions. Tunaley (1972) invoked physical arguments to suggest that if the frequence distributions in metallic films are stable then the observed noise characteristics in them may be understood. Based only on the experimental observations that near second order phase transitions where long tail spatial order develops, Jona-Lasinio (1975) considered stable distributions as a basic ingredient in understanding renormalization group notions in explaining such phenomena. Scher and Montroll (1975) connect intermittant currents in certain xerographic films to a stable distribition of waiting times for the jumping of charges out of a distribution of deep traps. This provided a basic theoretical model for dispersive transport in amorphous materials.

As examples of the exploration of the stable processes models in physical contexts, we may cite a few very interesting papers. Doob

(1942), West and Seshadri (1982) examined the response of a linear system driven by stable fluctuations. Mandelbrot and van Ness (1968) used Gaussian and stable fractional stochastic processes in several interesting situations. Montroll and West (1979), see also references there, Hughes, Shlesinger, and Montroll (1981), and Montroll and Shlesinger (1982) examined random walks with self-similar clusters leading to "Lévy flights" and "1/f noise". If the diffusion of defects in a medium containing many polar molecules is executed as a continuous-time random walk composed of an alternation of steps and passes and the pausing-time distribution function has a long tail, then Montroll and Bendler (1984) have obtained the Williams-Watts form of dielectric relaxation.

2. Decay theory of quantum systems.

The quantum description of decaying systems has been the subject of many investigations since the early days of quantum mechanics. For a review of different attempts to solve this interesting problem of quantum physics, see, Fonda et al, (1978). Let us recall only that the discovery of natural radioactivity Becquerel (1896) and the identification of two new radioactive elements, i.e., polonium and radium by Mme Sklodowska-Curie (1898) marks the beginning of the studies of decay processes. The classical theory of the decay is based on the assumption that radioactive nuclei have a certain probability of undergoing decay and that this probability does not depend on the past history of the individual decaying nuclei. From which one gets the exponential decay law

$$N(t) = N(0)\exp(-t/\tau), \tag{2}$$

where $N(t)$ is the number of radioactive nuclei which are present at time t and τ is the lifetime of a radioactive nucleus.

In quantum description of the decay process, one determines the probability $P(t)$ of finding, for a measurement at time t, the quantum system in the same physical situation, i.e., in the same state ψ, in which it was at the initial time $t=0$. The mathematical quantity which is relevant for this problem is then

$$P(t) = |A(t)|^2, \qquad t \geq 0 \tag{3}$$

where

$$A(t) = (\psi, \exp(-Dt/\hbar)\psi), \tag{4}$$

D being the development operator governing the dynamical evolution of the quantum system under investigation, and $h = 2\pi\hbar$ is Planck's constant. Originally, Krylov-Fock (1947) used in formula (4) Hamiltonian operator L, but their model is not now accepted, see Fonda et al. (1978), since it is known that a decaying system cannot be described by the unitary evolution $U_t = \exp(-iLt)$.

When an ensemble of identical quantum systems is considered, the number N(t) of systems which are found in the original state at time t is given by

$$N(t) = N(0)P(t). \tag{5}$$

Equation (5) is then the quantum analogue of the classical equation (2). Several authors have studied in various contexts the behaviour of P(t).

We shall derive the nonexponential form of P(t) for many-body systems from a completely asymmetric α-stable energy distribution of the decaying system. Let us mention that nonexponential decay law obtained first asymptotically for large times by Khalfin (1957), still attracts interest, see for example Bunimovich-Sinai (1981), Hack (1982), Lee (1983), and Hart-Girardeau (1983).

THEOREM 1.

The non-decay probability function for many-body weakly interacting quantum systems has the form

$$P(t) = \exp(-ct^{\alpha}), \qquad c > 0,\ 0 < \alpha < 1. \tag{6}$$

Proof. In the quantum statistical mechanics the time evolution of a physical system in equilibrium is given by a dynamical group $U_t = \exp(-iLt)$, which is uniquely determined by its generator defined by the Hamiltonian L of the system. In order to handle a decaying system this time evolution has to be generalized, since the decaying systems are not relevant to the discussion of equilibrium.

The time evolution of decaying system is described by dynamical semigroup cf. Davis (1976), Fonda et al. (1978), and Blum (1981). For this consider a continuous one-parameter semigroup T_t of contractions on the Hilbert space L(H) of all Hilbert-Schmidt operators on the Hilbert space H associated with a quantum mechanical system.

By Nagy-Foias (1960) theorem this semigroup uniquely splits into the orthogonal sum of a unitary semigroup and of a completely non-unitary (c.n.u.) semigroup

$$T_t = T_t^u \oplus T_t^{cnu}.$$

Hille-Yosida theorem gives a form of infinitesimal generators

$$T_t^u = \exp\{(-it\,\tilde{L})/\hbar\}$$

and

$$T_t^{cnu} = \exp\{(-t\tilde{D} - it\tilde{L}_1)/\hbar\},$$

where $\tilde{L}$, $\tilde{L}_1$, $\tilde{D}$ are self-adjoint operators on L(H) and $\tilde{D}$ has positive spectrum.

If $\rho(t)$ is a density operator of the system i.e., self-adjoint, positive with finite trace (see Blum (1981)), then $\rho(t) = T_t\rho(o)$ and

$$\rho(t) = (T_t^u \oplus T_t^{cnu})\ \rho(t) = \rho^u(t) \oplus \rho^{cnu}(t).$$

Moreover,

$$i\hbar \frac{d}{dt}\ \rho^u(t) = \tilde{L}\ \rho^u(t) = [L,\ \rho^u(t)] \tag{7}$$

and

$$i\hbar \frac{d}{dt}\ \rho^{cnu}(t) = (\tilde{L}_1 - i\ \tilde{D})\ \rho^{cnu}(t) = [L_1, \rho^{cnu}(t)]\ - [D, \rho^{cnu}(t)]_+ \tag{8}$$

where L, L_1, D are self-adjoint operator on H, L, L_1 are Hamiltonians and D a new development operator with positive spectrum. Formula (7) is a classical von Neumann equation and (8) its analogue for c.n.u. part. [,] denotes here comutator and $[\ ,\ ]_+$ anticomutator, for more details cf. Weron, Rajagopal, and Weron (1984).

In particular, assuming for simplicity that $\tilde{L}_1 = 0$, we have

$$\rho^{cnu}(t) = e^{-t\tilde{D}/\hbar}\rho^{cnu}(0) = e^{-tD/\hbar}\rho^{cnu}(0)e^{-tD/\hbar}, \tag{9}$$

which shows that D governs the dynamical evolution of the decaying quantum system.

Introducing probability density $p(\varepsilon)$ of the state ψ associated with the continuous spectrum of the development operator D one can write (see (3) and (4))

$$A(t) = (\psi,\ \exp(-Dt/\hbar)\psi) = \int_0^\infty \exp(-\varepsilon t/\hbar)\ (\psi, E(d\varepsilon)\psi) =$$

$$= \int_0^\infty \exp(-\varepsilon t/\hbar)\ p(\varepsilon)d\varepsilon, \tag{10}$$

where $E(\cdot)$ is the spectral measure of the development operator D. Thus A(t) is the Laplace transform of the probability density $p(\varepsilon)$ of the decaying state ψ.

Observe that there is an arbitrariness in the specification of ψ and $p(\varepsilon)$. In general one considers ψ to represent a decaying state for a many-body system, and therefore the number of components in the system should not influence the decay. In other words the same decaying law should be obtained for one portion or several portions of the system. Consequently, in a weakly interacting quantum system microscopic energies can be considered as independent identically distributed energy random variables. The macroscopic energy distribution $p(\varepsilon)d\varepsilon$ associated with the decaying system is identified to be the limit distribution of normalized sums of the microscopic energy random variables. By the limit theorem, Gnedenko and Kolmogorov (1954), it is well known that the limit $p(\varepsilon)d\varepsilon$ has α-stable distribution $0 < \alpha \leq 2$. Since $p(\varepsilon)$ is associated from the above construction with the development operator D, it has to have positive support. This holds only when $p(\varepsilon)d\varepsilon$ has a completely asymmetric ($\beta = 1$, $0 < \alpha < 1$) stable distribution. Thus by (10) it is enough to evaluate its Laplace transform. In the case at hand, formula (1) can be rewritten, if we put $\gamma = 0$, in the following form

$$\operatorname{Log} \phi(t) = -\sigma_1 t^{\alpha} (\cos(\pi\alpha/2) - i \sin(\pi\alpha/2)) = -\sigma_1(-it)^{\alpha},$$

where $\sigma_1 = \sigma/\cos(\pi\alpha/2)$ and $t \geq 0$.

Consequently, the Fourier transform of $p(\varepsilon)d\varepsilon$ has the form $\exp(-\sigma_1(t/i)^{\alpha})$. By the well known relation between Fourier and Laplace transforms, $F(f(x); t) = L(f(x); -it)$, when f(x) has positive support, see Gradshteyn and Ryzhik (1980) p. 1153. Hence we get that the Laplace transform of $p(\varepsilon)d\varepsilon$

$$L(p(\varepsilon);t) = \exp(-\sigma_1 t^{\alpha}) \tag{11}$$

Finally, by (3), (4), (10) and (11)

$$P(t) = [\exp(-\sigma_1 t^{\alpha})]^2 = \exp(-2\sigma_1 t^{\alpha}),$$

which gives formula (6) with $c = 2\sigma_1$ and $0 < \alpha < 1$. □

3. Ergodic properties of stable dynamical systems.

According to the theory of ensembles, cf. Arnold and Avez (1968), an isolated system is in equilibrium when it is represented by a "microcanonical ensemble" i.e., when all points on the surface of given energy have the same probability. This means that the energy must be the only invariant. But for many physical systems energy is far from being the only invariant and consequently a system is wandering on a very small fraction of the constant energy.

Boltzmann introduced a new (ergodic) type of dynamics system for which the energy is the only invariant. In its modern form Boltzmann's Ergodic Hypothesis: *"the point of phase space representing the state of Hamiltonian systems wanders everywhere on its hypersurface of constant energy"* is replaced by the notion of metric transitivity. It says that every subset of a hypersurface of constant energy that is carried into itself by the time development of the system is either of measure zero or is the complement of a subset of measure zero.

The measure referred to is given by the so-called micro-canonical ensemble

$$\mu(S) = \int_S \delta(E - H(q,p))\,dq\,dp,$$

where S is a subset of the hypersurface of given energy E, $q = \{q_1,\ldots,q_m\}$ and $p = \{p_1,\ldots,p_m\}$ are the canonical variables and H is the Hamiltonian of the system. The time evolution of the system is given by a *flow* i.e., a *measure preserving family of mappings* $T^t\{q,p\} = \{q(t), p(t)\}$ such that

$$T^o = I,\ T^nT^k = T^{n+k},\ \mu(T^nS) = \mu(S),\quad n,k \in \mathbb{Z}.$$

Any flow induces a one parameter group of transformations of functions f's defined on the hypersurface Ω_E of given energy E:

$$X_n \equiv (U^nf)\ (q,p) \equiv f(T^n\{q,p\}),\quad n \in \mathbb{Z}.$$

Now a flow is *metrically transitive* if the only functions satisfying $U^nf = f$ for all n, except possibly on a set of μ-measure zero, are constant almost everywhere.

Observe that $\Omega_E \subset \mathbb{R}^{2m}$ with the micro-canonical measure μ plays a role of probability space adequate to our problem. Thus all probabilistic characteristics of the flow can be expressed in terms of the measure μ. See Cornfeld-Fomin-Sinai (1982).

When the functions f's are chosen in the Hilbert space $L^2(\Omega_E,\mu)$, for example when μ is Gaussian, then $\{U^n,\ n \in \mathbb{Z}\}$ turns out to be a unitary one-parameter group $U^nU^k = U^{n+k}$, $U^o = I$ and

$(U^n)^* = U^{-n}$. Von Neumann's ergodic theorem says that

$$\lim_{N\to\infty} \frac{1}{N} \sum_{n=1}^{N} f(T^n x)$$

exists in $L^2(\Omega_E,\mu)$ and equals

$$\int_{\Omega_E} f(x)\, d\mu\, (x),$$

i.e., time average equals phase average.

When the functions f's are chosen in the Banach space $L^p(\Omega_E,\mu)$, for example when μ is a symmetric α-stable measure and $1 \leq p < \alpha < 2$, then $\{U^n,\ n \in \mathbb{Z}\}$ becomes a group of isometries on $L^p(\Omega_E,\mu)$ and similar ergodic behaviour follows from Bellow's (1964) ergodic theorem.

Professional statistical mechanicians are not much impressed. They ask: *How does one verify that a concretely given dynamical system is metrically transitive?* The answer is that it isn't so easy to do. However, it is well known that there are many ergodic ($\equiv$ metrically transitive) flows and also many which are not ergodic, cf. Cornfeld, Fomin and Sinai (1982), where Gaussian case is studied in detail. Since α-stable distributions form a universal class of limit distributions, the systematic use of limit theorems for rigorous proofs in statistical mechanics originated by Khinchin (1949), motivated us to study symmetric α-stable flows.

In order to define them consider the space S of all real sequences $\{x(n),\ n \in \mathbb{Z}\}$ with the minimal σ-algebra A containing all the finite-dimensional cylinders. The probability measure m on A is said to be symmetric α-stable if the joint distribution of any vector

$$X = (x(n_1),\ x(n_2),\dots,x(n_r))$$

is an r-dimensional symmetric α-stable distribution i.e., if its characteristic function has a form

$$\exp\,(-\int_{S_r} |\langle t,x\rangle|^{\alpha} d\Gamma_X(dx)),$$

where $t = (t_1,\dots,t_r)$, $x = (x_1,\dots,x_r) \in \mathbb{R}^r$, $\langle t,x\rangle = t_1x_1+\dots+t_rx_r$, and Γ_X is a symmetric finite measure on the unit sphere S_r of $\mathbb{R}^r$ - called the spectral measure of the vector X. Denote by T^n the shift transformation in the space S, i.e., $T^n x(k) = x(n+k)$. If the measure m is invariant w.r.t. T^n then the group $\{T^n, n \in \mathbb{Z}\}$ of shifts on the space S is said to be symmetric α-stable flow.

As it follows from Hardin (1982) any such flow (or equivalently its induced group U^n of transformations) can be represented in law by

$$X_n = \int_M (U^n f)(x) Z(dx), \tag{12}$$

where (M,Σ,ν) is a measure space, $f \in L^\alpha(M,\Sigma,\nu) \equiv L^\alpha(\nu)$ is a fixed function, $\{U^n, n \in \mathbb{Z}\}$ is a group of isometries on $L^\alpha(\nu)$ and Z is the canonical independly scattered symmetric α-stable measure on (M,Σ,ν) i.e., for all disjoint sets $M_1,\ldots,M_n \in \Sigma$ of finite ν-measure the random variables $Z(M_1),\ldots,Z(M_n)$ are independent with

$$\mathbb{E} \exp(itZ(M_k)) = \exp(-|t|^\alpha \nu(M_k)).$$

Observe that a mean zero Gaussian ($\alpha = 2$) flow on (Ω_E,μ) can be trivially represented in form (12). To see this, let Z be the canonical independently scattered Gaussian measure on (Ω_E,μ). Then by checking characteristic functions we see that the Gaussian flow $U^n f$ has the same distribution as

$$\int_{\Omega_E} U^n f(x) Z(dx)$$

The following general answer for the above discussed question can be immediately obtained from the recent result of Cambanis, Hardin and Weron (1984) on stationary stable processes.

THEOREM 2.

A symmetric α-stable flow, $0 < \alpha \le 2$ with the spectral representation (12) *is metrically transitive iff for each* $h \in \overline{sp}\{U^n f, n \in \mathbb{Z}\}_{L^\alpha(\nu)}$ *the following two conditions hold*

$$\lim_{N\to\infty} \frac{1}{N} \sum_{n=1}^{N} ||U^n h - h||_\alpha^\alpha = 2||h||_\alpha^\alpha . \tag{13}$$

and

$$\lim_{N\to\infty} \frac{1}{N} \sum_{n=1}^{N} ||U^n h - h||_\alpha^{2\alpha} = 4||h||_\alpha^{2\alpha}. \tag{14}$$

It turns out that these conditions provide a useful criterion for metric transitivity. For example moving average α-stable flows are metrically transitive and α-sub-Gaussian flows are never metrically transive. For the proof of Theorem 2 and more results we refer to Cambanis, Hardin, and Weron (1984). Here we will discuss only one example.

EXAMPLE

A real symmetric α-stable flow is called *harmonizable* if its induced group of transformations $X_n = U^n f$ has the following representation:

$$X_n = \mathrm{Re} \int_0^{2\Pi} e^{in\lambda} dW(\lambda), \quad n \in \mathbb{Z}, \tag{15}$$

where W is a rotationally invariant, i.e., the distribution of $\{e^{iv} W(\Delta),\ \Delta \in B[0,2\pi)\}$ does not depend on v, independently scattered complex symmetric α-stable measure on $([0,2\pi), B, \nu)$ and ν is finite, see Cambanis (1983). Of course, any real Gaussian flow has a harmonic representation (15), where $W(\cdot)$ is an independently scattered complex Gaussian measure such that $\mathbb{E} \exp\{i \mathrm{Re} \int u dW\} = \exp(-||u||^2)$, $u \in L^2(\nu)$. However, α-stable flows with $0 < \alpha < 2$ do not have in general harmonic representation cf. Cambanis and Soltani (1984).

It is well known, Maruyama (1949), Grenander (1950), and Fomin (1950), that a Gaussian flow is metrically transitive iff the spectral measure F ($F(\Delta) = \mathbb{E}|W(\Delta)|^2$) has no atoms. For $\alpha<2$ it is enough to check conditions (13) and (14) in Th. 2. Note that the left hand side of the formula (13) takes the form

$$\frac{1}{N} \sum_{n=1}^{N} ||(e^{in\lambda}-1)h(\lambda)||_\alpha^\alpha = \frac{1}{N} \sum_{n=1}^{N} \int_0^{2\pi} |e^{in\lambda}-1|^\alpha \, |h(\lambda)|^\alpha \nu(d\lambda) =$$

$$= \frac{2^\alpha}{N} \int_0^{2\pi} \sum_{n=1}^{N} |\sin \tfrac{n\lambda}{2}|^\alpha \, |h(\lambda)|^\alpha \nu(d\lambda)$$

$$\xrightarrow[N\to\infty]{} 2^\alpha/\pi \int_0^\pi |\sin x|^\alpha dx \int_{(0,2\pi)} |h(\lambda)|^\alpha \nu(d\lambda) \equiv C_\alpha ||h||_\alpha^\alpha$$

Observe that when $\alpha = 2$, $C_2 = 2$ and thus (13) holds provided $\nu\{0\}=0$. However, when $\alpha < 2$ then $C_\alpha < 2$ and (13) is not satisfied. Consequently by Th. 2 symmetric α-stable harmonizable flow is *never metrically transitive* for $\alpha < 2$. This fact has been established also by LePage (1980) by a different method. □

REFERENCES

[1] V. Arnold and A. Avez, Ergodic Problems of Classical Mechanics, Benjamin/Cummings Publishing Co. Reading, Mass. 1978.

[2] A. Bellow, Ergodic properties of isometries in L^p spaces, Bull. Amer. Math. Soc. 70 (1966), 366-371.

[3] K. Blum, Density Matrix Theory and Applications, Plenum Press, New York 1981.

[4] L.A. Bunimovich and Ya. G. Sinai, Statistical properties of Lorentz gas with periodic configuration of scatteres, Commun. Math. Phys. 78 (1981), 479-497.

[5] S. Cambanis, Complex symmetric stable variables and processes, Contributions to Statistics: Essays in Honour of Norman L. Johnson, P.K. Sen, Ed., North-Holland, New York 1983, 63-79.

[6] S. Cambanis, C.D. Hardin, Jr., and A. Weron, Ergodic properties of stationary stable processes, Center for Stochastic Processes Tech. Rept. No. 59, Univ. of North Carolina, Chapel Hill, 1984.

[7] S. Cambanis and A.R. Soltani, Prediction of stable processes: Spectral and moving average representations. Z. Wahrsch. verw. Geb. 66 (1984), 593-612.

[8] I.P. Cornfeld, S.V. Fomin, and Ya. G. Sinai, Ergodic Theory, Springer-Verlag, New York 1982.

[9] E.B. Davies, Quantum Theory of Open Systems, Academic Press, London, 1976.

[10] J. Doob, The Brownian movement and stochastic equations, Ann. Math. 43 (1942), 351-369.

[11] S.V. Fomin, Normal dynamical systems, (in Russian), Ukr. Mat. J. 2 (1950) no. 2., 25-47.

[12] L. Fonda, G.C. Ghirardi, and A. Rimini, Decay theory of unstable quantum systems, Rept. Progr. Phys. 41 (1978) 587-631.

[13] B.V. Gnedenko and A.N. Kolmogorov, Limit Distributions for Sums of Independent Random Variables, Adison-Wesley, Reading, Mass. 1954.

[14] I.S. Gradshteyn and I.M. Ryzhik, Table of Integrals, Series and Products, Academic Press, New York 1980.

[15] U. Grenander, Stochastic processes and statistical inference, Ark. Math. 1 (1950), 195-277.

[16] M.N. Hack, Long tame tails in decay theory, Phys. Lett. 90A (1982), 220-221.

[17] C.D. Hardin, Jr., On the spectral representation of symmetric stable processes, J. Multivariate Anal. 12 (1982) 385-401.

[18] C.F. Hart and M.D. Girardeau, New variational principle for decaying states, Phys. Rev. Lett. 51 (1983), 1725-1728.

[19] B.D. Hughes, M.F. Shlesinger, and E.W. Montroll, Random walks with self-similar clusters, Proc. Natl. Acad. Sci. USA 78 (1981), 3287-3291.

[20] G. Jona-Lasinio, The renormalization group: a probabilistic view, Il. Nuovo Cim. 26 (1975), 99-137.

[21] L.A. Khalfin, Contribution to the decay theory of a quasi stationary state, Zh. Eksp. Teor. Fiz. 33 (1957), 1371-1382.

[22] A.I. Khinchin, Mathematical Foundation of Statistical Mechanics, Dover Publ., Inc., New York 1949.

[23] N.S. Krylov and V.A. Fock, The uncertainty relation for energy and time, Zh. Eksp. Teor. Fiz. 17 (1947), 93-107.

[24] M.H. Lee, Can the velocity autocorrelation function decay exponentially?, Phys. Rev. Lett. 51 (1983), 1227-1230.

[25] R. LePage, Multidimensional infinitely divisible variables and processes. Part I: Stable case. Tech. Rept. No. 292, Dept. of Statistics, Stanford Univ. 1980.

[26] P. Lévy, Theorie des erreurs. La loi de Gauss et les lois exceptionelles, Bull. Soc. Math. France 52 (1924), 49-85.

[27] B.B. Mandelbrot and J.W. Van Ness, Fractional Brownian motions, fractional noises and applications., SIAM Review 10 (1968), 422-437.

[28] G. Maruyama, The harmonic analysis of stationary stochastic processes, Mem. Fac. Sci. Kyushu Univ. A4 (1949) 45-106.

[29] E.W. Montroll and J.T. Bendler, On Lévy (or stable) distributions and the Williams-Watts model of dielectric relaxation, J. Stat. Phys. 34 (1984), 129-162.

[30] E.W. Montroll and M.F. Shlessinger, On 1/f noise and other distributions with long tails, Proc. Natl. Acad. Sci. USA 79 (1982), 3380-3383.

[31] E.W. Montroll and B.J. West, On an enriched collection of stochastic processes, in Fluctuation Phenomena, Eds. E.W. Montroll and J.L. Lebowitz, North-Holland, New York 1979, 61-175.

[32] B. Sz.-Nagy and C. Foias, Sur les contractions de l'espace de Hilbert. IV, Acta Sci. Math. 21 (1960), 251-259.

[33] H. Scher and E.W. Montroll, Anomalous transit-time dispersion in amorphous solids, Phys. Rev. 12B (1975), 2455-2477.

[34] J.K.E. Tunaley, Conduction in a random lattice under a potential gradient, J. Appl. Phys. 43 (1972), 4783-4786.

[35] A. Weron, Stable processes and measures: A survey, Lecture Notes in Math. 1080, 306-364, Springer-Verlag 1984.

[36] A. Weron, A.K. Rajagopal, and K. Weron, Canonical decomposition of dynamical semigroups; Irreversibility and entropy, Preprint 1984.

[37] B.J. West and V. Seshadri, Linear systems with Lévy fluctuations, Physics 113A (1982), 203-216.

[38] V.M. Zolotarev, One-dimensional Stable Distributions, (in Russian), Nauka, Moscow 1983.

Comparison of Martingale Difference Sequences

Joel Zinn*
Texas A & M University
College Station, TX 77843/USA

§0. Let $\{d_j\}$ be a martingale difference sequence (mds) with respect to the increasing sequence of σ-fields $\{\mathcal{A}_j\}$. In ([1], p. 40), Burkholder obtained the following generalization of an inequality of Rosenthal [5].

0.1 **Theorem.** Let $2 < p < \infty$. Then there exists $K = K_p$ such that

$$\|\sum d_j\|_p \approx_K \|(\sum E(d_j^2 \mid \mathcal{A}_{j-1}))^{1/2}\|_p + (\sum \|d_j\|_p^p)^{1/p}$$

Here and in what follows $A \approx_K B$ means that $K^{-1}B \leqslant A \leqslant KB$

As an immediate (trivial) corollary we have

0.2 **Corollary.** Let $2 < p < \infty$ and $\{d_j\}$ and $\{e_j\}$ be mds's with respect to the same σ-fields $\{\mathcal{A}_j\}$. If

$$\|\sum E(d_j^2 \mid \mathcal{A}_{j-1})\|_{p/2} \leqslant \|\sum E(e_j^2 \mid \mathcal{A}_{j-1})\|_{p/2}$$

and

$$\sum \|d_j\|_p^p \leqslant \sum \|e_j\|_p^p ,$$

then

$$\|\sum d_j\|_p \leqslant L \|\sum e_j\|_p$$

where $L = K^2$, K as in Theorem 0.1.

Remark: This corollary plus a duality argument already yields most ($1 < p < \infty$) of the results (on "decoupling") which originally motivated us (see Examples 0.3 and 1.6 and Theorems 1.7 and 1.8). However, we then wondered whether a comparision theorem similar to corollary 0.2 holds for $0 < p < 2$. One version of such a comparison theorem is given in Theorem 1.4 and Corollary 1.5 below. Unfortunately, our comparison hypothesis (see Definition 1.1) seems rather strong. Nonetheless it still enables us to retrieve the "decoupling" results.

0.3 **Example.** Let $f_{ij} : \mathbf{R}^2 \to \mathbf{R}$ be Borel functions, $\{X_k\}$ a sequence of independent random variables, $\{X_k'\}$ an independent copy of $\{X_k\}$ and $\{\epsilon_k, \epsilon_k'\}$ a Rademacher sequence independent of $\{X_k, X_l'\}$. Further let

$$\mathcal{A}_j = \sigma\{X_i, X_i', \epsilon_i, \epsilon_i' : i \leqslant j\}$$

$$d_j = \epsilon_j \sum_{i<j} \epsilon_i f_{ij}(X_i, X_j) \text{ and}$$

$$e_j = \epsilon_j' \sum_{i<j} \epsilon_i f_{ij}(X_i, X_j').$$

*Research partially supported by NSF grant no. MCS-83-01367.

Then, $E(d_j^2 \mid A_{j-1}) = E(e_j^2 \mid A_{j-1})$, a.s.

and $\|d_j\|_p = \|e_j\|_p$.

Hence by Corollary 0.2 for $2 < p < \infty$ (and for $1 < p < 2$, by Theorem 1.7 and a duality argument)

$$(DC_p) \qquad \|\sum_j \epsilon_j \sum_{i<j} \epsilon_i f_{ij}(X_i, X_j)\|_p \approx_L \|\sum_j \epsilon_j' \sum_{i<j} \epsilon_i f_{ij}(X_i, X_j')\|_p .$$

In [3] McConnell and Taqqu show that in the case $f_{ij}(s,t) = a_{ij} st$ (DC_p) holds, in particular, for $1 \leq p < \infty$ and call this decoupling. Their interest (see [4]) in decoupling was motivated by the study of multiple stochastic integration

$$\int_0^1 \int_0^t f(s,t)\, dX_s\, dX_t$$

where X is an independent increment, symmetric stable process of index $1 < \alpha < 2$. The object was to compare the L_r norms $(r < \alpha)$ of such a stochastic integral with the L_r-norms of

$$\int_0^1 \int_0^t f(s,t)\, dX_s\, dX_t' ,$$

where (X_t') is an independent copy of (X_t).

One advantage of decoupling is that via Fubini's theorem we may fix the sequences $\{X_i\}$ and $\{\epsilon_i\}$ on the right hand side of (DC_p), in which case we have a sum of independent random variables.

§1. Our method of comparison is given in the

1.1 Definition. Let both $x = \{x_j\}$ and $y = \{y_j\}$ be adapted to $\{A_j\}$ and assume $x_j, y_j \geq 0$. We write $x \leq_c y$ if for all $t > 0$

$$(1.1) \qquad \sum_j E(x_j \wedge t \mid A_{j-1}) \leq \sum_j E(y_j \wedge t \mid A_{j-1}) \quad \text{a.s.}$$

The following simple lemma is a corollary of a result of Doob. It is basic to much of what follows.

1.2 Lemma. Let $\{x_j\}_{j=1}^{\infty}$ be adapted to the σ-fields $\{A_j\}_{j=1}^{\infty}$, which are increasing. If $0 \leq x_j \leq 1$ for all $j \geq 1$, then

$$\sum x_j < \infty \text{ a.s.} \quad \text{iff } \sum E(x_j \mid A_{j-1}) < \infty \text{ a.s.}$$

Proof. As in the proof of the conditional Borel-Cantelli lemma, one considers $\{d_j\} =: \{x_j - E(x_j \mid A_{j-1})\}$, a mds. Now $\sup_N \sum_{j=1}^N d_j \leq \sup_N \sum_{j=1}^N x_j$. Hence $\sum x_j < \infty$ a.s. implies $\sup_N \sum_{j=1}^N d_j < \infty$ a.s. Since we clearly have $E \sup_j |d_j| \leq 1$, we may apply Theorem 4.1(iv) in [Doob, p. 319] to obtain $\sum_j [x_j - E(x_j \mid A_{j-1})]$ converges a.s. and hence $\sum E(x_j \mid A_{j-1}) < \infty$ a.s. For the other implication we use $d_j = E(x_j \mid A_{j-1}) - x_j$. ∎

As an immediate corollary we obtain:

1.3 Corollary. If $\{x_j\}$ is adapted to $\{\ _j\}$, and $x_j \geqslant 0$

$$\sum x_j < \infty \text{ a.s.} \quad \text{iff for all (or for some) } t > 0 \quad \sum E(x_j \wedge t \mid \ _{j-1}) < \infty \text{ a.s.}$$

Proof. $\sum x_j < \infty$ a.s. iff $t^{-1}\sum(x_j \wedge t) < \infty$ a.s. which by Lemma 1.2 is equivalent to

$$\sum E(x_j \wedge t \mid \ _{j-1}) < \infty \quad \text{a.s.} \quad \blacksquare$$

We can now state the "non-negative" version of our comparison theorem.

1.4 Theorem. Assume $x \leqslant_c y$, $x_j, y_j \geqslant 0$, $j \geqslant 1$, and $0 < p < 1$. Then
(i) $\sum y_j < \infty$ a.s. implies $\sum x_j < \infty$ a.s.
and
(ii) there exists $C_p < \infty$ such that

$$\| \sum x_j \|_p \leqslant C_p \| \sum y_j \|_p \quad .$$

Proof. (i) is obvious from Corollary 1.3. To prove (ii) we first note that for $0 < p < 1$ a random variable $\xi \geqslant 0$ is in L_p iff $\sum \xi_k / k^{1/p} < \infty$ a.s., where $\{\xi, \xi_1, \xi_2, \cdots\}$ are i.i.d.

Now assume $Y = \sum y_j \in L_p$. Let $\pi_k : \Omega^{\mathbf{N}} \to \Omega$ be the coordinate maps, and consider

$$\ _{k,j} = \ _j \otimes \cdots \otimes \ _j \otimes \ _{j-1} \otimes \ _{j-1} \otimes \cdots$$

where there are k $\ _j$'s. Hence $\sum_j \sum_k (y_j \circ \pi_k) / k^{1/p} < \infty$ a.s. But this is equivalent to: for all $n_j \quad \infty$, $\sum_j \sum_{k \leqslant n_j} (y_j \circ \pi_k) / k^{1/p} < \infty$ a.s. Ordering things lexicographically, this last condition is by Corollary 1.3 equivalent to: for all $n_j \quad \infty$, using $\ _{k,j} = \ _{k,j}$ $k \geqslant 1$ and $\ _{0,j} = \ _{n_{j-1},j-1}$,

$$\sum_j \sum_{k \leqslant n_j} E[(y_j \circ \pi_k) / k^{1/p} \wedge 1 \mid \ _{k-1,j}]$$

$$= \sum_j \sum_{k \leqslant n_j} E\left[\frac{y_j}{k^{1/p}} \wedge 1 \mid \ _{j-1}\right] \circ \pi_k < \infty \text{ a.s.}$$

And this is equivalent to

$$\sum_{k=1}^{\infty} \sum_{j=1}^{\infty} E\left[\frac{y_j}{k^{1/p}} \wedge 1 \mid \ _{j-1}\right] \circ \pi_k$$

$$= \sum_{j=1}^{\infty} \sum_{k=1}^{\infty} E\left[\frac{y_j}{k^{1/p}} \wedge 1 \mid \ _{j-1}\right] \circ \pi_k < \infty .$$

Since $x \leqslant_c y$, the left land side in the equation above dominates

$$\sum_{k=1}^{\infty} \sum_{j=1}^{\infty} E\left[\frac{x_j}{k^{1/p}} \wedge 1 \mid \quad _{j-1}\right] \circ \pi_k$$

which by a repetition of the above argument is equivalent to $\sum x_j \in L_p$.

We now show the existence of the constant C_p . If no such constant exists, then for each $r \geqslant 1$, there exists n_r , $x^{(r)} = \{x_j^{(r)}\}_{j=1}^{n_r}$, $y^{(r)} = \{y_j^{(r)}\}_{j=1}^{n_r}$, and σ-fields $\{ \ \ _j^{(r)}\}_{j=1}^{n_r}$ with $x^{(r)} \leqslant_c y^{(r)}$ and such that

$$\| \sum_{j=1}^{n_r} y_j^{(r)} \|_p \leqslant 2^{-r/p} \ , \quad \| \sum_{j=1}^{n_r} x_j^{(r)} \|_p \geqslant 2^r \ .$$

We now consider the sequences of random variables $x = \{x_j^{(r)} \circ \pi_r : 1 \leqslant j \leqslant n_r \, , 1 \leqslant r < \infty\}$ and $y = \{y_j^{(r)} \circ \pi_r : 1 \leqslant j \leqslant n_r \, , 1 \leqslant r < \infty\}$ and σ-fields

$$_{j,r} = \ \ _{n_1}^{(1)} \otimes \cdots \otimes \ \ _{n_{r-1}}^{(r-1)} \otimes \ \ _j^{(r)} \otimes \ \ _0^{(r+1)} \otimes \ \ _0^{(r+2)} \otimes \cdots$$

lexicographically ordered. Then, since $0 < p < 1$.

$$\| \sum_{r=1}^{\infty} \sum_{j=1}^{n_r} y_j^{(r)} \circ \pi_r \|_p^p \leqslant \sum_{r=1}^{\infty} \| \sum_{j=1}^{n_r} y_j^{(r)} \circ \pi_r \|_p^p \leqslant 1$$

and by non-negativity of the $x_j^{(r)}$'s,

$$\| \sum_{r=1}^{\infty} \sum_{j=1}^{n_r} x_j^{(r)} \circ \pi_r \|_p \geqslant \| \sum_{r=1}^{n_r} x_j^{(r)} \circ \pi_r \|_p \geqslant 2^r \ ,$$

for all $r \geqslant 1$. Further, since $x \leqslant_c y$, we have obtained a contradiction. ■

Notation. Let $d^2 = \{d_j^2\}$, the sequence of squares.

We now give the analogous results for the "mean-zero" case. Again $\{\epsilon_j\}$ denotes a Rademacher sequence independent of all other random variables in the given context.

1.5 Corollary. Assume $d^2 \leqslant_c e^2$ (in particular, d_j^2, e_j^2 are $\ \ _j$-measurable). Then

(i) $\sum \epsilon_j d_j$ converges a. s. iff for all (some) $t > 0$ $\sum E(d_j^2 \wedge t \mid \ \ _{j-1}) < \infty$ a. s.

(ii) For each $0 < p < 2$ there exists a constant C_p such that

$$\| \sum \epsilon_j d_j \|_p \leqslant C_p \| \sum \epsilon_j e_j \|_p \ .$$

(iii) For each $1 < p < 2$ there exists a constant C_p such that for all mds's $\{d_j\}$ and $\{e_j\}$ with respect to the same $\{ \ \ _j\}$, $\| \sum d_j \|_p \leqslant C_p \| \sum e_j \|_p$.

Proof. (i). This follows from Corollary 1.3 and the fact that $\sum \epsilon_j d_j$ converges a. s. iff $\sum d_j^2 < \infty$ a. s.

(ii). This follows from Khintchine's inequality and Theorem 1.4.

(iii). By Burkholder ([1], Theorem 3.2), $\| \sum d_j \|_p \approx \| (\sum d_j^2)^{1/2} \|_p$ and similarly for $\{e_j\}$. Now apply Theorem 1.4(ii).

We now consider a version of decoupling.

1.6 Example. Let $\phi_j : \mathbf{R}^j \to \mathbf{R}$ be Borel measurable and let $\{X_j\}$ and $\{X_j'\}$ be independent copies of a sequence of independent random variables. Then, for example, if $\quad_j = \sigma(X_i, X_i' : 1 \leqslant i \leqslant j)$, we have for all $t > 0$

$$E(\phi_j^2((X_i)_{i \leqslant j}) \wedge t \mid \quad_{j-1}) = E(\phi_j^2((X_i)_{i<j}, X_j') \wedge t \mid \quad_{j-1}) \text{ a.s.}$$

Hence the preceeding theorems apply with equivalence. In particular, we have

1.7 Theorem. In the situation of Example 1.6 we have for $0 < p < 2$, there exists $C = C_p$ such that

$$\| \sum \epsilon_j \phi_j((X_i)_{i \leqslant j}) \|_p \approx_C \| \sum \epsilon_j \phi_j((X_i)_{i<j}, X_j') \|_p \ .$$

We also have, for example,

1.8 Theorem. In the situation of Example 0.3 we have for $0 < p < 2$ a constant $C = C_p$ such that

$$\| \sum_j \epsilon_j \sum_{i<j} \epsilon_i f_{ij}(X_i, X_j) \|_p \approx_C \| \sum_j \epsilon_j' \sum_{i<j} \epsilon_i f_{ij}(X_i, X_j') \|_p \ .$$

Proof. It is well known that there exists $D = D_p$ such that $\| \sum_j \epsilon_j \sum_{i<j} \epsilon_i \quad_{ij} \|_p \approx_D (\sum_j \sum_{i<j} \quad_{ij}^2)^{1/2} \approx_D \| \sum_j \epsilon_j' \sum_{i<j} \epsilon_i \quad_{ij} \|_p$. We, again, may use (for $i < j$)

$$E(f_{ij}^2(X_i, X_j) \mid X_k, X_k', k \leqslant j-1) = E(f_{ij}^2(X_i, X_j') \mid X_k, X_k', k \leqslant j-1) \text{ a.s.},$$

and apply Theorem 1.4. ■

Acknowledgements. I would like to thank Gilles Pisier for many important comments and suggestions made during the course of this investigation. I would also like to thank E. Giné, M. Marcus and W. Johnson for several helpful discussions.

[1] Burkholder, D. L. (1973). Distribution function inequalities for martingales. *Ann. of Probability*, **1**, 19-42.

[2] Doob, J. L. (1953). *Stochastic Processes.* J. Wiley and Sons, New York.

[3] McConnell, T. and Taqqu, M. (1984). Decoupling inequalities for multilinear forms in independent symmetric random variables. Preprint.

[4] __________ and __________ (1984). Double integration with respect to symmetric stable processes. Preprint.

[5] Rosenthal, H. P. (1970). On the subspaces of L^p $(p > 2)$ spanned by sequences of independent random variables. *Israel J. Math.* **8**, 273-303.